Progress in Computational Physics

Novel Trends in Lattice-Boltzmann Methods

Volume 3

Edited by:

Matthias Ehrhardt
Bergische Universität Wuppertal
Fachbereich Mathematik und Naturwissenschaften
Lehrstuhl für Angewandte Mathematik und Numerische Mathematik
Gaußstrasse 20, 42119 Wuppertal
Germany

CONTENTS

FOREWORD

Lattice Boltzmann Methods (LBM) emerged over twenty years ago as a new way of analyzing physical systems, mostly of the fluid flow type. While historically LBM evolved from lattice gas automata models, it is now better interpreted as a systematic approximation to the Boltzmann kinetic theory. Unlike a real microscopic many-body system, LBM represents a fluid system as a simple dynamical model residing in some special discrete phase space. The governing equation resulting from such formulation is called lattice Boltzmann equation. Although an LBM model is a drastic simplification to the corresponding realistic fluid system at the micro-dynamical level, correct physical properties at the macroscopic level can be shown to be recovered with proper formulation. This fact allows using LBM as an alternative computational fluid dynamics (CFD) approach for studying realistic physical phenomena. The fundamental difference between LBM and conventional CFD is that the former is based on a microscopic (or perhaps "mesoscopic") level description *via* Boltzmann - like kinetic equation, instead of solving hydrodynamic equations such as the Navier-Stokes equation. This difference has enabled a number of key advantages of LBM over the conventional approach. In addition to the familiar benefits such as being an attractive computational method for efficient and robust fluid simulation, LBM, due to its underlying kinetic theory basis also opens a way to simulate a wider variety of hydrodynamic phenomena for a broader range of physical regimes than the conventional computational methods. Many of such problems known to be extremely difficult or impossible to treat by the conventional techniques can be now approached.

Besides the basic scientific interest associated with LBM, this new field has a significant potential in real world applications. With the rapid progress in the information and computer technologies, computer aided engineering (CAE) has now become the leading trend in modern industrial engineering processes. The computer - based process allows a virtual platform for studying physical problems often impossible in real experiment, thus enabling deeper understanding of the underlying phenomena. Consequently, the new process greatly enhances and empowers innovation. In this context, LBM is not only offering great opportunities in the fluid dynamics area but is already making substantial impact in the CAE - based real world engineering process.

In the recent years LBM has become a very active and fast growing research field, as evident from the richness of topics covered in this eBook. This method now is not only standing on a much more solid theoretical foundation but also has a substantially expanded domain of applications. Comparing to the early years when LBM only described the simple Navier-Stokes fluids, it is now used to handle a wide variety of complex fluids and flows, as well as physical systems beyond fluid dynamics. As indicated in the title, this eBook covers a set of important extensions in LBM ranging from reacting flows to transport phenomena to fluid-solid interactions. Such problems are of central importance in many real world applications. LBM models of these physical systems open a way for deeper understanding of their essential physical properties. In addition to these extensions, the eBook also includes some of the state of the art theoretical underpinnings in LBM fundamentals and formulations. This eBook edited by Professor Matthias Ehrhardt will certainly be useful for a large audience of scholars.

Hudong Chen
Exa Corporation
55 Network Drive
Burlington, MA 01803
USA

PREFACE

This is the third volume of a new eBook series that is devoted to very recent research trends in computational physics. Hereby, it focus on the computational perspectives of current physical challenges, publishing new numerical techniques for the solution of mathematical equations including chapters describing certain real-world applications concisely. The goal of this series is to emphasize especially approaches that are of interdisciplinary nature. The scientific topics in the fields of modeling, numerical methods and practical applications include e.g. the coupling between free and porous media flow, coupling of flow and transport models, coupling of atmospheric and ground water models, etc.

This volume contains 9 chapters devoted to mathematical analysis of different issues related to tailor-made Lattice Boltzmann Methods (LBMs), advanced numerical techniques for physicochemical flows and fluid structure interaction and practical applications to real world problems. This eBook consists of 9 invited chapters that are structured in the five parts introduction, regularization, asymptotic analysis and lifting of Lattice Boltzmann Methods, reactive flow and physicochemical transport, Lattice Boltzmann Methods for fluid-structure interaction (FSI) and finally practical applications.

In the first part we first present a troughout introduction into the general topic and review the known derivation of the Lattice Boltzmann Method from the Boltzmann equation using finite differences and approximate the collision term with a single relaxation time model (BGK) Also, a multi-relaxation-time model is shortly presented. By applying a multiscale expansion (Chapman-Enskog), the solution of the numerical method is verified as a meaningful approximation of the solution of the Navier-Stokes equations. Next, the LBM is extended to handle coupled problems, like the movement of objects in a flow, the coupling to heat transport and the coupling of electric circuits with power dissipation (as heat) and heat transport.

In the second part consisting of Chapters 2–5, on regularization, asymptotic analysis and lifting of Lattice Boltzmann Methods, the authors first describe how regularization of LBMs can be achieved by dampening systems using controlled numerical dissipation. Where effective, each of the proposed techniques corresponds to an additional injection of dissipation compared with the standard LBGK model. Using some standard 1D and 2D benchmarks including the shock tube and lid driven cavity, the authors show effectiveness and accuracy penalties of the proposed methods.

In Chapter 3 the authors present discrete-velocity models and LBMs for solving convection-radiation effects in thermal fluid flows. The discrete-velocity equations are derived from the continuous Boltzmann equation with appropriate scaling suitable for incompressible flows. The radiative heat flux in the energy equation is obtained using the discrete-ordinates solution of the radiative transfer equation. Numerical results for several test examples on coupled convection-radiation flows in two dimensional dimensions show that the developed models are competitive tools for convection-radiation problems.

In Chapter 4 the authors perform a detailed asymptotic analysis of different numerical schemes for the interaction between an incompressible fluid and a rigid structure within a Lattice Boltzmann (LB) framework. Hereby they focus on moving boundary LB schemes and on force computation through a momentum exchange algorithm (MEA). Again, all theoretical results are supported by several numerical experiments, considering both adhoc designed benchmarks, to validate accuracy properties of the schemes, and more general test setups.

In Chapter 5 the authors present an overview of the various lifting strategies for LBMs which is useful in coupled LBM and PDE models, where one part of the domain is described by a PDE while another part is modeled by an LBM. At the interface between these models the lifting operator provides the correct boundary conditions for the LBM domain. The authors discuss the accuracy, computational cost and convergence rate of analytical and numerical lifting procedures.

This topic is a good bridge to the third part on reactive flow and physicochemical transport, consisting of Chapters 6–7. The author of Chapter 6 explains the recent progress of the LBMs applied to diffusion-reaction processes for chemical and environmental applications, in particular to physicochemical processes that take place in environmental systems, such as aquatic systems, porous media, sediment, soils and biofilm layer on inert substrate. Furthermore, the role of Michaelis-Menten boundary condition at a consuming interface will be investigated. These results have been obtained by using a new computer program MHEDYN developed by the author to compute metal fluxes at planar consuming surfaces in multiligand, chemically heterogeneous environmental systems.

In Chapter 7 the authors present a LBM for modeling coupled fluid flow, solute transport, and chemical reaction at a fundamental scale where the flow is governed by continuum fluid equations. This numerical model accounts for multiple processes, including fluid flow, diffusion and advection of species, ion-exchange and mineral precipitation/dissolution reactions, as well as the evolution of pore geometry due to dissolution/precipitation. Homogeneous reactions are described either kinetically or through local equilibrium mass action relations. Heterogeneous reactions are incorporated into the LBM through boundary conditions imposed at the mineral surface.

In the fourth part on LBMs for FSI, the authors investigate the validity and efficiency of the coupling of the LBM with finite element methods (FEMs) as well as rigid body approaches to model FSI. The results using the fluid solver VirtualFluids on two- and three-dimensional benchmark configurations show that an explicit coupling scheme is able to produce accurate results which agree with reference solutions very well. Furthermore, the coupling to a rigid body dynamics engine (PhysicsEngine-pe) leads to the possiblity to compute FSI problems with a huge number of particles which is the basis for numerical simulations in geothermic drilling.

Finally, Chapter 9 entitled LBM for MILD Oxy-fuel Combustion Research: A Potential Powerful Tool Responding to the Man-made Global Warming considers a hot topic with socio-economic relevance. It deals with the numerical modeling and simulation of multicomponent combustion under MILD oxy-fuel operation by the LBM to deepen our understanding on this novel combustion technology. The main objective of this final chapter is to demonstrate the opportunities to build a new modeling framework to accelerate scientific discovery in this topic by utilizing unique features of the LBM and indicate the challenges that we should overcome since until recently the LBM could not be employed in such research field due to combustion's inherent complexity.

We would like to thank Dr. Hudong Chen (Exa Corporation) for writing the foreword and for providing the figures for the title page (generated with Exa PowerFLOW/PowerVIZ) and Bentham Science Publishers, particularly the Manager Ms. Sana Mokarram for support and efforts.

Matthias Ehrhardt

Bergische Universität Wuppertal
Fachbereich Mathematik und Naturwissenschaften
Lehrstuhl für Angewandte Mathematik und Numerische Mathematik
Gaußstrasse 20, 42119 Wuppertal
Germany

CONTRIBUTORS

Davide Alemani	EPFL, Lausanne, Switzerland
Mapundi K. Banda	School of Computational and Applied Mathematics, University of the Witwatersrand, South Africa
Andreas Bartel	Lehrstuhl für Angewandte Mathematik und Numerische Analysis, Bergische Universität Wuppertal, Germany
Robert A. Brownlee	Department of Mathematics, University of Leicester, United Kingdom
Alfonso Caiazzo	Weierstrass Institute for Applied Analysis and Stochastics, Berlin, Germany
Sheng Chen	US-China Clean Energy Research Center, China-EU Institute for Clean and Renewable Energy, State Key Laboratory of Coal Combustion, Huazhong University of Science and Technology, China
Matthias Ehrhardt	Lehrstuhl für Angewandte Mathematik und Numerische Analysis, Bergische Universität Wuppertal, Germany
Sebastian Geller	Institut für rechnergestützte Modellierung im Bauingenieurwesen, Technische Universität Braunschweig, Germany
Alexander N. Gorban	Department of Mathematics, University of Leicester, United Kingdom
Daniel Heubes	Lehrstuhl für Angewandte Mathematik und Numerische Analysis, Bergische Universität Wuppertal, Germany
Christian Janßen	Institut für rechnergestützte Modellierung im Bauingenieurwesen, Technische Universität Braunschweig, Germany
Michael Junk	Fachbereich Mathematik und Statistik, Universität Konstanz, Germany
Qinjun Kang	Computational Earth Science Group (EES-16), Earth and Environmental Sciences Division, Los Alamos National Laboratory, Los Alamos, USA.
Manfred Krafczyk	Institut für rechnergestützte Modellierung im Bauingenieurwesen, Technische Universität Braunschweig, Germany.
Jeremy Levesley	Department of Mathematics, University of Leicester, United Kingdom
Peter Lichtner	Computational Earth Science Group (EES-16), Earth and Environmental Sciences Division, Los Alamos National Laboratory, Los Alamos, USA
David Packwood	Department of Mathematics, University of Leicester, United Kingdom
Dirk Roose	Department of Computer Science, Faculty of Applied Sciences, Katholieke Universiteit Leuven, Belgium
Mohammed Seaïd	School of Engineering and Computing Sciences, University of Durham, United Kingdom
Christophe Vandekerckhove	ArcelorMittal, Belgium
Ynte Vanderhoydonc	Universiteit Antwerpen, Belgium
Pieter Van Leemput	Department of Computer Science, Faculty of Applied Sciences, Katholieke Universiteit Leuven, Belgium
Wim Vanroose	Universiteit Antwerpen, Belgium

2

Progress in Computational Physics, **Vol. *3*, 2013, 3-30**

CHAPTER 1

An Introduction to the Lattice Boltzmann Method for Coupled Problems

Daniel Heubes*, Andreas Bartel, Matthias Ehrhardt

Bergische Universität Wuppertal, Fachbereich Mathematik und Naturwissenschaften, Lehrstuhl für Angewandte Mathematik und Numerische Mathematik, Gaußstrasse 20, 42119 Wuppertal, Germany

Abstract: The first part of this introduction is devoted to the known derivation of the lattice Boltzmann method (LBM): We track two different derivations, a historical one (via lattice gas automata) and a theoretical version (via a discretization of the Boltzmann equation). Thereby the collision term is approximated with a single relaxation time model (BGK) and we motivate the introduction of this common approximation. By applying a multi-scale expansion (Chapman-Enskog), the solution of the numerical method is verified as a meaningful approximation of the solution of the Navier-Stokes equations. To state a well posed problem, common boundary conditions are introduced and their realization within a LBM is discussed.

In the second part, the LBM is extended to handle coupled problems. Four cases are investigated: (i) multiphase and multicomponent flow, (ii) additional forces, (iii) the coupling to heat transport, (iv) coupling of electric circuits with power dissipation (as heat) and heat transport.

Keywords: BBGKY hierarchy, BGK approximation, boundary conditions, Chapman-Enskog expansion, circuit coupling, D3Q19, discrete velocity space, Gauß-Hermite quadrature, lattice gas automata, Navier-Stokes equations, thermal coupling.

1. INTRODUCTION

The circulation of blood in human blood vessels, the river flow at bridge pillars, the air flow passing a car are typical examples of complex problems from fluid mechanics that can only be solved numerically. Due to the today's demanding complexity of structures and models, such a fluid simulation must be very efficient. Often computations are based on continuous models for mass and momentum conservation as the Navier-Stokes equations operating on the macroscopic level. More recently, kinetic models and the lattice Boltzmann method (LBM) compete against standard finite element based simulations. The LBM is quite attractive, since an explicit update scheme is not only extremely efficient in computation, but also fits to modern computer architecture, especially graphical processing units (GPUs). Its drawback and its advantage at the same time is the microscopical description on the particle level, i.e. each interaction, boundary conditions etc. have to be formulated on this basis. For an extended introduction to the field of the LBM, there are several reviews and books available. We refer the interested reader to [9, 29, 36] and the references therein. This chapter is organized as follows: In the remainder of this section, we briefly state the Navier-Stokes equations and we discuss the basics of the Boltzmann equation, the collision integral and its approximation. In Section 2 we focus on the derivation of the LBM. First, we study the historical derivation via lattice gas automata. Next we show how the Boltzmann equation can be derived from a particle ensemble description and then we present the more recent derivation via discretization and Gauß-Hermite quadrature. For the latter part, we discuss in particular the D3Q19 model. At the end of Section 2, we show using a Chapman-Enskog expansion the relation to the Navier-Stokes equations. Section 3 is devoted to to a concise discussion of boundary conditions for the LBM. Afterwards, we discuss enhancements as additional forces, temperature and an electric circuit coupling in Section 4.

*Address correspondence to: Daniel Heubes, Bergische Universität Wuppertal, Fachbereich Mathematik und Naturwissenschaften, Lehrstuhl für Angewandte Mathematik und Numerische Mathematik, Gaußstrasse 20, 42119 Wuppertal, Germany; Tel: +49 202 439 4781; Fax: +49 202 439 4770; E-mail: heubes@math.uni-wuppertal.de

Matthias Ehrhardt (Ed.)

1.1. The Navier-Stokes Equations

The macroscopic motion of a Newtonian fluid can be described by a balance equation of momenta. To this end, let $\boldsymbol{u} = \boldsymbol{u}(\boldsymbol{x},t)$ denote the fluid velocity at the spatial location $\boldsymbol{x}$ and time t, $\rho = \rho(\boldsymbol{x},t)$ the local mass density, $p = p(\boldsymbol{x},t)$ the local pressure and η the dynamic viscosity. For a constant mass density $\rho \equiv$ const., the Navier-Stokes equations read [37]:

$$\nabla \cdot \boldsymbol{u} = 0, \tag{1}$$

$$\rho \frac{\partial \boldsymbol{u}}{\partial t} + \rho\,(\boldsymbol{u} \cdot \nabla)\,\boldsymbol{u} = -\nabla p + \eta \Delta \boldsymbol{u}. \tag{2}$$

The first equation (1) states the incompressibility of the fluid (conservation of mass). The second equation (2) is the balance equation for momenta, where the fluid transport is driven by the pressure p and the stress term $\eta \Delta \boldsymbol{u}$. This description yields the fundamental macroscopic fluid model, which we want to simulate. Clearly, if additional body forces are present, a further addend is needed.Furthermore if we have $\rho \neq$ const., the fluid is described by the compressible version of the Navier-Stokes equations:

$$\frac{\partial \rho}{\partial t} + \nabla \cdot (\rho \boldsymbol{u}) = 0,$$

$$\frac{\partial \rho \boldsymbol{u}}{\partial t} + \nabla \cdot \left(\rho \boldsymbol{u}\boldsymbol{u}^\top\right) = -\nabla p + \nabla \cdot \eta \left[\nabla \boldsymbol{u} + \nabla \boldsymbol{u}^\top - \frac{2}{3} \nabla \cdot \boldsymbol{u}\right].$$

For further details see, e.g., [2].

1.2. The Boltzmann Equation

In this section, we state the microscopic view on fluid modeling. Here, the Boltzmann equation without additional forces is an evolution equation in time (t) for the single particle distribution $f(\boldsymbol{x},\boldsymbol{v},t)$ at space position $\boldsymbol{x}$ with microscopic velocity $\boldsymbol{v} \in \mathbb{R}^3$:

$$\frac{\partial f(\boldsymbol{x},\boldsymbol{v},t)}{\partial t} + \boldsymbol{v} \cdot \nabla f(\boldsymbol{x},\boldsymbol{v},t) = Q(f) \tag{3}$$

The value $f(\boldsymbol{x},\boldsymbol{v},t)\,\mathrm{d}\boldsymbol{x}\,\mathrm{d}\boldsymbol{v}$ for all infinitesimal small $\mathrm{d}\boldsymbol{x}$ and $\mathrm{d}\boldsymbol{v}$ represents the probability to find a particle at time t having the position $\boldsymbol{x}$ and the velocity $\boldsymbol{v}$. In fact, this value is scaled with the constant particle mass m, thus the mass density is given by the integral (zeroth moment)

$$\rho(\boldsymbol{x},t) = \int_{\mathbb{R}^3} f(\boldsymbol{x},\boldsymbol{v},t)\,\mathrm{d}\boldsymbol{v} \tag{4}$$

in three dimensional space. The term $Q(f)$ in (3) expresses the change in the distribution f due to the collision of particles with different velocities. It is referred to as collision term. In Section 1.3., we state more details of this quantity. Moreover, a rough derivation of the Boltzmann equation (3) can be found in Section 2.2.1. In addition to (4), the other macroscopic fluid quantities as the velocity $\boldsymbol{u}$ and the temperature T can be determined by higher moments of the single particle distribution:

$$\rho(\boldsymbol{x},t)\boldsymbol{u}(\boldsymbol{x},t) = \int_{\mathbb{R}^3} \boldsymbol{v} f(\boldsymbol{x},\boldsymbol{v},t)\,\mathrm{d}\boldsymbol{v}, \tag{5}$$

$$\rho(\boldsymbol{x},t)\frac{3}{2}\frac{k_B T(\boldsymbol{x},t)}{m} = \int_{\mathbb{R}^3} \frac{|\boldsymbol{v} - \boldsymbol{u}(\boldsymbol{x},t)|^2}{2} f(\boldsymbol{x},\boldsymbol{v},t)\,\mathrm{d}\boldsymbol{v}, \tag{6}$$

with Boltzmann constant k_B. Integrating the Boltzmann equation (3) over the velocity space, that is, the zeroth moment, results in the continuity equation

$$\frac{\partial}{\partial t}\rho(\boldsymbol{x},t) + \nabla \cdot (\rho(\boldsymbol{x},t)\boldsymbol{u}(\boldsymbol{x},t)) = 0. \tag{7}$$

Here we used the fact that the zeroth moment of the collision term vanishes [5]. Furthermore, the first moment of the Boltzmann equation is computed by multiplying this equation by $\boldsymbol{v}$ and then by integrating over velocity space. This yields

$$\frac{\partial}{\partial t}(\rho(\boldsymbol{x},t)\boldsymbol{u}(\boldsymbol{x},t)) + \boldsymbol{\nabla}\cdot\left(\rho(\boldsymbol{x},t)\boldsymbol{u}(\boldsymbol{x},t)\boldsymbol{u}(\boldsymbol{x},t)^\top\right) = -\boldsymbol{\nabla}\cdot\int_{\mathbb{R}^3}(\boldsymbol{v}-\boldsymbol{u}(\boldsymbol{x},t))(\boldsymbol{v}-\boldsymbol{u}(\boldsymbol{x},t))^\top f(\boldsymbol{x},\boldsymbol{v},t)\,\mathrm{d}\boldsymbol{v}, \quad (8)$$

where the integral on the right hand side is the momentum flux tensor, which exhibits the form [18]

$$\int_{\mathbb{R}^3}(\boldsymbol{v}-\boldsymbol{u}(\boldsymbol{x},t))(\boldsymbol{v}-\boldsymbol{u}(\boldsymbol{x},t))^\top f(\boldsymbol{x},\boldsymbol{v},t)\,\mathrm{d}\boldsymbol{v} = p(\boldsymbol{x},t)I - \sigma(\boldsymbol{x},t)$$

with the pressure $p(\boldsymbol{x},t)$, the identity matrix I and

$$\sigma(\boldsymbol{x},t) = \rho(\boldsymbol{x},t)\nu\left[\boldsymbol{\nabla u}(\boldsymbol{x},t) + (\boldsymbol{\nabla u}(\boldsymbol{x},t))^\top\right].$$

Here ν denotes the kinematic viscosity, and the complete term σ expresses the dynamic portion in the momentum flux tensor. Hence, the equation (8) becomes

$$\frac{\partial}{\partial t}(\rho(\boldsymbol{x},t)\boldsymbol{u}(\boldsymbol{x},t)) + \boldsymbol{\nabla}\cdot\left(\rho(\boldsymbol{x},t)\boldsymbol{u}(\boldsymbol{x},t)\boldsymbol{u}(\boldsymbol{x},t)^\top\right) = -\boldsymbol{\nabla}p(\boldsymbol{x},t) + \boldsymbol{\nabla}\cdot\sigma(\boldsymbol{x},t). \quad (9)$$

Now, if ρ is constant, (7) yields (1) and (9) gives (2), where the dynamic viscosity is $\eta = \rho\nu$. Finally, we note that the temperature is derived a posteriori by (6) from the distribution f.

1.3. The Collision Term and BGK Approximation

Next, we derive the Boltzmann integral expression for $Q(f)$ making several assumptions [32]: One condition is that the particles interact only in two-particle collisions, i.e. we assume that interactions involving more than two particles can be neglected. For all two-particle collisions we assume that they appear locally in the sense that they take place at a single point $\boldsymbol{x}$. A similar condition holds for the time t; it is assumed that the duration of a collision is negligible. Particles involved in a collision are assumed to be uncorrelated, and the collision itself is modeled as an elastic collision, meaning that kinetic energy and especially momentum are conserved. For a collision of two particles, we have the velocities $\boldsymbol{v}$, $\boldsymbol{w}$ before and the velocities $\boldsymbol{v}'$, $\boldsymbol{w}'$ after the collision, the collision integral due to Boltzmann reads

$$Q(f) = \int_{\mathbb{R}^3}\int_{\mathbb{S}^2}\sigma(\Omega)|\boldsymbol{v}-\boldsymbol{w}|\left[f(\boldsymbol{x},\boldsymbol{v}',t)f(\boldsymbol{x},\boldsymbol{w}',t) - f(\boldsymbol{x},\boldsymbol{v},t)f(\boldsymbol{x},\boldsymbol{w},t)\right]\,\mathrm{d}\Omega\,\mathrm{d}\boldsymbol{w}, \quad (10)$$

where $\sigma(\Omega)$ denotes the differential collision cross section and the inner integration is done over all possible solid angles $\Omega \in \mathbb{S}^2$. Moreover, the post-collision velocities can be computed depending on the pre-collision velocities and the impact angle.

The collision integral (10) possesses a rather complicated form. In the following, we motivate the approximation introduced by Bhatnagar, Gross and Krook (BGK) [3] which simplifies the collision term and is used frequently nowadays. Let us note that, besides the BGK term, there exist different alternatives for the collision integral. One of them are multi-relaxation time models, see, e.g., [12]. In BGK schemes the collision integral is replaced by a single time relaxation term. This term is chosen in such a way that it emulates specific properties of the original collision integral, in particular those which were necessary to derive equations (7) and (9). It can be shown [5], that there exist five elementary invariants which read

$$\psi_1 = 1, \quad (\psi_2,\psi_3,\psi_4) = \boldsymbol{v} \quad \text{and} \quad \psi_5 = |\boldsymbol{v}|^2$$

and lead to vanishing integrals

$$\int_{\mathbb{R}^3}\psi_i Q(f)\,\mathrm{d}\boldsymbol{v} = 0, \quad i = 1,\dots,5.$$

It even holds the following equivalence

$$\int_{\mathbb{R}^3} Q(f)\varphi(\mathbf{v})\,\mathrm{d}\mathbf{v} = 0 \quad \Leftrightarrow \quad \varphi(\mathbf{v}) = \sum_{i=1}^{5} s_i \psi_i = \alpha + \boldsymbol{\beta}\cdot\mathbf{v} + \gamma|\mathbf{v}|^2,$$

for any real scalars $\alpha, \gamma, s_i \in \mathbb{R}$ $(i = 1, \ldots, 5)$ and any real vector $\boldsymbol{\beta} \in \mathbb{R}^3$. Since a Maxwellian distribution describes the equilibrium state for the Boltzmann equation, the BGK approximation is modeled as a relaxation towards a Maxwellian distribution

$$M(\mathbf{v};\rho,\mathbf{u},T) := \rho\left(\frac{m}{2\pi k_B T}\right)^{3/2} \exp\left(\frac{-m}{2k_B T}|\mathbf{v}-\mathbf{u}|^2\right). \tag{11}$$

We remark the relation of the Boltzmann constant k_B to the specific gas constant $R = k_B/m$. Substituting the Maxwellian distribution for f into (10), the collision term vanishes. In the BGK approximation the collision term reads

$$\Omega(f) := -\frac{1}{\tau_c}\left[f(\mathbf{x},\mathbf{v},t) - f^{(eq)}(\mathbf{x},\mathbf{v},t)\right],$$

where the local equilibrium distribution $f^{(eq)}(\mathbf{x},\mathbf{v},t) = M(\mathbf{v};\rho,\mathbf{u},T)$ is Maxwellian with corresponding ρ, $\mathbf{u}$ and T. The values are chosen such that the aforementioned properties are emulated. Therefore they have to be calculated in terms of $f(\mathbf{x},\mathbf{v},t)$ as given in (4)–(6).
The Boltzmann equation with BGK approximation for the collision term then reads

$$\frac{\partial f(\mathbf{x},\mathbf{v},t)}{\partial t} + \mathbf{v}\cdot\nabla f(\mathbf{x},\mathbf{v},t) = -\frac{1}{\tau_c}\left[f(\mathbf{x},\mathbf{v},t) - f^{(eq)}(\mathbf{x},\mathbf{v},t)\right]. \tag{12}$$

As for the Boltzmann equation (3) the zeroth and first moment of the BGK approximated version (12) lead also to equations (7) and (9).

2. THE LATTICE BOLTZMANN METHOD: BASIC CONCEPTS

The current section revolves completely around the LBM

$$f_i(\mathbf{x}+\mathbf{c}_i\Delta t, t+\Delta t) - f_i(\mathbf{x},t) = -\frac{\Delta t}{\tau_c}\left[f_i(\mathbf{x},t) - f_i^{(eq)}(\mathbf{x},t)\right]. \tag{13}$$

We explain two ways which lead to this equation. On the one hand, in Section 2.1. we demonstrate the historical development from very simple computational concepts to the LBM. The former could already be implemented on the first computing devices in the 1960s, and the LBM can be implemented efficiently on most recent computer architecture, for example a cluster of GPUs. On the other hand, in Section 2.2., we derive the LBM from the Boltzmann equation. This connection to the Boltzmann equation was not known directly with the appearance of the first lattice Boltzmann models in literature. Furthermore we also give a short idea how the Boltzmann equation can be derived from an exact microscopical description of a fluid.
After these sections, which both end with the LBM equation (13), we verify the physical fundament of it by a technique called Chapman-Enskog expansion. We complete Section 2. with providing a basis for beginners who get in touch with the LBM for the first time by briefly explaining how the LBM works and how it can be implemented.

2.1. Classical Approach

The historical development of the LBM started with an adaption of lattice gas automata, which itself can be understood as modified cellular automata. We track this development in the current section by briefly describing cellular automata and lattice gas automata, afterwards we explain the adaption of the lattice gas automata which lead to the LBM. Finally we demonstrate the relation of the LBM to the discrete velocity Boltzmann equation.

2.1.1. *Cellular Automata*

As a universal concept for computation, John von Neumann [38] introduced the cellular automaton (CA) as a discrete model. Imaging a CA with cells $\mathscr{C}$ formally placed on a D-dimensional lattice, i.e., we have $\mathscr{C} \subset \mathbb{Z}^D$. Each cell holds a state from a finite set of possible states $\mathscr{S} \subset \mathbb{Z}$ [25]. At discrete time levels the states are updated simultaneously by prescribed deterministic rules. These rules operate locally, that is, a new state of any interior cell $\boldsymbol{C}_i \in \mathscr{C}$ possibly depends on all old states of neighboring cells including the cell itself. Due to the regular structure of the lattice, we can refer to the neighboring cells by local coordinates $\mathscr{N}$, such that the set of neighbors is given by $\boldsymbol{C}_i + \mathscr{N}$. Boundary cells are updated due to some appropriate update rules.

As an illustrating example, we consider a one-dimensional CA consisting of 400 cells with only two possible states for each cell: $\mathscr{S} = \{0,1\}$. States of cell C_1 and C_{400} are set to 0 and are not updated. The updating rule for cells in the interior shall only depend on the states of the two adjacent cells and the cell itself, i.e., $\mathscr{N} = \{-1,0,1\}$. Hence, any update rule r must assign a new configuration to a local configuration (which are the corresponding three states), that is,

$$\begin{array}{llll} r(0,0,0) = r_0, & r(0,0,1) = r_1, & r(0,1,0) = r_2, & r(0,1,1) = r_3, \\ r(1,0,0) = r_4, & r(1,0,1) = r_5, & r(1,1,0) = r_6, & r(1,1,1) = r_7 \end{array}$$

with corresponding values $r_i \in \mathscr{S}$. This amounts to $2^8 = 256$ different possible update rules for this one-dimensional CA. Then the state s_i of the ith interior cell is updated to s_i':

$$s_i' = r(s_{i-1}, s_i, s_{i+1}), \qquad i = 2, \ldots, 399.$$

2.1.2. *Lattice Gas Automata*

Lattice gas automata (LGA) are derived from classical CA by some modifications, which simplify the construction and application of automata to given physical processes [39]. LGA are capable to simulate fluid flows successfully and lead to the Navier-Stokes equations in the macroscopic limit [41]. And despite of their relative simple nature, they can be applied in less simple themes such as for instance in the simulation of flows through porous media [8].

The first introduced LGA, known as HPP due to Hardy, de Pazzis and Pomeau [19], was proposed as a new technique for the numerical study of the Navier-Stokes equations. In this technique the direct integration of the Navier-Stokes equations is replaced by the simulation of a very simple microscopic system: particles of the same mass are allowed to move on a regular lattice, and local collision rules are introduced on the nodes which conserve the number of particles and momentum. However the momentum flux tensor in the resulting equations has not the correct form such as required by the Navier-Stokes equations.

The first LGA which overcome this problem were introduced by Frisch, Hasslacher and Pomeau in 1986 [16], the so called FHP model. The major difference between the HPP and FHP model consists in the fact that the HPP uses a square lattice and in the FHP model a triangular lattice with more velocities is used. The employment of a triangular lattice leads to a specific isotropic tensor of rank four which is necessary to achieve the correct form of the momentum flux tensor in sense of the Navier-Stokes equations, see [39] for details. The first step towards a formal description of LGA is the definition of a regular lattice. Depending on the lattice, a number of k different lattice vectors $\boldsymbol{d}_i$ $(i = 1, \ldots, k)$ are introduced which connect nearest neighbors. For the FHP model (more precisely, the FHP-I model) in two dimensions with $k = 6$, they read

$$\boldsymbol{d}_i = \left(\cos\left(\frac{\pi}{6}(2i-1)\right), \sin\left(\frac{\pi}{6}(2i-1)\right)\right) \qquad (i = 1, \ldots, 6). \tag{14}$$

The lattice vectors describe the possible particle velocities (and thus possible momenta). Now at certain discrete positions $\boldsymbol{r} \in \mathbb{R}^2$, lattice nodes are located. Each lattice node consists of k cells, which describe whether a particle with corresponding velocity is present or not. Thus we have binary states n_i $(i = 1, \ldots, k)$ with

$$n_i(\boldsymbol{r}, t) = \begin{cases} 0, & \text{cell } i \text{ is not occupied by a particle with velocity } \boldsymbol{d}_i, \\ 1, & \text{cell } i \text{ is occupied by a particle with velocity } \boldsymbol{d}_i. \end{cases}$$

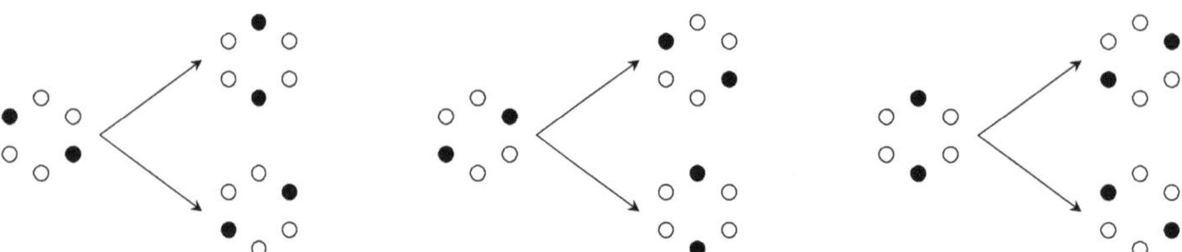

(a) Two particle collisions; conservation of mass and momentum yields two possible configurations after collision. Thus a random choice is applied (with probability one half).

(b) Three particle collisions.

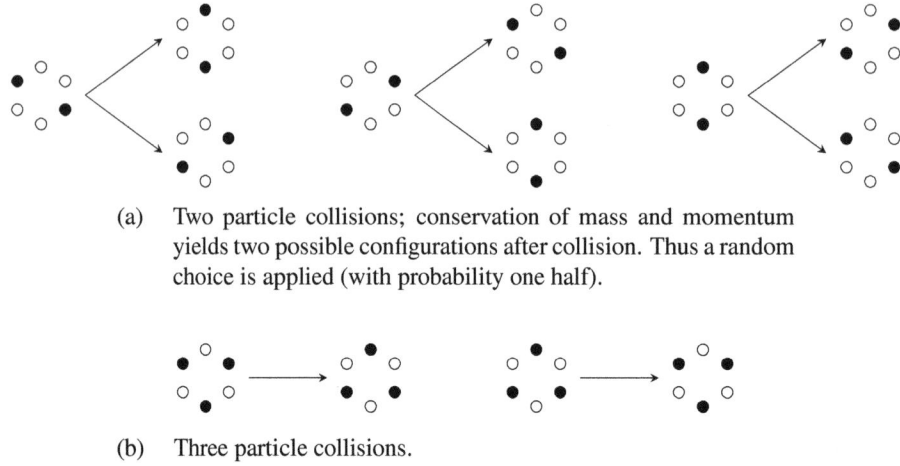

Fig. **1**: Collision rules used in the FHP model. Empty circles indicate a state of 0, whereas solid circles are occupied cells of state 1.

Here, t indicates the time. Due to an exclusion principle each cell is either occupied by one particle or holds no particle at all. All states are updated at discrete time levels simultaneously by an evaluation of a collision rule Ψ_i and a streaming:

$$n_i(\boldsymbol{r}+\boldsymbol{d}_i, t+\Delta t) = n_i(\boldsymbol{r},t) + \Psi_i(\boldsymbol{n}(\boldsymbol{r},t)) \quad \text{with} \quad \boldsymbol{n}(\boldsymbol{r},t) = \left(n_1(\boldsymbol{r},t),\, n_2(\boldsymbol{r},t),\, \ldots,\, n_k(\boldsymbol{r},t)\right)^{\top}. \tag{15}$$

The streaming is encoded in the location of the left-hand side $\boldsymbol{r}+\boldsymbol{d}_i$.
The collision rules $\Psi_i(\boldsymbol{n}(\boldsymbol{r},t))$ should be chosen such that they conserve the number of particles and momentum. We illustrate the collision rules of the FHP model in Fig. **1**. For the two particle head on collision there are two possible outcomes which conserve the number of particles and momentum and one has to take a choice randomly. Configurations not listed in Fig. **1** are not affected due to collision, i.e., only two and three particle collisions are considered.

2.1.3. *Translation to Continuous Variables*

In the following, our main focus is the historical development of LBM from LGA. Using a Chapman-Enskog expansion [6] up to second order terms, the original FHP model was shown to yield the continuous mathematical model [16]:

$$\begin{aligned} \frac{\partial \rho}{\partial t} + \nabla \cdot (\rho \boldsymbol{u}) &= 0, \\ \frac{\partial \rho u_\alpha}{\partial t} + \sum_\beta \frac{\partial}{\partial x_\beta}\left[\gamma(\rho)\rho u_\alpha u_\beta\right] &= -\frac{\partial}{\partial x_\alpha} p + \eta \Delta u_\alpha + \xi \frac{\partial}{\partial x_\alpha}(\nabla \cdot \boldsymbol{u}) \end{aligned}$$

with $\gamma(\rho) = \frac{\rho-3}{\rho-6}$. This result differs from the compressible variant of the Navier-Stokes equations mainly by the presence of the term $\gamma(\rho)$. If this term is a constant, for $0 < \rho < 3$, it can be absorbed in a rescaled time $t \rightsquigarrow \frac{t}{\gamma(\rho)}$. Thus in the incompressible limit, where all density variations are neglected except density variations in the pressure term, the latter equations become

$$\begin{aligned} \nabla \cdot \boldsymbol{u} &= 0, \\ \rho \frac{\partial \boldsymbol{u}}{\partial t} + \rho\gamma(\rho)(\boldsymbol{u} \cdot \nabla)\boldsymbol{u} &= -\nabla p + \eta \Delta \boldsymbol{u}. \end{aligned} \tag{16}$$

Now, the generalized substantial derivative

$$\frac{\partial \boldsymbol{u}}{\partial t} + \gamma \cdot (\boldsymbol{u} \cdot \nabla)\boldsymbol{u}$$

fulfills a Galilean invariance if and only if $\gamma \equiv 1$ (e.g., [39]), which is impossible in the original FHP model. This lack of Galilean invariance in this FHP model is due to the exclusion principle in combination with a finite set of allowed velocities. Though, there exist modifications of this model, in which the Galilean invariance holds [13].

What we have not yet dealt with is the question how the macroscopic quantities in (16) can be derived from LGA. This is specially demanding, since the quantities appearing in (15) are only binary ones. Both, mass density and fluid velocity are computed from local average populations N_i by

$$\rho(\boldsymbol{x},t) = \sum_i N_i(\boldsymbol{x},t), \qquad \rho(\boldsymbol{x},t)\boldsymbol{u}(\boldsymbol{x},t) = \sum_i \boldsymbol{d}_i N_i(\boldsymbol{x},t).$$

The notation N_i indicates an averaging of n_i that corresponds to the lattice vector $\boldsymbol{d}_i$. One procedure to compute N_i is called coarse graining; here one calculates the mean values over large subregions. Another possibility is ensemble averaging. In this approach one simulates the evolution of LGA many times with different initial configurations, and computes the average over all simulations. It could also be used in combination with coarse graining. Although the evolution equation (15) for LGA is quite simple, one needs to spend a high computational effort to obtain reasonable macroscopic quantities by the averaging procedure.

Two years after the introduction of the FHP McNamara and Zanetti [27] suggested to change the binary populations of LGA into real values representing the average values and thus to avoid the averaging process. Nowadays, this paper is seen as the origin of the LBM, even though similar probabilistic views of LGA were common in the literature before [15, 40]. Due to this suggestions, not only the n_i are replaced by continuous variables f_i, but also an arithmetic treatment of the collision term replaces the boolean collision term. Still, the f_i are referred to as *populations*. The BGK model, see Section 1.3., was independently introduced for lattices by [31] and [7] in 1992. Using the discretized BGK model for collision, the LGA model changed into the lattice Boltzmann equation

$$f_i(\boldsymbol{x} + \boldsymbol{c}_i \Delta t, t + \Delta t) - f_i(\boldsymbol{x},t) = -\frac{\Delta t}{\tau_c}\left[f_i(\boldsymbol{x},t) - f_i^{(eq)}(\boldsymbol{x},t)\right], \tag{17}$$

with lattice vectors $\boldsymbol{c}_i$ and an local equilibrium distribution

$$f_i^{(eq)}(\boldsymbol{x},t) = w_i \rho(\boldsymbol{x},t)\left[1 + 3(\boldsymbol{c}_i \cdot \boldsymbol{u}(\boldsymbol{x},t)) + \frac{9}{2}(\boldsymbol{c}_i \cdot \boldsymbol{u}(\boldsymbol{x},t))^2 - \frac{3}{2}|\boldsymbol{u}|^2\right],$$

where the weights w_i depend on the chosen lattice model and the macroscopic quantities are computed by moments

$$\rho(\boldsymbol{x},t) = \sum_i f_i(\boldsymbol{x},t), \qquad \rho(\boldsymbol{x},t)\boldsymbol{u}(\boldsymbol{x},t) = \sum_i \boldsymbol{c}_i f_i(\boldsymbol{x},t).$$

The LBM closes not only the lack of Galilean invariance, see above, but also another drawback of LGA not considered here which is statistical noise [11]. We will consider different lattices below.

2.1.4. *Relation to the Discrete Velocity Boltzmann Equation and Standard Discretizations*

The LBM (17) uses a discrete velocity space, say:

$$\mathcal{V} = \left\{\boldsymbol{c}_i \in \mathbb{R}^3 : i = 0, \ldots, n_v\right\}.$$

Thus it can be understood as a certain discretization of the Boltzmann equation (3). Applying this velocity discretization directly to the Boltzmann equation gives a discrete velocity Boltzmann equation [35]:

$$\frac{\partial f_i(\boldsymbol{x},t)}{\partial t} + \boldsymbol{c}_i \cdot \nabla f_i(\boldsymbol{x},t) = -\frac{1}{\tau_c}\left[f_i(\boldsymbol{x},t) - f_i^{(eq)}(\boldsymbol{x},t)\right], \qquad i = 0, \ldots, n_v, \tag{18}$$

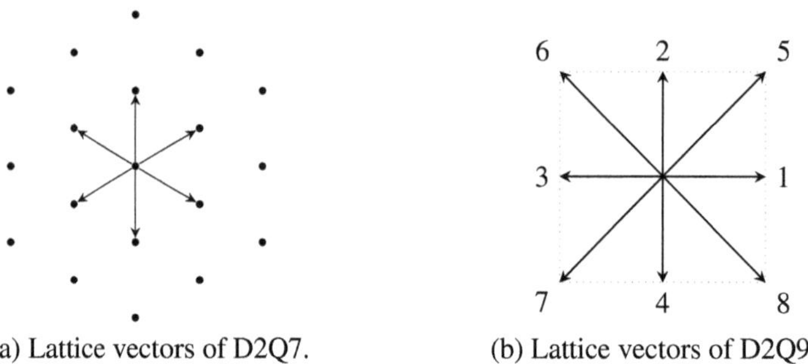

(a) Lattice vectors of D2Q7. (b) Lattice vectors of D2Q9.

Fig. **2**: Standard two dimensional lattice models. – Both cases include the rest velocity $\boldsymbol{c} = \boldsymbol{0}$.

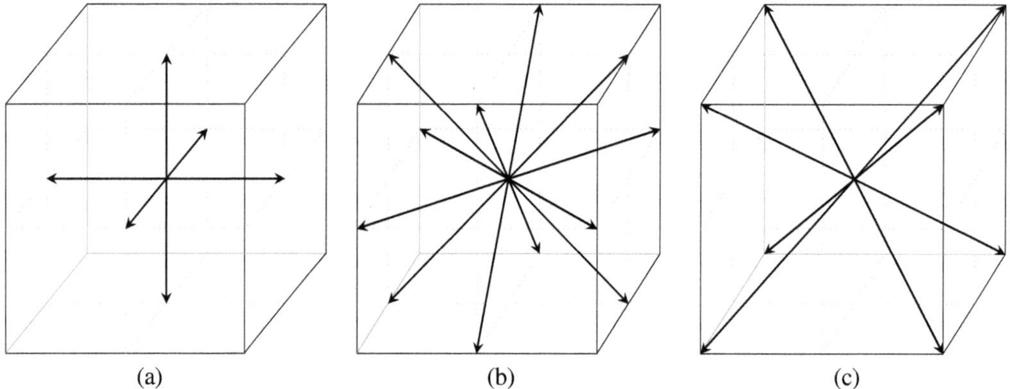

(a) (b) (c)

Fig. **3**: Lattice velocities for cubic lattice in three dimensions.

where the discrete velocity space $\mathscr{V}$ replaces $\mathbb{R}^3$, such that $f(\boldsymbol{x}, \boldsymbol{c}_i, t) \to f_i(\boldsymbol{x}, t)$ and $f^{(eq)}(\boldsymbol{x}, \boldsymbol{c}_i, t) \to f_i^{(eq)}(\boldsymbol{x}, t)$. The set $\mathscr{V}$ has to be interpreted here as a set of vectors defining the lattice. Frequently used lattices in LBM are the D2Q7 and D2Q9 in two dimensions (Fig. **2**) and D3Q15, D3Q19, as well as D3Q27 in three dimensions (Fig. **3**). In the notation DxQy, which goes back to [31], x denotes the spatial dimension and $y(= n_v + 1)$ denotes the number of lattice velocities. D2Q7 without rest velocity is the lattice used in the FHP model; the corresponding velocities are given in (14). Given the reference lattice velocity $c = \frac{\Delta x}{\Delta t}$, the lattice vectors of the more usual D2Q9 model read

$$\boldsymbol{c}_0 = \boldsymbol{0}, \qquad \boldsymbol{c}_i = c\begin{pmatrix}\cos\left(\frac{\pi}{2}(i-1)\right)\\ \sin\left(\frac{\pi}{2}(i-1)\right)\end{pmatrix}, \; i = 1,2,3,4 \qquad \boldsymbol{c}_j = c\begin{pmatrix}\sqrt{2}\cos\left(\frac{\pi}{2}(j-\frac{1}{2})\right)\\ \sqrt{2}\sin\left(\frac{\pi}{2}(j-\frac{1}{2})\right)\end{pmatrix}, \; j = 5,6,7,8.$$

For the three dimensional case, we consider a cubic lattice with cubes of edge length c and lattice nodes located at all vertices. Connecting the nearest neighbors, we achieve six lattice velocities of length c, twelve velocity vectors of length $c\sqrt{2}$ and eight of length $c\sqrt{3}$. See Fig. **3** for an illustration. In addition, we have the zero velocity, which belongs to any discrete velocity set. Different combinations of these lattice velocities yield the well-known three dimensional models: We obtain the D3Q15 model by taking the velocity vectors of length c and $c\sqrt{3}$. Extending this model by the twelve $c\sqrt{2}$-velocity vectors, we get the D3Q27 model. And the popular D3Q19 model is constructed of lattice velocities with length c and $c\sqrt{2}$, see Fig **4**, i.e., the lattice vectors read in this case:

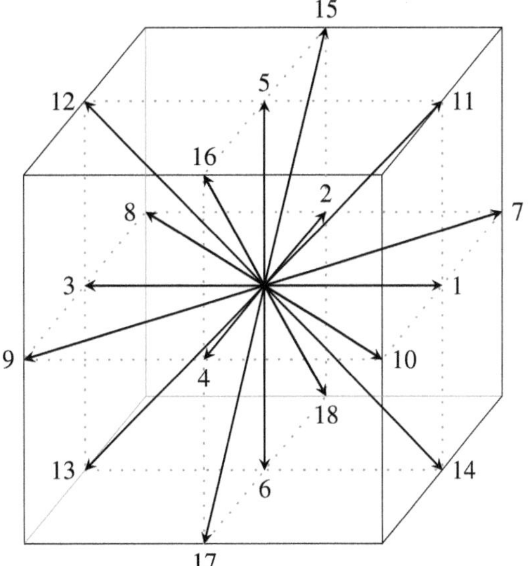

Fig. **4**: The lattice vectors of the D3Q19 model.

$$\begin{aligned}
&\boldsymbol{c}_0 = (0,0,0)^\top, && \boldsymbol{c}_1 = c(1,0,0)^\top, && \boldsymbol{c}_2 = c(0,1,0)^\top && \boldsymbol{c}_3 = c(-1,0,0)^\top, \\
&\boldsymbol{c}_4 = c(0,-1,0)^\top, && \boldsymbol{c}_5 = c(0,0,1)^\top, && \boldsymbol{c}_6 = c(0,0,-1)^\top && \boldsymbol{c}_7 = c(1,1,0)^\top, \\
&\boldsymbol{c}_8 = c(-1,1,0)^\top, && \boldsymbol{c}_9 = c(-1,-1,0)^\top, && \boldsymbol{c}_{10} = c(1,-1,0)^\top, && \boldsymbol{c}_{11} = c(1,0,1)^\top, \\
&\boldsymbol{c}_{12} = c(-1,0,1)^\top, && \boldsymbol{c}_{13} = c(-1,0,-1)^\top, && \boldsymbol{c}_{14} = c(1,0,-1)^\top, && \boldsymbol{c}_{15} = c(0,1,1)^\top, \\
&\boldsymbol{c}_{16} = c(0,-1,1)^\top, && \boldsymbol{c}_{17} = c(0,-1,-1)^\top, && \boldsymbol{c}_{18} = c(0,1,-1)^\top.
\end{aligned} \tag{19}$$

We come now to the space and time discretization of the discrete velocity Boltzmann equation (18). This discretization must be suitable for the velocity space, that is, for all velocity vectors $\boldsymbol{c}_i$ at an interior grid point $\boldsymbol{x}_j$, the sums $\boldsymbol{x}_j + \boldsymbol{c}_i\Delta t$ describe a further grid point, where Δt denotes the time step. Hence, we have a link between space discretization and time discretization. Then the advection term $\boldsymbol{c}_i \cdot \nabla f_i(\boldsymbol{x},t)$ in (18) can be discretized by finite differences as

$$\boldsymbol{c}_i \cdot \nabla f_i(\boldsymbol{x},t) = \frac{f_i(\boldsymbol{x}+\boldsymbol{c}_i\Delta t, t+\Delta t) - f_i(\boldsymbol{x}, t+\Delta t)}{\Delta t} + \mathcal{O}(\Delta t).$$

Using the forward Euler to approximate the time derivative, the fully discretized Boltzmann equation with BGK approximation gives the lattice Boltzmann equation (13). This completes the derivation.

2.2. Theoretical Approach

In the Section 2.1. we have outlined the historical development of the lattice Boltzmann method. In contrast the Section 2.2. explains roughly the theoretical background. First, we give an idea of how the Boltzmann equation can be derived from a microscopical description. Then we explain the derivation of LBM via suitable discretizations with know approximation error.

2.2.1. *Sketched Derivation of the Boltzmann Equation*

For a microscopical view of a fluid, we consider a system consisting of N particles (e.g., molecules) with the mass m each. Let the Hamiltonian $\mathscr{H} = \mathscr{H}(t,\boldsymbol{q},\boldsymbol{p})$ expresses the energy of the system as a function of generalized position coordinates $\boldsymbol{q} = \boldsymbol{q}(t) = (\boldsymbol{q}_1,\dots,\boldsymbol{q}_N)$ and generalized momenta $\boldsymbol{p} = \boldsymbol{p}(t) = (\boldsymbol{p}_1,\dots,\boldsymbol{p}_N)$, then the evolution of that dynamic system can be described by the Hamilton equations [5]:

$$\dot{\boldsymbol{q}}_k = \frac{\partial \mathscr{H}}{\partial \boldsymbol{p}_k}, \quad \dot{\boldsymbol{p}}_k = -\frac{\partial \mathscr{H}}{\partial \boldsymbol{q}_k}, \quad k = 1,\dots,N. \tag{20}$$

A direct consideration of all N particles is ruled out, because in practically all interesting cases the number of particles is too huge, $\mathscr{O}(10^k)$, $k > 20$. Fortunately, in general one is not directly interested in the movement of each specific particle, instead it is often enough to know how many particles with specific velocities are in average present. Therefore, one considers an N-particle distribution function P_N, where

$$P_N(\boldsymbol{q}_1,\boldsymbol{p}_1,\boldsymbol{q}_2,\boldsymbol{p}_2,\dots,\boldsymbol{q}_N,\boldsymbol{p}_N,t)\,\mathrm{d}\boldsymbol{q}_1\dots\mathrm{d}\boldsymbol{q}_N\,\mathrm{d}\boldsymbol{p}_1\dots\mathrm{d}\boldsymbol{p}_N$$

gives the probability to find this N particle ensemble with the ith particle in an infinitesimal small space volume $\mathrm{d}\boldsymbol{q}_i$ around $\boldsymbol{q}_i$ and with momentum in the infinitesimal small volume $\mathrm{d}\boldsymbol{p}_i$ around $\boldsymbol{p}_i$ at time t. The evolution of this probability distribution is expressed by the Liouville equation (see, e.g., [5])

$$\frac{\mathrm{d}P_N}{\mathrm{d}t} = \frac{\partial P_N}{\partial t} + \sum_{k=1}^{N} \dot{\boldsymbol{q}}_k \frac{\partial P_N}{\partial \boldsymbol{q}_k} + \sum_{k=1}^{N} \dot{\boldsymbol{p}}_k \frac{\partial P_N}{\partial \boldsymbol{p}_k} = 0, \tag{21}$$

where the time derivatives $\dot{\boldsymbol{q}}$ and $\dot{\boldsymbol{p}}$ are given by the Hamilton equations (20). The Liouville equation can be transformed into an equivalent system of coupled equations, which successively involves more and more particles. To this end, we introduce s-particle distribution functions by integration with respect to all particles from $s+1$ till N:

$$P_N^{(s)}(\boldsymbol{q}_1,\boldsymbol{p}_1,\boldsymbol{q}_2,\boldsymbol{p}_2,\dots,\boldsymbol{q}_s,\boldsymbol{p}_s,t) = \int P_N(\boldsymbol{q}_1,\boldsymbol{p}_1,\boldsymbol{q}_2,\boldsymbol{p}_2,\dots,\boldsymbol{q}_N,\boldsymbol{p}_N,t)\,\mathrm{d}\boldsymbol{q}_{s+1}\dots\mathrm{d}\boldsymbol{q}_N\,\mathrm{d}\boldsymbol{p}_{s+1}\dots\mathrm{d}\boldsymbol{p}_N,$$

$s = 1,\dots,N-1$. Moreover, we split the force terms for the kth particle (Hamilton equation)

$$\dot{\boldsymbol{p}}_k = -\frac{\partial \mathscr{H}}{\partial \boldsymbol{q}_k} = \boldsymbol{k}_{k,ex} + \sum_{j=1,\,j\neq k}^{N} \boldsymbol{k}_{k,j}$$

into external forces $\boldsymbol{k}_{k,ex}$ and forces coming from interactions with the jth particle $\boldsymbol{k}_{k,j}$. Now, we plan to successively remove the dependency on particles by integration, to this end we write the substitution of these split terms in (21) as follows:

$$\frac{\partial P_N}{\partial t} + \sum_{k=1}^{N} \dot{\boldsymbol{q}}_k \frac{\partial P_N}{\partial \boldsymbol{q}_k} + \sum_{k=1}^{N-1} \left[\left(\boldsymbol{k}_{k,ex} + \sum_{j=1,\,j\neq k}^{N-1} \boldsymbol{k}_{k,j} \right) \frac{\partial P_N}{\partial \boldsymbol{p}_k} + \boldsymbol{k}_{k,N} \frac{\partial P_N}{\partial \boldsymbol{p}_k} \right] + \boldsymbol{k}_{N,ex} \frac{\partial P_N}{\partial \boldsymbol{p}_N} + \sum_{j=1}^{N-1} \boldsymbol{k}_{N,j} \frac{\partial P_N}{\partial \boldsymbol{p}_N} = 0. \tag{22}$$

Each integration of (22) with respect to the coordinates and momenta of the last $N-1$, $N-2$, …, 1 particles yields one equation of the following equivalent system:

$$\begin{aligned}
\frac{\partial P_N^{(1)}}{\partial t} + \dot{\boldsymbol{q}}_1 \cdot \frac{\partial P_N^{(1)}}{\partial \boldsymbol{q}_1} + \boldsymbol{k}_1^{(1)} \cdot \frac{\partial P_N^{(1)}}{\partial \boldsymbol{p}_1} &= (1-N) \int \boldsymbol{k}_{1,2} \cdot \frac{\partial P_N^{(2)}}{\partial \boldsymbol{p}_1}\,\mathrm{d}\boldsymbol{q}_2\,\mathrm{d}\boldsymbol{p}_2 \\
&\vdots \\
\frac{\partial P_N^{(s)}}{\partial t} + \sum_{k=1}^{s} \dot{\boldsymbol{q}}_k \cdot \frac{\partial P_N^{(s)}}{\partial \boldsymbol{q}_k} + \sum_{k=1}^{s} \boldsymbol{k}_k^{(s)} \cdot \frac{\partial P_N^{(s)}}{\partial \boldsymbol{p}_k} &= (s-N) \int \sum_{k=1}^{s} \boldsymbol{k}_{k,s+1} \cdot \frac{\partial P_N^{(s+1)}}{\partial \boldsymbol{p}_k}\,\mathrm{d}\boldsymbol{q}_{s+1}\,\mathrm{d}\boldsymbol{p}_{s+1} \\
&\vdots \\
\frac{\partial P_N^{(N-1)}}{\partial t} + \sum_{k=1}^{N-1} \dot{\boldsymbol{q}}_k \cdot \frac{\partial P_N^{(N-1)}}{\partial \boldsymbol{q}_k} + \sum_{k=1}^{N-1} \boldsymbol{k}_k^{(N-1)} \cdot \frac{\partial P_N^{(N-1)}}{\partial \boldsymbol{p}_k} &= -\int \sum_{k=1}^{N-1} \boldsymbol{k}_{k,N} \cdot \frac{\partial P_N^{(N)}}{\partial \boldsymbol{p}_k}\,\mathrm{d}\boldsymbol{q}_N\,\mathrm{d}\boldsymbol{p}_N
\end{aligned} \tag{23}$$

with abbreviations for the truncated forces

$$\boldsymbol{k}_k^{(s)} := \boldsymbol{k}_{k,ex} + \sum_{j=1,\,j\neq k}^{s} \boldsymbol{k}_{k,j}.$$

E.g., for the last equation of (23) the derivatives with respect to the Nth particle in (22) balance and therefore these terms drop out of the equation. The system (23) is the so-called BBGKY-hierarchy named after Bogoliubov, Born, Green, Kirkwood and Yvon (see, e.g., [5]). The labeling of the particles was arbitrary, but the system is given for the specific particle distribution functions. For our purpose, it would be enough to know if there are particles at given positions, rather than having also the information which specific particles they are. Therefore we scale the specific particle distribution functions to obtain general particle distribution functions $\tilde{P}_N^{(s)}$:

$$\tilde{P}_N^{(s)} = \frac{N!}{(N-s)!} P_N^{(s)}.$$

By this, from (23) we get

$$\frac{\partial \tilde{P}_N^{(s)}}{\partial t} + \sum_{k=1}^{s} \dot{\boldsymbol{q}}_k \cdot \frac{\partial \tilde{P}_N^{(s)}}{\partial \boldsymbol{q}_k} + \sum_{k=1}^{s} \boldsymbol{k}_k^{(s)} \cdot \frac{\partial \tilde{P}_N^{(s)}}{\partial \boldsymbol{p}_k} = -\int \sum_{k=1}^{s} \boldsymbol{k}_{k,s+1} \cdot \frac{\partial \tilde{P}_N^{(s+1)}}{\partial \boldsymbol{p}_k} \,\mathrm{d}\boldsymbol{q}_{s+1}\,\mathrm{d}\boldsymbol{p}_{s+1}, \quad s = 1,\ldots,N-1. \tag{24}$$

The structure of the BBGKY hierarchy is simple, the first equation (i.e., $s=1$) is an equation for the single particle distribution function and it is coupled to the two particle distribution function by the integral on the right hand side. Then the evolution equation for the two particle distribution ($s=2$) couples to the three particle distribution function, and so on. The full system (24) is equivalent to the Liouville equation, however we pay special attention to the first equation, since it will yield the Boltzmann equation: its term on the right hand side covers all the coupling to the other particles and thus describes collisions. An exact modeling of collision as (24) has to incorporate a correlation of particle velocities due to preceding collisions. Now, the Boltzmann equation is based on this first equation with a certain approximation of the right-hand side term $Q(f)$ and the scaled unknown $f(\boldsymbol{x},\boldsymbol{v},t) := m\tilde{P}_N^{(1)}(\boldsymbol{q}_1,\boldsymbol{p}_1,t)$. This reads for the external force density $\boldsymbol{K} = \boldsymbol{k}_k^{(1)}$:

$$\frac{\partial f(\boldsymbol{x},\boldsymbol{v},t)}{\partial t} + \boldsymbol{v} \cdot \frac{\partial f(\boldsymbol{x},\boldsymbol{v},t)}{\partial \boldsymbol{x}} + \boldsymbol{K} \cdot \frac{\partial f(\boldsymbol{x},\boldsymbol{v},t)}{\partial \boldsymbol{v}} = Q(f).$$

This is the Boltzmann equation with external forces incorporated.

2.2.2. *Time Discretization*

For the sake of simplicity we consider the Boltzmann equation without external forces. Later, we incorporate external forces to the LBM again in Section 4.2. Also, we substitute the collision integral by the BGK approximation in the following. Hence, we have

$$\frac{\partial f(\boldsymbol{x},\boldsymbol{v},t)}{\partial t} + \boldsymbol{v} \cdot \boldsymbol{\nabla} f(\boldsymbol{x},\boldsymbol{v},t) = -\frac{1}{\tau_c}\left[f(\boldsymbol{x},\boldsymbol{v},t) - f^{(eq)}(\boldsymbol{x},\boldsymbol{v},t)\right], \tag{25}$$

with single time relaxation parameter τ_c. For the discretization, we follow the strategy from He and Luo [21]. By introducing the advective differential operator $\frac{\mathrm{d}}{\mathrm{d}t} = \frac{\partial}{\partial t} + \boldsymbol{v} \cdot \boldsymbol{\nabla}$, equation (25) can formally be written as a linear first order ordinary differential equation (ODE):

$$\frac{\mathrm{d}}{\mathrm{d}t} f(\boldsymbol{x},\boldsymbol{v},t) + \frac{1}{\tau_c} f(\boldsymbol{x},\boldsymbol{v},t) = \frac{1}{\tau_c} f^{(eq)}(\boldsymbol{x},\boldsymbol{v},t).$$

Formally, its solution at $t+\Delta t$ (with $\Delta t \geq 0$) reads

$$f(\boldsymbol{x}+\boldsymbol{v}\Delta t,\boldsymbol{v},t+\Delta t) = \exp\left(-\frac{\Delta t}{\tau_c}\right) f(\boldsymbol{x},\boldsymbol{v},t) + \frac{1}{\tau_c}\exp\left(-\frac{\Delta t}{\tau_c}\right)\int_0^{\Delta t} \exp\left(\frac{s}{\tau_c}\right) f^{(eq)}(\boldsymbol{x}+\boldsymbol{v}s,\boldsymbol{v},t+s)\,\mathrm{d}s.$$

A Taylor expansion for the integrand (around $s=0$)

$$\exp\left(\frac{s}{\tau_c}\right) f^{(eq)}(\boldsymbol{x}+\boldsymbol{v}s,\boldsymbol{v},t+s) = f^{(eq)}(x,t) + \mathcal{O}(s),$$

Taylor expansion for all exponential functions as well as neglecting all terms of order $\mathscr{O}(\Delta t^2)$ yields the following approximation of the Boltzmann equation as a time discrete formulation:

$$f(\boldsymbol{x}+\boldsymbol{v}\Delta t, \boldsymbol{v}, t+\Delta t) - f(\boldsymbol{x}, \boldsymbol{v}, t) = -\frac{\Delta t}{\tau_c}\left[f(\boldsymbol{x}, \boldsymbol{v}, t) - f^{(eq)}(\boldsymbol{x}, \boldsymbol{v}, t)\right]. \tag{26}$$

This equation already resembles the LBM, but with continuous velocity space. By the lattice construction, a velocity space discretization will imply a space discretization. This is our next topic.

2.2.3. *Finding Appropriate Discrete Velocity Spaces*

A discrete velocity space $\mathscr{V} = \{\boldsymbol{c}_i \in \mathbb{R}^3 : i = 0, \ldots, n_v\}$ is also used to compute the momenta of the distribution f for certain polynomials $\psi(\boldsymbol{v})$, i.e.,

$$I = \int_{\mathbb{R}^3} \psi(\boldsymbol{v}) f(\boldsymbol{x}, \boldsymbol{v}, t)\, \mathrm{d}\boldsymbol{v}$$

via a quadrature rule. Now, an appropriate quadrature rule is required to deliver the necessary momenta sufficiently accurate. In performing a Chapman-Enskog expansion to recover the Navier-Stokes equations, one derives that the first two moments of $f^{(0)}$ and $f^{(1)}$ have to be computed exactly in the numerical integration. The zeroth order term $f^{(0)}$ is equal to the Maxwellian distribution $f^{(eq)}$, the first order term is approximately given by

$$f^{(1)} \approx -\tau_c \left(\frac{\partial^{(1)}}{\partial t} + \boldsymbol{v} \cdot \boldsymbol{\nabla}^{(1)}\right) f^{(eq)}$$

and thus also related to the equilibrium distribution. Finally, one derives from the Chapman-Enskog expansion that the integral I is also given by

$$I = \int_{\mathbb{R}^3} \psi(\boldsymbol{v}) f^{(eq)}(\boldsymbol{x}, \boldsymbol{v}, t)\, \mathrm{d}\boldsymbol{v} \tag{27}$$

for all polynomials $\psi(\boldsymbol{v}) = v_x^m v_y^n v_z^p$ of up to third degree (i.e., $m+n+p \leq 3$). These integrals are needed for the computation of the macroscopic variables ρ and $\boldsymbol{u}$ [21].
Now, a Taylor expansion of the Maxwellian distribution $f^{(eq)}$ around $\boldsymbol{u} = \boldsymbol{0}$ up to second order yields the following approximation

$$f^{(eq)} \approx \rho \left(\frac{1}{2\pi RT}\right)^{3/2} \exp\left(-\frac{|\boldsymbol{v}|^2}{2RT}\right)\left[1 + \frac{1}{RT}(\boldsymbol{v}\cdot\boldsymbol{u}) + \frac{1}{2(RT)^2}(\boldsymbol{v}\cdot\boldsymbol{u})^2 - \frac{1}{2RT}|\boldsymbol{u}|^2\right]. \tag{28}$$

Consequently, we have to evaluate integrals of type

$$\widehat{I} = \int_{\mathbb{R}^3} \exp\left(-\frac{|\boldsymbol{v}|^2}{2RT}\right) \Psi(\boldsymbol{v})\, \mathrm{d}\boldsymbol{v}$$

to recover the Navier-Stokes equations, where $\Psi(\boldsymbol{v}) = v_x^m v_y^n v_z^p$ is a polynomial of degree five at most, i.e., $m+n+p \leq 5$. In order to evaluate the integral numerically but without further approximation error, we have to use a quadrature formula which is exact up to this degree. Using a quadrature formula leads to a discrete equilibrium distribution, which contains the weights and lattice velocities. He and Luo [21] separated the integral into products of one dimensional integrals, where they used 3-point Gauß-Hermite quadrature formula, thus they obtained the D3Q27 lattice Boltzmann model. Our purpose here is to introduce a quadrature formula which yields the D3Q19 model. Using the approximation (28) in the integral (27), we

Table **1**: Index mapping for α.

α	0	1	2	3	4	5	6	7	8	9	10	11	12	13	14	15	16	17	18
i	2	3	2	1	2	2	2	3	1	1	3	3	1	1	3	2	2	2	2
j	2	2	3	2	1	2	2	3	3	1	1	2	2	2	2	3	1	1	3
k	2	2	2	2	2	3	1	2	2	2	2	3	3	1	1	3	3	1	1

obtain

$$I \approx \frac{\rho}{(2\pi RT)^{3/2}} \Bigg[\left(1 - \frac{|\boldsymbol{u}|^2}{2RT}\right) I_{m,n,p} + \frac{1}{RT} \left(u_x I_{m+1,n,p} + u_y I_{m,n+1,p} + u_z I_{m,n,p+1}\right) + \frac{1}{2(RT)^2} \left(u_x^2 I_{m+2,n,p} + u_y^2 I_{m,n+2,p} + u_z^2 I_{m,n,p+2} + 2u_x u_y I_{m+1,n+1,p} + 2u_x u_z I_{m+1,n,p+1} + 2u_y u_z I_{m,n+1,p+1}\right) \Bigg] \tag{29}$$

with

$$I_{m,n,p} := \int_{\mathbb{R}^3} \exp\left(-\frac{|\boldsymbol{v}|^2}{2RT}\right) v_x^m v_y^n v_z^p \, \mathrm{d}\boldsymbol{v}.$$

Approximating the integrals $I_{m,n,p}$ by a Gauß-Hermite type formula on the velocity set of D3Q19, we have

$$I_{m,n,p} = (2RT)^{\frac{m+n+p+3}{2}} \int_{\mathbb{R}^3} \exp\left(-|\boldsymbol{\zeta}|^2\right) \zeta_x^m \zeta_y^n \zeta_z^p \, \mathrm{d}\boldsymbol{\zeta} \approx (2RT)^{\frac{m+n+p+3}{2}} \sum_{(i,j,k)\in J} w_{i,j,k} \zeta_i^m \zeta_j^n \zeta_k^p =: Q_{m,n,p}, \tag{30}$$

with nodes $\zeta_{1/3} = \mp\sqrt{3/2}$, $\zeta_2 = 0$, index set

$$J = \Big\{(1,1,2);\ (1,2,1);\ (1,2,2);\ (1,2,3);\ (1,3,2);\ (2,1,1);\ (2,1,2);\ (2,1,3);\ (2,2,1);\ (2,2,2); (2,2,3);\ (2,3,1);\ (2,3,2);\ (2,3,3);\ (3,1,2);\ (3,2,1);\ (3,2,2);\ (3,2,3);\ (3,3,2)\Big\},$$

and weights

$$\frac{1}{\pi^{3/2}} w_{i,j,k} = \begin{cases} \frac{1}{3} & \text{for } (i,j,k) = (2,2,2) \\ \frac{1}{18} & \text{for } (i,j,k) \in \big\{(1,2,2)\,;\,(2,1,2)\,;\,(2,2,1)\,;\,(2,2,3)\,;\,(2,3,2)\,;\,(3,2,2)\big\} \\ \frac{1}{36} & \text{otherwise.} \end{cases}$$

The quadrature $Q_{m,n,p}$ yields an exact integration for $I_{m,n,p}$ for $m+n+p \leq 5$. Thus, the substitution of these $Q_{m,n,p}$ in (29) gives

$$I = \sum_{\alpha=0}^{18} \rho w_\alpha \xi_i^m \xi_j^n \xi_k^p \left\{1 + \frac{1}{RT}(\boldsymbol{u}\cdot\boldsymbol{\xi}_\alpha) + \frac{1}{2(RT)^2}(\boldsymbol{u}\cdot\boldsymbol{\xi}_\alpha)^2 - \frac{|\boldsymbol{u}|^2}{2RT}\right\} + \mathcal{O}\left(|\boldsymbol{u}|^3\right)$$

with $\xi_i = \sqrt{2RT}\,\zeta_i$, $w_\alpha = \frac{1}{\pi^{3/2}} w_{i,j,k}$ and $\boldsymbol{\xi}_\alpha = (\xi_i, \xi_j, \xi_k)^\top$ corresponding to the index mapping defined by Table **1**. For the discrete equilibrium distribution it follows from (28):

$$f_\alpha^{(eq)} = \rho w_\alpha \left\{1 + \frac{3(\boldsymbol{c}_\alpha\cdot\boldsymbol{u})}{c^2} + \frac{9(\boldsymbol{c}_\alpha\cdot\boldsymbol{u})^2}{2c^4} - \frac{3|\boldsymbol{u}|^2}{2c^2}\right\}, \tag{31}$$

with $c = \|\sqrt{2RT}\zeta_1\| = \sqrt{3RT}$ and corresponding velocities $\boldsymbol{c}_\alpha = \boldsymbol{\xi}_\alpha$, which are exactly the velocities given in (19). In summary, we have

$$\int_{\mathbb{R}^3} \psi(\boldsymbol{v}) f^{(eq)}(\boldsymbol{x},\boldsymbol{v},t)\,\mathrm{d}\boldsymbol{v} = \sum_{\alpha=0}^{18} \psi(\boldsymbol{c}_\alpha) f_\alpha^{(eq)}(\boldsymbol{x},t) = \sum_{\alpha=0}^{18} \psi(\boldsymbol{c}_\alpha)(2\pi RT)^{3/2} \exp\left(\frac{|\boldsymbol{c}_\alpha|^2}{2RT}\right) w_\alpha f^{(eq)}(\boldsymbol{x},\boldsymbol{c}_\alpha,t),$$

and we conclude that moments of $f(\boldsymbol{x},\boldsymbol{v},t)$ can be computed by

$$\int_{\mathbb{R}^3} \psi(\boldsymbol{v}) f(\boldsymbol{x},\boldsymbol{v},t)\,\mathrm{d}\boldsymbol{v} = \sum_{\alpha=0}^{18} \psi(\boldsymbol{c}_\alpha) f_\alpha(\boldsymbol{x},t) \quad \text{with} \quad f_\alpha(\boldsymbol{x},t) := (2\pi RT)^{3/2} \exp\left(\frac{|\boldsymbol{c}_\alpha|^2}{2RT}\right) w_\alpha\, f(\boldsymbol{x},\boldsymbol{c}_\alpha,t).$$

Since the discrete velocity set is determined, equation (26) can be fully discretized, and it follows the system

$$f_\alpha(\boldsymbol{x}+\boldsymbol{c}_\alpha\Delta t, t+\Delta t) - f_\alpha(\boldsymbol{x},t) = -\frac{\Delta t}{\tau_c}\left[f_\alpha(\boldsymbol{x},t) - f_\alpha^{(eq)}(\boldsymbol{x},t)\right], \quad \alpha = 0,\dots,n_v. \tag{32}$$

In particular, it follows that the mass density ρ and fluid velocity $\boldsymbol{u}$ are computed as

$$\rho(\boldsymbol{x},t) = \sum_{\alpha=0}^{18} f_\alpha(\boldsymbol{x},t), \qquad \rho(\boldsymbol{x},t)\boldsymbol{u}(\boldsymbol{x},t) = \sum_{\alpha=0}^{18} \boldsymbol{c}_\alpha f_\alpha(\boldsymbol{x},t). \tag{33}$$

This completes the theoretical derivation. One can show that the LBM can achieve a quadratic convergence order with respect to the grid size, see [35]. For a further finite difference interpretation of the method, we refer to [24].

2.3. Verification of the Physical Fundament: Chapman-Enskog-Expansion

In this section, we draw the relation of the lattice Boltzmann equation (13) to the Navier-Stokes equations for incompressible fluids. To this end, a Taylor expansion in Δt of the lattice Boltzmann equation (32) up to second order terms is performed. This gives

$$f_\alpha(\boldsymbol{x}+\boldsymbol{c}_\alpha\Delta t, t+\Delta t) - f_\alpha(\boldsymbol{x},t) = \Delta t\left(\frac{\partial}{\partial t} + \boldsymbol{c}_\alpha\cdot\nabla\right) f_\alpha(\boldsymbol{x},t) + \frac{1}{2}\Delta t^2\left(\frac{\partial}{\partial t} + \boldsymbol{c}_\alpha\cdot\nabla\right)^2 f_\alpha(\boldsymbol{x},t) + \mathcal{O}(\Delta t^3).$$

Neglecting the third order terms in this expansion and inserting the result into (32) gives:

$$\Delta t\left(\frac{\partial}{\partial t} + \boldsymbol{c}_\alpha\cdot\nabla\right) f_\alpha(\boldsymbol{x},t) + \frac{1}{2}\Delta t^2\left(\frac{\partial}{\partial t} + \boldsymbol{c}_\alpha\cdot\nabla\right)^2 f_\alpha(\boldsymbol{x},t) = -\frac{\Delta t}{\tau_c}\left[f_\alpha(\boldsymbol{x},t) - f_\alpha^{(eq)}(\boldsymbol{x},t)\right]. \tag{34}$$

Next, we expand the distribution $f_\alpha(\boldsymbol{x},t)$ around the equilibrium distribution:

$$f_\alpha(\boldsymbol{x},t) = f_\alpha^{(0)}(\boldsymbol{x},t) + \varepsilon f_\alpha^{(1)}(\boldsymbol{x},t) + \varepsilon^2 f_\alpha^{(2)}(\boldsymbol{x},t) + \mathcal{O}(\varepsilon^3), \tag{35}$$

where the zeroth order $f_\alpha^{(0)}(\boldsymbol{x},t)$ represents the local equilibrium distribution $f_\alpha^{(eq)}(\boldsymbol{x},t)$ and ε is a small parameter. The macroscopic density and velocity compute as follows

$$\begin{aligned}\rho(\boldsymbol{x},t) &= \sum_{\alpha=0}^{n_v} f_\alpha(\boldsymbol{x},t) = \sum_{\alpha=0}^{n_v} f_\alpha^{(0)}(\boldsymbol{x},t),\\ \rho(\boldsymbol{x},t)\boldsymbol{u}(\boldsymbol{x},t) &= \sum_{\alpha=0}^{n_v} \boldsymbol{c}_\alpha f_\alpha(\boldsymbol{x},t) = \sum_{\alpha=0}^{n_v} \boldsymbol{c}_\alpha f_\alpha^{(0)}(\boldsymbol{x},t).\end{aligned} \tag{36}$$

Hence, we can conclude

$$\sum_{\alpha=0}^{n_v} f_\alpha^{(k)}(\boldsymbol{x},t) = 0, \qquad \sum_{\alpha=0}^{n_v} \boldsymbol{c}_\alpha f_\alpha^{(k)}(\boldsymbol{x},t) = 0. \tag{37}$$

for any $k \geq 1$. The differential operator with respect to time is expanded as well

$$\frac{\partial}{\partial t} = \varepsilon \frac{\partial^{(1)}}{\partial t} + \varepsilon^2 \frac{\partial^{(2)}}{\partial t} + \mathcal{O}(\varepsilon^3), \tag{38}$$

whereas the spatial differential operator obeys

$$\boldsymbol{\nabla} = \varepsilon \boldsymbol{\nabla}^{(1)}. \tag{39}$$

Substituting the expansions (35), (38) and (39) in (34) and splitting terms with respect to the order of ε yield the first order equation

$$\varepsilon \left(\frac{\partial^{(1)}}{\partial t} + \boldsymbol{c}_\alpha \cdot \boldsymbol{\nabla}^{(1)} \right) f_\alpha^{(0)}(\boldsymbol{x},t) \approx -\frac{1}{\tau_c} \varepsilon f_\alpha^{(1)}(\boldsymbol{x},t), \tag{40}$$

and the second order equation

$$\varepsilon^2 \left[\left(1 - \frac{\Delta t}{2\tau_c}\right) \left(\frac{\partial^{(1)}}{\partial t} + \boldsymbol{c}_\alpha \cdot \boldsymbol{\nabla}^{(1)} \right) f_\alpha^{(1)}(\boldsymbol{x},t) + \frac{\partial^{(2)}}{\partial t} f_\alpha^{(0)}(\boldsymbol{x},t) \right] \approx -\frac{1}{\tau_c} \varepsilon^2 f_\alpha^{(2)}(\boldsymbol{x},t) \tag{41}$$

which was simplified by using (40). Summing up (40) for $\alpha = 0, \ldots, n_v$ yields

$$\frac{\partial^{(1)}}{\partial t} \rho(\boldsymbol{x},t) + \boldsymbol{\nabla}^{(1)} \cdot (\rho(\boldsymbol{x},t)\boldsymbol{u}(\boldsymbol{x},t)) = 0 \tag{42}$$

by (36) and (37). This corresponds to the zeroth moment. Correspondingly, the first moment results in

$$\frac{\partial^{(1)}}{\partial t} (\rho(\boldsymbol{x},t)\boldsymbol{u}(\boldsymbol{x},t)) + \boldsymbol{\nabla}^{(1)} \cdot P^{(0)}(\boldsymbol{x},t) = 0, \tag{43}$$

with the zeroth order term

$$P^{(0)}(\boldsymbol{x},t) = \sum_{\alpha=0}^{n_v} \boldsymbol{c}_\alpha \boldsymbol{c}_\alpha^\top f_\alpha^{(0)}(\boldsymbol{x},t)$$

of the expression

$$P(\boldsymbol{x},t) = \sum_{\alpha=0}^{n_v} \boldsymbol{c}_\alpha \boldsymbol{c}_\alpha^\top f_\alpha(\boldsymbol{x},t)$$

in the expansion

$$P(\boldsymbol{x},t) = P^{(0)}(\boldsymbol{x},t) + \varepsilon P^{(1)}(\boldsymbol{x},t) + \varepsilon^2 P^{(2)}(\boldsymbol{x},t) + \mathcal{O}(\varepsilon^3).$$

The given form of $f^{(0)}(\boldsymbol{x},t)$ yields

$$P^{(0)}(\boldsymbol{x},t) = \rho(\boldsymbol{x},t)\boldsymbol{u}(\boldsymbol{x},t)\boldsymbol{u}(\boldsymbol{x},t)^\top + p(\boldsymbol{x},t)I$$

with identity matrix I and the pressure in LBMs

$$p(\boldsymbol{x},t) = \frac{c^2}{3} \rho(\boldsymbol{x},t). \tag{44}$$

Hence, equation (43) becomes

$$\frac{\partial^{(1)}}{\partial t} (\rho(\boldsymbol{x},t)\boldsymbol{u}(\boldsymbol{x},t)) + \boldsymbol{\nabla}^{(1)} \cdot \left(\rho(\boldsymbol{x},t)\boldsymbol{u}(\boldsymbol{x},t)\boldsymbol{u}(\boldsymbol{x},t)^\top \right) = -\boldsymbol{\nabla}^{(1)} p(\boldsymbol{x},t). \tag{45}$$

Accordingly, the zeroth and first moment of (41) have to be computed. Like above, the zeroth moment is readily given as

$$\frac{\partial^{(2)}}{\partial t}\rho(\boldsymbol{x},t) = 0. \tag{46}$$

Multiplying (41) with $\boldsymbol{c}_\alpha$ and summing up gives the terms of the momentum equation of second order. It follows

$$\boldsymbol{\nabla}^{(1)} \cdot \left[\left(1 - \frac{\Delta t}{2\tau_c}\right) P^{(1)}(\boldsymbol{x},t)\right] + \frac{\partial^{(2)}}{\partial t}(\rho(\boldsymbol{x},t)\boldsymbol{u}(\boldsymbol{x},t)) = 0. \tag{47}$$

The crucial part is now the computation of the first order term:

$$P^{(1)}(\boldsymbol{x},t) = \sum_{\alpha=0}^{n_v} \boldsymbol{c}_\alpha \boldsymbol{c}_\alpha^\top f_\alpha^{(1)}(\boldsymbol{x},t).$$

Using (40) makes the computation straightforward, a detailed calculation is given for instance in [39]. By using the previous results of lower order, equation (44) and neglecting terms of $\mathcal{O}(|\boldsymbol{u}|^2)$ one obtains

$$P^{(1)}(\boldsymbol{x},t) = -\tau_c \frac{c^2}{3}\rho(\boldsymbol{x},t)\left(\boldsymbol{\nabla}^{(1)}\boldsymbol{u}(\boldsymbol{x},t) + \left[\boldsymbol{\nabla}^{(1)}\boldsymbol{u}(\boldsymbol{x},t)\right]^\top\right).$$

Eventually, by substituting the latter equation into (47), the first moment of (41) is given by

$$\frac{\partial^{(2)}}{\partial t}(\rho(\boldsymbol{x},t)\boldsymbol{u}(\boldsymbol{x},t)) = \left(\frac{2\tau_c - \Delta t}{6}c^2\right)\boldsymbol{\nabla}^{(1)} \cdot \left[\rho(\boldsymbol{x},t)\boldsymbol{\nabla}^{(1)}\boldsymbol{u}(\boldsymbol{x},t) + \rho(\boldsymbol{x},t)\left(\boldsymbol{\nabla}^{(1)}\boldsymbol{u}(\boldsymbol{x},t)\right)^\top\right]. \tag{48}$$

Adding the equations of the zeroth moments, (42) and (46), leads to an approximation up to second order of the continuity equation

$$\frac{\partial}{\partial t}\rho(\boldsymbol{x},t) + \boldsymbol{\nabla}\cdot(\rho(\boldsymbol{x},t)\boldsymbol{u}(\boldsymbol{x},t)) + \mathcal{O}(\varepsilon^3) = 0.$$

Similarly, by addition of the first moments, both first and second order in ε, i.e., (45) and (48), yields

$$\begin{aligned}
&\frac{\partial}{\partial t}(\rho(\boldsymbol{x},t)\boldsymbol{u}(\boldsymbol{x},t)) + \boldsymbol{\nabla}\cdot\left(\rho(\boldsymbol{x},t)\boldsymbol{u}(\boldsymbol{x},t)\boldsymbol{u}(\boldsymbol{x},t)^\top\right) + \mathcal{O}(\varepsilon^3)\\
&= -\boldsymbol{\nabla}p(\boldsymbol{x},t) + \left(\frac{2\tau_c - \Delta t}{6}c^2\right)\boldsymbol{\nabla}\cdot\left[\rho(\boldsymbol{x},t)\boldsymbol{\nabla}\boldsymbol{u}(\boldsymbol{x},t) + \rho(\boldsymbol{x},t)(\boldsymbol{\nabla}\boldsymbol{u}(\boldsymbol{x},t))^\top\right].
\end{aligned}$$

Under the assumption that density fluctuations are negligible these equations coincide with the Navier-Stokes equations for incompressible fluids (1-2). We derive that the kinematic viscosity ν in the LBM is tuned by varying the relaxation parameter τ_c as:

$$\nu = \frac{\eta}{\rho} = \frac{2\tau_c - \Delta t}{6}c^2.$$

This completes the link of LBM to the Navier-Stokes equations.

2.4. The Lattice Boltzmann Method: How It Works

Basically, the LBM can be divided in two separate steps, one called the streaming or propagation step and the other called the collision step. These terms are frequently used when dealing with the LBM, and we will explain them here now.

Each population f_i is linked with the lattice vector $\boldsymbol{c}_i$. Fig. **5** shows the 8 populations for one location in space in the D2Q9 model, where the different lengths of the arrows symbolize the different magnitudes of

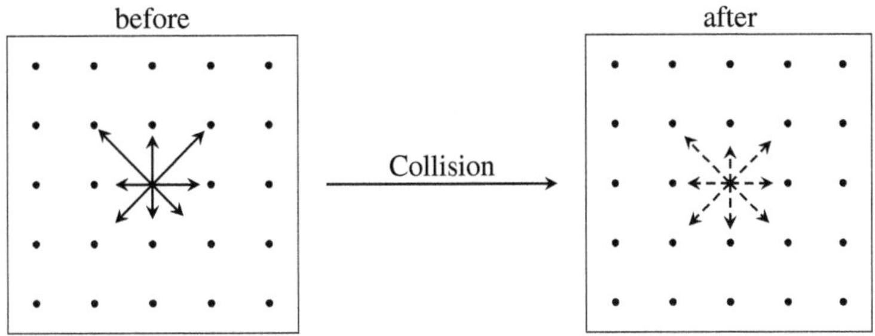

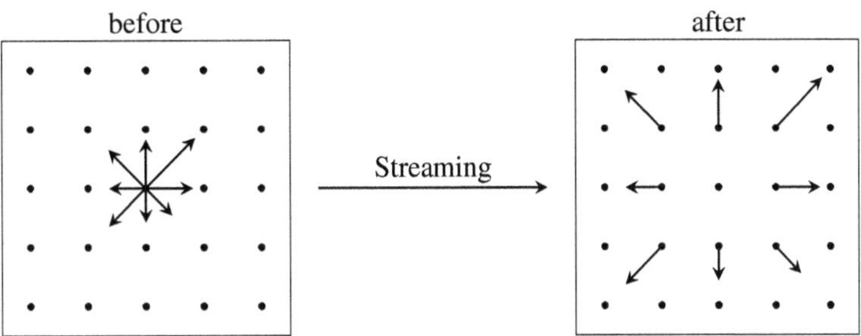

Fig. **5**: Illustration of streaming and collision.

the populations. The magnitude of the population corresponding to the rest velocity is omitted in this picture. Writing the lattice Boltzmann equation (13) as

$$f_i(\boldsymbol{x}+\boldsymbol{c}_i\Delta t, t+\Delta t) = f_i(\boldsymbol{x},t) - \frac{\Delta t}{\tau_c}\left[f_i(\boldsymbol{x},t) - f_i^{(eq)}(\boldsymbol{x},t)\right]$$

separates the two steps. An evaluation of the right hand side, which is done at a fixed time point t yields new values for the populations at any location $\boldsymbol{x}$:

$$\tilde{f}_i(\boldsymbol{x},t) = f_i(\boldsymbol{x},t) - \frac{\Delta t}{\tau_c}\left[f_i(\boldsymbol{x},t) - f_i^{(eq)}(\boldsymbol{x},t)\right].$$

This process is called *collision*. It is illustrated in the first line of Fig. **5**, the dashed arrows correspond to the post-collision populations $\tilde{f}_i(\boldsymbol{x},t)$. The so-called *streaming* or *propagation* step is described by the following equation:

$$f_i(\boldsymbol{x}+\boldsymbol{c}_i\Delta t, t+\Delta t) = \tilde{f}_i(\boldsymbol{x},t).$$

It is noteworthy that the time evolution is exclusively specified by this step. In the streaming step, the post-collision population at position $\boldsymbol{x}$ corresponding to direction $\boldsymbol{c}_i$ becomes the pre-collision population at position $\boldsymbol{x}+\boldsymbol{c}_i\Delta t$, see also the second line of Fig. **5**.

We finish this section with a pseudocode for the LBM on D3Q19 with BGK collision term in Algorithm 1. For sake of simplicity the constant lattice velocity c is set to 1. The algorithm illustrates the principal idea. For an implementation it is necessary to restrict the fluid to a bounded domain, which is the topic of the following section.

***Initialization*:;**
  Set time horizon T_{end} for simulation ;
  Define lattice velocities:;

$$\begin{array}{ll} c_0 = [\ 0;\ 0;\ 0], & c_1 = [\ 1;\ 0;\ 0], \\ c_2 = [\ 0;\ 1;\ 0], & c_3 = [-1;\ 0;\ 0], \\ c_4 = [\ 0;-1;\ 0], & c_5 = [\ 0;\ 0;\ 1], \\ c_6 = [\ 0;\ 0;-1], & c_7 = [\ 1;\ 1;\ 0], \\ c_8 = [-1;\ 1;\ 0], & c_9 = [-1;-1;\ 0], \\ c_{10} = [\ 1;-1;\ 0], & c_{11} = [\ 1;\ 0;\ 1], \\ c_{12} = [-1;\ 0;\ 1], & c_{13} = [-1;\ 0;-1], \\ c_{14} = [\ 1;\ 0;-1], & c_{15} = [\ 0;\ 1;\ 1], \\ c_{16} = [\ 0;-1;\ 1], & c_{17} = [\ 0;-1;-1], \\ c_{18} = [\ 0;\ 1;-1] & \end{array};$$

  Set $t = 0$;
  Choose relaxation parameter τ_c;
  Determine time step Δt;
  Set initial values:;
    foreach *lattice point* $\boldsymbol{x}$ **do**
      for *i=0* **to***18* **do**
        Set initial values $f_i(\boldsymbol{x}, 0)$;
      end
    end

***Time Evolution*:;**
  while $t < T_{end}$ **do**
    ***Collision*:;**
    foreach *lattice point* $\boldsymbol{x}$ **do**
      Compute density: $\rho = \sum_{i=0}^{18} f_i(\boldsymbol{x}, t)$;
      Compute velocity: $u = \frac{1}{\rho} \sum_{i=0}^{18} c_i f_i(\boldsymbol{x}, t)$;
      Compute equilibrium distribution:;
        $f_0^{(eq)} = \frac{1}{3}\rho[1 - \frac{3}{2}|u|^2]$;
      for *i=1* **to***6* **do**
        $f_i^{(eq)} = \frac{1}{18}\rho \left[1 + 3(c_i \cdot u) + \frac{9}{2}(c_i \cdot u)^2 - \frac{3}{2}|u|^2\right]$;
      end
      for *i=7* **to***18* **do**
        $f_i^{(eq)} = \frac{1}{36}\rho \left[1 + 3(c_i \cdot u) + \frac{9}{2}(c_i \cdot u)^2 - \frac{3}{2}|u|^2\right]$;
      end
      for *i=0* **to***18* **do**
        $f_i(\boldsymbol{x}, t) = f_i(\boldsymbol{x}, t) - \frac{\Delta t}{\tau_c}\left[f_i(\boldsymbol{x}, t) - f_i^{(eq)}\right]$;
      end
    end
    ***Streaming*:;**
    foreach *lattice point* $\boldsymbol{x}$ **do**
      for *i=0* **to***18* **do**
        $f_i(\boldsymbol{x} + \boldsymbol{c}_i\Delta t, t + \Delta t) = f_i(\boldsymbol{x}, t)$;
      end
    end
    ***Update*:;**
    $t = t + \Delta t$;
  end

Algorithm 1: Pseudocode for LBM with D3Q19 model.

3. STANDARD BOUNDARY CONDITIONS

For an implementation of the LBM, a restriction to a bounded domain is necessary. An interior lattice point has the full number of neighbors. For these points the main concept of the LBM holds. Nevertheless, there are lattice points which do not have the full number of neighbors, we call these points boundary lattice points.

For interior points, after the streaming step all populations of the density distribution are determined by pre-streaming values of adjacent points with the only exception of the value for the rest particle. For boundary lattice points, there are unknown populations after the streaming step due to missing neighbors. Now, boundary conditions in lattice Boltzmann models have to define the missing populations, in a way which leads to a desired macroscopic behavior at the boundary. There are various possibilities. In general, we find local and nonlocal conditions. A condition is nonlocal if information from neighboring lattice points is used. This can be combined with conditions which solely compute the unknown populations at the boundary or with conditions which additionally modify the known populations at boundary lattice points.

If a local boundary condition only determines the unknown populations, then the action of a boundary condition can be viewed as an additional streaming step, where the unknown populations are streamed from imaginary lattice points located outside of the actual numerical domain.

In the following subsections we will give a short description of the most popular boundary conditions when dealing with the LBM. These are: periodic, bounceback, Inamuro, and Zou/He conditions. We focus on, how they are applied and their relation to physical conditions as no slip, pressure and velocity conditions. For this discussion, we restrict the description of boundary conditions to straight boundaries which are aligned with the main lattice directions. For boundaries not aligned with lattice directions, schemes based on extrapolations like in [10] can be used. A consideration of curved boundaries can be found, e.g., in [28].

3.1. Periodic Boundary Conditions

The boundary conditions which are simplest implemented are the so-called periodic boundary conditions. Here the domain is wrapped around, i.e., corresponding boundary counterparts are identified. The missing neighboring lattice points are replaced with boundary lattice points of the corresponding boundary counterpart, and vice versa. Thus, populations leaving the domain on one side in the streaming step, re-enter the domain at the other side.

This kind of boundary condition is usually applied if the domain of interest and the physical process has a regular symmetry. They are also typically applied in order to get rid of actual physical boundaries, e.g., if one is interested in the behavior of a multi-component mixture for a given initial state where surface effects play a negligible role [36].

3.2. Bounceback Boundary Conditions

This condition is often applied for solid walls and obstacles. Its implementation is also not difficult, because one does not need to consider the orientation of the boundary. For those lattice points where a bounceback condition is assigned, the collision step, i.e., the relaxation towards equilibrium in the BGK model, is omitted. Now all unknown populations are adopted from those of opposite directions. A lattice point with bounceback condition normally does not belong to the fluid domain, but the effective physical boundary lies half way between these bounceback lattice points and the neighboring fluid lattice points, see Fig. **6**. Certainly, the macroscopic quantities (33) cannot be computed in these points, but since they do not belong to the fluid domain, that is not a problem. This kind of boundary condition is often implemented if no-slip boundaries are present, however they do not model a no-slip condition exactly [22]. In fact, in the time average, bounceback yields the no-slip condition. An extended bounceback scheme for moving boundaries was introduced in [4].

3.3. Inamuro No-Slip Boundary Conditions

Inamuro et al. [23] presented an approach such that no-slip boundaries can be realized exactly. To this end, we consider a wall which may move; we denote its velocity and thus the fluid velocity by $\boldsymbol{u}_{wall}$ and the outer

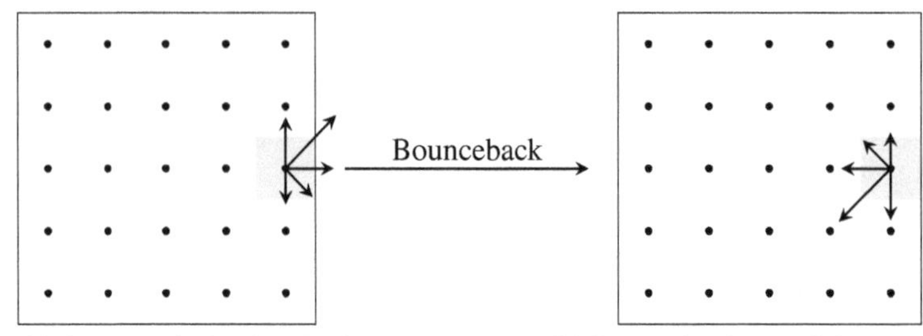

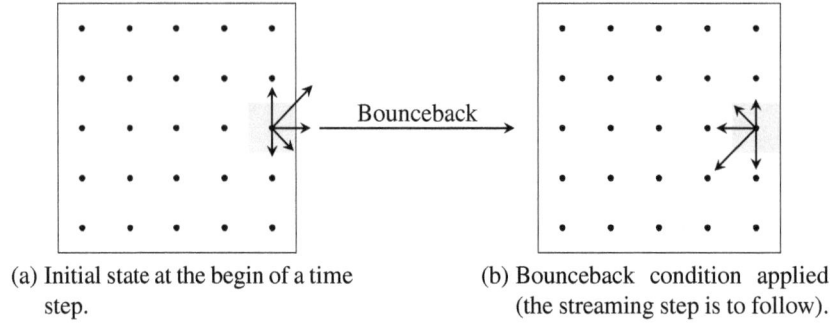

(a) Initial state at the begin of a time step.

(b) Bounceback condition applied (the streaming step is to follow).

Fig. **6**: Illustration of Bounceback Condition for one lattice point (the shaded area represents a boundary or an obstacle).

normal (of the wall as a boundary) is denoted by $\boldsymbol{n}$. For a stationary wall, we simply have to set $\boldsymbol{u}_{wall} = \mathbf{0}$. Moreover, the effective boundary of this wall shall pass through lattice points, such that $\boldsymbol{n}$ is aligned with the main lattice directions. Let us consider a lattice point on this wall, then the macroscopic velocity in the normal direction $u_n := \boldsymbol{u}_{wall} \cdot \boldsymbol{n}$ is fixed in that lattice point (by the wall velocity).
In the following, we aim at assigning only the unknown populations. Before we introduce the needed additional degrees of freedom, we show that the known value u_n already fixes the macroscopic density at the wall. Now, let $\mathscr{I}_- = \{i \in \{0,\ldots,n_v\} \mid f_i \text{ unknown after streaming step}\}$ be the set of indices corresponding to unknown populations. Then we abbreviate the corresponding sum by

$$\rho_- = \sum_{i \in \mathscr{I}_-} f_i.$$

Moreover, for a given lattice velocity index i let i^- be the index of the opposite lattice velocity, i.e., $\boldsymbol{c}_{i^-} = -\boldsymbol{c}_i$. Analogously, we have ρ_+ as

$$\rho_+ = \sum_{i \in \mathscr{I}_+} f_i \quad \text{with} \quad \mathscr{I}_+ = \left\{ i \in \{0,\ldots,n_v\} \mid i^- \in \mathscr{I}_- \right\}.$$

Denoting the sum of remaining populations by ρ_0, the macroscopic quantities (33) are given by

$$\rho = \rho_- + \rho_+ + \rho_0, \qquad \rho u_n = \rho_+ - \rho_-. \tag{49}$$

The equations (49) can be combined to

$$\rho_{wall} := \rho = \frac{2\rho_+ + \rho_0}{1 + \frac{u_n}{c}}, \tag{50}$$

which shows that the density is fixed exclusively by the normal velocity u_n and the known populations.
When modeling the collision as a relaxation towards equilibrium, like in the BGK model, then the discrete equilibrium distribution can be computed already by (31). By simply replacing all populations, also the given ones, by their corresponding equilibrium values would result in a less accurate method, cf. [26]. The idea of the so-called Inamuro boundary conditions is to replace the unknown populations by equilibrium values corresponding to (31) with $\rho \to \rho'$ and velocity $\boldsymbol{u} \to \boldsymbol{u}'$, such that, we satisfy the velocity condition from (33) at the wall:

$$\boldsymbol{u}_{wall} \stackrel{!}{=} \frac{1}{\rho_{wall}} \sum_i \boldsymbol{c}_i f_i(\boldsymbol{x},t). \tag{51}$$

The density ρ' is a free parameter and the velocity $\boldsymbol{u}'$ is the given wall velocity plus a so-called counterslip velocity $\boldsymbol{w}'$, i.e., $\boldsymbol{u}' = \boldsymbol{u}_{wall} + \boldsymbol{w}'$. For $\boldsymbol{w}'$ only tangential wall components are non-zero, that is, $\boldsymbol{w}' \cdot \boldsymbol{n} = 0$. Hence the number of unknown components in $\boldsymbol{u}'$ is the space dimension minus one. For example in a 2D simulation with a bottom wall, it would read $\boldsymbol{u}' = (u_{x,wall} + w'_x, u_{y,wall})^\top$ with the unknown w'_x. Thus, the number of free velocity component(s) in $\boldsymbol{w}'$ plus one unknown density ρ' matches the number of constraints in (51). We note that the wall density is then also automatically fulfilled:

$$\rho_{wall} = \sum_i f_i(\boldsymbol{x},t).$$

This also justifies the need of the free density parameter ρ'. For the D2Q9 model, the solution of (51) can be stated explicitly. For instance at a bottom wall we have

$$\rho' = 6\frac{\rho_{wall}\frac{u_{y,wall}}{c} + \rho_+}{1 + \frac{u_{y,wall}}{c} + 3(\frac{u_{y,wall}}{c})^2}, \qquad w'_x = \frac{6}{\rho'} \cdot \frac{\rho_{wall}\frac{u_{x,wall}}{c} - f_1 + f_3 + f_7 - f_8}{1 + 3\frac{u_{y,wall}}{c}} - \frac{u_{x,wall}}{c}$$

with populations labeled as indicated in Fig. **2**(b). In the D3Q19 model, this procedure yields a nonlinear system of equations, which can be solved for instance by the Newton-Raphson method.
Finally we note that Inamuro boundary conditions can also be used to assign velocity boundary conditions (e.g., for an inlet or outlet) by setting artificial wall velocities.

3.4. Zou/He Boundary Conditions

Zou and He [42] present a possibility to implement both a pressure boundary condition and a velocity boundary condition in lattice Boltzmann simulations. Next, we address both conditions.

3.4.1. *Velocity Condition*

Again, the aim is to find the unknown populations, such that the macroscopic velocity matches a prescribed velocity $\boldsymbol{u}_{wall}$, that is, (51) shall hold. Consider a simulation in d space dimensions and suppose a velocity is prescribed for a boundary lattice point. Then this will result in d constraints for the missing populations. For instance using the D2Q9 model, we have three unknowns (inward directed populations), but only two constraints. This similarly holds for the D3Q19 model. There we have five missing populations but only three relations to fulfill. Hence additional relations are required. Now, Zou and He suggest that the non-equilibrium part of the population normal to the boundary is bounced back. Following this suggestion, we get

$$f_n = f_{n^-} - f_{n^-}^{(eq)} + f_n^{(eq)},$$

where the index n corresponds to an inward normal direction and n^- corresponds to the opposite direction (outer normal). Note that the discrete equilibrium distribution can be computed when a velocity is prescribed, since the consideration leading to (50) is also valid here. Thereby, we now have for the D2Q9 model three equations, which determine the three unknown populations. Unfortunately, the three-dimensional case of D3Q19 is not covered by this consideration, because there are still more unknowns than equations. One way to overcome this lack of constraints is the following. First one computes proposals $f_i^\star$ for the still unknown populations by extending the idea of bouncing back the non-equilibrium part to all unknown populations. Thus for D3Q19 and a boundary with outer normal vector $\boldsymbol{c}_1 = (1,0,0)^\top$ (i.e., $n = 3$, see Fig. **4**), we obtain

$$f_8^\star = f_8 = f_{10} - f_{10}^{(eq)} + f_8^{(eq)}, \qquad f_9^\star = f_9 = f_7 - f_7^{(eq)} + f_9^{(eq)},$$
$$f_{12}^\star = f_{12} = f_{14} - f_{14}^{(eq)} + f_{12}^{(eq)}, \qquad f_{13}^\star = f_{13} = f_{11} - f_{11}^{(eq)} + f_{13}^{(eq)}.$$

By this, the velocity normal to the boundary computed by (33) with the proposals matches the prescribed one. However, generally there is a mismatch for the velocity components tangential to the boundary, i.e., the term

$$\Delta\boldsymbol{j} := \sum_i \boldsymbol{c}_i f_i - \rho_{wall}\boldsymbol{u}_{wall}$$

is in general not equal to the zero vector. Therefore the proposals are adapted to match these velocities as well. This is done by redistributing $\boldsymbol{\Delta j}$ over the proposed populations. In the example above this reads

$$f_i = f_i^\star - \frac{1}{2c^2}\boldsymbol{c}_i \cdot \boldsymbol{\Delta j}, \tag{52}$$

with $i = 8, 9, 12, 13$. Note that in the current example $(\Delta j)_1 = 0$ and therefore does not contribute to the equation above. Also note the coefficient being here $\frac{1}{2}$, because in each case only two populations contribute to a direction, hence the momentum excess $\boldsymbol{\Delta j}$ is split in portions of $\frac{1}{2}$. In D3Q19, the formula (52) can be used for all walls aligned with the main lattice direction. When using the D3Q15 model, the coefficient would read $\frac{1}{4}$ instead.
We conclude this paragraph with some remarks: (i) As we have seen above, a prescribed velocity fixes the density in the boundary lattice points by equation (50). (ii) In principle the statement of part a) can also be used to model a no-slip boundary by setting the velocity appropriately. (iii) Unlike Inamuro boundary conditions, the Zou/He boundary conditions are explicit also in three dimensions.

3.4.2. *Pressure Condition*

In LBMs the pressure is proportional to the density, see (44). Hence, prescribing a pressure in boundary lattice points is equivalent to prescribing a density ρ_{wall}. Writing equation (50) as

$$u_n = \left(\frac{2\rho_+ + \rho_0}{\rho_{wall}} - 1\right) \cdot c$$

shows that a prescribed pressure determines the normal velocity. Now, the easiest way to continue is to additionally enforce the magnitude of the tangential velocity. This leads to the same situation as described for velocity boundary conditions and the missing populations can be found as stated above. A typical choice for the tangential velocity is setting it equal to zero, which gives a no-slip condition.
This completes the standard boundary conditions. We consider now a collection of enhancements of the basic LBM concept.

4. COMMON ENHANCEMENTS OF THE BASIC CONCEPT

So far, we have introduced the basic version of LBM. Many applications, e.g., from computational physics, mechanics and life sciences need enhancements and further couplings to other models. We discuss here multiphase and multicomponent flows and the incorporation of additional forces (as gravitation). Furthermore, we address couplings to other phenomena as temperature and we roughly present an application and a coupling for chip design. There are other important couplings: for example applications in fluid structure interaction (from civil engineering), coupling to chemical reaction kinetics (for, e.g., the behavior of the atmosphere), which we just name here.

4.1. Multiphase/Multicomponent Flows

Up to now we have considered fluids consisting of a single component only. Also our consideration was limited to a single phase of the fluid. However, in practical applications one often has a mixture of several components. The most common way to incorporate these extensions to the LBM is the multiphase/multicomponent model of Shan and Chen [33]. In this model nonlocal interactions among the different species are incorporated.
Next, we shortly summarize how a fluid consisting of C components is modeled in a LBM. Each component evolves, correspondingly to (13), as follows

$$f_i^m(\boldsymbol{x} + \boldsymbol{c}_i\Delta t, t + \Delta t) - f_i^m(\boldsymbol{x}, t) = -\frac{\Delta t}{\tau_m}\left[f_i^m(\boldsymbol{x}, t) - f_i^{m(eq)}(\boldsymbol{x}, t)\right], \quad m = 1, \ldots, C$$

where f_i^m is the single-particle distribution for the mth component. Note that there are C independent relaxation parameter τ_m, which control the viscosity of each component. As usual the $f_i^{m(eq)}$, $m = 1, \ldots, C$, are identical to the equilibrium distribution of a single component simulation, i.e. (31), with accordingly computed macroscopic quantities ρ_m and $\boldsymbol{u}'_m$ [34]:

$$\rho_m(\boldsymbol{x},t) = \sum_i f_i^m(\boldsymbol{x},t),$$

$$\boldsymbol{u}'_m(\boldsymbol{x},t) = \left(\sum_{j=1}^{C} \frac{\rho_j(\boldsymbol{x},t)\boldsymbol{u}_j(\boldsymbol{x},t)}{\tau_j}\right)\left(\sum_{j=1}^{C} \frac{\rho_j(\boldsymbol{x},t)}{\tau_j}\right)^{-1} + \frac{\tau_m}{\rho_m(\boldsymbol{x},t)} F_m(\boldsymbol{x},t)$$

where $F_m(\boldsymbol{x},t)$ models the momentum change of the mth component due to interactions. The velocities $\boldsymbol{u}_m(\boldsymbol{x},t)$ for $m = 1,\ldots,C$ are given by

$$\rho_m(\boldsymbol{x},t)\boldsymbol{u}_m(\boldsymbol{x},t) = \sum_i \boldsymbol{c}_i f_i^m(\boldsymbol{x},t).$$

The interaction force $F_m(\boldsymbol{x},t)$ can be expressed by (e.g., $\Psi = \rho$)

$$F_m(\boldsymbol{x},t) = -\Psi(\boldsymbol{x},t) \sum_{j=1}^{C} G_{m,j} \sum_i \Psi(\boldsymbol{x} + \boldsymbol{c}_i\Delta t, t)\boldsymbol{c}_i,$$

with a symmetric matrix $G = (G_{m,j}) \in \mathbb{R}^{C\times C}$ having diagonal elements equal to zero. The entry $G_{m,j}$ controls the strength of the interaction between the components m and j.

4.2. Additional Forces

Typically in real experimentsm one has to incorporate external forces, which act on the fluid. One example is gravity. There are many other circumstances which make it desirable to be able to incorporate internal or external forces in numerical simulations.

On the macroscopic scale, where the Navier-Stokes equations describe the motion of the fluid, forces can be incorporated by simply adding an acceleration term corresponding to the desired force. Thus, in the lattice Boltzmann sense it is the task to find a modification which leads to

$$\begin{aligned} \boldsymbol{\nabla}\cdot\boldsymbol{u} &= 0, \\ \frac{\partial \boldsymbol{u}}{\partial t} + (\boldsymbol{u}\cdot\boldsymbol{\nabla})\boldsymbol{u} &= -\boldsymbol{\nabla}\hat{p} + \nu\Delta\boldsymbol{u} + \boldsymbol{k}, \end{aligned} \tag{53}$$

in the multiscale expansion where $\boldsymbol{k}$ is the acceleration due to the force density $\boldsymbol{K} = \rho\boldsymbol{k}$ (and $\hat{p} = p/\rho$). One possibility for realization is a modification proposed by Guo et al. [17]. We briefly explain the necessary modifications. Let $\boldsymbol{K}(\boldsymbol{x},t)$ be a given force density, then a corresponding discrete quantity reads

$$K_i(\boldsymbol{x},t) = \left(1 - \frac{\Delta t}{2\tau_c}\right) w_i \left[3\frac{(\boldsymbol{c}_i - \boldsymbol{u}^\star(\boldsymbol{x},t))}{c^2} + 9\frac{(\boldsymbol{c}_i\cdot\boldsymbol{u}^\star(\boldsymbol{x},t))}{c^4}\boldsymbol{c}_i\right]\cdot\boldsymbol{K}(\boldsymbol{x},t), \tag{54}$$

where the adapted velocity $\boldsymbol{u}^\star$ is now computed by

$$\boldsymbol{u}^\star(\boldsymbol{x},t) = \frac{1}{\rho(\boldsymbol{x},t)}\left(\sum_{i=0}^{n_v} \boldsymbol{c}_i f_i(\boldsymbol{x},t) + \frac{\boldsymbol{K}(\boldsymbol{x},t)}{2}\right) \tag{55}$$

replacing (33). The discrete force density (54) is used in a modified collision step. The modified lattice Boltzmann method satisfying (53) reads

$$f_i(\boldsymbol{x} + \boldsymbol{c}_i\Delta t, t + \Delta t) - f_i(\boldsymbol{x},t) = -\frac{\Delta t}{\tau_c}\left[f_i(\boldsymbol{x},t) - f_i^{(eq)}(\boldsymbol{x},t)\right] + \Delta t\, K_i(\boldsymbol{x},t).$$

4.3. Thermal Coupling

The goal of this subsection is to develop a numerical method to simulate the temperature evolution coupled with fluid simulation. Therefore we review a method which extends the lattice Boltzmann method by a temperature T (respectively an internal energy E), which is advected by the fluid's flow and diffuses in it. This is to say, the temperature/energy obeys an advection-diffusion equation

$$\frac{\partial}{\partial t}(\rho(\boldsymbol{x},t)E(\boldsymbol{x},t)) + \boldsymbol{\nabla}\cdot(\rho(\boldsymbol{x},t)\boldsymbol{u}(\boldsymbol{x},t)E(\boldsymbol{x},t)) = \chi\Delta(\rho(\boldsymbol{x},t)E(\boldsymbol{x},t)), \tag{56}$$

with thermal diffusivity χ. The temperature T is related to the internal energy by

$$E(\boldsymbol{x},t) = \frac{3}{2}\frac{k_B T(\boldsymbol{x},t)}{m}$$

in $\mathbb{R}^3$. Like viscosity in the LBM, a parameter which can be chosen to tune the thermal diffusivity is available in the numerical scheme.

Basically, the advection-diffusion equation (56) is coupled to the Navier-Stokes equations only in one way. The fluid is not affected due to the temperature, this means it behaves like an isothermal fluid, which especially means that the viscosity is temperature independent. Consequently, this limits the application of the proposed method to those problems where the influence of the temperature on the viscosity of the fluid is small. However, an external force dependent on the temperature or more precisely temperature variations can be incorporated to the basic LBM. By this we indirectly achieve a two way coupling. Furthermore, the compression work done by the pressure and the viscous heat dissipation are neglected in the method we state here. When simulating the motion of real incompressible fluids in applications these terms are (very often) negligible. A method which incorporates also the compression work done by the pressure and the viscous heat dissipation is presented in [20].

Towards the LBM, we first derive an analog to the Boltzmann equation for the temperature. Recalling (6), we define a new distribution, calling it the energy distribution, by

$$g(\boldsymbol{x},\boldsymbol{v},t) = \frac{|\boldsymbol{v}-\boldsymbol{u}(\boldsymbol{x},t)|^2}{2} f(\boldsymbol{x},\boldsymbol{v},t).$$

The derivatives of g can be computed in terms of the single particle distribution f. The latter is a solution of the Boltzmann equation, therefore we get by straightforward calculations

$$\frac{\partial g(\boldsymbol{x},\boldsymbol{v},t)}{\partial t} + \boldsymbol{v}\cdot\boldsymbol{\nabla} g(\boldsymbol{x},\boldsymbol{v},t) = \frac{|\boldsymbol{v}-\boldsymbol{u}(\boldsymbol{x},t)|^2}{2} Q(f) - f(\boldsymbol{x},\boldsymbol{v},t)q(\boldsymbol{x},\boldsymbol{v},t), \tag{57}$$

with

$$q(\boldsymbol{x},\boldsymbol{v},t) = (\boldsymbol{v}-\boldsymbol{u}(\boldsymbol{x},t))\cdot\left[\frac{\partial \boldsymbol{u}(\boldsymbol{x},t)}{\partial t} + (\boldsymbol{v}\cdot\boldsymbol{\nabla})\boldsymbol{u}(\boldsymbol{x},t)\right].$$

The term $f(\boldsymbol{x},\boldsymbol{v},t)q(\boldsymbol{x},\boldsymbol{v},t)$ only contributes to the compression work and the heat dissipation. As explained above these ingredients are neglected here, therefore we can neglect this term in (57) as well. Furthermore we approximate

$$\frac{|\boldsymbol{v}-\boldsymbol{u}(\boldsymbol{x},t)|^2}{2} Q(f) \approx -\frac{1}{\tau_d}\left[g(\boldsymbol{x},\boldsymbol{v},t) - g^{(eq)}(\boldsymbol{x},\boldsymbol{v},t)\right]$$

in a BGK consistent way with

$$g^{(eq)}(\boldsymbol{x},\boldsymbol{v},t) := \frac{|\boldsymbol{v}-\boldsymbol{u}(\boldsymbol{x},t)|^2}{2} f^{(eq)}(\boldsymbol{x},\boldsymbol{v},t). \tag{58}$$

Thus, we end up with an evolution equation for the energy distribution

$$\frac{\partial g(\boldsymbol{x},\boldsymbol{v},t)}{\partial t} + \boldsymbol{v}\cdot\boldsymbol{\nabla} g(\boldsymbol{x},\boldsymbol{v},t) = -\frac{1}{\tau_d}\left[g(\boldsymbol{x},\boldsymbol{v},t) - g^{(eq)}(\boldsymbol{x},\boldsymbol{v},t)\right]. \tag{59}$$

Due to (6) it holds

$$\int_{\mathbb{R}^3} g(\boldsymbol{x},\boldsymbol{v},t)\,\mathrm{d}\boldsymbol{v} = \rho(\boldsymbol{x},t)E(\boldsymbol{x},t),$$

and the first two moments of the energy equilibrium distribution can be computed explicitly

$$\begin{aligned}\int_{\mathbb{R}^3} g^{(eq)}(\boldsymbol{x},\boldsymbol{v},t)\,\mathrm{d}\boldsymbol{v} &= \rho(\boldsymbol{x},t)E(\boldsymbol{x},t),\\ \int_{\mathbb{R}^3} \boldsymbol{v}g^{(eq)}(\boldsymbol{x},\boldsymbol{v},t)\,\mathrm{d}\boldsymbol{v} &= \rho(\boldsymbol{x},t)\boldsymbol{u}(\boldsymbol{x},t)E(\boldsymbol{x},t).\end{aligned} \tag{60}$$

The Chapman-Enskog expansion of (59) verifies that the outcome matches the advection-diffusion equation (56) using (60).
A comparison of the structure of equations (59) and (12) shows that they differ only in a renaming of the distribution $f \to g$. Hence a scheme like (13) can be used, it follows

$$g_\alpha(\boldsymbol{x}+\boldsymbol{c}_\alpha\Delta t, t+\Delta t) - g_\alpha(\boldsymbol{x},t) = -\frac{\Delta t}{\tau_d}\left[g_\alpha(\boldsymbol{x},t) - g_\alpha^{(eq)}(\boldsymbol{x},t)\right]. \tag{61}$$

The parameter τ_d in this equation can be used to tune the thermal diffusivity [30]. The equation (61) is not well defined, since the discrete energy equilibrium distribution is not determined. However, using the definitions of the energy equilibrium distribution (58) and the Maxwellian distribution (11), a Taylor expansion like in the derivation of (28) yields

$$\begin{aligned}g^{(eq)} = \rho E\left(\frac{1}{2\pi RT}\right)^{3/2}&\exp\left(-\frac{|\boldsymbol{v}|^2}{2RT}\right)\left[\frac{|\boldsymbol{v}|^2}{3RT} + \left(\frac{|\boldsymbol{v}|^2}{3RT} - \frac{2}{3}\right)\frac{(\boldsymbol{v}\cdot\boldsymbol{u})}{RT} + \frac{(\boldsymbol{v}\cdot\boldsymbol{u})^2}{2(RT)^2} - \frac{|\boldsymbol{u}|^2}{2RT}\right]\\ +\rho E\left(\frac{1}{2\pi RT}\right)^{3/2}&\exp\left(-\frac{|\boldsymbol{v}|^2}{2RT}\right)\left[\left(\frac{|\boldsymbol{v}|^2}{3RT} - \frac{7}{3}\right)\frac{(\boldsymbol{v}\cdot\boldsymbol{u})^2}{2(RT)^2} - \left(\frac{|\boldsymbol{v}|^2}{3RT} - \frac{5}{3}\right)\frac{|\boldsymbol{u}|^2}{2RT}\right] + \mathcal{O}(|\boldsymbol{u}|^3).\end{aligned}$$

The second line has no contribution to the integrals (60), hence this addend can be neglected. Thus the quadrature formula (30) can be used for the discrete computation of the integrals (60):

$$\int_{\mathbb{R}^3} \psi(\boldsymbol{v})g^{(eq)}(\boldsymbol{x},\boldsymbol{v},t)\,\mathrm{d}\boldsymbol{v} \approx \sum_\alpha \psi(\boldsymbol{c}_\alpha)g_\alpha^{(eq)},$$

with the discrete energy equilibrium distribution

$$g_\alpha^{(eq)} = \rho E w_\alpha\left\{\frac{|\boldsymbol{c}_\alpha|^2}{c^2} + \left(\frac{|\boldsymbol{c}_\alpha|^2}{c^2} - \frac{2}{3}\right)\frac{3(\boldsymbol{c}_\alpha\cdot\boldsymbol{u})}{c^2} + \frac{9(\boldsymbol{c}_\alpha\cdot\boldsymbol{u})^2}{2c^4} - \frac{3|\boldsymbol{u}|^2}{2c^2}\right\}. \tag{62}$$

This completes the thermal extension.

4.4. Circuit Coupling

As one specific example, we want to model the coupling of integrated electric circuits using a fluid (air) current produced for instance by some fan. Let $\Omega \subset \mathbb{R}^3$ denote the fluid region, where air can flow. Due to (12) and (59) the fluid and temperature inside Ω is governed by

$$\begin{aligned}\frac{\partial f(\boldsymbol{x},\boldsymbol{v},t)}{\partial t} + \boldsymbol{v}\cdot\boldsymbol{\nabla}f(\boldsymbol{x},\boldsymbol{v},t) &= -\frac{1}{\tau_c}\left[f(\boldsymbol{x},\boldsymbol{v},t) - f^{(eq)}(\boldsymbol{x},\boldsymbol{v},t)\right],\\ \frac{\partial g(\boldsymbol{x},\boldsymbol{v},t)}{\partial t} + \boldsymbol{v}\cdot\boldsymbol{\nabla}g(\boldsymbol{x},\boldsymbol{v},t) &= -\frac{1}{\tau_d}\left[g(\boldsymbol{x},\boldsymbol{v},t) - g^{(eq)}(\boldsymbol{x},\boldsymbol{v},t)\right].\end{aligned}$$

The relation to the temperature is as follows (cf. (6)):

$$T(\boldsymbol{x},t) = \frac{2}{3\rho(\boldsymbol{x},t)R} \int_{\mathbb{R}^3} g(\boldsymbol{x},\boldsymbol{v},t)\,\mathrm{d}\boldsymbol{v}.$$

The electric network model shall be described by the modified nodal analysis (MNA), see for instance [14]:

$$\frac{\mathrm{d}}{\mathrm{d}t}(A\boldsymbol{q}(\boldsymbol{e},\mathbf{j})) = \boldsymbol{h}(\boldsymbol{e},\mathbf{j},t)$$

with unknowns $\boldsymbol{e}$ and $\mathbf{j}$ being the node voltages and currents through voltage defining elements (voltage sources, inductors), respectively. The term $\boldsymbol{q}$ models charges and fluxes, $\boldsymbol{h}$ is the so-called static part including independent sources. A describes the incidence matrix, that matches the components of $\boldsymbol{q}$ to the corresponding nodes.
We notice that the term $\boldsymbol{h}$ includes thermally relevant resistors with incidence matrix A_T. We separate $\boldsymbol{h}$ for the following using the corresponding conduction matrix G_T:

$$\boldsymbol{h}(\boldsymbol{e},\mathbf{j},t) = \tilde{\boldsymbol{h}}(\boldsymbol{e},\mathbf{j},t) + A_T G_T A_T^\top \boldsymbol{e}$$

We remark that within MNA no geometry information is present, which we need here to supply such that the physical location of thermally relevant elements is defined.
The coupling of the models for fluid/temperature and the circuit consists in two ways. In the circuit model devices depend on an environment temperature, and on the other hand, the dissipated power of the electric circuit is an energy input for the temperature model. Let the ith thermal active element show a common interface Γ_i with the fluid, the environment temperature $T_{\mathrm{env},i}$ of this ith thermal element can be obtained directly out of the fluid/temperature model, e.g., by averaging:

$$T_{\mathrm{env},i} = \frac{1}{||\Gamma_i||} \int_{\Gamma_i} T(\boldsymbol{x},t)\,\mathrm{d}x = \frac{2}{3R||\Gamma_i||} \int_{\Gamma_i} \frac{1}{\rho} \int_{\mathbb{R}^3} g(\boldsymbol{x},\boldsymbol{v},t)\,\mathrm{d}\boldsymbol{v}\,\mathrm{d}x.$$

To set up a coupling in the other direction, we need to compute the dissipated power of the electric circuit. Using all thermally relevant resistors for heat dissipation we obtain the dissipated power for the ith element as [1]

$$P_i = e_i \cdot j_i = \frac{1}{R_i} e_i^2$$

(with resistance R_i). This is a production term for the energy balance of the ith thermal element. Furthermore, we model the energy transfer to the fluid via Newton cooling on the active boundary Γ_i, such that we obtain

$$c_v \frac{\mathrm{d}}{\mathrm{d}t} T_i = P_i - \gamma ||\Gamma_i|| (T_i - T_{\mathrm{env},i})$$

with the lumped heat capacity c_v and the parameter γ denoting the transfer coefficient. Then the boundary condition can be formulated for T (for the heat flux):

$$\lambda \left(\boldsymbol{n} \cdot \frac{\partial T(\boldsymbol{x},t)}{\partial \boldsymbol{x}} \right) = -\gamma (T_i - T_{\mathrm{env},i}) \quad \text{on} \quad \Gamma_i, \tag{63}$$

where $\boldsymbol{n}$ is the unit outer normal vector of the boundary Γ_i and λ denotes the heat conductivity (plus further boundary conditions). Assuming the boundary conditions for the fluid, i.e., for f, and the conditions for temperature on remaining boundaries are determined, the continuous problem is fully defined. Next we have to consider how the continuous description can be transferred to the discrete model. We again restrict this consideration to boundaries aligned with the main lattice direction. We focus only on the transfer of the boundary condition (63) to the LBM (61). Due to the assumption of alignment of the boundaries with the main lattice direction, except for sign the normal vector is a unit vector, hence the left hand side of (63) is simply a partial derivative. Especially, there is a lattice velocity with $\boldsymbol{c}_\beta = c\boldsymbol{n}$. The non-central finite difference quotient of second order can be used to compute a temperature on the boundary, which embodies the von Neumann condition (63):

$$T(\boldsymbol{x},t) = \frac{1}{3}\left[4T(\boldsymbol{x}+\boldsymbol{c}_\beta\Delta t,t) - T(\boldsymbol{x}+2\boldsymbol{c}_\beta\Delta t,t) - 2c\Delta t\gamma\frac{1}{\lambda}(T_i - T_{\mathrm{env},i})\right]. \tag{64}$$

Therefore we have transferred the von Neumann boundary condition to a Dirichlet boundary condition. With the temperature (64) an equilibrium distribution according to (62) can be computed.

CONCLUSIONS

In this chapter we have recapitulated the derivation of LBM from two perspectives: one was the historical development from lattice gas automata and the other was a more recent derivation using certain discretizations and Gauß-Hermite quadrature. The second approach was here transferred to the D3Q19 lattice. A pseudocode for LBM using D3Q19 concluded this section. Then the basic and standard boundary conditions were explained with remarks for implementation and applicability. After this part, we have discussed a couple of enhancements for LBM and further trends. We have also added a rough mathematical modeling for a coupled system from chip design, where the fluid and temperature could be simulated via LBM. This thermal electric coupling becomes more and more important due to the increasing power challenges in chip design.

ACKNOWLEDGEMENT

Declared none.

CONFLICT OF INTEREST

The authors confirm that this chapter content has no conflict of interest.

REFERENCES

[1] Bartel A. Partial Differential-Algebraic Models in Chip Design-Thermal and Semiconductor Problems. VDI Verlag, Düsseldorf, Germany, 2004.
[2] Batchelor, GK. An Introduction to Fluid Dynamics. Cambridge University Press, New York, USA, 2000.
[3] Bhatnagar PL, Gross EP, Krook M. A Model for Collision Processes in Gases. I. Small Amplitude Processes in Charged and Neutral One-Component Systems. Phys Rev 1954; 94: 511-525.
[4] Bouzidi, M, Firdaouss, M, Lallemand, P. Momentum transfer of a Boltzmann-lattice fluid with boundaries. Phys Fluids 2001; 13: 3452-3459.
[5] Cercignani C. The Boltzmann equation and its applications. Springer Verlag, New York, USA, 1988.
[6] Chapman S, Cowling T. The Mathematical Theory of Non-Uniform Gases. Cambridge University Press, London, UK, 1970.
[7] Chen H, Chen S, Matthaeus WH. Recovery of the Navier-Stokes equations using a lattice-gas Boltzmann method. Phys Rev A 1992; 45: R5339-R5342.
[8] Chen S, Diemer K, Doolen GD, Eggert K, Fu C, Gutman S, Travis BJ. Lattice gas automata for flow through porous media. Physica D: Nonlin Phen 1991; 47: 72-84.
[9] Chen S, Doolen GD. Lattice Boltzmann Method for Fluid Flows. Annu Rev Fluid Mechanics 1998; 30: 329-364.
[10] Chen S, Martinez D, Mei R. On boundary conditions in lattice Boltzmann methods. Phys Fluids 1996; 8: 2527-2536.
[11] Dahlburg JP, Montgomery D, Doolen GD. Noise and compressibility in lattice-gas fluids. Phys Rev A 1987; 36: 2471-2474.
[12] d'Humieres D, Ginzburg I, Krafczyk M, Lallemand P, Luo L-S. Multiple-relaxation-time lattice Boltzmann models in three dimensions. Phil Trans Royal Soc A 2002; 360: 437-451.
[13] d'Humieres D, Lallemand P, Searby G. Numerical Experiments on Lattice Gases: Mixtures and Galilean Invariance. Complex Systems 1987; 1: 633-647.
[14] Feldmann U, Günther M. CAD-based electric-circuit modeling in industry I: mathematical structure and index of network equations. Surveys Math Industry 1999; 8: 97-129.
[15] Frisch U, d'Humieres D, Hasslacher B, Lallemand P, Pomeau Y, Rivet J-P. Lattice gas hydrodynamics in two and three dimensions. Complex Systems 1987; 1: 649-707.
[16] Frisch U, Hasslacher B, Pomeau Y. Lattice-Gas Automata for the Navier-Stokes Equation. Phys Rev Lett 1986; 56: 1505-1508.
[17] Guo Z, Zheng C, Shi B. Discrete lattice effects on the forcing term in the lattice Boltzmann method. Phys Rev E 2002; 65: 046308.
[18] Hänel D. Molekulare Gasdynamik. Springer Verlag, 2004.
[19] Hardy J, Pomeau Y, de Pazzis O. Time evolution of a two-dimensional model system. I. Invariant states and time correlation functions. J Math Phys 1973; 14: 1746-1759.
[20] He X, Chen S, Doolen GD. A novel thermal model for the lattice Boltzmann method in incompressible limit. J Comput Phys 1998; 146: 282-300.
[21] He X, Luo L-S. Theory of the lattice Boltzmann method: From the Boltzmann equation to the lattice Boltzmann equation. Phys Rev E 1997; 56: 6811-6817.
[22] He X, Zou Q, Luo L-S, Dembo M. Analytic solutions of simple flows and analysis of nonslip boundary conditions for the lattice Boltzmann BGK model. J Stat Phys 1997; 87: 115-136.
[23] Inamuro T, Yoshino M, Ogino F. A non-slip boundary condition for lattice Boltzmann simulations. Phys Fluids 1995; 7: 2928-2930.
[24] Junk M. A finite difference interpretation of the lattice Boltzmann method. Numer Meth Part Diff Eq 2001; 17: 383-402.

[25] Kutrib M, Vollmar R, Worsch T. Introduction to the special issue on cellular automata. Parallel Computing 1997; 23: 1567-1576.
[26] Latt J, Chopard B, Malaspinas O, Deville M, Michler A. Straight velocity boundaries in the lattice Boltzmann method. Phys Rev E 2008; 77: 056703.
[27] McNamara GR, Zanetti G. Use of the Boltzmann Equation to Simulate Lattice-Gas Automata. Phys Rev Lett 1988; 61: 2332-2335.
[28] Mei R, Shyy W, Yu D. Lattice Boltzmann Method for 3-D Flows with Curved Boundary. J Comput Phys 2000; 161: 680-699.
[29] Nourgaliev R, Dinh T, Theofanous T, Joseph D. The lattice Boltzmann equation method: theoretical interpretation, numerics and implications. Int J Multiphase Flow 2003; 29: 117-169.
[30] Peng Y, Shu C, Chew YT. A 3D incompressible thermal lattice Boltzmann model and its application to simulate natural convection in a cubic cavity. J Comput Phys 2004; 193: 260-274.
[31] Qian Y, d'Humieres D, Lallemand P. Lattice BGK Models for Navier-Stokes Equation. Europhys Lett 1992; 17: 479-484.
[32] Saint-Raymond L. Hydrodynamic Limits of the Boltzmann Equation. Springer Verlag, Berlin, Germany, 2009.
[33] Shan X, Chen H. Lattice Boltzmann model for simulating flows with multiple phases and components. Phys Rev E 1993; 47: 1815-1819.
[34] Shan X, Doolen G. Multicomponent lattice-Boltzmann model with interparticle interaction. J Stat Phys 1995; 81: 379-393.
[35] Sterling JD, Chen S. Stability Analysis of Lattice Boltzmann Methods. J Comput Phys 1996; 123: 196-206.
[36] Succi S. The Lattice Boltzmann Equation for Fluid Dynamics and Beyond. Oxford University Press, Oxford, UK, 2001.
[37] Temam R. Navier-Stokes equations: theory and numerical analysis. AMS Chelsea Publishing, Providence, RI, USA, 2001.
[38] von Neumann J. Theory of Self-Reproducing Automata. University of Illinois Press, Champaign, IL, USA, 1966.
[39] Wolf-Gladrow D. Lattice-Gas Cellular Automata and Lattice Boltzmann Models. Springer Verlag, Berlin, Germany, 2000.
[40] Wolfram S. Cellular automaton fluids 1: Basic theory. J Stat Phys 1986; 45: 471-526.
[41] Zanetti G. Hydrodynamics of lattice-gas automata. Phys Rev A 1989; 40: 1539-1548.
[42] Zou Q, He X. On pressure and velocity boundary conditions for the lattice Boltzmann BGK model. Phys Fluids 1997; 9: 1591-1598.

CHAPTER 2

Add-ons for Lattice Boltzmann Methods: Regularization, Filtering and Limiters

Robert A. Brownlee, Jeremy Levesley, David Packwood, Alexander N. Gorban*

Department of Mathematics, University of Leicester, Leicester LE1 7RH, UK

Abstract: We describe how regularization of lattice Boltzmann methods can be achieved by modifying dissipation. Classes of techniques used to try to improve regularization of LBMs include flux limiters, enforcing the exact correct production of entropy and manipulating non-hydrodynamic modes of the system in relaxation. Each of these techniques corresponds to an additional modification of dissipation compared with the standard LBGK model. Using some standard 1D and 2D benchmarks including the shock tube and lid driven cavity, we explore the effectiveness of these classes of methods.

Keywords: Lattice Boltzmann, Dissipation, Stability, Entropy, Filtering, Multiple Relaxation Times (MRT), Shock Tube, Lid-Driven Cavity.

1. INTRODUCTION

Lattice Boltzmann Methods (LBM) are a class of discrete computational schemes which can be used to simulate hydrodynamics and more[34]. They have been proposed as a discretization of Boltzmann's kinetic equation. Instead of fields of moments M, the LBM operates with fields of discrete distributions f.

All computational methods for continuum dynamics meet some troubles with stability when the gradients of the flows become too sharp. In Computational Fluid Dynamics (CFD) such situations occur when the Mach number Ma is not small, or the Reynolds number Re is too large. The possibility for using grid refinement is bounded by computational time and memory restrictions. Moreover, for nonlinear systems with shocks then grid refinement does not guarantee convergence to the proper solution. Methods of choice to remedy this are based on the *modification of dissipation* with limiters, additional viscosity, and so on [37]. All these approaches combine high order methods in relatively quiet regions with low order regularized methods in the regions with large gradients. The areas of the high-slope flows are assumed to be small but the loss of the order of accuracy in a small region may affect the accuracy in the whole domain because of the phenomenon of error propagation. Nevertheless, this loss of accuracy for systems with high gradients seems to be unavoidable.

It is impossible to successfully struggle with some spurious effects without a local decrease in the order of accuracy. In a more formal setting, this has been proven. In 1959, Godunov [14] proved that a (linear) scheme for a partial differintial equation could not, at the same time, be monotone and second-order accurate. Hence, we should choose between spurious oscillations in high order non-monotone schemes and additional dissipation in first-order schemes. Lax [23] demonstrated that un-physical dispersive oscillations in areas with high slopes are unavoidable due to discretization. For hydrodynamic simulations using the standard LBGK model (Sec.) such oscillations become prevalent especially at high Re and non-small Ma. Levermore and Liu used differential approximation to produce the "modulation equation" for the dispersive oscillation in the simple initial-value Hopf problem [24] and demonstrated directly how for a nonlinear problem a solution of the discretized equation does not converge to the solution of the continuous model with high slope when the step $h \to 0$.

Some authors expressed a hope that precisely keeping the entropy balance can make the computation more "physical", and that thermodynamics can help to suppress nonphysical effects. Tadmor and Zhong constructed a new family of entropy stable difference schemes which retain the precise entropy decay of the Navier–Stokes equations and demonstrated that this precise keeping of the entropy balance does not help to avoid the nonphysical dispersive oscillations [35].

Address correspondence to: Alexander Gorban, Department of Mathematics, University of Leicester, University Road, Leicester LE1 7RH, United Kingdom; Tel: ++44 116 223 14 33; E-mail: ag153@leicester.ac.uk

Matthias Ehrhardt (Ed.)

To prevent nonphysical oscillations, most upwind schemes employ limiters that reduce the spatial accuracy to first order through shock waves. A mixed-order scheme may be defined as a numerical method where the formal order of the truncation error varies either spatially, for example, at a shock wave, or for different terms in the governing equations, for example, third-order convection with second-order diffusion [32].
Several techniques have been proposed to help suppress these pollutive oscillations in LBM, the three which we deal with in this work are entropic lattice Boltzman (ELBM), entropic limiters and generalized lattice Boltzmann, also known as multiple relaxation time lattice Boltzmann (MRT). Where effective each of these techniques corresponds to an additional degree of complexity in the dissipation to the system, above that which exists in the LBGK model.
The Entropic lattice Boltzmann method (ELBM) was invented first in 1998 as a tool for the construction of single relaxation time LBM which respect the H-theorem [19]. For this purpose, instead of the mirror image with a local equilibrium as the reflection center, the entropic involution was proposed, which preserves the entropy value. Later, it was called the *Karlin-Succi involution* [15]. In 2000, it was reported that exact implementation of the Karlin-Succi involution (which keeps the entropy balance) significantly regularizes the post-shock dispersive oscillations [1]. This regularization seems very surprising, because the entropic lattice BGK (ELBGK) model gives a second-order approximation to the Navier–Stokes equation similarly to the LBGK model (different proofs of that degree of approximation were given in [34] and [4]).
Entropic limiters [4] are an example of flux limiter schemes [4, 21, 30], which are invented to combine high resolution schemes in areas with smooth fields and first order schemes in areas with sharp gradients. The idea of flux limiters can be illustrated by the computation of the flux $F_{0,1}$ of the conserved quantity u between a cell marked by 0 and one of two its neighbour cells marked by ± 1:

$$F_{0,1} = (1-\phi(r))f_{0,1}^{\text{low}} + \phi(r)f_{0,1}^{\text{high}}, \tag{1}$$

where $f_{0,1}^{\text{low}}$, $f_{0,1}^{\text{high}}$ are low and high resolution scheme fluxes, respectively, $r = (u_0 - u_{-1})/(u_1 - u_0)$, and $\phi(r) \geq 0$ is a flux limiter function. For r close to 1, the flux limiter function $\phi(r)$ should be also close to 1. Many flux limiter schemes have been invented during the last two decades [37]. No particular limiter works well for all problems, and a choice is usually made on a trial and error basis. Particular examples of the limiters we use are introduced in Section 3.3..
MRT has been developed as a true generalization of the collisions in the lattice Bhatnagar–Gross–Krook (LBGK) scheme [11, 22] from a one parameter diagonal relaxation matrix, to a general linear operation with more free parameters, the number of which is dependent on the particular discrete velocity set used and the number of conserved macroscopic variables. Different variants of MRT have been shown to improve accuracy and stability, including in our benchmark examples [26] in comparison with the standard LBGK systems.
The lattice Boltzmann paradigm is now mature, and explanations for some of its successes are available. However, in its applications it approaches the boundaries of the applicability and need special additional tools to extend the area of applications. It is well-understood that near to shocks, for instance, special and specific attention must be paid to avoid unphysical effects. In this paper then, we will discuss a variety of *add-ons* for LBM and apply them to a variety of standard 1D and 2D problems to test their effectiveness. In particular we will describe a family of entropic filters and show that we can use them to signficantly expand the effective range of operation of the LBM.

2. BACKGROUND

Lattice Boltzmann methods can be derived independently by a discretization Boltzmann's equation for kinetic transport or by naively creating a discrete scheme which matches moments with the Maxwellian distribution up to some finite order.
In each case the final discrete algorithm consists of two alternating steps, advection and collision, which are applied to m single particle distribution functions $f_i \equiv f_i(\mathbf{x},t), (i = 1 \ldots m)$, each of which corresponds with a discrete velocity vector $\mathbf{v}_i, (i = 1 \ldots m)$. The values f_i are also sometimes known as *populations* or *densities* as they can be thought of as representative of the densities of particles moving in the direction of the corresponding discrete velocities.

The advection operation is simply free flight for the discrete time step ε in the direction of the corresponding velocity vector,

$$f_i(\mathbf{x}, t+\varepsilon) = f_i(\mathbf{x} - \varepsilon \mathbf{v}_i, t). \tag{2}$$

The collision operation is instantaneous and can be different for each distribution function but depends on every distribution function, this might be written,

$$f_i(\mathbf{x}) \rightarrow F_i(\{f_i(\mathbf{x})\}). \tag{3}$$

In order to have a slightly more compact notation we can write these operations in vector form, in the below equation it should be inferred that the ith distribution function is advecting along its corresponding discrete velocity,

$$\mathbf{f}(\mathbf{x}, t+\varepsilon) = \mathbf{f}(\mathbf{x} - \varepsilon \mathbf{v}_i, t), \tag{4}$$

$$\mathbf{f}(\mathbf{x}) \rightarrow F(\mathbf{f}(\mathbf{x})). \tag{5}$$

To transform our vector of microscopic variables at a point in space $\mathbf{f}(\mathbf{x})$ to a vector of macroscopic variables $M(\mathbf{x})$ we use a vector of linear functions $M(\mathbf{x}) = m(\mathbf{f}(\mathbf{x}))$. In the athermal hydrodynamic systems we consider in this work the momemts are density ρ and momentum density $\rho\mathbf{u}$, $\{\rho, \rho\mathbf{u}\}(\mathbf{x}) = \sum_i \{1, \mathbf{v}_i\} f_i(\mathbf{x})$. These macroscopic moments are conserved by the collision operation, $m(\mathbf{f}) = m(F(\mathbf{f}))$.
The simplest and most common choice for the collision operation F is the Bhatnagar-Gross-Krook(BGK) [7, 9, 17, 34] operator with over-relaxation

$$F(\mathbf{f}) = \mathbf{f} + \alpha\beta(\mathbf{f}^{\mathrm{eq}} - \mathbf{f}). \tag{6}$$

For the standard LBGK method $\alpha = 2$ and $\beta \in [0,1]$ (usually, $\beta \in [1/2, 1]$) is the over-relaxation coefficient used to control viscosity. For $\beta = 1/2$ the collision operator returns the *local equilibrium* $\mathbf{f}^{\mathrm{eq}}$ and $\beta = 1$ (the *mirror reflection*) returns the collision for a liquid at the zero viscosity limit. The definition of $\mathbf{f}^{\mathrm{eq}}$ defines the dynamics of the system, often it chosen as an approximation to the continuous Maxwellian distribution. An equilibrium can also be independently derived by constructing a discrete system which matches moments of the Maxwellian up to some finite order. For hydrodynamic systems often this finite order is chosen to be 2, as this is sufficient to accurately replicate the Euler(non dissipative) component of the Navier Stokes equations. For a dissipative fluid with viscosity ν the parameter β is chosen by $\beta = \varepsilon/(2\nu + \varepsilon)$.
Each of the techniques we test in this paper can be introduced as developments of the generic LBGK system and such a presentation follows in the next sections.

3. ENTROPIC LATTICE BOLTZMANN

[illegible] *H* [illegible]

In the continuous case the Maxwellian distribution maximizes entropy, as measured by the Boltzmann H function, and therefore also has zero entropy production. In the context of lattice Boltzmann methods a discrete form of the H-theorem has been suggested as a way to introduce thermodynamic control to the system [18, 1].
From this perspective the goal is to find an equilibrium state equivalent to the Maxwellian in the continuum which will similarly maximize entropy. Before the equilibrium can be found an appropriate H function must be known for a given lattice. These functions have been constructed in a lattice dependent fashion in [18], and $H = -S$ with S from (7) is an example of a H function constructed in this way.
One way to implement an ELBM is as a variation on the LBGK, known as the ELBGK [1]. In this case α is varied to ensure a constant entropy condition according to the discrete H-theorem. In general the entropy function is based upon the lattice and cannot always be found explicitly. However for some examples such as the simple one dimensional lattice with velocities $\mathbf{v} = (-c, 0, c)$ and corresponding populations $\mathbf{f} = (f_-, f_0, f_+)$ an explicit Boltzmann style entropy function is known [18]:

$$S(\mathbf{f}) = -f_- \log(f_-) - f_0 \log(f_0/4) - f_+ \log(f_+). \tag{7}$$

With knowledge of such a function α is found as the non-trivial root of the equation

$$S(\mathbf{f}) = S(\mathbf{f} + \alpha(\mathbf{f}^* - \mathbf{f})). \tag{8}$$

The trivial root $\alpha = 0$ returns the entropy value of the original populations. ELBGK then finds the non-trivial α such that (8) holds. This version of the BGK collision one calls entropic BGK (or EBGK) collision. A solution of (8) must be found at every time step and lattice site. Entropic equilibria (also derived from the H-theorem) are always used for ELBGK.

[illegible]

The definition of ELBM for a given entropy equation (8) is incomplete. First of all, it is possible that the non-trivial solution does not exist. Moreover, for most of the known entropies (like the perfect entropy [18]) there always exist such f that the equation (8) for the ELBM collision has no non-trivial solutions. These f should be sufficiently far from equilibrium. For completeness, every user of ELBM should define collisions when the non-trivial root of (8) does not exist. We know and tried two rules for this situation:

1. The most radical approach gives the the Ehrenfest rule [16, 3]: "if the solution does not exist then go to equilibrium", i.e. if the solution does not exists then take $\alpha = 1$.

2. The most gentle solution gives the "positivity rule" [3, 25, 36, 33]: to take the maximal value of α that guarantees $f_i + \alpha(f_i^* - f_i) \geq 0$ for all i.

In general, the Ehrenfest rule prescribes to send the most non-equilibrium sites to equilibrium and the positivity rule is applied for any LBM as a recommendation to substitute the non-positive vectors $\mathbf{f}$ by the closest non-negative on the interval of the straight line $[\mathbf{f}, \mathbf{f}^*]$ that connects $\mathbf{f}$ to equilibrium. These rules give the examples of the pointwise LBM limiter and we discuss them separately.
By its nature, the ELBM adds more dissipation than the positivity rule when the non-trivial root of (8) does not exist. It does not always keep the entropy balance but increases dissipation for highly nonequilibrium sites.

[illegible]

Another source for additional dissipation in the ELBM may be the numerical method used for the solution of (8). For the full description of ELBM we have to select a numerical method for this equation. This method has to have an uniform accuracy in the wide range of parameters, for all possible deviation from equilibrium (distribution of these deviations has "heavy tails" [5]).
In order to investigate the stabilization properties of ELBGK it is necessary to craft a numerical method capable of finding the non-trivial root in (8). In this section we fix the population vectors $\mathbf{f}$ and $\mathbf{f}^*$, and are concerned only with this root finding algorithm. We recast (8) as a function of α only:

$$S_f(\alpha) = S(\mathbf{f} + \alpha(\mathbf{f}^{\mathrm{eq}} - \mathbf{f})) - S(\mathbf{f}). \tag{9}$$

In this setting we attempt to find the non-trivial root r of (9) such that $S_f(r) = 0$. It should be noted that as we search for r numerically we should always take care that the approximation we use is less than r itself. An upper approximation could result in negative entropy production. A simple algorithm for finding the roots of a concave function, based on local quadratic approximations to the target function, has cubic convergence order. Assume that we are operating in a neighbourhood $r \in N$, in which S_f' is negative (as well of course S_f'' is negative). At each iteration the new estimate for r is the greater root of the parabola P, the second order Taylor polynomial at the current estimate. Analogously to the case for Newton iteration, the constant in the estimate is the ratio of third and first derivatives in the interval of iteration:

$$|(r - \alpha_{n+1})| \leq C|\alpha_n - r|^3,$$
$$\text{where } C = \tfrac{1}{6}\sup_{a\in N}|S_f'''(a)| \Big/ \inf_{b\in N}|S_f'(b)|\,,$$

where α_n is the evaluation of r on the nth iteration.
We use a Newton step to estimate the accuracy of the method at each iteration: because of the concavity of S

$$|\alpha_n - r| \lesssim \left|S_f(\alpha_n)/S_f'(\alpha_n)\right|. \tag{10}$$

In fact we use a convergence criteria based not solely on α but on $\alpha||\mathbf{f}^* - \mathbf{f}||$, this has the intuitive appeal that in the case where the populations are close to the local equilibrium $\Delta S = S(\mathbf{f}^*) - S(\mathbf{f})$ will be small and a very precise estimate of α is unnecessary. We have some freedom in the choice of the norm used and we select between the standard L_1 norm and the entropic norm. The entropic norm is defined as

$$||\mathbf{f}^{\mathrm{eq}} - \mathbf{f}||_{\mathbf{f}^{\mathrm{eq}}} = -((\mathbf{f}^{\mathrm{eq}} - \mathbf{f}), D^2S\big|_{\mathbf{f}^{\mathrm{eq}}}(\mathbf{f}^{\mathrm{eq}} - \mathbf{f})),$$

where $D^2S\big|_{\mathbf{f}^{\mathrm{eq}}}$ is the second differential of entropy at point $\mathbf{f}^{\mathrm{eq}}$, and (x,y) is the standard scalar product.
The final root finding algorithm then is beginning with the LBGK estimate $x_0 = 2$ to iterate using the roots of successive parabolas. We stop the method at the point,

$$|\alpha_n - r| \cdot ||\mathbf{f}^* - \mathbf{f}|| < \varepsilon. \tag{11}$$

To ensure that we use an estimate that is less than the root, at the point where the method has converged we check the sign of $S_f(\alpha_n)$. If $S_f(\alpha_n) > 0$ then we have achieved a lower estimate, if $S_f(\alpha_n) < 0$ we correct the estimate to the other side of the root with a double length Newton step,

$$\alpha_n = \alpha_n - 2\frac{S_f(\alpha_n)}{S_f'(\alpha_n)}. \tag{12}$$

At each time step before we begin root finding we eliminate all sites with $\Delta S < 10^{-15}$. For these sites we make a simple LBGK step. At such sites we find that round off error in the calculation of S_f by solution of equation (8) can result in the root of the parabola becoming imaginary. In such cases a mirror image given by LBGK is effectively indistinct from the exact ELBGK collision. In the numerical examples given in this work the case where the non-trivial root of the entropy parabola does not exist was not encountered.

4. ENTROPIC FILTERING

All the specific LBM limiters [5, 30] are based on a representation of distributions f in the form:

$$\mathbf{f} = \mathbf{f}^{\mathrm{eq}} + ||\mathbf{f} - \mathbf{f}^{\mathrm{eq}}|| \frac{\mathbf{f} - \mathbf{f}^{\mathrm{eq}}}{||\mathbf{f} - \mathbf{f}^{\mathrm{eq}}||}, \tag{13}$$

where $\mathbf{f}^{\mathrm{eq}}$ is the corresponding quasiequilibrium (conditional equilibrium) for given moments M, $\mathbf{f} - \mathbf{f}^{\mathrm{eq}}$ is the nonequilibrium "part" of the distribution, which is represented in the form "norm×direction" and $||f - f^*||$ is the norm of that nonequilibrium component (usually this is the entropic norm).
All limiters we use change the norm of the nonequilibrium component $\mathbf{f} - \mathbf{f}^{\mathrm{eq}}$, but do not touch its direction or the quasiequilibrium. In particular, limiters do not change the macroscopic variables, because moments for $\mathbf{f}$ and $\mathbf{f}^{\mathrm{eq}}$ coincide. These limiters are transformations of the form

$$\mathbf{f} \mapsto \mathbf{f}^{\mathrm{eq}} + \phi \times (\mathbf{f} - \mathbf{f}^{\mathrm{eq}}) \tag{14}$$

with $\phi > 0$. If $\mathbf{f} - \mathbf{f}^{\mathrm{eq}}$ is too big, then the limiter should decrease its norm.
For the first example of the realization of this *pointwise filtering* we use the kinetic idea of the *positivity rule*, the prescription is simple [3, 25, 36, 33]: to substitute nonpositive $F(\mathbf{f})$ by the closest nonnegative state that belongs to the straight line

$$\left\{\lambda\mathbf{f} + (1-\lambda)\mathbf{f}^{\mathrm{eq}} \middle| \lambda \in \mathbb{R}\right\} \tag{15}$$

defined by the two points, **f** and the corresponding quasiequilibrium **1**. This operation is to be applied pointwise, at points of the lattice where positivity is violated. This technique preserves the positivity of populations, but can affect the accuracy of the approximation. This rule is necessary for ELBM when the positive "mirror state" with the same entropy as **f** does not exists on the straight line (15).

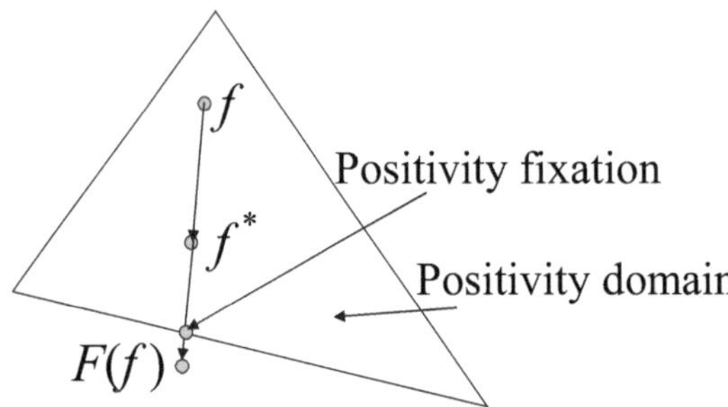

Fig. **1**: Positivity rule in action. The motions stops at the positivity boundary.

The positivity rule measures the deviation $f - f^*$ by a binary measure: if all components of this vector $f - f^*$ are non-negative then it is not too large. If some of them are negative that this deviation is too large and needs corrections.

To construct pointwise flux limiters for LBM, based on dissipation, the entropic approach remains very convenient. The local nonequilibrium entropy for each site is defined

$$\Delta S(\mathbf{f}) := S(\mathbf{f}^{\mathrm{eq}}) - S(\mathbf{f}). \tag{16}$$

The positivity limiter was targeted wherever population densities became negative. Entropic limiters are targeted wherever non-equilibrium entropy becomes large.

The first limiter is *Ehrenfests' regularisation* [4, 3], it provides "entropy trimming": we monitor local deviation of $\mathbf{f}$ from the corresponding quasiequilibrium, and when $\Delta S(\mathbf{f})(\mathbf{x})$ exceeds a pre-specified threshold value δ, perform local Ehrenfests' steps to the corresponding equilibrium: $\mathbf{f} \mapsto \mathbf{f}^{\mathrm{eq}}$ at those points.

Not all lattice Boltzmann models are entropic, and an important question arises: "how can we create nonequilibrium entropy limiters for LBM with non-entropic (quasi)equilibria?". We propose a solution of this problem based on the discrete Kullback entropy [20]:

$$S_{\mathrm{K}}(\mathbf{f}) = -\sum_i f_i \ln\left(\frac{f_i}{f_i^{\mathrm{eq}}}\right). \tag{17}$$

For entropic quasiequilibria with perfect entropy the discrete Kullback entropy gives the same ΔS: $-S_{\mathrm{K}}(f) = \Delta S(f)$. Let the discrete entropy have the standard form for an ideal (perfect) mixture [18]:

$$S(\mathbf{f}) = -\sum_i f_i \ln\left(\frac{f_i}{W_i}\right).$$

In quadratic approximation,

$$-S_{\mathrm{K}}(\mathbf{f}) = \sum_i f_i \ln\left(\frac{f_i}{f_i^{\mathrm{eq}}}\right) \approx \sum_i \frac{(f_i - f_i^{\mathrm{eq}})^2}{f_i^{\mathrm{eq}}}. \tag{18}$$

If we define $\mathbf{f}$ as the conditional entropy maximum for given $M_j = \sum_k m_{jk} f_k$, then

$$\ln f_k^{\mathrm{eq}} = \sum_j \mu_j m_{jk},$$

where $\mu_j(M)$ are the Lagrange multipliers (or "potentials"). For this entropy and conditional equilibrium we find

$$\Delta S = S(\mathbf{f}^{\mathrm{eq}}) - S(\mathbf{f}) = \sum_i f_i \ln\left(\frac{f_i}{f_i^{\mathrm{eq}}}\right) = -S_{\mathrm{K}}(\mathbf{f}), \tag{19}$$

if $\mathbf{f}$ and $\mathbf{f}^{\mathrm{eq}}$ have the same moments, $m(\mathbf{f}) = m(\mathbf{f}^{\mathrm{eq}})$.

In what follows, ΔS is the Kullback distance $-S_K(\mathbf{f}^{\mathrm{eq}})$ (19) for general (positive) quasiequilibria $\mathbf{f}^{\mathrm{eq}}$, or simply $S(\mathbf{f}^{\mathrm{eq}}) - S(\mathbf{f})$ for entropic quasiequilibria (or second approximations for these quantities (18)).
So that the Ehrenfests' steps are not allowed to degrade the accuracy of LBGK it is pertinent to select the k sites with highest $\Delta S > \delta$. An a posteriori estimate of added dissipation could easily be performed by the analysis of entropy production in the Ehrenfests' steps. Numerical experiments show (see, e.g., [4, 3]) that even a small number of such steps drastically improves stability.
The positivity rule and Ehrenfests' regularisation provide rare, intense and localised corrections. Of course, it is easy and also computationally cheap to organise more gentle transformations with a smooth shift of highly nonequilibrium states to quasiequilibrium. The following regularisation transformation with a smooths function ϕ distributes its action smoothly:

$$\mathbf{f} \mapsto \mathbf{f}^{\mathrm{eq}} + \phi(\Delta S(\mathbf{f}))(\mathbf{f} - \mathbf{f}^{\mathrm{eq}}). \tag{20}$$

The choice of function ϕ is highly ambiguous, for example, $\phi = 1/(1+\alpha\Delta S^k)$ for some $\alpha > 0$ and $k > 0$. There are two significantly different choices: (i) ensemble-independent ϕ (i.e., the value of ϕ depends on local value of ΔS only) and (ii) ensemble-dependent ϕ, for example

$$\phi(\Delta S) = \frac{1 + (\Delta S/(\alpha \mathrm{E}(\Delta S)))^{k-1/2}}{1 + (\Delta S/(\alpha \mathrm{E}(\Delta S)))^{k}}, \tag{21}$$

where $\mathrm{E}(\Delta S)$ is the average value of ΔS in the computational area, $k \geq 1$, and $\alpha \gtrsim 1$. For small ΔS, $\phi(\Delta S) \approx 1$ and for $\Delta S \gg \alpha \mathrm{E}(\Delta S)$ it tends to $\sqrt{\alpha \mathrm{E}(\Delta S)/\Delta S}$. It is easy to select an ensemble-dependent ϕ with control of total additional dissipation.

[illegible]

Two monotonicity properties are important in the theory of nonequilibrium entropy limiters:

1. A limiter should move the distribution to equilibrium: in all cases of (14) $0 \leq \phi \leq 1$. This is the *dissipativity* condition which means that limiters never produce negative entropy.

2. A limiter should not change the order of states on the line: if for two distributions with the same moments, $\mathbf{f}$ and $\mathbf{f}'$, $\mathbf{f}' - \mathbf{f}^{\mathrm{eq}} = x(\mathbf{f} - \mathbf{f}^{\mathrm{eq}})$ and $\Delta S(\mathbf{f}) > \Delta S(\mathbf{f}')$ before the limiter transformation, then the same inequality should hold after the limiter transformation too. For example, for the limiter (20) it means that $\Delta S(\mathbf{f}^{\mathrm{eq}} + x\phi(\Delta S(\mathbf{f}^{\mathrm{eq}} + x(\mathbf{f} - \mathbf{f}^{\mathrm{eq}}))(\mathbf{f} - \mathbf{f}^{\mathrm{eq}}))$ is a monotonically increasing function of $x > 0$.

In quadratic approximation,

$$\begin{aligned} \Delta S(\mathbf{f}^{\mathrm{eq}} + x(\mathbf{f} - \mathbf{f}^{\mathrm{eq}})) &= x^2 \Delta S(\mathbf{f}), \\ \Delta S(\mathbf{f}^{\mathrm{eq}} + x\phi(\Delta S(\mathbf{f}^{\mathrm{eq}} + x(\mathbf{f} - \mathbf{f}^{\mathrm{eq}}))(\mathbf{f} - \mathbf{f}^{\mathrm{eq}})) &= x^2 \phi^2 (x^2 \Delta S(\mathbf{f})), \end{aligned}$$

and the second monotonicity condition transforms into the following requirement: $y\phi(y^2 s)$ is a monotonically increasing (not decreasing) function of $y > 0$ for any $s > 0$.
If a limiter satisfies both monotonicity conditions, we call it "double monotonic". For example, Ehrenfests' regularisation satisfies the first monotonicity condition, but violates the second one. The limiter (21) violates the first condition for small ΔS, but is dissipative and satisfies the second one in quadratic approximation for large ΔS. The limiter with $\phi = 1/(1+\alpha\Delta S^k)$ always satisfies the first monotonicity condition, violates the second if $k > 1/2$, and is double monotonic (in quadratic approximation for the second condition), if $0 < k \leq 1/2$. The threshold limiter (26) is also double monotonic.
For smooth functions, the condition of double monotonicity (in quadratic approximation) is equivalent to the system of differential inequalities:

$$\begin{aligned} \phi(x) + 2x\phi'(x) &\geq 0; \\ \phi'(x) &\leq 0. \end{aligned}$$

The initial condition $\phi(0)=1$ means that in the limit $\Delta S \to 0$ limiters do not affect the flow. Following these inequalities we can write: $2x\phi'(x) = -\eta(x)\phi(x)$, where $0 \le \eta(x) \le 1$. The solution of these inequalities with initial condition is

$$\phi(x) = \exp\left(-\frac{1}{2}\int_0^x \frac{\eta(\chi)}{\chi}\mathrm{d}\chi\right), \tag{22}$$

if the integral on the right-hand side exists. This is a general solution for double monotonic limiters (in the second approximation for entropy). If $\eta(x)$ is the Heaviside step function, $\eta(x) = H(x - \Delta S_{\mathrm{t}})$ with threshold value ΔS_{t}, then the general solution (22) gives us the threshold limiter. If, for example, $\eta(x) = x^k/(\Delta S_{\mathrm{t}}^k + x^k)$, then

$$\phi(x) = \left(1 + \frac{x^k}{\Delta S_{\mathrm{t}}^k}\right)^{-\frac{1}{2k}}. \tag{23}$$

This special form of limiter function is attractive because for small x it gives

$$\phi(x) = 1 - \frac{1}{2k}\frac{x^k}{\Delta S_{\mathrm{t}}^k} + o(x^k).$$

Thus, the limiter does not affect the motion up to the $(k+1)$st order, and the macroscopic equations coincide with the macroscopic equations for LBM without limiters up to the $(k+1)$st order in powers of deviation from quasiequilibrium. Furthermore, for large x we get the kth order approximation to the threshold limiter (26):

$$\phi(x) = \sqrt{\frac{\Delta S_{\mathrm{t}}}{x}} + o(x^{-k}).$$

Of course, it is not forbidden to use any type of limiters under the local and global control of dissipation, but double monotonic limiters provide some natural properties automatically, without additional care.

[illegible]

For given β, the entropy production in one LBGK step in quadratic approximation for ΔS is:

$$\delta_{\mathrm{LBGK}} S \approx [1 - (2\beta - 1)^2] \sum_x \Delta S(\mathbf{x}),$$

where $\mathbf{x}$ is the grid point, $\Delta S(\mathbf{x})$ is nonequilibrium entropy (16) at point $\mathbf{x}$, $\delta_{\mathrm{LBGK}} S$ is the total entropy production in a single LBGK step. It would be desirable if the total entropy production for the limiter $\delta_{\mathrm{lim}} S$ was small relative to $\delta_{\mathrm{LBGK}} S$:

$$\delta_{\mathrm{lim}} S < \delta_0 \delta_{\mathrm{LBGK}} S. \tag{24}$$

A simple ensemble-dependent limiter (perhaps, the simplest one) for a given δ_0 operates as follows. Let us collect the histogram of the $\Delta S(\mathbf{x})$ distribution, and estimate the distribution density, $p(\Delta S)$. We have to estimate a value ΔS_0 that satisfies the following equation:

$$\int_{\Delta S_0}^{\infty} p(\Delta S)(\Delta S - \Delta S_0)\,\mathrm{d}\Delta S = \delta_0 [1 - (2\beta - 1)^2] \int_0^{\infty} p(\Delta S)\Delta S\,\mathrm{d}\Delta S. \tag{25}$$

In order not to affect distributions with a small expectation of ΔS, we choose a threshold $\Delta S_{\mathrm{t}} = \max\{\Delta S_0, \delta\}$, where δ is some predefined value (as in the Ehrenfests' regularization). For states at sites with $\Delta S \ge \Delta S_{\mathrm{t}}$ we provide homothety with equilibrium center $\mathbf{f}^{\mathrm{eq}}$ and coefficient $\sqrt{\Delta S_{\mathrm{t}}/\Delta S}$ (in quadratic approximation for nonequilibrium entropy):

$$\mathbf{f}(\mathbf{x}) \mapsto \mathbf{f}^{\mathrm{eq}}(\mathbf{x}) + \sqrt{\frac{\Delta S_{\mathrm{t}}}{\Delta S}}(\mathbf{f}(\mathbf{x}) - \mathbf{f}^{\mathrm{eq}}(\mathbf{x})). \tag{26}$$

To avoid the change of accuracy order "on average", the number of sites with this step should be $\le \mathcal{O}(Nh/L)$ where N is the total number of sites, h is the step of the space discretization and L is the macroscopic characteristic length. But this rough estimate of accuracy across the system might be destroyed by a concentration of Ehrenfests' steps in the most nonequilibrium areas, for example, in boundary layers. In that case, instead of the total number of sites N in $\mathcal{O}(Nh/L)$ we should take the number of sites in a specific region. The effects of such concentration could be analysed a posteriori.

[illegible]

The Ehrenfest step described above provides pointwise correction of nonequilibrium entropy at the "most nonequilibrium" points. Due to the pointwise nature, the technique does not introduce any nonisotropic effects, and provides some other benefits. But if we involve local structure, we can correct local non-monotone irregularities without touching regular fragments. For example, we can discuss monotone increase or decrease of nonequilibrium entropy as regular fragments and concentrate our efforts on reduction of "speckle noise" or "salt and pepper noise". This approach allows us to use the accessible resource of entropy change (24) more thriftily. Salt and pepper noise is a form of noise typically observed in images. It represents itself as randomly occurring white (maximal brightness) and black pixels. For this kind of noise, conventional low-pass filtering, e.g., mean filtering or Gaussian smoothing is unsuccessful because the perturbed pixel value can vary significantly both from the original and mean value. For this type of noise, *median filtering* is a common and effective noise reduction method. Median filtering is a common step in image processing [31] for the smoothing of signals and the suppression of impulse noise with preservation of edges.

The median is a more robust average than the mean (or the weighted mean) and so a single very unrepresentative value in a neighbourhood will not affect the median value significantly. Hence, we suppose that the median entropy filter will work better than entropy convolution filters.

For the nonequilibrium entropy field, the median filter considers each site in turn and looks at its nearby neighbours. It replaces the nonequilibrium entropy value ΔS at the point with the median of those values ΔS_{med}, then updates f by the transformation (26) with the homothety coefficient $\sqrt{\Delta S_{\text{med}}/\Delta S}$. The median, ΔS_{med}, is calculated by first sorting all the values from the surrounding neighbourhood into numerical order and then replacing that being considered with the middle value. For example, if a point has 3 nearest neighbours including itself, then after sorting we have 3 values ΔS: $\Delta S_1 \leq \Delta S_2 \leq \Delta S_3$. The median value is $\Delta S_{\text{med}} = \Delta S_2$. For 9 nearest neighbours (including itself) we have after sorting $\Delta S_{\text{med}} = \Delta S_5$. For 27 nearest neighbours $\Delta S_{\text{med}} = \Delta S_{14}$.

We accept only dissipative corrections (those resulting in a decrease of ΔS, $\Delta S_{\text{med}} < \Delta S$) because of the second law of thermodynamics. The analogue of (25) is also useful for the acceptance of the most significant corrections. In "salt and pepper" terms, we correct the salt (where ΔS exceeds the median value) and do not touch the pepper.

[illegible]

The separation of $\mathbf{f}$ in equilibrium and nonequilibrium parts (13) allows one to use any nonlocal filtering procedure. Let $\mathbf{f}^{\text{neq}} = \mathbf{f} - \mathbf{f}^{\text{eq}}$. The values of moments for $\mathbf{f}$ and $\mathbf{f}^{\text{eq}}$ coincide, hence we can apply any transformation of the form

$$\mathbf{f}^{\text{neq}}(\mathbf{x}) \mapsto \sum_{y} d(\mathbf{y})\mathbf{f}^{\text{neq}}(\mathbf{x}+\mathbf{y})$$

for any family of vectors $\mathbf{y}$ that shift the grid into itself and any coefficients $d(\mathbf{y})$. If we apply this transformation, the macroscopic variables do not change but their time derivatives may change. We can control the values of some higher moments in order not to perturb significantly some macroscopic parameters, the shear viscosity, for example [10]. Several local (but not pointwise) filters of this type have been proposed and tested recently [30].

5. MULTIPLE RELAXATION TIMES

The MRT lattice Boltzmann system [22, 26, 11] generalizes the BGK collision into a more general linear transformation of the population functions,

$$F(\mathbf{f}) = \mathbf{f} + A(\mathbf{f}^{\text{eq}} - \mathbf{f}), \tag{27}$$

where A is a square matrix of size m. The use of this more general operator allows more different parameters to be used within the collision, to manipulate different physical properties, or for stability purposes.

To facilitate this a change of basis matrix can be used to switch the space of the collision to the moment space. Since the moment space of the system may be several dimensions smaller than the population space, to complete the basis linear combinations of higher order polynomials of the discrete velocity vectors may be used. For our later experiments we will use the D2Q9 system, we should select a particular enumeration of the discrete velocity vectors for the system, the zero velocity is numbered one and the positive x velocity is numbered 2, the remainder are numbered clockwise from this system,

$$\begin{matrix} 7 & 8 & 9 \\ 6 & 1 & 2 \\ 5 & 4 & 3 \end{matrix}. \tag{28}$$

The change of basis matrix given in some of the literature [22] is chosen to represent specific macroscopic quantities and in our velocity system is as follows,

$$M_1 = \begin{pmatrix} 1 & 1 & 1 & 1 & 1 & 1 & 1 & 1 & 1 \\ -4 & -1 & 2 & -1 & 2 & -1 & 2 & -1 & 2 \\ 4 & -2 & 1 & -2 & 1 & -2 & 1 & -2 & 1 \\ 0 & 1 & 1 & 0 & -1 & -1 & -1 & 0 & 1 \\ 0 & -2 & 1 & 0 & -1 & 2 & -1 & 0 & 1 \\ 0 & 0 & -1 & -1 & -1 & 0 & 1 & 1 & 1 \\ 0 & 0 & -1 & 2 & -1 & 0 & 1 & -2 & 1 \\ 0 & 1 & 0 & -1 & 0 & 1 & 0 & -1 & 0 \\ 0 & 0 & -1 & 0 & 1 & 0 & -1 & 0 & 1 \end{pmatrix} \tag{29}$$

As an alternative we could complete the basis simply using higher powers of the velocity vectors, the basis would be $1, v_x, v_y, v_x^2+v_y^2, v_x^2-v_y^2, v_xv_y, v_x^2v_y, v_xv_y^2, v_x^2v_y^2$, in our velocity system then this change of basis matrix is as follows,

$$M_2 = \begin{pmatrix} 1 & 1 & 1 & 1 & 1 & 1 & 1 & 1 & 1 \\ 0 & 1 & 0 & -1 & 0 & 1 & -1 & -1 & 1 \\ 0 & 0 & 1 & 0 & -1 & 1 & 1 & -1 & -1 \\ 0 & 1 & 1 & 1 & 1 & 2 & 2 & 2 & 2 \\ 0 & 1 & -1 & 1 & -1 & 0 & 0 & 0 & 0 \\ 0 & 0 & 0 & 0 & 0 & 1 & -1 & 1 & -1 \\ 0 & 0 & 0 & 0 & 0 & 1 & -1 & -1 & 1 \\ 0 & 0 & 0 & 0 & 0 & 1 & 1 & -1 & -1 \\ 0 & 0 & 0 & 0 & 0 & 1 & 1 & 1 & 1 \end{pmatrix} \tag{30}$$

When utilized properly any basis should be equivalent (although with different rates). In any case in this system there are 3 conserved moments and 9 population functions. Altogether then there are 6 degrees of freedom in relaxation in this system. We need some of these degrees of freedom to implement hydrodynamic rates such as shear viscosity or to force isotropy and some are 'spare' and can be manipulated to improve accuracy or stability. Typically these spare relaxation modes are sent closer to equilibrium than the standard BGK relaxation rate. This is in effect an additional contraction in the finite dimensional non-equilibrium population function space, corresponding to an increase in dissipation.

Once in moment space we can apply a diagonal relaxation matrix C_1 to the populations and then the inverse moment transformation matrix M_1^{-1} to switch back into population space, altogether $A_1 = M_1^{-1}C_1M_1$. If we use the standard athermal polynomial equilibria then three entries on the diagonal of C_1 are actually not important as the moments will be automatically conserved since $m(\mathbf{f}) = m(\mathbf{f}^{\mathrm{eq}})$, for simplicity we set them equal to 0 or 1 to reduce the complexity of the terms in the collision matrix. There are 6 more parameters on the diagonal matrix C which we can set. Three of these correspond to second order moments, one each is required for shear and bulk viscosity which are called s_e and s_v respectively and one for isotropy. Two correspond to third order moments, one gives a relaxation rate s_q and again one is needed for isotropy. Finally one is used to give a relaxation rate s_ε for the single fourth order moment. We have then in total four relaxation parameters which appear on the diagonal matrix in the following form:

$$C_1 = \begin{pmatrix} 1 & 0 & 0 & 0 & 0 & 0 & 0 & 0 & 0 \\ 0 & s_e & 0 & 0 & 0 & 0 & 0 & 0 & 0 \\ 0 & 0 & s_\varepsilon & 0 & 0 & 0 & 0 & 0 & 0 \\ 0 & 0 & 0 & 1 & 0 & 0 & 0 & 0 & 0 \\ 0 & 0 & 0 & 0 & s_q & 0 & 0 & 0 & 0 \\ 0 & 0 & 0 & 0 & 0 & 1 & 0 & 0 & 0 \\ 0 & 0 & 0 & 0 & 0 & 0 & s_q & 0 & 0 \\ 0 & 0 & 0 & 0 & 0 & 0 & 0 & s_\nu & 0 \\ 0 & 0 & 0 & 0 & 0 & 0 & 0 & 0 & s_\nu \end{pmatrix}. \tag{31}$$

Apart from the parameter s_ν which is used to control shear viscosity, in an incompressible system the other properties can be varied to improve accuracy and stability. In particular, there exists a variant of MRT known as TRT (two relaxation time) [13] where the relaxation rates s_e, s_ε are made equal to s_ν. In a system with boundaries the final rate is calculated $s_q = 8(2 = s_\nu)/(8 - s_\nu)$, this is done, in particular, to combat numerical slip on the boundaries of the system.

We should say that in some of the literature regarding MRT the equilibrium is actually built in moment space, that is the collision operation would be written,

$$F(\mathbf{f}) = \mathbf{f} + M_1^{-1} C_1 (\mathbf{m}^{\text{eq}} - M_1 \mathbf{f}). \tag{32}$$

This could be done to increase efficiency, depending on the implementation, however each moment equilibrium $\mathbf{m}^{\text{eq}}$ has an equivalent population space equilibrium $\mathbf{f}^{\text{eq}} = M_1^{-1}\mathbf{m}^{\text{eq}}$, so the results of implementing either system should be the same up to rounding error.

We can also conceive of using an MRT type collision as a limiter, that is to apply it only on a small number of points on the lattice where non equilibrium entropy passes a certain threshold. This answers a criticism of the single relaxation time limiters that they fail to preserve dissipation on physical modes. As well as using the standard MRT form given above we can build an MRT limiter using the alternative change of basis matrix M_2.

The limiter in this case is based on the idea of sending every mode except shear viscosity directly to equilibrium again the complete relaxation matrix is given by $A_2 = M_2^{-1} C_2 M_2$ where,

$$C_2 = \begin{pmatrix} 1 & 0 & 0 & 0 & 0 & 0 & 0 & 0 & 0 \\ 0 & 1 & 0 & 0 & 0 & 0 & 0 & 0 & 0 \\ 0 & 0 & 1 & 0 & 0 & 0 & 0 & 0 & 0 \\ 0 & 0 & 0 & 1 & 0 & 0 & 0 & 0 & 0 \\ 0 & 0 & 0 & 0 & s_\nu & 0 & 0 & 0 & 0 \\ 0 & 0 & 0 & 0 & 0 & s_\nu & 0 & 0 & 0 \\ 0 & 0 & 0 & 0 & 0 & 0 & 1 & 0 & 0 \\ 0 & 0 & 0 & 0 & 0 & 0 & 0 & 1 & 0 \\ 0 & 0 & 0 & 0 & 0 & 0 & 0 & 0 & 1 \end{pmatrix}. \tag{33}$$

This could be considered a very *aggressive* form of the MRT which maximizes regularization on every mode except shear viscosity and would not be appropriate for general use in a system, especially as most systems violate the incompressibility assumption and hence bulk viscosity is not small. The advantage of using the different change of basis matrix is that the complete collision matrix A_2 is relatively sparse with just twelve off diagonal elements and hence is easy to implement and not too expensive to compute with.

6. 1D SHOCK TUBE

A standard experiment for the testing of LBMs is the one-dimensional shock tube problem. The lattice velocities used are $\mathbf{v} = (-1, 0, 1)$, so that space shifts of the velocities give lattice sites separated by the unit distance. 800 lattice sites are used and are initialized with the density distribution

$$\rho(x) = \begin{cases} 1, & 1 \leq x \leq 400, \\ 0.5, & 401 \leq x \leq 800. \end{cases}$$

Initially all velocities are set to zero. We compare the ELBGK equipped with the parabola based root finding algorithm using the entropic norm with the standard LBGK method using both standard polynomial and entropic equilibria. The polynomial equilibria are given in [7, 34]:

$$f_-^* = \frac{\rho}{6}\left(1-3u+3u^2\right), \quad f_0^* = \frac{2\rho}{3}\left(1-\frac{3u^2}{2}\right),$$
$$f_+^* = \frac{\rho}{6}\left(1+3u+3u^2\right).$$

The entropic equilibria also used by the ELBGK are available explicitly as the maximum of the entropy function (7),

$$f_-^* = \frac{\rho}{6}(-3u-1+2\sqrt{1+3u^2}), \quad f_0^* = \frac{2\rho}{3}(2-\sqrt{1+3u^2}),$$
$$f_+^* = \frac{\rho}{6}(3u-1+2\sqrt{1+3u^2}).$$

Now following the prescription fromm Sec. the governing equations for the simulation are

$$f_-(x,t+1)=f_-(x+1,t)+\alpha\beta(f_-^*(x+1,t)-f_-(x+1,t)),$$
$$f_0(x,t+1)=f_0(x,t)+\alpha\beta(f_0^*(x,t)-f_0(x,t)),$$
$$f_+(x,t+1)=f_+(x-1,t)+\alpha\beta(f_+^*(x-1,t)-f_+(x-1,t)).$$

From this experiment we observe no benefit in terms of regularization in using the ELBGK rather than the

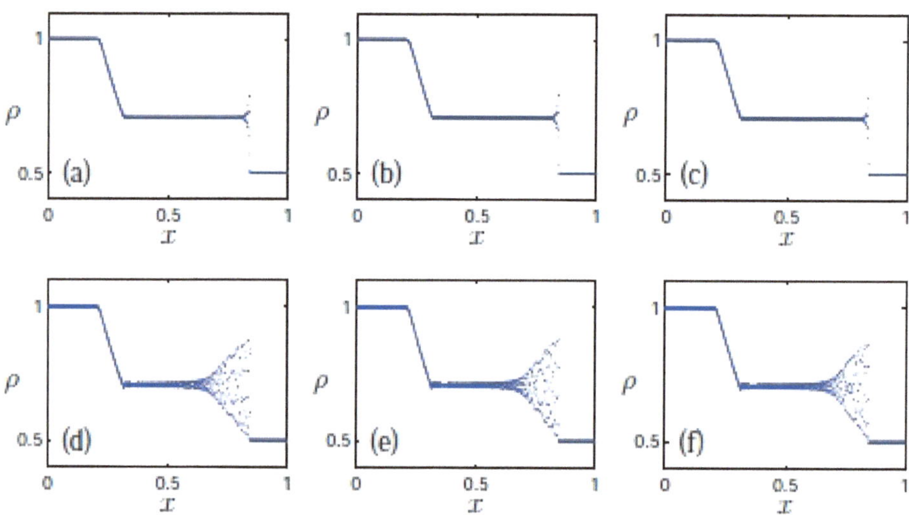

Fig. **2**: Density profile of the simulation of the shock tube problem following 400 time steps using **(a)** LBGK with polynomial equilibria $[\nu = (1/3)\cdot 10^{-1}]$; **(b)** LBGK with entropic equilibria $[\nu = (1/3)\cdot 10^{-1}]$; **(c)** ELBGK $[\nu = (1/3)\cdot 10^{-1}]$; **(d)** LBGK with polynomial equilibria $[\nu = 10^{-9}]$; **(e)** LBGK with entropic equilibria $[\nu = 10^{-9}]$; **(f)** ELBGK $[\nu = 10^{-9}]$.

standard LBGK method (Fig. **2**). In both the medium and low viscosity regimes ELBGK does not supress the spurious oscillations found in the standard LBGK method. The observation is in full agreement with the Tadmor and Zhong [35] experiments for schemes with precise entropy balance.

Entropy balance gives a nice additional possibility to monitor the accuracy and the basic physics but does not give an omnipotent tool for regularization.

7. 2D SHEAR DECAY

For the second test we use a simple test proposed to measure the observable viscosity of a lattice Boltzmann simulation to validate the shear viscosity production of the MRT models. We take the 2D isothermal nine-velocity model with standard polynomial equilibria. Our computational domain will a square which we discretize with $L+1 \times L+1$ uniformly spaced points and periodic boundary conditions. The initial condition is $\rho(x,y)=1, u_x(x,y)=0$ and $u_y(x,y)=u_0 sin(2\pi x/L)$, with $u_0=0.05$. The exact velocity solution to this problem is an exponential decay of the initial condition: $u_x(x,y,t)=0, u_y(x,y,t)=u_0\exp(-\lambda u_0 t/\mathrm{Re}L)\sin(2\pi x/L)$, where λ is some constant and $Re=u_0L/\nu$ is the Reynolds number of the flow. Here, ν is the theoretical shear viscosity of the fluid due to the relaxation parameters of the collision operation.

Now, we simulate the flow over L/u_0 time steps and measure the constant λ from the numerical solution. We do this for LBGK, the 'aggressive' MRT with collision matrix A_2 and the MRT system with collision matrix A_1 with additional parameters $s_e=1.64, s_\varepsilon=1.54, s_q=3(2-s_\nu)/(3-s_\nu)$ [22]. The shear viscosity relaxation parameter s_ν is varied to give different viscosities and therefore Reynolds numbers for $L=50$ and for $L=100$.

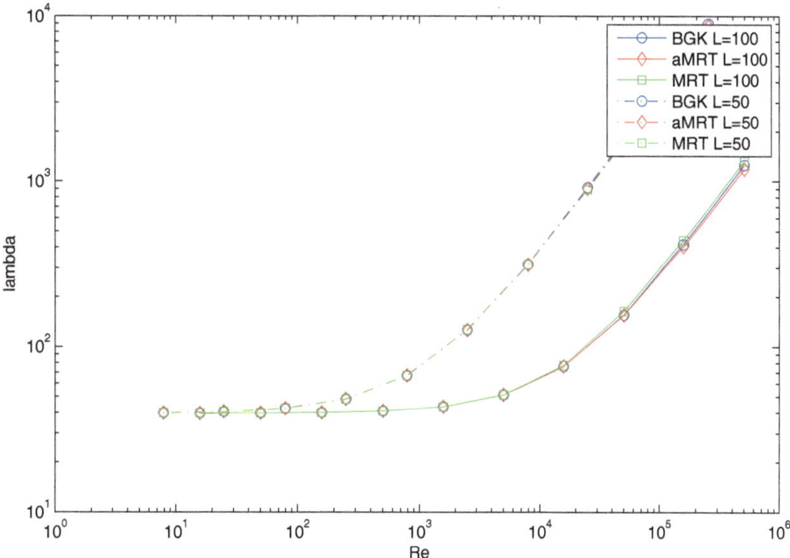

Fig. **3**: Observed λ in the shear decay experiment

From Figure **3** it can be seen that in all the systems, for increasing theoretical Reynolds numbers (decreasing viscosity coefficient) following a certain point numerical dissipation due to the lattice begins growing. The most important observation from this system is that our MRT models do indeed produce shear viscosity at almost the same rate as the BGK model.

In particular, the bulk viscosity in this system is zero, so all dissipation is given by shear viscosity or higher order modes. Since the space derivatives of the velocity modes are well bounded, as is the magnitude of the velocity itself, the proper asymptotic decay of higher order modes is observed and the varying higher order relaxation coefficients have only a very marginal effect.

We should reiterate that we while we use the 'aggressive' MRT across the system, this is only appropriate as bulk viscosity is zero and in fact the selection of the bulk viscosity coefficient makes no difference in this example. This example is a special case in this regard.

8. LID DRIVEN CAVITY

Our next 2D example is the benchmark 2D lid driven cavity. In this case this is a square system of side length 129. Bounce back boundary conditions are used and the top boundary imposes a constant velocity of $u_{\text{wall}} = 0.1$. For a variety of Reynolds numbers we run experiments for up to 10000000 time steps and check which methods have remained stable up until that time step.

The methods which we test are the standard BGK system, the BGK system equipped with Ehrenfest steps, the BGK system equipped with the MRT limiter, the TRT system, an MRT system with the TRT relaxation rate for the third order moment and the other rates $s_e = 1.64, s_\varepsilon = 1.54$ and finally an MRT system which we call MRT1 with the rates $s_q = 1.9 s_e = 1.64, s_\varepsilon = 1.54$ [26]. In each case of the system equipped with limiters the maximum number of sites where the limiter is used is 9.

All methods are equipped with the standard 2nd order compressible quasi-equilibrium, which is available as the product of the 1D equilibria 4.4..

When calculating the stream functions of the final states of these simulations we use Simpson integration in first the x and then y directions.

Additionally we measure Enstrophy $\mathscr{E}$ in each system over time. Enstrophy is calculated as the sum of vorticity squared across the system, normalized by the number of lattice sites. This statistic is useful as vorticity is theoretically only dissipated due to shear viscosity, at the same time in the lid driven system vorticity is produced by the moving boundary. For these systems $\mathscr{E}$ becomes constant as the vorticity field becomes steady. The value of this constant indicates where the *balance* between dissipation and production of vorticity is found. The lower the final value of $\mathscr{E}$ the more dissipation produced in the system.

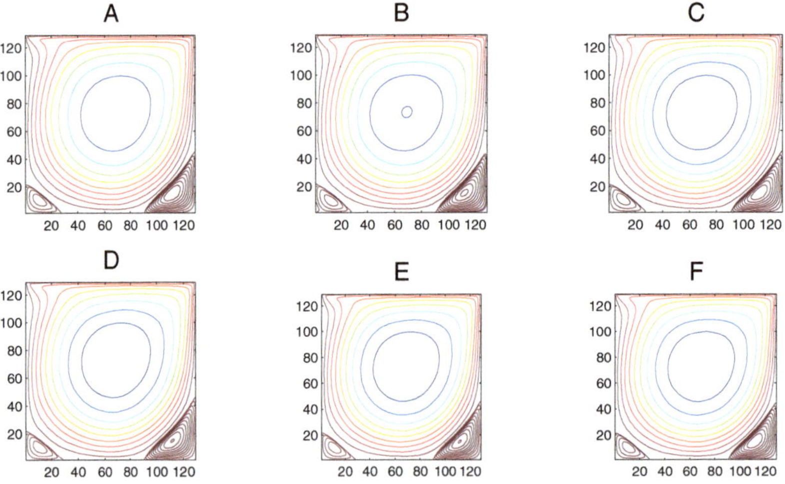

Fig. **4**: Contour plots of stream functions of A: BGK, B: BGK + Ehrenfest Steps, C: BGK + MRT Limiter, D: TRT, E: MRT, F: MRT1 following 10000000 time steps at Re1000.

All of these systems are stable for Re1000, the contour plots of the final state are given in Figure **4** and there appears only small differences. We calculate the average enstrophy in each system and plot it as a function of time in Figure **5**. We can see that in the different systems that enstrophy and hence dissipation varies. Compared with the BGK system all the other systems except MRT1 exhibit a lower level of enstrophy indicating a higher rate of dissipation. For MRT1 the fixed relaxation rate of the third order mode is actually less dissipative than the BGK relaxation rate for this Reynolds number, hence the increased enstrophy. An artifact of using the pointwise filtering techniques is that they introduce small scale local oscillations in the modes that they regularize, therefore the system seems not to be asymptotically stable. This might be

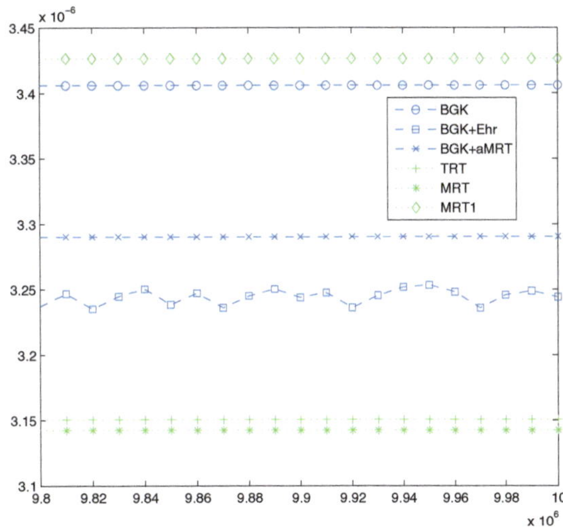

Fig. **5**: Enstrophy in the Re1000 systems during the final $2 \cdot 10^5$ time steps

remedied by increasing the threshold of ΔS below which no regularization is performed. Nevertheless in these experiments after sufficient time the enstrophy values remain within a small enough boundary for the results to be useful.

The next Reynolds number we choose is Re2500. Only 4 of the original 6 systems complete the full number of time steps for this Reynolds number, the contour plots of the final stream functions are given in Figure **6**. Of the systems which did not complete the simulation it should be said that the MRT system survived a few 10s of thousands of time steps while the BGK system diverged almost immediately, indicating that it does provide stability benefits which are not apparent at the coarse granularity of Reynolds numbers used in this study. One feature to observe in the stream function plots is the absence of an upper left vortex in the Ehrenfest limiter. This system selects the "most non-equilibrium" sites to apply the filter. These typically occur near the corners of the moving lid. It seems here that the local increase in shear viscosity is enough to prevent this vortex forming. This problem does not seem to affect the MRT limiter which preserves the correct production of shear viscosity.

Again we check the enstrophy of the systems and give the results for the final timesteps in Figure **7**. Due to the failure of the BGK system to complete this simulation there is no "standard" result to compare the improved methods with. The surviving methods maintain their relative positions with respect to enstrophy production.

For the theoretical Reynolds number of 5000 only two systems remain, their streamfunction plots are given in Figure **8**. At this Reynolds number the upper left vortex has appeared in the Ehrenfest limited simulation, however a new discrepancy has arisen. The lower right corner exhibits a very low level of streaming.

In Figure **9** the enstrophy during the final parts of the simulation is given. We note that for the first time the MRT1 system produces less enstrophy (is more dissipative) than the BGK system with Ehrenfest limiter.

For the final two Reynolds numbers we use, 7500 and 10000, only the BGK system with the Ehrenfest limiter completes the simulation. The corresponding streamfunction plots are given in Figure **10** and they exhibit multiple vortices in the corners of the domain.

The enstrophy plots are given in Figure **11**, as the theoretical Reynolds number increases so does the level of enstrophy.

[illegible]

As Reynolds number increases the flow in the cavity is no longer steady and a more complicated flow pattern emerges. On the way to a fully developed turbulent flow, the lid-driven cavity flow is known to undergo a

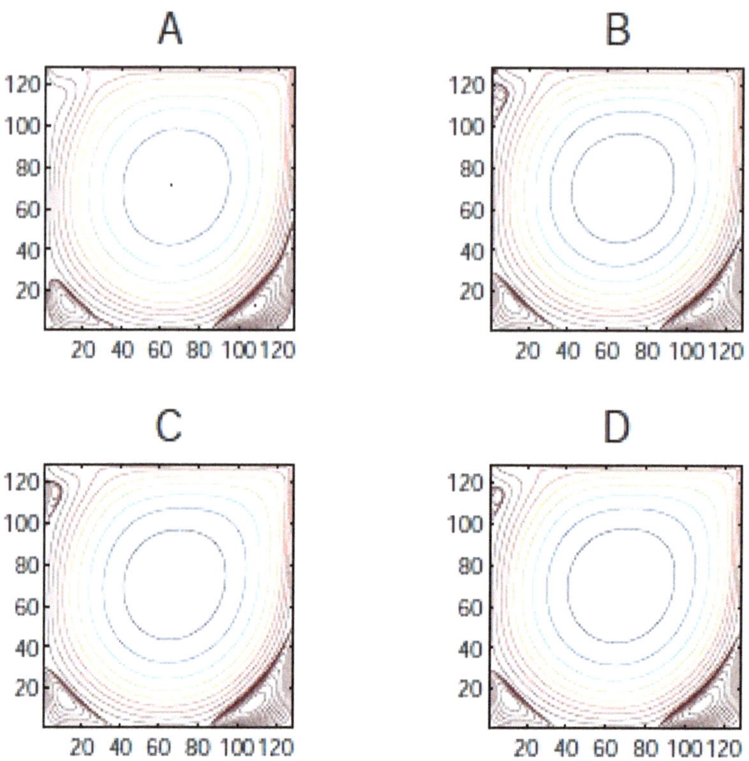

Fig. 6: Contour plots of stream functions of A: BGK + Ehrenfest Steps, B: BGK + MRT Limiter, C: TRT, D: MRT1,following 10000000 time steps at Re2500.

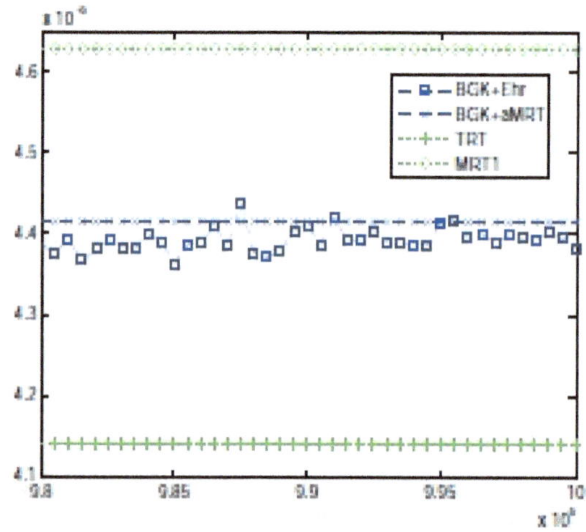

Fig. 7: Enstrophy in the Re2500 systems during the final $2 \cdot 10^5$ time steps

series of period doubling Hopf bifurcations.

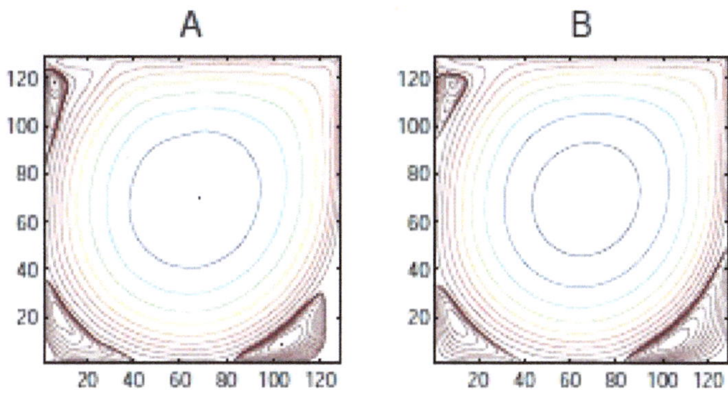

Fig. 8: Contour plots of stream functions of A: BGK + Ehrenfest Steps, B: MRT1,following 10000000 time steps at Re5000.

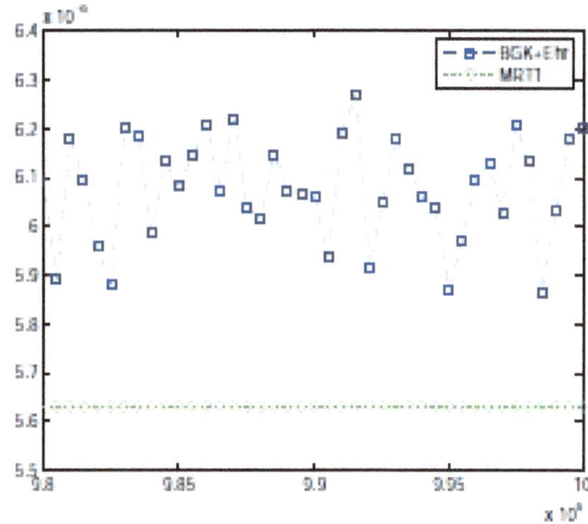

Fig. 9: Enstrophy in the Re5000 systems during the final $2 \cdot 10^5$ time steps

A survey of available literature reveals that the precise value of Re at which the first Hopf bifurcation occurs is somewhat contentious, with most current studies (all of which are for incompressible flow) ranging from around Re $= 7400$–8500 [6, 28, 29]. Here, we do not intend to give a precise value because it is a well observed grid effect that the critical Reynolds number increases (shifts to the right) with refinement (see, e.g., Fig. 3 in [29]). Rather, we will be content to localise the first bifurcation and, in doing so, demonstrate that limiters are capable of regularising without effecting fundamental flow features.

To localise the first bifurcation we take the following algorithmic approach. Entropic equilibria are in use. The initial uniform fluid density profile is $\rho = 1.0$ and the velocity of the lid is $u_0 = 1/10$ (in lattice units). We record the unsteady velocity data at a single control point with coordinates $(L/16, 13L/16)$ and run the simulation for $5000L/u_0$ time steps. Let us denote the final 1% of this signal by $(u_{\mathrm{sig}}, v_{\mathrm{sig}})$. We then compute the *energy* E_u (ℓ_2-norm normalised by non-dimensional signal duration) of the deviation of u_{sig} from its mean:

$$E_u := \left\| \sqrt{\frac{L}{u_0 |u_{\mathrm{sig}}|}} (u_{\mathrm{sig}} - \overline{u_{\mathrm{sig}}}) \right\|_{\ell_2}, \tag{34}$$

where $|u_{\mathrm{sig}}|$ and $\overline{u_{\mathrm{sig}}}$ denote the length and mean of u_{sig}, respectively. We choose this robust statistic instead

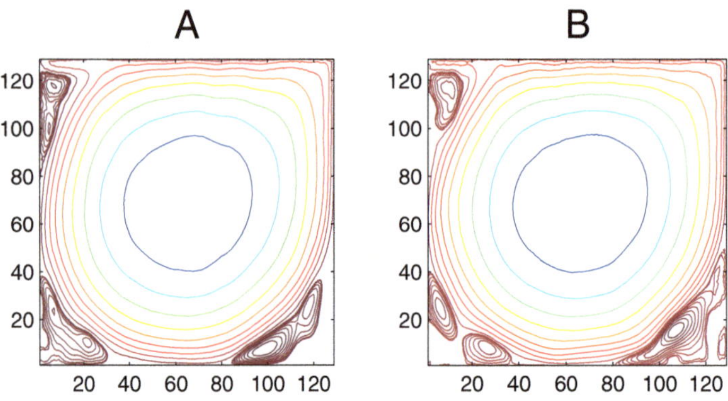

Fig. **10**: Contour plots of stream functions of A: Re7500 and B: Re10000 BGK + Ehrenfest systems,following 10000000 time steps at Re5000.

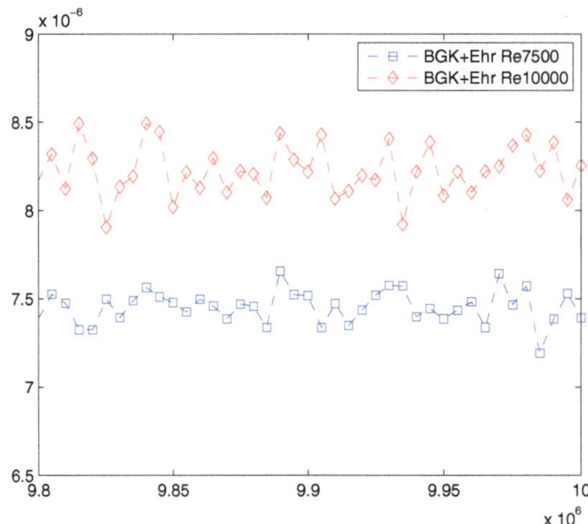

Fig. **11**: Enstrophy in the Re7500 and Re10000 BGK + Ehrenfest systems during the final $2 \cdot 10^5$ time steps

of attempting to measure signal amplitude because of numerical noise in the LBM simulation. The source of noise in LBM is attributed to the existence of an inherently unavoidable neutral stability direction in the numerical scheme (see, e.g., [3]).

We opt not to employ the "bounce-back" boundary condition used in the previous steady state study. Instead we will use the diffusive Maxwell boundary condition (see, e.g., [8]), which was first applied to LBM in [2]. The essence of the condition is that populations reaching a boundary are reflected, proportional to equilibrium, such that mass-balance (in the bulk) and detail-balance are achieved. The boundary condition coincides with "bounce-back" in each corner of the cavity.

To illustrate, immediately following the advection of populations consider the situation of a wall, aligned with the lattice, moving with velocity u_{wall} and with outward pointing normal to the wall in the negative y-direction (this is the situation on the lid of the cavity with $u_{\text{wall}} = u_0$). The implementation of the diffusive Maxwell boundary condition at a boundary site (x,y) on this wall consists of the update

$$f_i(x,y,t+1) = \gamma f_i^*(u_{\text{wall}}), \qquad i = 4,7,8,$$

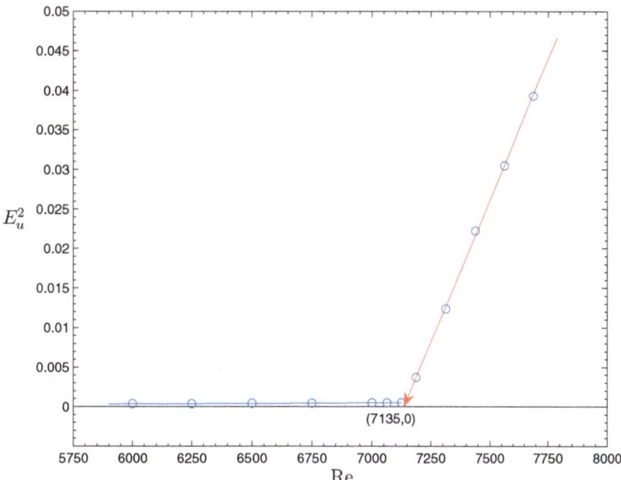

Fig. **12**: Plot of energy squared, E_u^2 (34), as a function of Reynolds number, Re, using LBGK regularised with the median filter limiter with $\delta = 10^{-3}$ on a 100×100 grid. Straight lines are lines of best fit. The intersection of the sloping line with the x-axis occurs close to $\mathrm{Re} = 7135$.

with

$$\gamma = \frac{f_2(x,y,t) + f_5(x,y,t) + f_6(x,y,t)}{f_4^*(u_{\mathrm{wall}}) + f_7^*(u_{\mathrm{wall}}) + f_8^*(u_{\mathrm{wall}})}.$$

Observe that, because density is a linear factor of the equilibria, the density of the wall is inconsequential in the boundary condition and can therefore be taken as unity for convenience. As is usual, only those populations pointing in to the fluid at a boundary site are updated. Boundary sites do not undergo the collisional step that the bulk of the sites are subjected to.

We prefer the diffusive boundary condition over the often preferred "bounce-back" boundary condition with constant lid profile. This is because we have experienced difficulty in separating the aforementioned numerical noise from the genuine signal at a single control point using "bounce-back". We remark that the diffusive boundary condition does not prevent unregularised LBGK from failing at some critical Reynolds number.

Now, we conduct an experiment and record (34) over a range of Reynolds numbers. In each case the median filter limiter is employed with parameter $\delta = 10^{-3}$. Since the transition between steady and periodic flow in the lid-driven cavity is known to belong to the class of standard Hopf bifurcations we are assured that $E_u^2 \propto \mathrm{Re}$ [12]. Fitting a line of best fit to the resulting data localises the first bifurcation in the lid-driven cavity flow to $\mathrm{Re} = 7135$ (Fig. **12**). This value is within the tolerance of $\mathrm{Re} = 7402 \pm 4\%$ given in [29] for a 100×100 grid. We also provide a (time averaged) phase space trajectory and Fourier spectrum for $\mathrm{Re} = 7375$ at the monitoring point (Fig. **13** and Fig. **14**) which clearly indicate that the first bifurcation has been observed.

9. CONCLUSION

In the 1D shock tube we do not find any evidence that maintaining the proper balance of entropy (implementing ELBM) regularizes spurious oscillations in the LBM. We note that entropy production controlled by α and viscosity controlled by β are composite in the collision integral (6). A weak lower approximation to α would lead effectively to addition of dissipation at the mostly far from equilibrium sites and therefore would locally increase viscosity. Therefore the choice of the method to implement the entropic involution is crucial in this scheme. Any method which is not sufficiently accurate could give a misleading result.

In the 2D lid driven cavity test we observe that implementing TRT[13] or MRT[22] with certain relaxation rates can improve stability. The increase in stability from using TRT can be attributed to the correction of the numerical slip on the boundary, as well as increasing dissipation. What is the best set of parameters to

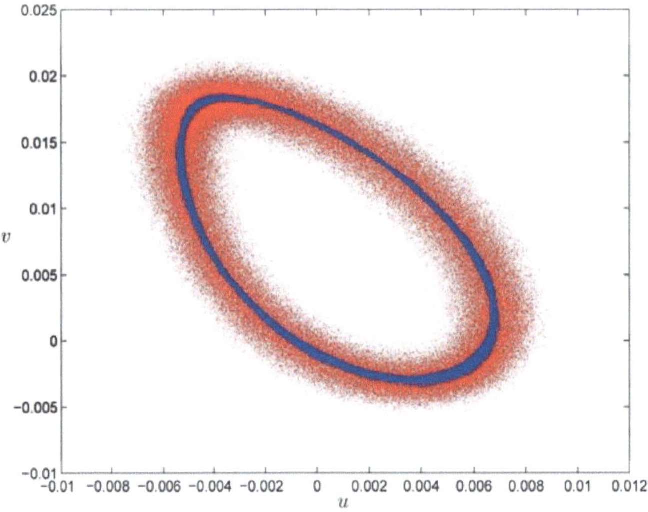

Fig. **13**: A phase trajectory for velocity components for the signal $(u_{\mathrm{sig}}, v_{\mathrm{sig}})$ at the monitoring point $(L/16, 13L/16)$ using LBGK regularised with the median filter limiter with $\delta = 10^{-3}$ on a 100×100 grid $(\mathrm{Re} = 7375)$. Dots represent simulation results at various time moments and the solid line is a 100 step time average of the signal.

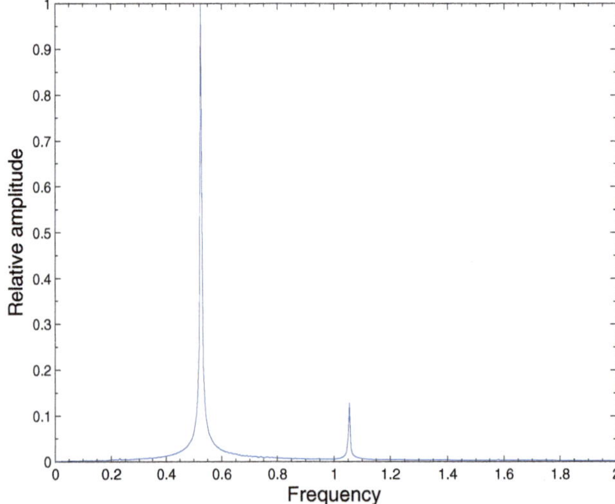

Fig. **14**: Relative amplitude spectrum for the signal u_{sig} at the monitoring point $(L/16, 13L/16)$ using LBGK regularised with the median filter limiter with $\delta = 10^{-3}$ on a 100×100 grid $(\mathrm{Re} = 7375)$. We measure a dominant frequency of $\omega = 0.525$.

choose for MRT is not a closed question. The parameters used in this work originally proposed by Lallemand and Luo [22] are based on a linear stability analysis. Certain choices of relaxation parameters may improve stability while qualitatively changing the flow, so parameter choices should be justified theoretically, or alternatively the results of simulations should be somehow validated. Nevertheless the parameters used in this work exhibit an improvement in stability over the standard BGK system.

Modifying the relaxation rates of the different modes changes the production of dissipation of different components at different orders of the dynamics. The higher order dynamics of latttice Boltzmann methods include higher order space derivatives of the distribution functions. MRT could exhibit the very nice property that where these derivatives are near to zero that MRT has little effect, while where these derivatives are large (near shocks and oscillations) that additional dissipation could be added, regularizing the system.

Using entropic limiters explicitly adds dissipation locally[4]. The Ehrenfest steps succeed to stabilize the system at Reynolds numbers where other tested methods fail, at the cost of the smoothness of the flow. We also implemented an entropic limiter using MRT technology. This also succeeded in stabilizing the system to a degree, however the amount of dissipation added is less than an Ehrenfest step and hence it is less effective. The particular advantage of a limiter of this type over the Ehrenfest step is that it can preserve the correct production of dissipation on physical modes across the system. Other MRT type limiters can easily be invented by simply varying the relaxation parameters.

As previously mentioned there have been more filtering operations proposed[30]. These have a similar idea of local (but not pointwise) filtering of lattice Boltzmann simulations. A greater variety of variables to filter have been examined, for example the macroscopic field can be filtered rather than the mesoscopic population functions.

We can use the Enstrophy statistic to measure effective dissipation in the system. The results from the lid-driven cavity experiment indicate that increased total dissipation does not necessarily increase stability. The increase in dissipation needs to be targeted onto specific parts of the domain or specific modes of the dynamics to be effective.

Using the global median filter in the lid-driven cavity we find that the expected Reynolds number of the first Hopf bifurcation seems to be preserved, despite the additional dissipation. This is an extremely positive result as it indicates that if the addition of dissipation needed to stabilize the system is added in an appropriate manner then qualitative features of the flow can be preserved.

Finally we should note that the stability of lattice Boltzmann systems depends on more than one parameter. In all these numerical tests the Reynolds number was modified by altering the rate of production of shear viscosity. In particular, for the lid driven cavity the Reynolds number could be varied by altering the lid speed, which was fixed at 0.1 in all of these simulations. Since the different modes of the dynamics include varying powers of velocity, this would affect the stability of the system in a different manner to simply changing the shear viscosity coefficient. In such systems the relative improvements offered by these methods over the BGK system could be different.

The various LBMs all work well in regimes where the macroscopic fields are smooth. Each method has its limitation as they attempt to simulate flows with, for example, shocks and turbulence, and it is clear that something needs to be done in order to simulate beyond these limitations. We have examined various add-ons to well-known implementations of the LBM, and explored their efficiency. We demonstrate that add-ons based on the gentle modification of dissipation can significantly expand the stable boundary of operation of the LBM.

ACKNOWLEDGEMENT

Declared none

CONFLICT OF INTEREST

The authors confirm that this chapter content has no conflict of interest.

REFERENCES

[1] Ansumali S, Karlin IV. Stabilization of the lattice Boltzmann method by the H-theorem: A numerical test. Phys Rev E 2000; 62: 7999-8003.

[2] Ansumali S, Karlin IV. Kinetic boundary conditions in the lattice Boltzmann method. Phys Rev E 2002; 66: 026311.
[3] Brownlee RA, Gorban AN, Levesley J. Stabilisation of the lattice-Boltzmann method using the Ehrenfests' coarse-graining. Phys Rev E 2006; 74: 037703.
[4] Brownlee RA, Gorban AN, Levesley J. Stability and stabilisation of the lattice Boltzmann method. Phys Rev E 2007; 75: 036711.
[5] Brownlee RA, Gorban AN, Levesley J. Nonequilibrium entropy limiters in lattice Boltzmann methods, Physica A 2008; 387: 385-406.
[6] Bruneau C-H, Saad M. The 2D lid-driven cavity problem revisited. Comput Fluids 2006; 35: 326-348.
[7] Benzi R., Succi S, Vergassola M. The lattice Boltzmann-equation – theory and applications. Phys Reports 1992; 222: 145-197.
[8] Cercignani C. Theory and application of the Boltzmann equation. Scottish Academic Press, Edinburgh 1975.
[9] Chen S, Doolen GD. Lattice Boltzmann method for fluid flows. Annu Rev Fluid Mech 1998; 30: 329-364.
[10] Dellar PJ. Bulck and shear viscosities in lattice Boltzmann equation. Phys Rev E 2001; 64: 031203.
[11] Dellar PJ. Incompressible limits of lattice Boltzmann equations using multiple relaxation times. J Comput Phys 2003; 190: 351-370.
[12] Ghaddar NK, Korczak KZ, Mikic BB, Patera AT. Numerical investigation of incompressible flow in grooved channels. Part 1. Stability and self-sustained oscillations. J Fluid Mech 1986; 163: 99-127.
[13] Ginzburg I. Generic boundary conditions for lattice Boltzmann models and their application to advection and anisotropic dispersion equations. Adv Water Res 2005; 28: 1196-1216.
[14] Godunov SK. A difference scheme for numerical solution of discontinuous solution of hydrodynamic equations. Math Sbornik 1959; 47: 271-306.
[15] Gorban AN. Basic type of coarse-graining. In: Gorban AN, Kazantzis N, Kevrekidis IG, Öttinger HC, Theodoropoulos C. (eds). Model Reduction and Coarse-Graining Approaches for Multiscale Phenomena. Springer, Berlin-Heidelberg, New York 2006: 117-176.
[16] Gorban AN, Karlin IV, Öttinger HC, Tatarinova LL. Ehrenfests' argument extended to a formalism of nonequilibrium thermodynamics. Phys Rev E 2001; 63: 066124.
[17] Higuera F, Succi S, Benzi R. Lattice gas – dynamics with enhanced collisions. Europhys Lett 1989; 9: 345–349.
[18] Karlin IV, Ferrante A, Öttinger HC. Perfect entropy functions of the lattice Boltzmann method. Europhys Lett 1999; 47: 182–188.
[19] Karlin IV, Gorban AN, Succi S, Boffi V. Maximum entropy principle for lattice kinetic equations. Phys Rev Lett 1998; 81:6–9 .
[20] Kullback S. Information theory and statistics. Wiley, New York 1959.
[21] Kuzmin D, Lohner R, Turek S. Flux corrected transport. Springer 2005.
[22] Lallemand P, Luo LS. Theory of the lattice Boltzmann method: Dispersion, dissipation, isotropy, Galilean invariance, and stability. Phys Rev E 2000; 61: 6546-6562.
[23] Lax PD. On dispersive difference schemes. Physica D 1986; 18: 250-254.
[24] Levermore CD, Liu J-G. Oscillations arising in numerical experiments. Physica D 1996; 99: 191-216.
[25] Li Y, Shock R, Zhang R, Chen H. Numerical study of flow past an impulsively started cylinder by the lattice-Boltzmann method. J Fluid Mech 2004; 519: 273-300.
[26] Luo LS, Liao W, Chen X, Peng Y, Zhang W. Numerics of the lattice Boltzmann method: Effects of collision models on the lattice Boltzmann simulations. Phys Rev E 2011; 83: 056710.
[27] Packwood D. Entropy balance and dispersive oscillations in lattice Boltzmann models. Phys Rev E 2009; 80: 067701.
[28] Pan TW, Glowinksi R. A projection/wave-like equation method for the numerical simulation of incompressible viscous fluid flow modeled by the Navier-Stokes equations. Comp Fluid Dyn J 2000; 9: 28-42.
[29] Peng Y-F, Shiau Y-H, Hwang RR. Transition in a 2-D lid-driven cavity flow. Comput Fluids 2003; 32: 337-352.
[30] Ricot D, Marie S, Sagaut P, Bailly C. Lattice Boltzmann method with selective viscosity filter. J Comput Phys 2009; 228: 4478-4490.
[31] Pratt WK. Digital Image Processing. Wiley, New York 1978.
[32] Roy CJ. Grid convergence error analysis for mixed-order numerical schemes. AIAA J 2003; 41: 595-604
[33] Servan-Camas B, Tsai FT-C. Non-negativity and stability analyses of lattice Boltzmann method for advection-diffusion equation. J Comput Phys 2009; 228: 236-256.
[34] Succi S. The lattice Boltzmann equation for fluid dynamics and beyond. Oxford University Press, New York 2001.
[35] Tadmor E, Zhong W. Entropy stable approximations of Navier-Stokes equations with no artificial numerical viscosity. J Hyperbolic DEs 2006; 3: 529-559.
[36] Tosi F, Ubertini S, Succi S, Chen H, Karlin IV. A comparison of single-time relaxation lattice Boltzmann schemes with enhanced stability. Math Comput Simulation 2006; 72: 227-231.
[37] Wesseling P. Principles of Computational Fluid Dynamics. Springer Series in Computational Mathematics 29, Springer, Berlin 2001.

CHAPTER 3

Discrete-Velocity Models and Lattice Boltzmann Methods for Convection-Radiation Problems

Mapundi K. Banda[1], **Mohammed Seaid**[2,*]

[1] *School of Computational and Applied Mathematics, University of the Witwatersrand, South Africa,*
[2] *School of Engineering and Computing Sciences, University of Durham, UK*

Abstract: Discrete-velocity models and lattice Boltzmann methods are presented for solving convection-radiation effects in thermal fluid flows. The discrete-velocity equations are derived from the continuous Boltzmann equation with appropriate scaling suitable for incompressible flows. The radiative heat flux in the energy equation is obtained using the discrete-ordinates solution of the radiative transfer equation. This chapter is structured in two parts. In the first part we investigate the derivation of the discrete-velocity models and we perform an asymptotic analysis to show at the leading order, the macroscopic models are recovered from these discrete-velocity models. In this part we also formulate the lattice Boltzmann method for solving the convection-radiation problems. The second part of this chapter is devoted to the computational aspects of the considered models. Consistent boundary and initial conditions in the lattice Boltzmann models are also discussed in this part. Numerical results are presented for several test examples on coupled convection-radiation flows in two dimensional enclosures. Detailed simulation results at different flow and radiative regimes, as well as benchmark solutions, are also presented and discussed. The obtained results show that the developed models are competitive tools for convection-radiation problems.

Keywords: Discrete-velocity models, forced convection, heat transfer, implicit-explicit schemes, incompressible Navier-Stokes equations, large-eddy simulation, Lattice-Boltzmann method, natural convection, radiative heat transfer, relaxation scheme.

1. INTRODUCTION

The aim of the present chapter is to present a methodology for developing computational schemes for incompressible Navier-Stokes equations coupled with heat energy transport equation and radiative transfer. Such computational schemes are developed from a relaxation system derived from a lattice-Boltzmann type discrete-velocity model with diffuse scaling. In particular two relaxation systems are developed: the first for the fluid flow and the second for energy transport. The lattice-Boltzmann method is used as a platform for developing models for thermal flows because of its physical underpinning. Such an approach gives relaxation systems that are based on physical considerations rather than a pure mathematical formulation. This provides a consistent tool for deriving relaxation systems especially for practitioners in the applied sciences. The schemes presented here consider an energy equation with all modes of energy transport namely, convection, conduction and radiation. The analytical derivations also demonstrate a relationship between the lattice-Boltzmann method and such relaxation-based schemes. Moreover, such a multiscale approach provides a natural tool for solving microscopic and macroscopic models with the same underlying model. It should also be pointed out that the work presented in this chapter is mainly based on results presented in [4, 8] and the references therein.

The work in this chapter was mainly motivated by the desire to contribute to the active research activities to solve many practical applications involving high temperature regimes, like convection in a glass melting furnace, crystal growth, the design of combustion chambers for gas turbines, cooling of electronic circuitry, nuclear reactor insulation and ventilation rooms. These require the extension of the underlying flow equations such that radiation effects are included. In the problems considered here fluid flow is mainly defined

Address correspondence to: Mohammed Seaid, School of Engineering and Computing Sciences, University of Durham, South Road, Durham DH1 3LE, UK; Tel: ++44 191 334 2476; Fax: ++44 191 334 2407; Email: m.seaid@durham.ac.uk

by the limit of the incompressible Navier-Stokes equations *i.e.*, the flow is driven by density differences at small Mach numbers. In general temperature variations are significant. As a result transport properties vary with temperature. Energy and momentum equations are hence coupled and need to be solved simultaneously. The second effect of temperature variation is that density variation interacts with gravity creating a body force that may modify the flow structures considerably and may be the main driving force in the flow. The latter is mainly referred to as buoyancy-driven or natural convection. The relative importance of forced convection and buoyancy effects is measured by the ratio of the Rayleigh number Ra and the Reynolds number Re. If the ratio Re/Ra is very large the effects of natural convection may be ignored. In purely buoyancy-driven flows it may be possible to ignore density variations in all terms except for the body force in the vertical momentum equation. In the case where such flow is turbulent one faces the challenge of designing numerical schemes that resolve the flow efficiently. Just a few remarks on turbulence are now in order. The turbulent flow will modelled based on the Large Eddy Simulation (LES) formulation. Recently, developments of the LES techniques, both from the theoretical and applications point of views have been made. The LES equations for incompressible thermal flows are obtained by applying a spatial filtering to the Navier-Stokes equations, see [12, 55, 22] and the references therein. However, this procedure introduces a term called a subgrid stress tensor which needs to be modelled. This term has to be seen as the interaction between the large and small scales in the system. For the work in this chapter a subgrid problem based on the Smagorinsky model [64] will be applied. The LES simulations are normally performed on grids that are just fine enough to resolve the large flow scales, and numerical errors on such grids can have large effects on the simulation results. Although in the literature (see [22], and references therein) many numerical schemes have been presented, it is still not clear what the effect of the numerical and modelling errors in an LES simulation is. This is still an area of active research. However, it is commonly agreed that high-order accurate methods offer a means to obtain accurate solutions efficiently in comparison to low-order accurate methods.
To derive the relaxation systems referred to above, one considers kinetic equations or discrete-velocity models. Such kinetic equations or discrete-velocity models of kinetic equations yield, in the limit for small Knudsen and Mach numbers, an approximation of the incompressible Navier-Stokes equations. A classical example is given by the discrete-velocity models used for lattice-Boltzmann methods, see [11, 16, 28, 30, 17, 14] among others. These discrete-velocity models can also be viewed as relaxation systems for the incompressible Navier-Stokes equations. For such flows, pressure variations occurring are small such that variations in density can be ignored. Density variations can also be caused by changes in temperature and such variations can not, in general, be neglected. Even at small velocities, density of non-uniformly heated fluid can not be assumed constant [42, 43]. In this chapter the development of a computational approach for fluid flows with temperature changes is presented. In this approach lattice-Boltzmann approaches are considered, see also [25, 27, 46, 62] and references therein. A relaxation system for incompressible Navier-Stokes equations coupled with heat transport and radiation is developed. An extension of such an approach to incompressible Navier-Stokes equations and turbulent flows based on large-eddy simulation is also presented [6, 9, 8].
To solve relaxation systems one can apply relaxation type schemes with special properties, for example, asymptotic preserving schemes. In particular, a large number of numerical methods for kinetic equations with stiff relaxation terms have been considered in fluid dynamics or diffusive limits. For such relaxation methods and asymptotic-preserving methods, the reader may refer to [1, 13, 35, 40] among others. It can also be mentioned that in the context of hyperbolic conservation laws, relaxation schemes are closely related to central schemes in the sense that both approaches provide efficient high resolution and Riemann-solver free numerical methods. The advantage of using such a kinetic or lattice Boltzmann approach is that it provides a simple, clear and consistent framework from which relaxation systems and their high-order relaxation schemes can be developed. High-order relaxation methods are developed in a straightforward manner by discretizing relaxation systems, see Section 3.2.. The main results presented in this chapter demonstrate how a numerical approach based on synthesizing different schemes can be applied to a coupled system of radiating thermal incompressible flow models. The numerical approach is designed by synthesizing different approaches for different models based on a common multi-scale Boltzmann-type model. This approach is simple, practical and very accurate for such problems as the test cases in Section 5.3. demonstrate.
Towards the end of the chapter a demonstration of the performance of this approach using a third-order relaxation scheme for the flow equations is presented. The scheme works with uniform accuracy with respect to the Knudsen and Mach numbers [6]. In the low-Mach number limit it reduces to a third-order

explicit scheme for the incompressible Navier-Stokes/Boussinesq equations. This is achieved by combining the ideas developed in [41] with a third-order non-oscillatory spatial discretization and an Implicit-Explicit (IMEX) Runge-Kutta time discretization [6]. To obtain a discretization of the limit incompressible Navier-Stokes/Boussinesq equations, one can use the above mentioned spatial discretization on the relaxation system and apply any high-order time discretization directly to the resulting semi-discrete relaxed schemes. Hence high-order incompressible Navier-Stokes solvers with better stability properties are obtained. To solve the radiative transfer equation, the well-established discrete ordinates method along with a diffusion-synthetic acceleration procedure is formulated. This approach can also be viewed as a discrete-velocity model for the kinetic radiative transfer.

The rest of the chapter is organized in the following way. Section contains the lattice-Boltzmann type discrete-velocity model and its equivalent associated closed moment system relaxing to the incompressible Navier-Stokes/Boussinesq equations. Some simplified relaxation systems are also presented. In particular we introduce a simplified relaxation system that is suitable to provide relaxed schemes for the thermal incompressible Navier-Stokes equations. Section 3.2. describes numerical methods and solution procedures for the flow system and radiative transfer equation. Finally, Section 5.3. presents a numerical investigation of the schemes for two test examples on natural and forced convection-radiation problems. Concluding remarks are presented in Section 6.5..

2. DISCRETE-VELOCITY MODELS FOR THERMAL FLOWS AND SIMPLIFIED RELAXATION SYSTEMS

Many models in fluid dynamics can be derived from discrete-velocity models of kinetic equations. In addition, some numerical methods for solving such fluid dynamic problems can be systematically derived from kinetic equations, for example, the lattice Boltzmann method [26] and relaxation schemes for the compressible and the incompressible flow problems [39, 5, 6]. In this section, this approach will be extended to the numerical solution of convection-radiation as well as turbulent thermal flows modelled with LES techniques. First, the discrete-velocity model for incompressible Navier-Stokes equations will be described, and later modified to incorporate equations for LES and heat transfer. Finally, a relaxation LES with heat transport is formulated. Details of the discrete-velocity LES for isothermal flows can also be found in [9]. Here the presentation mainly focuses on the extension of this method in order to couple it to the heat transport.

[illegible]

The two-dimensional kinetic equation

$$\frac{\partial f}{\partial t} + \mathbf{v} \cdot \nabla f = J(f) + F, \tag{1}$$

describes the evolution of a particle density $f(\mathbf{x}, \mathbf{v}, t)$, with $\mathbf{x} = (x, y)^T \in \mathbb{R}^2$ and $\mathbf{v} = (v_1, v_2)^T \in \mathbb{R}^2$. The left hand side of equation (1) represents free transport of the particles while the right hand side describes interactions through collisions, $J(f)$, and a forcing term, F. For the two-dimensional discrete models one has

$$\mathbf{v} \in \{\mathbf{c}_0, \ldots, \mathbf{c}_{N-1}\}, \qquad \mathbf{c}_i \in \mathbb{R}^2.$$

Here, a model with nine velocities ($N = 9$) is considered:

$$\begin{aligned}
\mathbf{c}_1 &= \begin{pmatrix} 1 \\ 0 \end{pmatrix}, & \mathbf{c}_2 &= \begin{pmatrix} 0 \\ 1 \end{pmatrix}, & \mathbf{c}_3 &= \begin{pmatrix} -1 \\ 0 \end{pmatrix}, & \mathbf{c}_4 &= \begin{pmatrix} 0 \\ -1 \end{pmatrix}, \\
\mathbf{c}_5 &= \begin{pmatrix} 1 \\ 1 \end{pmatrix}, & \mathbf{c}_6 &= \begin{pmatrix} -1 \\ 1 \end{pmatrix}, & \mathbf{c}_7 &= \begin{pmatrix} -1 \\ -1 \end{pmatrix}, & \mathbf{c}_8 &= \begin{pmatrix} 1 \\ -1 \end{pmatrix},
\end{aligned}$$

and $\mathbf{c}_0 = \mathbf{0}$. In the discrete case, the $\mathbf{v}$-dependence of the particle distribution $f(\mathbf{x}, \mathbf{v}, t)$ is uniquely determined through N functions

$$f_i(\mathbf{x}, t) = f(\mathbf{x}, \mathbf{c}_i, t), \qquad i = 0, \ldots, N-1.$$

Macroscopic quantities like mass-, momentum- or energy-density are obtained by taking velocity moments of f. If ζ is any $\mathbf{v}$-dependent function, the discrete-velocity integral is denoted by

$$\langle\zeta\rangle = \sum_{i=0}^{N-1} \zeta(\mathbf{c}_i).$$

Mass and momentum density are then given by

$$\rho(\mathbf{x},t) = \langle f(\mathbf{x},\mathbf{v},t)\rangle \quad \text{and} \quad \rho\mathbf{u}(\mathbf{x},t) = \langle \mathbf{v}f(\mathbf{x},\mathbf{v},t)\rangle. \tag{2}$$

In what follows the components of the velocity are denoted by $\mathbf{u} = (u_1,u_2)^T$. In most lattice-Boltzmann applications, the collision operator $J(f)$ in (1) is typically of BGK-type

$$J(f) = -\frac{1}{\tau_v}(f - f^{eq}). \tag{3}$$

The parameter $\tau_v > 0$ is called *relaxation time* and f^{eq} is the *equilibrium distribution*. In the isothermal case, the equilibrium distribution f^{eq} depends on f through the variables ρ and $\mathbf{u}$ which are calculated according to (2), see for example [28, 29, 63]. Note that a scaling $\varepsilon^2 F_\varepsilon$ has been used in order to obtain the correct Boussinesq approximation in the limit. This is a consequence of the transformation into an equivalent set of moment equations which will be undertaken below. In this case

$$F_\varepsilon = 3f^*\frac{\mathbf{v}}{\varepsilon}\cdot\mathbf{G},$$

where $\mathbf{G}$ is an external force acting per unit mass, compare [25, 36]. For the standard D2Q9-model with 9 velocities [54], we have

$$f^{eq}[\rho,\mathbf{u}](\mathbf{v}) = \rho\left(1 + 3\mathbf{u}\cdot\mathbf{v} - \frac{3}{2}|\mathbf{u}|^2 + \frac{9}{2}(\mathbf{u}\cdot\mathbf{v})^2\right)f^*(\mathbf{v}), \tag{4}$$

where f^* is defined by

$$f^*(\mathbf{c}_i) = \begin{cases} \frac{4}{9}, & i = 0, \\ \frac{1}{9}, & i = 1,\cdots,4, \\ \frac{1}{36}, & i = 5,\cdots,8. \end{cases}$$

The equilibrium distribution is constructed in such a way that

$$\langle J(f)\rangle = 0 \quad \text{and} \quad \langle \mathbf{v}J(f)\rangle = 0,$$

which reflects respectively, conservation of mass and conservation of momentum in the collision process. The forcing term F is defined as

$$F = 3f^*(\mathbf{v})\mathbf{v}\cdot\mathbf{G},$$

where $\mathbf{G}$ is an external force acting per unit mass [25, 36]. Since

$$\langle 1, f^*\rangle = 1, \qquad \left\langle 1, \mathbf{v}_\alpha\mathbf{v}_\beta f^*\right\rangle = \frac{1}{3}\delta_{\alpha\beta}, \qquad \left\langle 1, \mathbf{v}_\alpha\mathbf{v}_\beta\mathbf{v}_\gamma\mathbf{v}_\theta f^*\right\rangle = \frac{1}{9}(\delta_{\alpha\beta}\delta_{\gamma\theta} + \delta_{\alpha\gamma}\delta_{\beta\theta} + \delta_{\alpha\theta}\delta_{\beta\gamma}),$$

where $\delta_{\alpha\beta}$ denotes the Kronecker symbol. Hence, we obtain

$$\langle 1, 3f^*\mathbf{v}\cdot\mathbf{G}\rangle = 0 \qquad \text{and} \quad \langle 1, 3f^*\mathbf{v}\mathbf{v}\mathbf{G}\rangle = 3\langle 1, f^*\mathbf{v}\mathbf{v}\rangle\mathbf{G} = \mathbf{G}.$$

In order to obtain a relation between the kinetic equation (1) and the incompressible Navier-Stokes equations, the *diffusive scaling* $\mathbf{x} \to \mathbf{x}/\varepsilon$, $t \to t/\varepsilon^2$ together with a rescaling of velocity $\mathbf{u} \to \varepsilon\mathbf{u}$ is introduced. This scaling describes the small Knudsen and low Mach number limit of kinetic equations, see [65, 19, 10, 32] for details. Under these transformations equation (1) is rewritten as

$$\frac{\partial f}{\partial t} + \frac{1}{\varepsilon}\mathbf{v}\cdot\nabla f = -\frac{1}{\varepsilon^2\tau_v}\Big(f - f^{eq}[\rho,\varepsilon\mathbf{u}]\Big) + \varepsilon^2 F_\varepsilon. \tag{5}$$

Notice that, in our case, equation (5) consists of nine equations for the occupation numbers $f_0,\ldots,f_8$. Here one also observes that a scaling $\varepsilon^2 F_\varepsilon$ has been used in order to obtain the correct source term in the incompressible limit as a consequence of the transformation into an equivalent set of moment equations in (6) below. In this case

$$F_\varepsilon = 3f^*(\mathbf{v})\frac{\mathbf{v}}{\varepsilon}\cdot\mathbf{G}.$$

Equation (5) is transformed into an equivalent set of moment equations (see also [41, 31] for a similar approach) using moments based on the following **v**-polynomials [24]

$$\begin{aligned}
&P_0(\mathbf{v}) = 1, && && \\
&P_1(\mathbf{v}) = \frac{v_1}{\varepsilon}, && P_2(\mathbf{v}) = \frac{v_2}{\varepsilon}, && \\
&P_3(\mathbf{v}) = \frac{v_1^2}{\varepsilon^2} - \frac{1}{3\varepsilon^2}, && P_4(\mathbf{v}) = \frac{v_1 v_2}{\varepsilon^2}, && P_5(\mathbf{v}) = \frac{v_2^2}{\varepsilon^2} - \frac{1}{3\varepsilon^2}, \\
&P_6(\mathbf{v}) = \frac{(3|\mathbf{v}|^2-4)v_1}{\varepsilon^3}, && P_7(\mathbf{v}) = \frac{(3|\mathbf{v}|^2-4)v_2}{\varepsilon^3}, && P_8(\mathbf{v}) = \frac{9|\mathbf{v}|^4 - 15|\mathbf{v}|^2+2}{\varepsilon^4}.
\end{aligned} \tag{6}$$

Note that $\langle P_0 f\rangle = \rho$, $\langle P_1 f\rangle = \rho u_1$ and $\langle P_2 f\rangle = \rho u_2$. The second order moments form a symmetric tensor

$$\Theta = (\Theta^x, \Theta^y) = \begin{pmatrix} \theta_{11} & \theta_{12} \\ \theta_{12} & \theta_{22} \end{pmatrix} = \begin{pmatrix} \langle P_3 f\rangle & \langle P_4 f\rangle \\ \langle P_4 f\rangle & \langle P_5 f\rangle \end{pmatrix},$$

where $\Theta^x = (\theta_{11}, \theta_{12})^T$ and $\Theta^y = (\theta_{12}, \theta_{22})^T$. For the remaining moments one can set

$$\mathbf{q} = \begin{pmatrix} q_1 \\ q_2 \end{pmatrix} = \begin{pmatrix} \langle P_6 f\rangle \\ \langle P_7 f\rangle \end{pmatrix}, \qquad s = \langle P_8 f\rangle.$$

The equations of mass and momentum conservation are

$$\begin{aligned}
\partial_t \rho + \operatorname{div} \rho\mathbf{u} &= 0, \\
\partial_t \rho\mathbf{u} + \operatorname{div}\Theta + \frac{1}{3\varepsilon^2}\nabla\rho &= \mathbf{G}.
\end{aligned} \tag{7}$$

Here, the divergence is applied to the rows of Θ. The equation for Θ is

$$\partial_t\Theta + \frac{2}{3\varepsilon^2}\mathbf{S}[\rho\mathbf{u}] + \frac{1}{3}\mathbf{Q}[\mathbf{q}] = -\frac{1}{\varepsilon^2\tau_\nu}(\Theta - \rho\mathbf{u}\otimes\mathbf{u}), \tag{8}$$

where

$$\mathbf{S}[\mathbf{u}] = \frac{1}{2}\begin{pmatrix} 2\partial_x u_1 & \partial_y u_1 + \partial_x u_2 \\ \partial_y u_1 + \partial_x u_2 & 2\partial_y u_2 \end{pmatrix}, \qquad \mathbf{Q}[\mathbf{q}] = \begin{pmatrix} \partial_y q_2 & \partial_y q_1 + \partial_x q_2 \\ \partial_y q_1 + \partial_x q_2 & \partial_x q_1 \end{pmatrix}.$$

Finally, the third and fourth order moments satisfy

$$\begin{aligned}
\partial_t \mathbf{q} + \frac{1}{\varepsilon^2}\operatorname{div}\begin{pmatrix} \theta_{22} & 2\theta_{12} \\ 2\theta_{12} & \theta_{11} \end{pmatrix} + \frac{1}{6}\nabla s &= -\frac{1}{\varepsilon^2\tau_\nu}\mathbf{q}, \\
\partial_t s + \frac{4}{\varepsilon^2}\operatorname{div}\mathbf{q} &= -\frac{1}{\varepsilon^2\tau_\nu}s.
\end{aligned} \tag{9}$$

Altogether, a hyperbolic system with stiff relaxation terms is obtained. Below some characteristics of the above system are highlighted:

Remark 1. (i.) Using the lattice Boltzmann equation (5) and its equivalent moment system, a hyperbolic system with stiff relaxation terms is obtained. Hence higher-order schemes applied in solving hyperbolic systems with stiff relaxation terms can also be applied. In short the kinetic approach presents a clear and consistent framework in which one can derive such relaxation systems.

(ii.) Further the equations for $\mathbf{q}$ and s do not contribute to the hydrodynamics in the lowest ε-order (such variables are referred to as ghost variables in [11]). The implication is that the lattice Boltzmann equations provide more information than is necessary in the approximation of the Navier-Stokes system. There is a lattice Boltzmann approach based on six discrete velocities on a hexagonal space grid [16]. For this case the equivalent moment system consists only of equations of mass, momentum and Θ. The extra three variables in the case of nine velocities are due to the required symmetries of the velocity space on a square lattice.

(iii.) Relaxation schemes are based on the idea of replacing nonlinear relaxation terms. This means new variables are introduced. In a bounded domain it is not clear how the boundary conditions for the new variables, some of which have no physical meaning, are prescribed [66].

(iv.) As a consequence of the above three observations, one takes advantage of the relaxation hyperbolic structure derived in the framework of kinetic theory as above and rewrites it in a simplified form to which high-order relaxation schemes of the form in [35] can be applied in a straightforward manner. Such simplifications are made in such a way that the size of the system of equations is reduced and an accurate approximation of the hydrodynamic information contained in the model is retained. Thus the ghost variables are in turn replaced by carefully chosen hydrodynamic information (also based on the characteristic information of the hyperbolic system). Details of this procedure are presented below in section 3.2..

Now a demonstration of how the diffusion limit is obtained will be presented. From the momentum equation in (7) it can be concluded that $\nabla \rho$ tends to zero as $\varepsilon \to 0$. Hence, ρ approaches a constant ρ_∞ (which is the Boussinesq relation in the isothermal case). Writing $\rho = \rho_\infty(1 + 3\varepsilon^2 p)$, equations (7) transform into

$$\begin{aligned} \partial_t p + \frac{1}{3\varepsilon^2}\operatorname{div}\mathbf{u} &= -\operatorname{div} p\mathbf{u}, \\ \partial_t \mathbf{u} + \operatorname{div}\left(\frac{1}{\rho_\infty}\Theta\right) + \nabla p &= -3\varepsilon^2 \partial_t(p\mathbf{u}) + \mathbf{G}. \end{aligned} \tag{10}$$

For $\varepsilon \to 0$, equation (8) yields at the lowest order

$$\frac{1}{\rho_\infty}\Theta = \mathbf{u} \otimes \mathbf{u} - \frac{2\tau_\nu}{3}\mathbf{S}[\mathbf{u}]. \tag{11}$$

Since the system (9) decouples completely from the other equations (in the lowest order) and since $2\operatorname{div}\mathbf{S}[\mathbf{u}] = (\Delta + \nabla \operatorname{div})\mathbf{u}$, one obtains from (10) and (11) the incompressible Navier-Stokes equations as a limiting system

$$\begin{aligned} \operatorname{div}\mathbf{u} &= 0, \\ \partial_t \mathbf{u} + \operatorname{div}\mathbf{u} \otimes \mathbf{u} + \nabla p &= \frac{\tau_\nu}{3}\Delta \mathbf{u} + \mathbf{G}, \end{aligned} \tag{12}$$

where the kinematic viscosity is related to the relaxation time by $\nu = \tau_\nu/3$. The equations (7)-(9) can be viewed as a relaxation system for the incompressible Navier-Stokes equations (12).

[illegible]

In this section a presentation of a derivation of the thermal energy equation based on a lattice-Boltzmann formulation [27, 62] is undertaken. The case in which the temperature field is passively advected by fluid flow and obeys a simple passive scalar equation is considered. Let R denote the gas constant, ρ, $\mathbf{u}$, and T the macroscopic density, velocity and temperature, respectively. An energy moment system for two-dimensional problems can be derived from

$$\rho RT = \left\langle P_0(\mathbf{v})\frac{1}{2}f(\mathbf{v}-\mathbf{u})^2 \right\rangle.$$

A temperature distribution function can be defined as

$$g = \frac{(\mathbf{v}-\mathbf{u})^2}{2R} f$$

and the temperature is obtained from

$$\rho T = \langle P_0(\mathbf{v}) g \rangle . \tag{13}$$

An alternative formulation is presented in [25, 61]. Thus, it was shown that for small Mach number flows, the discrete-velocity BGK model can be formulated using

$$\tilde{g} = \frac{g}{\rho}.$$

Using the transformed distribution one obtains the following equilibrium distribution function

$$\tilde{g}^{eq}[T,\rho,\mathbf{u}](\mathbf{v}) = T\left(1 + 3\mathbf{u}\cdot\mathbf{v} - \frac{3}{2}|\mathbf{u}|^2 + \frac{9}{2}(\mathbf{u}\cdot\mathbf{v})^2\right) f^*(\mathbf{v}), \tag{14}$$

where

$$T = \langle P_0(\mathbf{v})\tilde{g} \rangle .$$

Also denote $\langle P_1\tilde{g}\rangle = \Psi_1$, $\langle P_2\tilde{g}\rangle = \Psi_2$, $\langle P_1\tilde{g}^{eq}\rangle = u_1 T$ and $\langle P_2\tilde{g}^{eq}\rangle = u_2 T$, where

$$\tilde{g}^{eq}[T,\rho,\mathbf{u}](\mathbf{v}) = T f^{eq}[\rho,\mathbf{u}](\mathbf{v}),$$

with f^{eq} denoting the equilibrium function given by (4). In the subsequent discussion $\tilde{g}$ is replaced by g. As usual the second-order moments form a symmetric tensor

$$\Pi = (\Pi^x, \Pi^y) = \begin{pmatrix} \Pi_{11} & \Pi_{12} \\ \Pi_{12} & \Pi_{22} \end{pmatrix} = \begin{pmatrix} \langle P_3 g\rangle & \langle P_4 g\rangle \\ \langle P_4 g\rangle & \langle P_5 g\rangle \end{pmatrix},$$

where $\Pi^x = (\Pi_{11}, \Pi_{12})^T$ and $\Pi^y = (\Pi_{12}, \Pi_{22})^T$. For the remaining moments set

$$\mathbf{q}_c = \begin{pmatrix} q_{c1} \\ q_{c2} \end{pmatrix} = \begin{pmatrix} \langle P_6 g\rangle \\ \langle P_7 g\rangle \end{pmatrix}, \qquad s_c = \langle P_8 g\rangle .$$

The evolution equation is defined from equation (5) and for small Mach numbers and cases where compression does not take place and viscous heat dissipation effects are negligibly small

$$\partial_t g + \frac{1}{\varepsilon}\mathbf{v}\cdot\nabla g = -\frac{1}{\varepsilon^2\tau_c}\left(g - g^{eq}[\rho,\varepsilon\mathbf{u}]\right). \tag{15}$$

is applied. If there is a heat source term, the above equation is extended similarly to equation (5). The thermal source term $\mathbf{Q}$ is defined using

$$q_\varepsilon = 3 f^*(\mathbf{v})\frac{\mathbf{v}}{\varepsilon}\cdot\mathbf{Q}. \tag{16}$$

The equation of the thermal distribution function can be written in the form

$$\partial_t g + \frac{1}{\varepsilon}\mathbf{v}\cdot\nabla g = -\frac{1}{\varepsilon^2\tau_c}\left(g - g^{eq}[\rho,\varepsilon\mathbf{u}]\right) + q_\varepsilon. \tag{17}$$

The relaxation system for the energy density and the energy flow can now be written as

$$\begin{aligned} \partial_t T + \operatorname{div}\Psi &= 0, \\ \partial_t \Psi + \operatorname{div}\Pi + \frac{1}{3\varepsilon^2}\nabla T &= -\frac{1}{\varepsilon^2\tau_c}(\Psi - \mathbf{u}T) + \frac{1}{\varepsilon^2}\mathbf{Q}. \end{aligned} \tag{18}$$

Taking higher-order moments completes the system. The details will be skipped but it should be noted that the remaining moment equations decouple from the ones derived above. The lowest-order term in the limit as $\varepsilon \to 0$ gives the following equation

$$
\begin{aligned}
\partial_t T + \operatorname{div} \Psi &= 0, \\
\Psi &= -\frac{\tau_c}{3} \nabla T + \mathbf{u} T + \frac{\tau_c}{3} \mathbf{Q},
\end{aligned}
\tag{19}
$$

or simply

$$
\partial_t T + \operatorname{div} \left(\mathbf{u} T - \frac{\tau_c}{3} \nabla T + \frac{\tau_c}{3} \mathbf{Q} \right) = 0 \tag{20}
$$

where the thermal diffusion coefficient is now $\kappa = \tau_c/3$. Also take note that in this case the source term q_ε in (17) provides an extra flux in the energy equation in the limit.

3. LARGE-EDDY SIMULATION AND RELAXATION SYSTEMS

The modelling technique used here to compute turbulent flows is the large-eddy simulation. This approach computes space averaged flow quantities. To derive the space-filtered Navier-Stokes equations, the Navier-Stokes equations are convoluted with a chosen filter function G_Δ (where Δ is the filter width) and then integrated over the spatial domain. For more details we refer to [55] and further references can be found therein. In this section we will present an LES formulation for the system of equations derived above. We will close by presenting a requisite relaxation system for LES of thermal flows.

[illegible]

In the case of turbulent flow one denotes the parameters $\tau_\nu = \tau_\nu(\mathbf{x}) > 0$ and $\tau_c = \tau_c(\mathbf{x}) > 0$ which are now called *effective relaxation times*. For turbulent models spatial dependence in τ_ν and τ_c are a consequence of spatial dependence of the viscosity ν and thermal diffusivity κ which are determined by the effective relaxation times τ_ν and τ_c, respectively. Thus, the total LES effective relaxation times should correspond to the modified effective viscosity ν and modified effective diffusivity κ which are dynamic and given by [67]

$$
\nu = \nu_0 + \nu_t, \qquad \kappa = \kappa_0 + \kappa_t, \tag{21}
$$

where the molecular viscosity is denoted by ν_0 and the turbulence or eddy viscosity by ν_t. Likewise, the molecular diffusivity is denoted by κ_0 and the turbulence or eddy diffusivity by κ_t. For the subgrid closure, the Reynolds stress is modeled using the Smagorinsky model [64] as discussed below.

The basic idea of LES is to compute a space averaged flow accurately. To achieve this, each flow variable ψ is decomposed into a large-scale component $\overline{\psi}$ and a subgrid scale component ψ' such that

$$
\psi = \overline{\psi} + \psi'.
$$

The large-scale component is obtained by an application of a filter operator. This operator is a convolution integral of the form

$$
\overline{\psi}(\mathbf{x},t) = \int_{\mathbb{R}^2} \psi(\mathbf{y},t) G_\Delta(\mathbf{x}-\mathbf{y})\, d\mathbf{y},
$$

where G_Δ is the filter such as the volume-averaged box-filter [55]. Note that, with the filter function one filters out the small scales from the flow. The space-filtered Navier-Stokes equations take the form

$$
\begin{aligned}
\nabla \cdot \overline{\mathbf{u}} &= 0, \\
\partial_t \overline{\mathbf{u}} + \nabla \cdot \left(\overline{\mathbf{u}} \otimes \overline{\mathbf{u}}^T \right) - \nabla \cdot \left(2\nu_0 \mathbf{S}[\overline{\mathbf{u}}] \right) + \nabla \overline{p} + \nabla \cdot \mathscr{T}(\mathbf{u}) &= \overline{\mathbf{G}}, \\
\partial_t \overline{T} + \nabla \cdot \left(\overline{\mathbf{u}} \otimes \overline{T} \right) - \nabla \cdot \left(\kappa \nabla \overline{T} \right) + \nabla \cdot \mathscr{K}(\mathbf{u},T) &= 0,
\end{aligned}
\tag{22}
$$

where $\mathscr{T}(\mathbf{u}) = \overline{\mathbf{u} \otimes \mathbf{u}^T} - \overline{\mathbf{u}} \otimes \overline{\mathbf{u}}^T$ and $\mathscr{K}(\mathbf{u},T) = \overline{\mathbf{u}T} - \overline{\mathbf{u}}\overline{T}$ represent the Reynolds subgrid-scale tensor, and the subgrid-scale turbulent heat flux, respectively. In order to close the system of equations the well known Smagorinsky model [64] for $\mathscr{T}(\mathbf{u})$ along with a Fickian approach for the subgrid-scale turbulent flux $\mathscr{K}(\mathbf{u},T)$ is used, see for instance [15]. Thus,

$$\mathscr{T}(\mathbf{u}) \approx -\nu_t(\overline{\mathbf{u}})\mathbf{S}[\overline{\mathbf{u}}] + \text{terms incorporated into } \overline{p}, \qquad \mathscr{K}(\mathbf{u},T) \approx -\kappa_t(\overline{\mathbf{u}})\nabla\overline{T}, \tag{23}$$

where the turbulent viscosity ν_t and the turbulent diffusivity κ_t are defined by

$$\nu_t(\overline{\mathbf{u}}) = (c_s\Delta)^2 \|\mathbf{S}[\overline{\mathbf{u}}]\|, \qquad \|\mathbf{S}[\overline{\mathbf{u}}]\| = \left(\mathbf{S}[\overline{\mathbf{u}}]:\mathbf{S}[\overline{\mathbf{u}}]\right)^{\frac{1}{2}}, \qquad \kappa_t = \frac{\nu_t}{\mathrm{Pr}_t}.$$

Substituting (23) into (22) gives

$$\begin{aligned} \nabla\cdot\overline{\mathbf{u}} &= 0, \\ \partial_t\overline{\mathbf{u}} + \nabla\cdot\left(\overline{\mathbf{u}}\otimes\overline{\mathbf{u}}^T\right) - \nabla\cdot\Big((2\nu_0+\nu_t)\mathbf{S}[\overline{\mathbf{u}}]\Big) + \nabla\overline{p} &= \overline{\mathbf{G}}, \\ \partial_t\overline{T} + \nabla\cdot\left(\overline{\mathbf{u}}\otimes\overline{T}\right) - \nabla\cdot\Big((\kappa_0+2\kappa_t)\nabla\overline{T}\Big) &= 0. \end{aligned} \tag{24}$$

The values for c_s which can be applied range from 0.01 to 0.1, see for example [55]. Further, in the simulations below Pr_t is kept constant and equal to 0.9 as suggested for direct numerical simulations in [15, 38, 45].

[illegible]

Flow problems in participating grey media that interact with radiative transfer through emission, absorption and scattering are considered. This means the source term in the flow equations contains a radiative heat flux which is denoted by $\mathbf{Q}_R$. Let $I(t,\mathbf{x},\omega)$ be the radiative intensity at position $\mathbf{x}$, time t, traveling in direction ω with speed c. Then, the isotropic radiative transfer equation is given by [49, 47]

$$\frac{1}{c}\partial_t I + \omega\cdot\nabla I = \sigma_s\left(\frac{1}{4\pi}\int_{S^2} I\,d\omega - I\right) + \sigma_a\Big(B(T) - I\Big), \tag{25}$$

where σ_s and σ_a are the scattering and absorption coefficients, respectively, which may depend on space $\mathbf{x}$. The integral in (25) is taken over all directions in the unit sphere S^2. The function $B(T)$ is the grey Planck's function which defines a nonlinear relationship between the participating medium and its temperature, and it is defined as

$$B(T) = \sigma_B T^4, \tag{26}$$

where σ_B is the Stefan-Boltzmann constant [47]. Note that in the above coupling a thermodynamic equilibrium is assumed such that the fluid temperature and the radiation temperature are equal. Since the photons travel with the speed of light, the term $1/c$ in (25) is negligible and is dropped in the computational examples below. The radiative heat flux $\mathbf{Q}_R$ is defined by

$$\mathbf{Q}_R = \int_{S^2} I\omega\,d\omega. \tag{27}$$

Hence, integrating equation (25) over the whole solid angle $\omega\in S^2$ yields a relation for the radiative heat source

$$\nabla\cdot\mathbf{Q}_R = \int_{S^2}\sigma_a\Big(B(T) - I\Big)\,d\omega, \tag{28}$$

which is substituted in the energy equation (20). The term $\tau_c/3$ is incorporated into σ_a. In the models considered here the flow is assumed to be in the low Mach number limit. For such flows there is a linear relationship between density and temperature defined as

$$\rho(T) = \rho_\infty\Big(1 - \beta(T - T_\infty)\Big),$$

where β is the coefficient of thermal expansion, ρ_∞ and T_∞ are reference density and reference temperature, respectively. The system of equations takes the form

$$\begin{aligned}
\operatorname{div}\mathbf{u} &= 0,\\
\partial_t\mathbf{u}+\operatorname{div}\mathbf{u}\otimes\mathbf{u}+\nabla p &= \nu\Delta\mathbf{u}+\mathbf{G},\\
\partial_t T+\operatorname{div}\big(\mathbf{u}T\big) &= \kappa\Delta T-\nabla\cdot\mathbf{Q}_R,\\
\omega\cdot\nabla I+(\sigma_a+\sigma_s)I &= \frac{\sigma_s}{4\pi}\int_{S^2} I\,d\omega+\sigma_a B(T),
\end{aligned} \tag{29}$$

where $\mathbf{G}=\rho\mathbf{g}$, $\mathbf{g}$ is the gravitational force and $\nabla\cdot\mathbf{Q}_R$ is given by (28).

Remark 2. Some remarks are now in order:

(i) In the microscopic setting the radiative heat transfer is included as a source term while the asymptotic derivations realize it as an extra flux term.

(ii) It is well-known that the radiative transfer is a non-local phenomenon; photons which are the main medium of transfer do not need a medium in order to propagate; and the temperature dependence of radiation is nonlinear. Conduction and convection have a linear temperature dependence (Fourier's law). Therefore, it makes sense to treat the radiative term as a source in the microscopic scale.

(iii) In addition, whether radiation is a minor or dominant mode of heat transfer depends on the temperature and the nature of the material.

For the computational examples dimensionless forms of the fluid and energy flow equations in a grey participating medium are presented.

4. SIMPLIFIED RELAXATION SYSTEMS

In this section a relaxation system that will be used in the numerical schemes in the next section will be derived. The system of equations in (8) and (10) with $\rho_\infty\equiv 1$ is considered. The system will be simplified in such a way that the same limit for lower order terms as the original system, (8) and (10), is preserved. Next, from equation (10) the terms $-\operatorname{div} p\mathbf{u}$ and $-3\varepsilon^2\partial_t p\mathbf{u}$ are neglected. From equation (8) the terms $\frac{1}{3}\mathbf{Q}[\mathbf{q}]$, $\nabla\cdot\Pi$ are neglected and new terms, $\mathbf{A}\nabla\mathbf{u}$, $\mathbf{B}\nabla T$, respectively, are introduced as follows:

For p, $\mathbf{u}$ and $\Theta=(\Theta^1,\Theta^2)=\begin{pmatrix}\Theta^{11} & \Theta^{12}\\ \Theta^{12} & \Theta^{22}\end{pmatrix}$ as defined above, the system

$$\begin{aligned}
\partial_t p+\frac{1}{\varepsilon^2}\nabla\cdot\mathbf{u} &= 0,\\
\partial_t\mathbf{u}+\nabla\cdot\Theta+\nabla p &= \mathbf{G},\\
\partial_t\Theta+\mathbf{A}\nabla\mathbf{u} &= -\frac{1}{\varepsilon^2\tau_\nu}\Big(\Theta-\mathbf{u}\otimes\mathbf{u}+2\tau_\nu\mathbf{S}^\varepsilon[\mathbf{u}]\Big),\\
\partial_t T+\nabla\cdot\Psi &= 0,\\
\partial_t\Psi+\mathbf{B}\nabla T &= -\frac{1}{\varepsilon^2\tau_c}\Big(\Psi-\mathbf{u}T+\tau_c\nabla T\Big),
\end{aligned} \tag{30}$$

where $\mathbf{S}^\varepsilon[\mathbf{u}]=\mathbf{S}[\mathbf{u}]-\dfrac{\varepsilon^2}{2}\mathbf{A}\nabla\mathbf{u}$. We have added and subtracted the term

$$\mathbf{A}\nabla\mathbf{u}=\left(A^1\partial_x\mathbf{u},A^2\partial_y\mathbf{u}\right)=\begin{pmatrix}a^{11}\partial_x u & a^{21}\partial_y u\\ a^{12}\partial_x v & a^{22}\partial_y v\end{pmatrix},$$

with $\mathbf{A}=(A^1,A^2)$, where A^1, A^2 and $\mathbf{B}$ are positive diagonal matrices with the same dimension as $\mathbf{u}$:

$$A^1=\begin{pmatrix}a^{11} & 0\\ 0 & a^{21}\end{pmatrix},\qquad A^2=\begin{pmatrix}a^{12} & 0\\ 0 & a^{22}\end{pmatrix}\qquad\text{and}\qquad \mathbf{B}=\begin{pmatrix}b^1 & 0\\ 0 & b^2\end{pmatrix}.$$

Obviously, the limit equations for this system are again the coupled equations with kinematic viscosity $\nu=\tau_\nu$ and energy diffusion coefficient $\kappa=\tau_c$.

Remark 3. Considering the nonstiff advection parts in (30) separately for $\mathbf{u}$, Θ, T and Ψ we obtain a hyperbolic system with characteristic speeds $\pm\sqrt{a^{11}}$, $\pm\sqrt{a^{21}}$, $\pm\sqrt{b^1}$ in the x-direction and $\pm\sqrt{a^{12}}$, $\pm\sqrt{a^{22}}$, $\pm\sqrt{b^2}$ in the y-direction of the form

$$\begin{aligned} \partial_t \mathbf{u} + \nabla\cdot\Theta &= \mathbf{G}, \\ \partial_t \Theta + \mathbf{A}\nabla\mathbf{u} &= \mathbf{0}, \\ \partial_t T + \nabla\cdot\Psi &= 0, \\ \partial_t \Psi + \mathbf{B}\nabla T &= \mathbf{0}. \end{aligned} \tag{31}$$

As will be seen in Section 3.2., the relaxation parameters $\mathbf{A}$ and $\mathbf{B}$ can be selected based on a rough estimation of the eigenvalues of the flux functions in (32). In the computations presented in Section 5.3., these parameters are chosen depending on the local characteristic speeds.

Clearly, one can develop another class of relaxation systems by letting the first equation in equation (30) relax to its limit in the lowest order terms as $\varepsilon \to 0$ without altering the others. This is practical if one considers an implementation in the vorticity-stream function formulation, compare [6] for a similar formulation in the isothermal case.

A relaxation system for LES can be formulated based on the equations (30) and application of filters to obtain a system of the form

$$\begin{aligned} \partial_t \overline{p} + \frac{1}{\varepsilon^2}\nabla\cdot\overline{\mathbf{u}} &= 0, \\ \partial_t \overline{\mathbf{u}} + \nabla\cdot\overline{\Theta} + \nabla\overline{p} &= \overline{\mathbf{G}}, \\ \partial_t \overline{\Theta} + \mathbf{A}\nabla\overline{\mathbf{u}} &= -\frac{1}{\varepsilon^2}\left(\overline{\Theta} - \overline{\mathbf{u}}\otimes\overline{\mathbf{u}}^T + (2\nu_0+\nu_t)\mathbf{S}[\overline{\mathbf{u}}]\right), \\ \partial_t \overline{T} + \nabla\cdot\overline{\Psi} &= 0, \\ \partial_t \overline{\Psi} + \mathbf{B}\nabla\overline{T} &= -\frac{1}{\varepsilon^2}\left(\overline{\Psi} - \overline{\mathbf{u}}\overline{T} + (\kappa_0+2\kappa_t)\nabla\overline{T}\right), \end{aligned} \tag{32}$$

where ε is the stress relaxation rate, $\overline{\Theta}$ is the stress relaxation tensor. Note that $\mathbf{A}$ and $\mathbf{B}$ are positive diagonal matrices with the same dimension as $\overline{\mathbf{u}}$. Formally, for leading order the equations (32) converge to the LES problem (24). In fact, when $\varepsilon \longrightarrow 0$, the third equation in (32) reduces to

$$\overline{\Theta} = \overline{\mathbf{u}}\otimes\overline{\mathbf{u}}^T - (2\nu_0+\nu_t)\mathbf{S}[\overline{\mathbf{u}}]. \tag{33}$$

and the last equation reduces to

$$\overline{\Psi} = \overline{\mathbf{u}}\overline{T} - (\kappa_0+2\kappa_t)\nabla\overline{T}. \tag{34}$$

Back-substituting (33) and (34) in the second and fourth equation in (32), respectively, one recovers the original equations (24) in the limit as $\varepsilon \longrightarrow 0$.

Remark 4. Some remarks are in order:

(i.) The main advantage of considering the relaxation systems in (30) and (32) over the original models in (12) and (20) as well as (24) is essentially the semi-linear structure of the system (30) and (32). Such a system can be solved numerically without relying on Riemann problem solvers or characteristics decomposition, see also [35, 56, 6, 7, 5, 58].

(ii.) Considering the nonstiff advection parts in (32) separately for $\overline{\mathbf{u}}$, $\overline{\Theta}$, $\overline{T}$ and $\overline{\Psi}$ a hyperbolic system with characteristic speeds $\pm\sqrt{a^{11}}$, $\pm\sqrt{a^{21}}$, $\pm\sqrt{b^1}$ in the x-direction and $\pm\sqrt{a^{12}}$, $\pm\sqrt{a^{22}}$, $\pm\sqrt{b^2}$ in the y-direction of the form

$$\begin{aligned} \partial_t \overline{\mathbf{u}} + \nabla\cdot\overline{\Theta} &= \overline{\mathbf{G}}, \\ \partial_t \overline{\Theta} + \mathbf{A}\nabla\overline{\mathbf{u}} &= \mathbf{0}, \\ \partial_t \overline{T} + \nabla\cdot\overline{\Psi} &= 0, \\ \partial_t \overline{\Psi} + \mathbf{B}\nabla\overline{T} &= \mathbf{0}. \end{aligned} \tag{35}$$

is obtained.

In the next section a relaxation scheme based on the relaxation system (30) or (32) are presented.

5. NUMERICAL SCHEMES

To develop numerical schemes for the relaxation systems developed in Section 3.2., equations (30) are considered. In the next subsection, a derivation for a numerical procedure applied to flow equations will be discussed. Thereafter a numerical approach for radiative transfer is presented.

[illegible]

In this section high-order upwind discretizations are developed for the nonstiff advection part in the relaxation system (30). The stiff part is treated by high-order centered differences, as in [6]. In the remainder of this section the time continuous version of the scheme is considered (method of lines). Also note that the non-stiff parts from momentum and energy transport equations are completely decoupled. Hence, they are treated separately in the ensuing discussion.
To discretize the equations in space, for the sake of simplicity, a uniform grid in x and y with gridpoints (x_i, y_j) and spacing $\Delta x = \Delta y = h$ is used. Consider the non-stiff linear part of the system (30) as presented in (31). One observes that for the x-direction, $\Theta^x \pm a_1\mathbf{u}$ and $\Psi_1 \pm b_1 T$ are the characteristic variables associated with the characteristic speeds $\pm a_1$ and $\pm b_1$, respectively. For the y-direction the characteristic variables associated with the characteristic speeds $\pm a_2$ and $\pm b_2$ are $\Theta^y \pm a_2\mathbf{u}$ and $\Psi_2 \pm b_2 T$, respectively. According to these considerations the values of the characteristic variables are determined at cell-boundaries following the approach in [35]. This can be done in a straightforward way for a second-order method. For a third-order method, for the reconstruction step a third-order CWENO interpolant is used, see for example [53, 44] and further references therein. Similar reconstructions were also applied in [6, 7, 60]. For the convenience of the reader the polynomials for the two-dimensional reconstruction in the x-direction will be discussed. These polynomials at the gridpoint (x_i, y_j) are constructed as

$$\mathbf{p}_{ij}(\mathbf{z};x) = \mathbf{z}_{ij} + \mathbf{s}_{ij}(x - x_i), \tag{36}$$

where the MinMod limiter

$$\mathbf{s}_{ij}(\mathbf{z}) = \frac{1}{h}\mathrm{MinMod}(\mathbf{z}_{ij} - \mathbf{z}_{i-1j}, \mathbf{z}_{i+1j} - \mathbf{z}_{ij}),$$

for the second-order MUSCL case and as

$$\mathbf{p}_{ij}(\mathbf{z};x) = w_L\mathbf{P}^L_{ij}(\mathbf{z};x) + w_R\mathbf{P}^R_{ij}(\mathbf{z};x) + w_C\mathbf{P}^C_{ij}(\mathbf{z};x), \tag{37}$$

with

$$\mathbf{P}^R_{ij}(\mathbf{z};x) = \mathbf{z}_{ij} + \frac{1}{h}(\mathbf{z}_{i+1j} - \mathbf{z}_{ij})(x - x_i), \qquad \mathbf{P}^L_{ij}(\mathbf{z};x) = \mathbf{z}_{ij} + \frac{1}{h}(\mathbf{z}_{ij} - \mathbf{z}_{i-1j})(x - x_i),$$
$$\mathbf{P}^C_{ij}(\mathbf{z};x) = \mathbf{z}_{ij} - \frac{1}{12}(\mathbf{z}_{i+1j} - 2\mathbf{z}_{ij} + \mathbf{z}_{i-1j}) - \frac{1}{12}(\mathbf{z}_{ij+1} - 2\mathbf{z}_{ij} + \mathbf{z}_{ij-1}) +$$
$$\frac{1}{2h}(\mathbf{z}_{i+1j} - \mathbf{z}_{i-1j})(x - x_i) + \frac{1}{h^2}(\mathbf{z}_{i+1j} - 2\mathbf{z}_{ij} + \mathbf{z}_{i-1j})(x - x_i)^2.$$

for the third-order CWENO case. The nonlinear CWENO weights w_k, $k = L, R, C$ in (37) are given as

$$w_k = \frac{\alpha_k}{\sum_l \alpha_l}, \qquad \alpha_k = \frac{c_k}{(\gamma + \mathrm{IS}_k)^\beta}, \qquad c_L = \frac{1}{4}, \qquad c_R = \frac{1}{4}, \qquad c_C = \frac{1}{2},$$

with $\gamma = 10^{-6}$ and $\beta = 2$. The smoothness indicators IS_k, $k = L, R, C$ are defined by

$$\begin{aligned} \mathrm{IS}_L &= (v_{ij} - v_{i-1j})^2, \qquad \mathrm{IS}_R = (v_{i+1j} - v_{ij})^2, \\ \mathrm{IS}_C &= \frac{13}{3}(v_{i+1j} - 2v_{ij} + v_{i-1j})^2 + \frac{1}{4}(v_{i+1j} - v_{i-1j})^2. \end{aligned}$$

Clearly any other high-order reconstruction procedure applies. To determine the characteristic variables at the boundary of the cells $[x_{i-1/2}, x_{i+1/2}] \times [y_{j-1/2}, y_{j+1/2}]$ we apply

$$\begin{aligned} (\Theta^x + a_1\mathbf{u})_{i+1/2j} &= \mathbf{p}_{ij}(\Theta^x + a_1\mathbf{u}; x_{i+1/2}), \qquad (\Theta^x - a_1\mathbf{u})_{i+1/2j} = \mathbf{p}_{i+1j}(\Theta^x - a_1\mathbf{u}; x_{i+1/2}), \\ (\Psi_1 + b_1 T)_{i+1/2j} &= \mathbf{p}_{ij}(\Psi_1 + b_1 T; x_{i+1/2}), \qquad (\Psi_1 - b_1 T)_{i+1/2j} = \mathbf{p}_{i+1j}(\Psi_1 - b_1 T; x_{i+1/2}). \end{aligned} \tag{38}$$

An analogous procedure is used for the y-direction by considering the characteristic variables $(\Theta^y \pm a_2\mathbf{u})_{ij+1/2}$ and $(\Psi_2 \pm b_2 T)_{ij+1/2}$. Denote the discretization of the convective parts in equations (30) $\operatorname{div}\Theta$, $\nabla^{\mathbf{a}}[\mathbf{u}]$, $\operatorname{div}\Psi$, and $\operatorname{div}\Pi_b$ by $\mathbf{F}_h^{(1)}$, $\mathbf{F}_h^{(2)}$, $\mathbf{F}_h^{(3)}$, and $\mathbf{F}_h^{(4)}$, respectively. Using the reconstruction polynomial given above componentwise one obtains

$$\begin{aligned} \mathbf{F}_h^{(1)}(\Theta,\mathbf{u}) &= \frac{1}{h}\left(\Theta^x_{i+1/2j} - \Theta^x_{i-1/2j}\right) + \frac{1}{h}\left(\Theta^y_{ij+1/2} - \Theta^y_{ij-1/2}\right), \\ \mathbf{F}_h^{(2)}(\Theta,\mathbf{u}) &= \left(\frac{a_1^2}{h}(\mathbf{u}_{i+1/2,j} - \mathbf{u}_{i-1/2j}), \frac{a_2^2}{h}(\mathbf{u}_{ij+1/2} - \mathbf{u}_{ij-1/2})\right)^T, \\ \mathbf{F}_h^{(3)}(\Psi,T) &= \frac{1}{h}(\Psi_{1i+1/2j} - \Psi_{1i-1/2j}) + \frac{1}{h}(\Psi_{2ij+1/2} - \Psi_{2ij-1/2}), \\ \mathbf{F}_h^{(4)}(\Psi,T) &= \left(\frac{b_1^2}{h}(T_{i+1/2,j} - b_1^2 T_{i-1/2j}), \frac{b_2^2}{h}(T_{ij+1/2} - T_{ij-1/2})\right)^T, \end{aligned}$$

where the numerical fluxes are given by

$$\begin{aligned} \Theta^x_{i+1/2j} &= \frac{1}{2}\left((\Theta^x + a_1\mathbf{u})_{i+1/2j} + (\Theta^x - a_1\mathbf{u})_{i+1/2j}\right), \\ \mathbf{u}_{i+1/2j} &= \frac{1}{2a_1}\left((\Theta^x + a_1\mathbf{u})_{i+1/2j} - (\Theta^x - a_1\mathbf{u})_{i+1/2j}\right), \\ \Psi_{1i+1/2j} &= \frac{1}{2}\left((\Psi_1 + b_1 T)_{i+1/2j} + (\Psi_1 - b_1 T)_{i+1/2j}\right), \\ T_{i+1/2j} &= \frac{1}{2b_1}\left((\Psi_1 + b_1 T)_{i+1/2j} - (\Psi_1 - b_1 T)_{i+1/2j}\right), \end{aligned}$$

where the terms on the right hand side are defined by the interpolants in (38). For more details, the reader may refer to [6]. To discretize the pressure variable and the stiff parts *i.e.*, terms with the coefficient $\frac{1}{\varepsilon^2}$ in equations (30), firstly, the discrete gradient is denoted by $\mathbf{G}_h$ and the discrete divergence by $\mathbf{D}_h$. They are given by second- or fourth-order centered differences for second-order or third-order approaches, respectively. $\mathbf{S}_h^\varepsilon$ and $\mathbf{S}_h$ denote second- or fourth-order centered difference approximations of $\mathbf{S}^\varepsilon$ and $\mathbf{S}$, respectively.
Finally, a high-order spatial discretization for the moment system characterized by

$$\begin{aligned} \dot{p} + \frac{1}{\varepsilon^2}\mathbf{D}_h\cdot\mathbf{u} &= 0, \\ \dot{\mathbf{u}} + \mathbf{F}_h^{(1)}(\Theta,\mathbf{u}) + \mathbf{G}_h p &= \mathbf{G}, \\ \dot{\Theta} + \mathbf{F}_h^{(2)}(\Theta,\mathbf{u}) + \frac{2}{\varepsilon^2}\mathbf{S}_h^\varepsilon(\mathbf{u}) &= -\frac{1}{\varepsilon^2\tau_v}(\Theta - \mathbf{u}\otimes\mathbf{u}), \\ \dot{T} + \mathbf{F}_h^{(3)}(\Psi,T) &= 0, \\ \dot{\Psi} + \mathbf{F}_h^{(4)}(\Psi,T) + \frac{1}{\varepsilon^2}G_h T &= -\frac{1}{\varepsilon^2\tau_c}(\Psi - \mathbf{u}T) + \frac{1}{\varepsilon^2}\mathbf{Q}, \end{aligned} \tag{39}$$

is obtained or equivalently

$$\begin{aligned}
\mathbf{D}_h \cdot \mathbf{G}_h p - 2\varepsilon^2 \ddot{p} &= -\mathbf{D}_h \cdot \mathbf{F}_h^{(1)}(\Theta, \mathbf{u}), \\
\dot{\mathbf{u}} + \mathbf{F}_h^{(1)}(\Theta, \mathbf{u}) + \mathbf{G}_h p &= \mathbf{G}, \\
\dot{\Theta} + \mathbf{F}_h^{(2)}(\Theta, \mathbf{u}) &= -\frac{1}{\varepsilon^2 \tau_v}\Big(\Theta - \mathbf{u}\otimes\mathbf{u} + 2\tau_v \mathbf{S}_h^{\varepsilon}(\mathbf{u})\Big), \\
\dot{T} + \mathbf{F}_h^{(3)}(\Psi, T) &= 0, \\
\dot{\Psi} + \mathbf{F}_h^{(4)}(\Psi, T) &= -\frac{1}{\varepsilon^2 \tau_c}(\Psi - \mathbf{u}T + \tau_c \mathbf{G}_h T - \tau_c \mathbf{Q}).
\end{aligned}$$

A corresponding high-order upwind-based space discretization for the relaxed incompressible Navier-Stokes/Boussinesq equations is obtained considering the limit of the above discretization as $\varepsilon \to 0$

$$\begin{aligned}
\mathbf{D}_h \cdot \mathbf{G}_h p &= -\mathbf{D}_h \cdot \mathbf{F}_h^{(1)}(\mathbf{u}), \\
\dot{\mathbf{u}} &= -\mathbf{F}_h^{(1)}(\mathbf{u}\otimes\mathbf{u} - 2\tau_v \mathbf{S}_h(\mathbf{u}), \mathbf{u}) - \mathbf{G}_h p + \mathbf{G}, \\
\dot{T} &= -\mathbf{F}_h^{(3)}(\mathbf{u}T - \tau_c G_h T + \tau_c \mathbf{Q}, T).
\end{aligned}$$

Remark 5. To treat only the limit equations ($\varepsilon = 0$) one could use any explicit high order Runge-Kutta method combined with a Poisson solver and the limiting (relaxed) spatial discretization. The Poisson equation is in this case only used to determine the divergence-free velocities via ∇p and not to advance the pressure for one time step.

Let us denote the time step by Δt and use superscript n to denote the time iterations. The non-stiff parts are treated explicitly and the stiff parts implicitly. The application of this approach is demonstrated using a first-order scheme. For the first-order IMEX method, one can use the following simple time discretization

$$\begin{aligned}
\mathbf{u}^{n+1} &= \mathbf{u}^n - \Delta t\left(\operatorname{div}\Theta^n + \nabla p^{n+1}\right) + \Delta t \mathbf{G}^n, \\
\Theta^{n+1} &= \Theta^n - \Delta t \nabla^{\mathbf{a}}[\mathbf{u}^n] - \frac{\Delta t}{\varepsilon^2 \tau_v}\left(\Theta^{n+1} + 2\tau \mathbf{S}^{\varepsilon}[\mathbf{u}^{n+1}] - \mathbf{u}^{n+1}\otimes\mathbf{u}^{n+1}\right), \\
T^{n+1} &= T^n - \Delta t \operatorname{div}\Psi^n, \\
\Psi^{n+1} &= \Psi^n - \Delta t \operatorname{div}\Pi_b{}^n - \frac{\Delta t}{\varepsilon^2 \tau_c}\left(\Psi^{n+1} - \mathbf{u}^{n+1} T^{n+1} + \tau_c \mathbf{G}_h T^{n+1} - \tau_c \mathbf{Q}^{n+1}\right), \\
p^{n+1} &= p^n - \frac{\Delta t}{\varepsilon^2}\operatorname{div}\mathbf{u}^{n+1}.
\end{aligned} \tag{40}$$

Substituting the last equation in (40) into the first equation in (40) yields a Helmholtz equation for the pressure

$$\Delta p^{n+1} - \frac{\varepsilon^2}{\Delta t^2} p^{n+1} = \frac{1}{\Delta t}\operatorname{div}\mathbf{u}^n - \operatorname{div}(\operatorname{div}\Theta^n) - \frac{\varepsilon^2}{\Delta t^2} p^n. \tag{41}$$

This equation can be solved by a suitable iterative method. The velocity, $\mathbf{u}^{n+1}$, is determined using the first equation in (40). Obviously, as $\varepsilon \to 0$ the time marching tends to a time discretization of the thermal incompressible Navier-Stokes equations. For $\varepsilon \to 0$ the Poisson equation is obtained for the pressure

$$\Delta p^{n+1} = \frac{1}{\Delta t}\operatorname{div}\mathbf{u}^n - \operatorname{div}(\operatorname{div}\Theta^n),$$

together with

$$\begin{aligned}
\mathbf{u}^{n+1} &= \mathbf{u}^n - \Delta t\left(\operatorname{div}\Theta^n + \nabla p^{n+1}\right) + \Delta t \mathbf{G}^n, \\
\Theta^{n+1} &= 2\tau_v \mathbf{S}[\mathbf{u}^{n+1}] - \mathbf{u}^{n+1}\otimes\mathbf{u}^{n+1}, \\
T^{n+1} &= T^n - \Delta t \operatorname{div}\Psi^n, \\
\Psi^{n+1} &= \mathbf{u}^{n+1} T^{n+1} - \tau_c \mathbf{G}_h T^{n+1} + \tau_c \mathbf{Q}^{n+1}.
\end{aligned}$$

Thus, in the limit, the usual projection method for the incompressible Navier-Stokes/Boussinesq equations is obtained. Take note that the incompressibility condition is fulfilled for the velocity field $\mathbf{u}$ at every time step. In the IMEX notation the above first-order scheme is given by the explicit and implicit *Butcher tableau*

$$
\begin{array}{c|cc}
0 & 0 & 0 \\
1 & 0 & 1 \\
\hline
 & 0 & 1
\end{array}
\qquad\qquad
\begin{array}{c|cc}
0 & 0 & 0 \\
1 & 1 & 0 \\
\hline
 & 1 & 0
\end{array}
$$

For the above semi-implicit time discretization the usual hyperbolic and parabolic CFL conditions have to be fulfilled to guarantee stability. For the third-order time discretization we choose a two stage IMEX Runge Kutta method [51]. Its associated explicit and implicit *Butcher tableau* are

$$
\begin{array}{c|ccc}
0 & 0 & 0 & 0 \\
\gamma & \gamma & 0 & 0 \\
1-\gamma & \gamma-1 & 2-2\gamma & 0 \\
\hline
 & 0 & \frac{1}{2} & \frac{1}{2}
\end{array}
\qquad\qquad
\begin{array}{c|ccc}
0 & 0 & 0 & 0 \\
\gamma & 0 & \gamma & 0 \\
1-\gamma & 0 & 1-2\gamma & \gamma \\
\hline
 & 0 & \frac{1}{2} & \frac{1}{2}
\end{array}
$$

where $\gamma = \frac{3+\sqrt{3}}{6}$. Its formulation can be performed as in (40) and details are omitted. Other IMEX schemes of third and higher order are also discussed in [51].

[illegible]

A numerical approach for the LES model is presented based on a high-order relaxation approach as well. The equations (32) are used to reconstruct a high-order relaxation scheme. The relaxation system (32) is numerically solved using a method of lines technique. For the temporal discretization, operator splitting is applied where the advection stage and relaxation stage are discretized separately. To discretize the system (32), an equally spaced control volume $\mathscr{C}_{ij} = [x_{i-1/2}, x_{i+1/2}] \times [y_{j-1/2}, y_{j+1/2}]$ with dimensions $\Delta x = x_{i+1/2} - x_{i-1/2}$ and $\Delta y = y_{j+1/2} - y_{j-1/2}$ in x- and y-directions, respectively, is assumed for simplicity. Integrating (32) over the volume and keeping the time continuous, the following semi-discrete relaxation system

$$
\begin{aligned}
\frac{d\overline{p}_{ij}}{dt} + \frac{1}{\varepsilon^2}\left(\mathscr{D}_x \overline{u}_{ij} + \mathscr{D}_y \overline{v}_{ij}\right) &= 0, \\
\frac{d\overline{u}_{ij}}{dt} + \mathscr{D}_x \overline{\Theta}^{11}_{ij} + \mathscr{D}_y \overline{\Theta}^{12}_{ij} + \mathscr{D}_x \overline{p}_{ij} &= 0, \\
\frac{d\overline{v}_{ij}}{dt} + \mathscr{D}_x \overline{\Theta}^{21}_{ij} + \mathscr{D}_y \overline{\Theta}^{22}_{ij} + \mathscr{D}_y \overline{p}_{ij} &= 0, \\
\frac{d\overline{\Theta}^{11}_{ij}}{dt} + a^{11}_{ij} \mathscr{D}_x \overline{u}_{ij} &= -\frac{1}{\varepsilon^2}\left(\overline{\Theta}^{11}_{ij} - \overline{u}_{ij}\overline{u}_{ij} + (2\nu_0 + \nu_t) S^{11}[\overline{\mathbf{u}}_{ij}]\right), \\
\frac{d\overline{\Theta}^{12}_{ij}}{dt} + a^{12}_{ij} \mathscr{D}_y \overline{u}_{ij} &= -\frac{1}{\varepsilon^2}\left(\overline{\Theta}^{12}_{ij} - \overline{u}_{ij}\overline{v}_{ij} + (2\nu_0 + \nu_t) S^{12}[\overline{\mathbf{u}}_{ij}]\right), \\
\frac{d\overline{\Theta}^{21}_{ij}}{dt} + a^{21}_{ij} \mathscr{D}_x \overline{v}_{ij} &= -\frac{1}{\varepsilon^2}\left(\overline{\Theta}^{21}_{ij} - \overline{u}_{ij}\overline{v}_{ij} + (2\nu_0 + \nu_t) S^{21}[\overline{\mathbf{u}}_{ij}]\right), \\
\frac{d\overline{\Theta}^{22}_{ij}}{dt} + a^{22}_{ij} \mathscr{D}_y \overline{v}_{ij} &= -\frac{1}{\varepsilon^2}\left(\overline{\Theta}^{22}_{ij} - \overline{v}_{ij}\overline{v}_{ij} + (2\nu_0 + \nu_t) S^{22}[\overline{\mathbf{u}}_{ij}]\right), \\
\frac{d\overline{T}_{ij}}{dt} + \mathscr{D}_x \overline{\Psi}^{1}_{ij} + \mathscr{D}_y \overline{\Psi}^{2}_{ij} &= 0, \\
\frac{d\overline{\Psi}^{1}_{ij}}{dt} + b^{1}_{ij} \mathscr{D}_x \overline{T}_{ij} &= -\frac{1}{\varepsilon^2}\left(\overline{\Psi}^{1}_{ij} - \overline{u}_{ij}\overline{T}_{ij} + (\kappa_0 + 2\kappa_t)\mathscr{G}_x \overline{T}_{ij}\right), \\
\frac{d\overline{\Psi}^{2}_{ij}}{dt} + b^{2}_{ij} \mathscr{D}_y \overline{T}_{ij} &= -\frac{1}{\varepsilon^2}\left(\overline{\Psi}^{2}_{ij} - \overline{v_{ij}}\overline{T}_{ij} + (\kappa_0 + 2\kappa_t)\mathscr{G}_y \overline{T}_{ij}\right),
\end{aligned}
\tag{42}
$$

is obtained. Here S^{11}, S^{12}, S^{21}, S^{22} are entries of the 2×2-matrix $\mathbf{S}$ in (32), $\mathscr{D}_x$ and $\mathscr{D}_y$ are difference quotients defined as

$$\mathscr{D}_x\psi_{ij} = \frac{\psi_{i+1/2j}-\psi_{i-1/2j}}{\Delta x}, \qquad \mathscr{D}_y\psi_{ij} = \frac{\psi_{ij+1/2}-\psi_{ij-1/2}}{\Delta y}, \tag{43}$$

with ψ_{ij} denoting the space average of a generic function ψ in the cell $\mathscr{C}_{ij}$, $\psi_{i+1/2j}$ and $\psi_{ij+1/2}$ are the numerical fluxes at gridpoints $(x_{i+1/2},y_j)$ and $(x_i,y_{j+1/2})$,

$$\psi_{ij} = \frac{1}{\Delta x}\frac{1}{\Delta y}\int_{x_{i-1/2}}^{x_{i+1/2}}\int_{y_{j-1/2}}^{y_{j+1/2}}\psi(x,y,t)dxdy,$$
$$\psi_{i+1/2j} = \psi(x_{i+1/2},y_j,t), \qquad \psi_{ij+1/2} = \psi(x_i,y_{j+1/2},t).$$

The terms $\mathscr{G}_x$ and $\mathscr{G}_y$ denote discrete gradients in the x- and y-directions, respectively. Notice that the source term $\mathbf{G}$ in (32) has been ignored for simplicity. Indeed, for forced flow problems at high Reynolds number, its role is insignificant. The spatial discretization of the relaxation LES system (42) is complete when a numerical reconstruction of the fluxes in (43) is chosen. Since the system (24) has linear characteristics given by

$$\begin{aligned} &\mathscr{F}^{\pm} = \overline{\Theta}^{11} \pm \sqrt{a^{11}}\overline{u}, \qquad &&\mathscr{H}^{\pm} = \overline{\Theta}^{12} \pm \sqrt{a^{12}}\overline{v}, \qquad &&\mathscr{P}^{\pm} = \overline{\Psi}^{1} \pm \sqrt{b^{1}}\overline{T}, \\ &\mathscr{G}^{\pm} = \overline{\Theta}^{12} \pm \sqrt{a^{12}}\overline{u}, &&\mathscr{L}^{\pm} = \overline{\Theta}^{22} \pm \sqrt{a^{22}}\overline{v}, &&\mathscr{Q}^{\pm} = \overline{\Psi}^{2} \pm \sqrt{b^{2}}\overline{T}, \end{aligned} \tag{44}$$

upwind schemes can be easily implemented. First-order and second-order upwind schemes have been proposed in [35], whereas a third-order reconstruction has been introduced in [6, 7, 58]. In the current work, a fifth-order WENO reconstruction [34] for the numerical fluxes in (43) is formulated. Higher-order reconstructions are also possible, such as those developed in [3]. The central idea is based on the fact that such reconstructions are applicable directly to the linear characteristic variables (44). In what follows we only formulate the numerical fluxes $\overline{u}_{i+1/2j}$ and $\overline{\Theta}^{11}_{i+1/2j}$, and the expressions for other fluxes are obtained in an entirely analogous manner.

Since $\mathscr{F}^{+}$ and $\mathscr{F}^{-}$ travel along constant characteristics with speed $+a^{11}$ and $-a^{11}$, respectively, WENO reconstructions can be easily applied to them. Thus, the numerical fluxes $\mathscr{F}^{\pm}_{i+1/2j}$ at the cell boundary $(x_{i+1/2},y_j)$ are defined as left and right extrapolated values, $\mathscr{F}^{+,L}_{i+1/2j}$ and $\mathscr{F}^{-,R}_{i+1/2j}$, by

$$\mathscr{F}^{+}_{i+1/2j} = \mathscr{F}^{+,L}_{i+1/2j}, \qquad \mathscr{F}^{-}_{i+1/2j} = \mathscr{F}^{-,R}_{i+1/2j}. \tag{45}$$

For a scalar flow function $\psi(x,y)$ the fifth-order accurate left boundary extrapolated value $\psi^{L}_{i+1/2j}$ is defined as

$$\psi^{L}_{i+1/2j} = \omega_0 v_0 + \omega_1 v_1 + \omega_2 v_2, \tag{46}$$

where v_r is the extrapolated value obtained from cell averages in the rth stencil $(i-r, i-r+1, i-r+2)$ in x-direction:

$$v_0 = \frac{1}{6}\left(-\psi_{i+2j}+5\psi_{i+1j}+2\psi_{ij}\right), \qquad v_1 = \frac{1}{6}\left(-\psi_{i-1j}+5\psi_{ij}+2\psi_{i+1j}\right),$$
$$v_2 = \frac{1}{6}\left(2\psi_{i-2j}-7\psi_{i-1j}+11\psi_{ij}\right),$$

and ω_r, $r=0,1,2$, are nonlinear WENO weights given by

$$\omega_k = \frac{\alpha_k}{\sum_{l=0}^{2}\alpha_l}, \qquad \alpha_0 = \frac{3}{10(\mathrm{IS}_0+\xi)^2}, \qquad \alpha_1 = \frac{3}{5(\mathrm{IS}_1+\xi)^2}, \qquad \alpha_2 = \frac{1}{10(\mathrm{IS}_2+\xi)^2},$$

where ξ is a small parameter introduced to guarantee that the denominator does not become zero and is empirically taken to be 10^{-6}. The smoothness indicators IS_r, $r = 0,1,2$, are given by [34]

$$\begin{aligned}
\mathrm{IS}_0 &= \frac{13}{12}\left(\psi_{ij} - 2\psi_{i+1j} + \psi_{i+2j}\right)^2 + \frac{1}{4}\left(3\psi_{ij} - 4\psi_{i+1j} + \psi_{i+2j}\right)^2, \\
\mathrm{IS}_1 &= \frac{13}{12}\left(\psi_{i-1j} - 2\psi_{ij} + \psi_{i+1j}\right)^2 + \frac{1}{4}\left(\psi_{i-1j} - \psi_{i+1j}\right)^2, \\
\mathrm{IS}_2 &= \frac{13}{12}\left(\psi_{i-2j} - 2\psi_{i-1j} + \psi_{ij}\right)^2 + \frac{1}{4}\left(\psi_{i-2j} - 4\psi_{i-1j} + 3\psi_{ij}\right)^2.
\end{aligned}$$

The right value $\psi^R_{i+1/2j}$ is obtained as $\psi^R_{i+1/2j} = \omega_0 v_0 + \omega_1 v_1 + \omega_2 v_2$ where

$$v_0 = \frac{1}{6}\left(2\psi_{i+3j} - 7\psi_{i+2j} + 11\psi_{i+1j}\right), \qquad v_1 = \frac{1}{6}\left(-\psi_{i+2j} + 5\psi_{i+1j} + 2\psi_{ij}\right),$$
$$v_2 = \frac{1}{6}\left(-\psi_{i-1j} + 5\psi_{ij} + 2\psi_{i+1j}\right).$$

The corresponding weights and smoothness indicators are defined by symmetry and their formulation is omitted. Once $\mathscr{F}^+_{i+1/2j}$ and $\mathscr{F}^-_{i+1/2j}$ are reconstructed in (45), the numerical fluxes $\overline{u}_{i+1/2j}$ and $\overline{\Theta}^{11}_{i+1/2j}$ in the relaxation system are obtained from (44) by

$$\overline{u}_{i+1/2j} = \mathscr{F}^+_{i+1/2j} + \mathscr{F}^-_{i+1/2j}, \qquad \overline{\Theta}^{11}_{i+1/2j} = a^{11}_{ij}\left(\mathscr{F}^+_{i+1/2j} - \mathscr{F}^-_{i+1/2j}\right). \tag{47}$$

The time integration of the semi-discrete relaxation system (42) is carried out using the implicit-explicit Runge-Kutta (IMEX) schemes, compare [35, 52, 39, 7] among others. As an example, we formulate the first-order IMEX scheme for the equations (32). Let Δt be the time step and ψ^n denote the approximation of a function ψ at time $t = n\Delta t$. The simple first-order IMEX scheme is implemented as

$$\begin{aligned}
\overline{\mathbf{u}}^{n+1} &= \overline{\mathbf{u}}^n - \Delta t\left(\nabla\cdot\overline{\Theta}^n + \nabla\overline{p}^{n+1}\right), \\
\overline{\Theta}^{n+1} &= \overline{\Theta}^n - \Delta t\mathbf{A}\nabla\overline{\mathbf{u}}^n - \frac{\Delta t}{\varepsilon^2}\left(\overline{\Theta}^{n+1} - \overline{\mathbf{u}}^{n+1}\otimes(\overline{\mathbf{u}}^{n+1})^T + (2\nu_0 + \nu_t)\mathbf{S}[\overline{\mathbf{u}}^{n+1}]\right), \\
\overline{T}^{n+1} &= \overline{T}^n - \Delta t\nabla\cdot\overline{\Psi}^n, \\
\overline{\Psi}^{n+1} &= \overline{\Psi}^n - \Delta t\mathbf{B}\nabla\overline{T} - \frac{\Delta t}{\varepsilon^2}\left(\overline{\Psi}^{n+1} - \overline{\mathbf{u}}^{n+1}\overline{T}^{n+1} + (\kappa_0 + 2\kappa_t)\nabla\overline{T}^{n+1}\right), \\
\overline{p}^{n+1} &= \overline{p}^n - \frac{\Delta t}{\varepsilon^2}\nabla\cdot\overline{\mathbf{u}}^{n+1}.
\end{aligned} \tag{48}$$

Substituting the first equation into the last equation in (48) yields a Helmholtz equation for the pressure

$$-\Delta\overline{p}^{n+1} + \frac{\varepsilon^2}{(\Delta t)^2}\overline{p}^{n+1} = \frac{\varepsilon^2}{(\Delta t)^2}\overline{p}^n - \frac{1}{\Delta t}\nabla\cdot\overline{\mathbf{u}}^n + \nabla\cdot\left(\nabla\cdot\overline{\Theta}^n\right). \tag{49}$$

This equation can be solved by a suitable iterative method. Thus, $\overline{\mathbf{u}}^{n+1}$ is determined using the first equation in (48). All the results presented in this paper are performed using a third-order IMEX scheme represented by the associated double Butcher's *tableau* as

$$\begin{array}{c|ccc}
0 & 0 & 0 & 0 \\
\gamma & \gamma & 0 & 0 \\
1-\gamma & \gamma-1 & 2-2\gamma & 0 \\
\hline
 & 0 & \frac{1}{2} & \frac{1}{2}
\end{array}
\qquad\qquad
\begin{array}{c|ccc}
0 & 0 & 0 & 0 \\
\gamma & 0 & \gamma & 0 \\
1-\gamma & 0 & 1-2\gamma & \gamma \\
\hline
 & 0 & \frac{1}{2} & \frac{1}{2}
\end{array} \tag{50}$$

where $\gamma = \frac{3+\sqrt{3}}{6}$. The left and right tables represent the explicit and the implicit Runge-Kutta methods, respectively.

Table **1**: Discrete ordinates and quadrature weights for the S_8 set (one octant only).

m	μ_m	ξ_m	η_m	w_m
1	0.97097459	0.16912768	0.16912768	0.14613894
2	0.79878814	0.16912768	0.57735027	0.15983890
3	0.79878814	0.57735027	0.16912768	0.15983890
4	0.57735027	0.16912768	0.79878814	0.15983890
5	0.57735027	0.57735027	0.57735027	0.17334611
6	0.57735027	0.79878814	0.16912768	0.15983890
7	0.16912768	0.16912768	0.97097459	0.14613894
8	0.16912768	0.57735027	0.79878814	0.15983890
9	0.16912768	0.79878814	0.57735027	0.15983890
10	0.16912768	0.97097459	0.16912768	0.14613894

Remark 6. Note that:

(i.) By using the IMEX procedures neither linear algebraic equations nor nonlinear source terms can arise.

(ii.) Since the advective part in (42) is treated explicitly, this relaxation scheme is conditionally stable so that the time step Δt has to satisfy the usual hyperbolic CFL condition.

(iii.) Obviously, as $\varepsilon \to 0$ the time integration scheme (32) tends to an explicit time discretization of the LES equations based on the left table in (50).

In the next section the accuracy and high-resolution of the scheme presented above will be tested.

[illegible]

The radiative transfer equation (25) can be solved using any existing code from computational radiative transfer such as Monte Carlo or discrete-ordinates methods. The later method is selected to be used in the present work mainly because the Monte Carlo methods are computationally very demanding. Other fast solvers as those recently developed in [57, 59] can also be used. Here, the angle and space variables are discretized using discrete-ordinates and finite volume methods. The outstanding aspect of discrete-ordinates method is that such an integral is approximated by a procedure with low computational cost, which is analogous to the Gauss-Legendre method. It consists of the substitution of the integral term in the radiative transfer equation (25) with a weighted summation of the integrand at selected ordinates of the unit sphere. This method is sometimes referred to as the S_N approximation, where N represents the number of discrete values of direction cosines to be considered. In general, the total number of ordinate directions M in a set S_N is given by $M = N(N+2)/2$.

The S_8 discretization is well designed for solving radiative transfer problems. For each direction ω_m of the quadrature, a specific weight w_m is associated with a set of three cosine angles μ_m, ξ_m and η_m in direct Cartesian grid (x,y,z). The cosine angles and the weights for S_8 quadrature are listed in Table **1**. Other discrete S_N sets from [49, 21] are also applicable. For the two-dimensional Cartesian coordinates, the radiative transfer equation (25) can be expressed for each individual ordinate direction, m, as

$$\mu_m \partial_x I_m + \xi_m \partial_y I_m + (\sigma_a + \sigma_s) I_m = \frac{\sigma_s}{4\pi} \sum_{k=1}^{M} w_k I_k + \sigma_a B(T), \qquad m = 1, 2, \ldots, M, \tag{51}$$

where I_m denotes the radiative intensity at the discrete ordinate ω_m. Next, the finite volume technique is applied, which consists of the integration of the S_N-equation (51) over a control volume. The result is an equation relating the value of an arbitrary function ψ at the nodal point, $\psi_{i+1/2j+1/2}$, to the value at each adjacent line ψ_{ij}, ψ_{i+1j}, ψ_{ij+1} and ψ_{i+1j+1}. The fully discrete problem corresponding to the equation (25) is written as

$$\mu_m \frac{I_{m,i+1j} - I_{m,ij}}{\Delta x} + \xi_m \frac{I_{m,ij+1} - I_{m,ij}}{\Delta y} + \left(\sigma_{a_{i+1/2j+1/2}} + \sigma_{s_{i+1/2j+1/2}}\right) I_{m,i+1/2j+1/2} = \frac{\sigma_{s_{i+1/2j+1/2}}}{4\pi} \sum_{k=1}^{M} w_k I_{k,i+1/2j+1/2} + \sigma_{a_{i+1/2j+1/2}} B_{i+1/2j+1/2}(T), \tag{52}$$

where the cell averages of a function ψ are given by

$$\psi_{i+1j} = \frac{1}{\Delta y} \int_{y_j}^{y_{j+1}} \psi(x_i, y) dy, \qquad \psi_{ij+1} = \frac{1}{\Delta x} \int_{x_i}^{x_{i+1}} \psi(x, y_j) dx, \qquad \psi_{ij} = \frac{1}{\Delta x \Delta y} \int_{x_i}^{x_{i+1}} \int_{y_j}^{y_{j+1}} \psi(x, y) dx dy, \tag{53}$$

The cell center and cell boundary function in fluxes (53) are related to each other by a selected interpolation scheme. Here, the diamond difference method which consists of approximating the function values at the cell centers by the average of their values at the neighboring nodes is used. Thus, the function value of $\psi_{i+1/2j+1/2}$ at the cell center is simply approximated by bilinear interpolation as

$$\psi_{i+1/2j+1/2} = \frac{\psi_{ij} + \psi_{i+ij} + \psi_{ij+1} + \psi_{i+1j+1}}{4}. \tag{54}$$

With the introduction of the above interpolation relations into (52), the resulting fully discrete equations can be solved by proceeding in the direction of photon travel on the spatial mesh, that is, sweeping away from the boundary conditions in the mesh. For problems involving scattering media or reflecting boundaries, the discrete-ordinates equations are coupled and therefore have to be solved iteratively.

In the present work, to solve the discretized equations (52), the intensity is eliminated from the resulting discretized equations (52) then a linear system of the form

$$\left(\mathbf{I} - \mathbf{A}\right)\phi = \mathbf{f}, \tag{55}$$

has to be solved for the mean intensity ϕ

$$\phi_{ij} = \sum_{k=1}^{M} w_k I_{k,ij}.$$

The Schur matrix **A** contains the discretized transport and integral operators from (52), **f** the right hand side and **I** is the identity matrix, see [57] for a detailed matrix formulation of (55). In order to construct the matrix **A** in (55), one has to use a Gaussian elimination known in computational radiative transfer as sweeping procedure. All the results given throughout this chapter were obtained using the diffusion-synthetic acceleration (DSA) method. The DSA method uses the diffusion approach as preconditioner for the source iteration applied to the linear system (55). For a detailed formulation of this algorithm the reader is referred to [57].

6. NUMERICAL TESTS AND DISCUSSIONS

In order to study the performance of the proposed numerical models, some test examples have been presented herein convection-radiation flows. The aim is to show that, for small relaxation rates, the discrete-velocity relaxation models reproduce the corresponding flow patterns, and also to show that the high-order relaxation schemes accurately capture the flow structures with very little numerical diffusion, even after long time simulations. All the results presented in this section are computed with variable time steps Δt adjusted at each step by

$$\Delta t = Cr \cdot \min\left(\frac{\Delta x}{\left|a_{i\pm1/2j}\right|}, \frac{\Delta y}{\left|b_{ij\pm1/2}\right|}\right), \tag{56}$$

where $a_{i\pm1/2j}$ and $b_{ij\pm1/2}$ are the characteristic speeds in the relaxation system selected locally in each grid cell $[x_{i-1/2}, x_{i+1/2}] \times [y_{j-1/2}, y_{j+1/2}]$ as

$$a_{i+1/2j} = 2\max\left\{\left|p_{ij}(u_1; x_{i+1/2})\right|, \left|p_{i+1j}(u_1; x_{i+1/2})\right|\right\},$$

$$b_{ij+1/2} = 2\max\left\{\left|p_{ij}(u_2; y_{j+1/2})\right|, \left|p_{ij+1}(u_2; y_{j+1/2})\right|\right\}.$$

The other characteristic speeds $a_{i-1/2j}$ and $b_{ij-1/2}$ are calculated analogously. Here, the polynomials p_{ij} are defined for scalar variables u_1 and u_2 analogous to the definition in (36) for second-order method or in (37) for the third-order method. In (56), Cr is the Courant number set to 0.8 for all test cases to ensure stability. The time integration process is stopped if

$$\frac{\left|\phi_{ij}^{n+1} - \phi_{ij}^{n}\right|}{\max\left|\phi_{ij}^{n+1}\right|} \le 10^{-6}, \tag{57}$$

where ϕ represents u, v or T. For the Large Eddy Simulation case three test examples are selected to check the accuracy and performance of the proposed numerical models. The first example assesses convergence rates for an incompressible flow problem with a known analytical solution. The second and third examples are well-known benchmark problems in computational fluid dynamics. These examples show that for small relaxation rates, the discrete-velocity relaxation models reproduce the corresponding LES flow patterns. In addition the high-order relaxation schemes accurately capture the flow structures with very little numerical diffusion, even after long time simulations. All the results presented in this section are computed with variable time steps Δt adjusted at each step by

$$\Delta t = \mathrm{Cr}\cdot\min\left(\frac{\Delta x}{\max\left(\left|a_{ij}^{11}\right|, \left|a_{ij}^{12}\right|, \left|b_{ij}^{1}\right|\right)}, \frac{\Delta y}{\max\left(\left|a_{ij}^{21}\right|, \left|a_{ij}^{22}\right|, \left|b_{ij}^{2}\right|\right)}\right), \tag{58}$$

where a_{ij} and b_{ij} are the characteristic speeds in the relaxation system selected locally at each grid cell, compare [6, 8] for more details and discussion on the selection of characteristic speeds in the relaxation approximations. In (56) and (58), the Courant number is set to 0.75 for all examples. To avoid initial and boundary layers in the relaxation system (32) or (42), initial and boundary conditions for variables Θ and Ψ are calculated from their associated local equilibrium, see for example [35, 56, 7, 58]. In these computations the relaxation rate $\varepsilon = 10^{-10}$ and all linear systems of algebraic equations are solved using a preconditioned conjugate gradient solver with a stopping criteria set to 10^{-6}.

[illegible]

This test example is the Taylor vortex problem used to quantify the errors in the relaxation method. It consists of a laminar isothermal incompressible Navier-Stokes equation with exact solution given in [18],

$$u(x,y,t) = -\cos(x)\sin(y)e^{-2t/Re}, \qquad v(x,y,t) = \sin(x)\cos(y)e^{-2t/Re},$$
$$p(x,y,t) = -\frac{1}{4}\left(\cos(2x) + \cos(2y)\right)e^{-4t/Re}.$$

The problem is solved in a squared domain $[-\pi/2, \pi/2] \times [-\pi/2, \pi/2]$ uniformly discretized with spacing grid $h = \Delta x = \Delta y$, and results are displayed at time $t = 2$ for $\mathrm{Re} = 10$ and at time $t = 4\times 10^4$ for $\mathrm{Re} = 10^5$. In Table **2**, the relative error-norms for the u-velocity varying the mesh size h with $\mathrm{Re} = 10$ and $\mathrm{Re} = 10^5$ are reported. All the errors are calculated by the difference between the point-values of the exact solution and the reconstructed point-values of the computed solution at the final time.

The clear indication from Table **2** is that the error-norms decay as the spatial step h decreases. For $\mathrm{Re} = 10^5$, a slower decay in all error-norms than those obtained for $\mathrm{Re} = 10$ has been detected. It is evident that the proposed relaxation method reaches the expected fifth-order accuracy in space for both Reynolds numbers. For instance, computing the approximate convergence rate between the last two consecutive space refinings, the results for the L^1-error are rates of 5.12 at $\mathrm{Re} = 10$ and 5.08 at $\mathrm{Re} = 10^5$. This clearly demonstrates that our relaxation scheme preserves the fifth-order accuracy for both small and large diffusion values. Similar behavior has been observed for the other flow variables.

[illegible]

The well-established problem of natural convection in a squared cavity subject to a horizontal temperature difference (applied at the walls), which in turn induces natural convection by the fluid within the cavity

Table **2**: Error-norms in u-velocity for the Taylor vortex problem at two different Reynolds numbers.

Re	h	L^∞-error	Rate	L^1-error	Rate	L^2-error	Rate
10	$1/32$	0.13715E-00	—	0.95018E-01	—	0.98936E-01	—
	$1/64$	0.48892E-02	4.810	0.29899E-02	4.990	0.33136E-02	4.900
	$1/128$	0.16375E-03	4.900	0.91513E-04	5.030	0.10355E-03	5.001
	$1/256$	0.50120E-05	5.029	0.26315E-05	5.120	0.31693E-05	5.030
10^5	$1/32$	0.31067E-00	—	0.26553E-00	—	0.29960E-00	—
	$1/64$	0.12043E-01	4.689	0.89676E-02	4.888	0.11003E-01	4.767
	$1/128$	0.40450E-03	4.895	0.28082E-03	4.997	0.36220E-03	4.924
	$1/256$	0.12614E-04	5.003	0.87090E-05	5.010	0.11264E-04	5.006

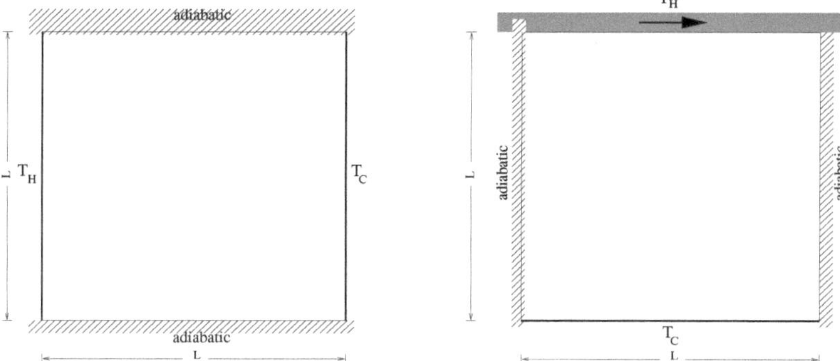

Fig. **1**: Geometry of test problem for the natural (left) and forced (right) convection-radiation.

is considered. The flow domain is a squared cavity with dimension L as shown in the left diagram of Fig. **1**. The left and right vertical walls are maintained at dimensionless hot temperature $T_H = 0.5$ and cold temperature $T_C = -0.5$, respectively. The bottom and top horizontal walls are insulated, whereas no-slip boundary conditions are imposed for the fluid flow on all cavity walls. The Prandtl number Pr and the Rayleigh number Ra are defined as

$$Pr = \frac{\nu}{k}, \qquad Ra = \frac{g\beta L^3 (T_H - T_C)}{\nu}.$$

This test problem has been extensively studied in the literature for non-radiating flows, see for example [20]. Radiation effects in a non-scattering participating fluid are considered. The left and right cavity walls are assumed to be diffusive to radiation such that boundary conditions for the radiative intensity are $I = B(T_H)$ and $I = B(T_C)$ on the left and right wall, respectively. Total reflection is imposed at the top and bottom cavity walls. Computational results for the Prandtl number $Pr = 0.71$ and absorption coefficient $\sigma_a = 1$ with Rayleigh numbers Ra ranging from 10^4 to 10^7 are presented. In order to evaluate the heat transfer rate along the hot wall, the local Nusselt number is calculated, Nu_H, and its averaged value, $\overline{\mathrm{Nu}}_H$, as

$$\mathrm{Nu}_H(y) = \frac{L}{T_H - T_C}\frac{\partial T}{\partial x}\bigg|_{x=0}, \qquad \overline{\mathrm{Nu}}_H = \frac{1}{L}\int_0^L \mathrm{Nu}_H(y)\, dy.$$

A series of computations was presented in [4] for the selected set of parameters to determine effects of the grid on the solution. Four meshes with 32×32, 64×64, 128×128 and 256×256 gridpoints were considered. In Fig. **2** the local Nusselt numbers at the hot wall and cross sections of the temperature at mid-height cavity ($y = L/2$) for $Ra = 10^4$ and 10^7. At $Ra = 10^4$ are presented. Differences in the obtained results are very small on the considered meshes. For this convection-radiation regime, a mesh of 64×64 can be considered grid independent. However, these differences become larger for the computations with $Ra = 10^7$, compare the plots of local Nusselt numbers in Fig. **2**. The discrepancies in the averaged Nusselt number $\overline{\mathrm{Nu}}_H$ on meshes 128×128 and 256×256 are less than 0.45% and 5% for $Ra = 10^4$ and $Ra = 10^7$, respectively. Therefore, bearing in mind the slight change in the results from a mesh of 128×128 and 256×256 at the expense of rather significant increase in CPU times, the mesh 128×128 is believed to be adequate to obtain the results free of grid effects. Hence, the results presented herein are based on the mesh with 128×128 gridpoints [4].

The next results for this test example examine the accuracy of the third-order relaxation method. To this end in Fig. **3** the isotherms and streamlines obtained using the second-order reconstruction (36) and the third-order reconstruction (37) for the two selected Rayleigh numbers $Ra = 10^4$ and 10^7 are presented. For better insight, in Fig. **4** cross sections of the temperature at mid-height cavity and the local Nusselt numbers at the hot wall for both reconstructions are presented. As can be observed from both figures, there is little differences between the second- and third-order results at $Ra = 10^4$, these accuracy differences become more pronounced in the test case with $Ra = 10^7$. This is attributed to the fact that at low Rayleigh numbers ($Ra = 10^4$) the physical diffusion present in the equations is higher than the numerical diffusion introduced by spatial and temporal discretizations such that second- and third-order reconstructions produce roughly the same results. At high Rayleigh numbers ($Ra = 10^7$) convective terms in the equations become dominant and for such flow regimes, numerical dissipation introduced by the second-order reconstruction becomes more visible. Similar behavior has been observed for other Rayleigh numbers. Unlike the second-order scheme, the third-order relaxation scheme accurately resolves this test example with very little numerical diffusion, even after long time simulations are carried out. In the sequel, all the presented results are computed using the third-order reconstruction (37).

In Fig. **5** the isothermal contours, velocity vectors and streamlines obtained for Rayleigh numbers $Ra = 10^4$, 10^5, 10^6 and 10^7 for the coupled convection-radiation test problem are displayed. For comparison the results obtained for the test problem without radiation are presented in Fig. **6**. As can be seen, for this later test case, the flow is symmetric with respect to the cavity center for all the Rayleigh numbers. At low Ra numbers, the flow exhibits a central vortex, and the heat transfer is dominated by conduction regime. At $Ra = 10^5$, the vortex breaks into two vortices moving towards the vertical walls. At high Rayleigh numbers, the flow becomes fully convection dominated, the cold fluid is entrained right to the hot wall where high temperature

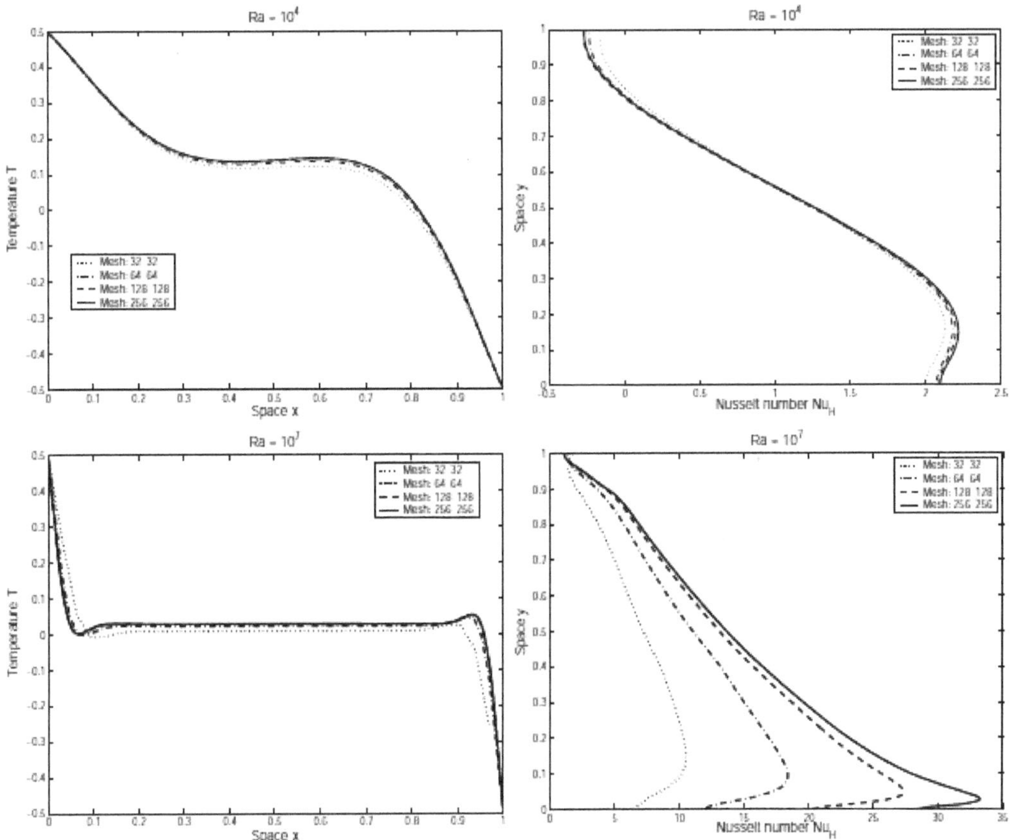

Fig. **2**: Local Nusselt numbers at the hot wall (right column) and cross sections of the temperature at mid-height cavity (left column) for $Ra = 10^4$ and $Ra = 10^7$.

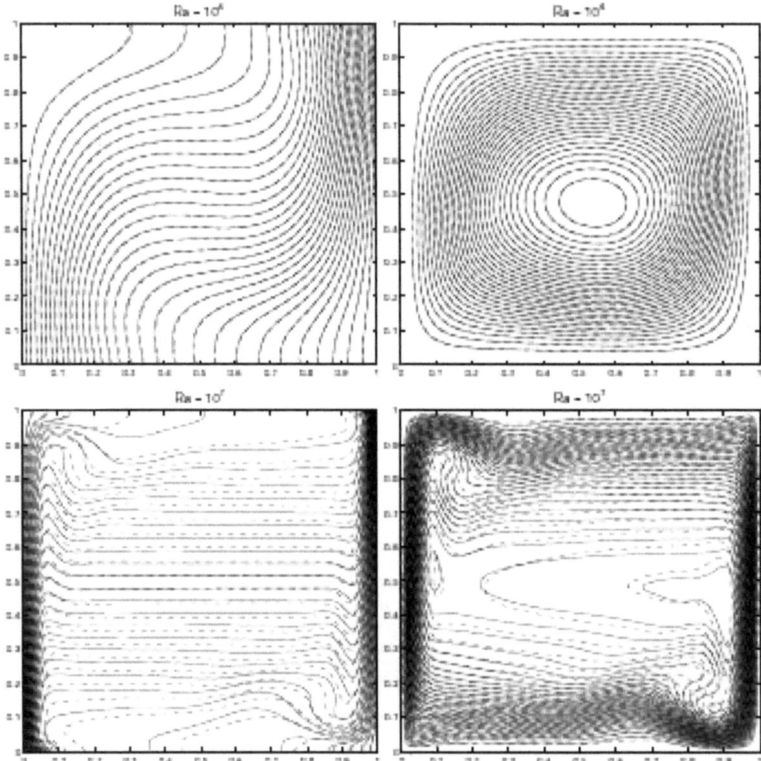

Fig. **3**: Accuracy comparisons between second-order reconstruction (dashed lines) and third-order reconstruction (solid lines) for $Ra = 10^4$ and $Ra = 10^7$.

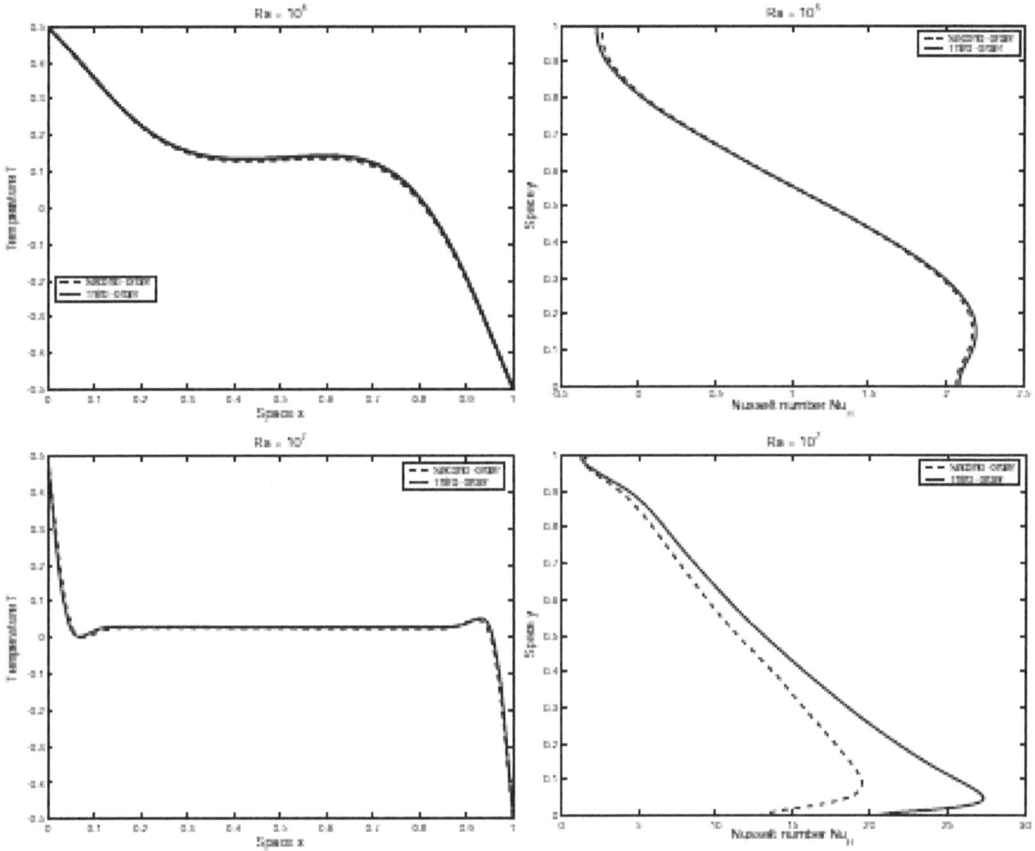

Fig. **4**: Accuracy comparisons between second-order reconstruction and third-order reconstruction.

gradients are created. There is excellent agreement between these results and those published in [20] for non-radiating natural convection in a squared cavity.

Accounting for radiation introduces another mode for heat transfer in the squared cavity. The radiation effects alter not only the temperature distribution but also the velocity field. For the considered radiative conditions, the effect of radiative heat transfer in the natural convection decreases as the Rayleigh numbers increase. For instance, at $Ra = 10^7$ the averaged Nusselt number $\overline{\mathrm{Nu}}_H$ for computations with radiation is 11% less than for computations without radiation. This deviation becomes more than 45% at $Ra = 10^4$. It should be stressed that the performance of the relaxation models is very attractive since the computed solutions remain stable and highly accurate without solving nonlinear problems or requiring Riemann solvers or special characteristic decomposition.

[illegible]

The next test example consists of a laminar scattering and absorbing driven flow in a squared cavity. The isothermal version of this test problem has widely served as a benchmark for validating numerical methods for the incompressible Navier-Stokes equations in the literature [23]. The non-radiating thermal driven flow in a squared cavity was also studied in [33]. According to this reference, the geometrical description of the problem is depicted in the right diagram of Fig. **1**. The top wall is moving with a velocity $u = 1$, for which the Reynolds number $Re = 1/\nu$. The problem is tested with $L = 1$, $Pr = 0.7$ and $\sigma_a = 1$ for three Reynolds numbers $Re = 100$, 400 and 1000, which corresponds to laminar regimes. The bottom wall is kept at cold

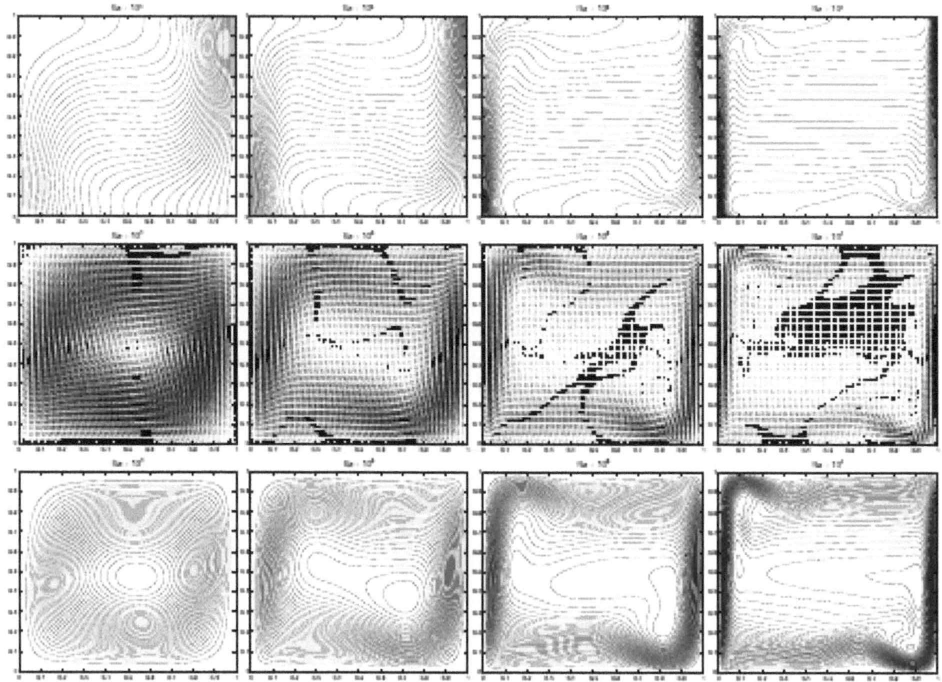

Fig. **5**: Isotherms (top row), velocity field (middle row) and streamlines (bottom row) for coupled natural convection-radiation. From left to right: $Ra = 10^4$, 10^5, 10^6 and 10^7.

temperature $T_C = 0$, the moving top wall is at hot temperature $T_H = 1$, while the left and right walls are adiabatic. No-slip boundary conditions are used for flow velocity, and radiation equilibrium is assumed at the bottom and top boundaries *i.e.*, $I = B(T_C)$ and $I = B(T_H)$ at the bottom and top wall, respectively. On the remaining boundaries total radiative reflection is used. The flow domain is discretized in a uniform mesh with 128×128 gridpoints. This mesh structure has been selected after a grid independence study assessed by comparing different meshes. In contrast to the previous test example where natural convection-radiation is solved in a cavity bounded by rigid and fixed walls, the considered mixed convective-radiation problem is solved in a cavity with moving top wall and scattering medium. As a consequence, the latter flow is more difficult to handle; the results shown here illustrate the robustness of the relaxation models.

In Fig. **7** temperature distributions, velocity vectors and streamlines for coupled convection-radiation in pure absorbing medium ($\sigma_a = 0$) are presented. The results for absorbing and scattering medium with $\sigma_a = 1$ and $\sigma_a = 10$ are displayed in Fig. **8** and Fig. **9**, respectively. For the sake of comparison, results obtained for the non-radiating medium are included in Fig. **10**. For this latter case, it is apparent that the flow structure is in good agreement with previous work [33]. The presented results give a clear view of the overall flow pattern and the effect of Reynolds numbers on the structure of the steady recirculating eddy in the cavity. In both calculations, with and without radiation at low Reynolds numbers, the center of the vortex is located at the mid-width and at about one-third of the cavity depth from the top. As the *Re* increases, the vortex center moves to the right and becomes increasingly circular. The contribution of radiation on both temperature distribution and flow field can also be seen from the presented results. It is noticeable that the thermal radiation at different scattering regimes alters the fluid motion inside the enclosure, compare the recirculation region and its location in Fig. **7**-Fig. **10**. For more comparisons, Fig. **11** shows profiles of the temperature and velocity components along the horizontal and vertical center lines of the cavity. There is

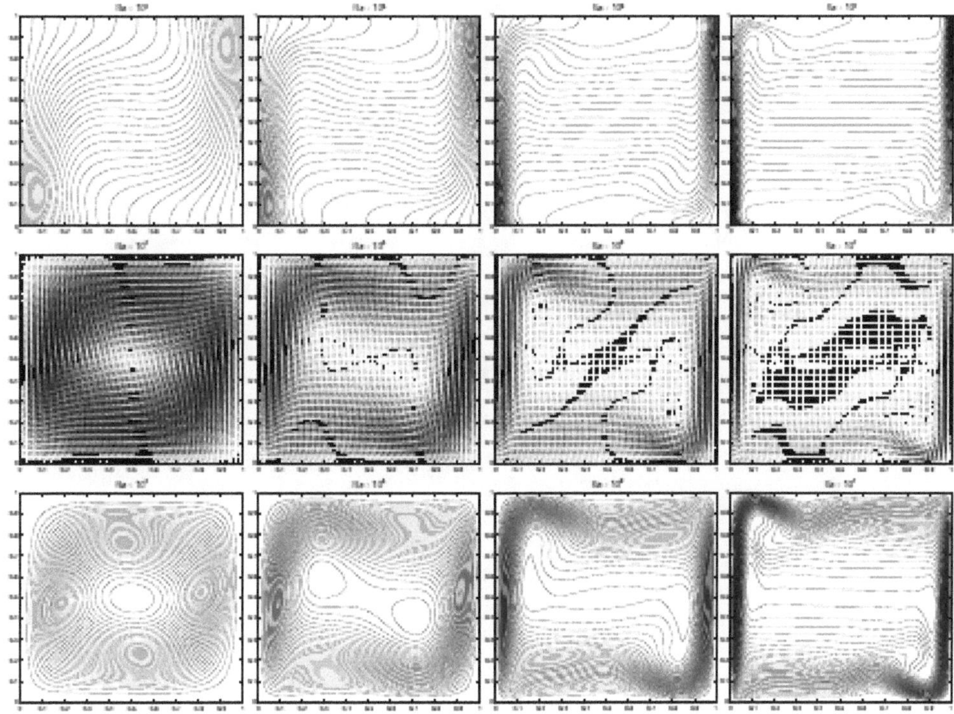

Fig. **6**: Isotherms (top row), velocity field (middle row) and streamlines (bottom row) for natural convection without radiation. From left to right: $Ra = 10^4$, 10^5, 10^6 and 10^7.

a large difference between the results obtained without accounting for radiation and results obtained with radiation for different scattering coefficients. These differences are more visible in the temperature profiles presented in Fig. **11**. The proposed discrete-velocity models and relaxation schemes accurately approximate the numerical solution of this forced convection-radiation problem.

To give an idea of the cost of the computation it can be said that in these implementations, the additional CPU time required to carry out a step in the coupled convection-radiation model with $\sigma_a = 0$, $\sigma_a = 1$ and $\sigma_a = 10$ is about 13%, 21% and 63% more costly than a computational step in no-radiation simulation [4]. It is also important to mention two points concerning the DSA method for solving radiative transfer. First, the DSA method requires a large number of iterations only in the scattering-dominated computations. Otherwise, five iterations were sufficient to reach the tolerance of 10^{-6} for low-scattering computations. Second, for total reflective boundary conditions, an extra iterative process has to be added to the DSA iteration, which may result in an increase of computational cost.

[illegible]

The problem of turbulent forced convection in a lid-driven squared cavity with length $L = 1$ is considered. The geometrical description of the problem and boundary conditions are depicted as in Fig. **12**. This test problem has been numerically studied in [50] and experimental data has also been provided in [48]. The left and right vertical walls are maintained at dimensionless hot temperature $T = 1$ and cold temperature $T = 0$, respectively. The bottom and top horizontal walls are insulated. According to reference [50], the left wall is moving with a velocity $v = 1$, for which the Reynolds number $\text{Re} = 1/\nu$. The problem is tested with the Prandtl number $\text{Pr} = 0.71$, the Smagorinsky constant $c_s = 0.1$, and for Reynolds numbers $\text{Re} = 5 \times 10^4$, 10^5,

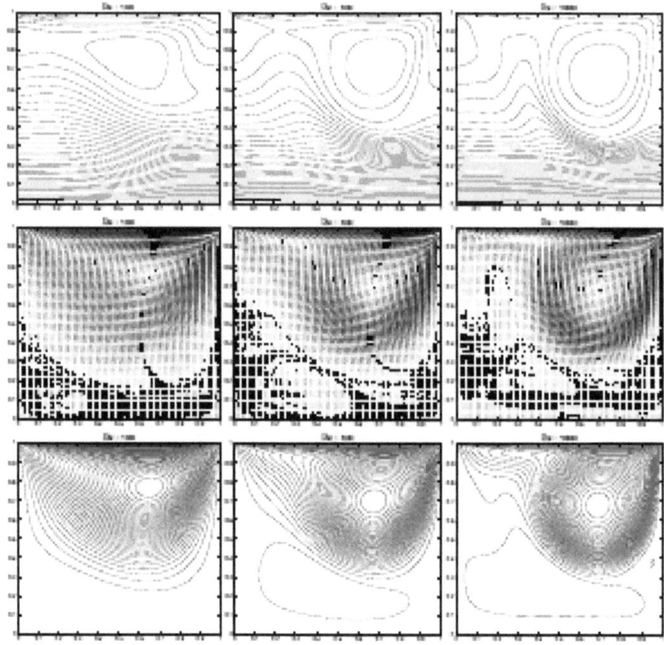

Fig. 7: Isotherms (top row), velocity field (middle row) and streamlines (bottom row) for coupled forced convection-radiation using $\sigma_a = 0$. From left to right: $Re = 100$, 400 and 1000.

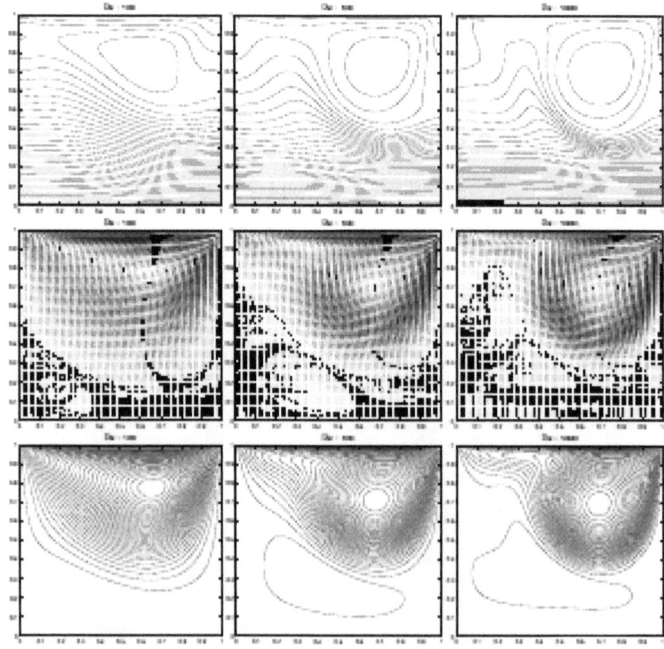

Fig. **8**: Isotherms (top row), velocity field (middle row) and streamlines (bottom row) for coupled forced convection-radiation using $\sigma_a = 1$. From left to right: $Re = 100$, 400 and 1000.

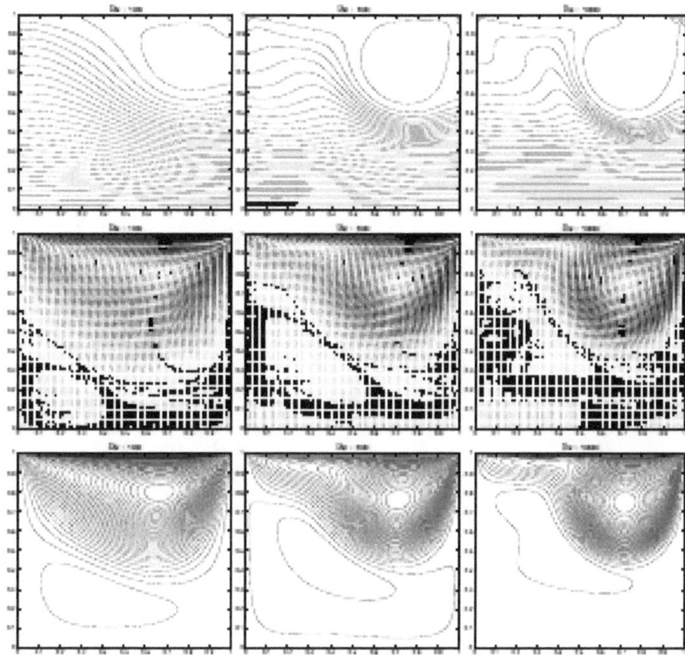

Fig. 9: Isotherms (top row), velocity field (middle row) and streamlines (bottom row) for coupled forced convection-radiation using $\sigma_a = 10$. From left to right: $Re = 100$, 400 and 1000.

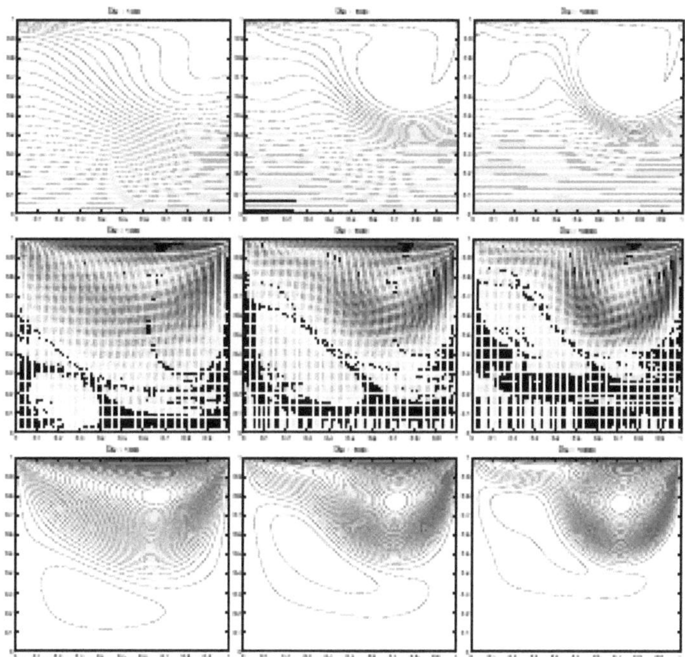

Fig. **10**: Isotherms (top row), velocity field (middle row) and streamlines (bottom row) for forced convection without radiation. From left to right: $Re = 100$, 400 and 1000.

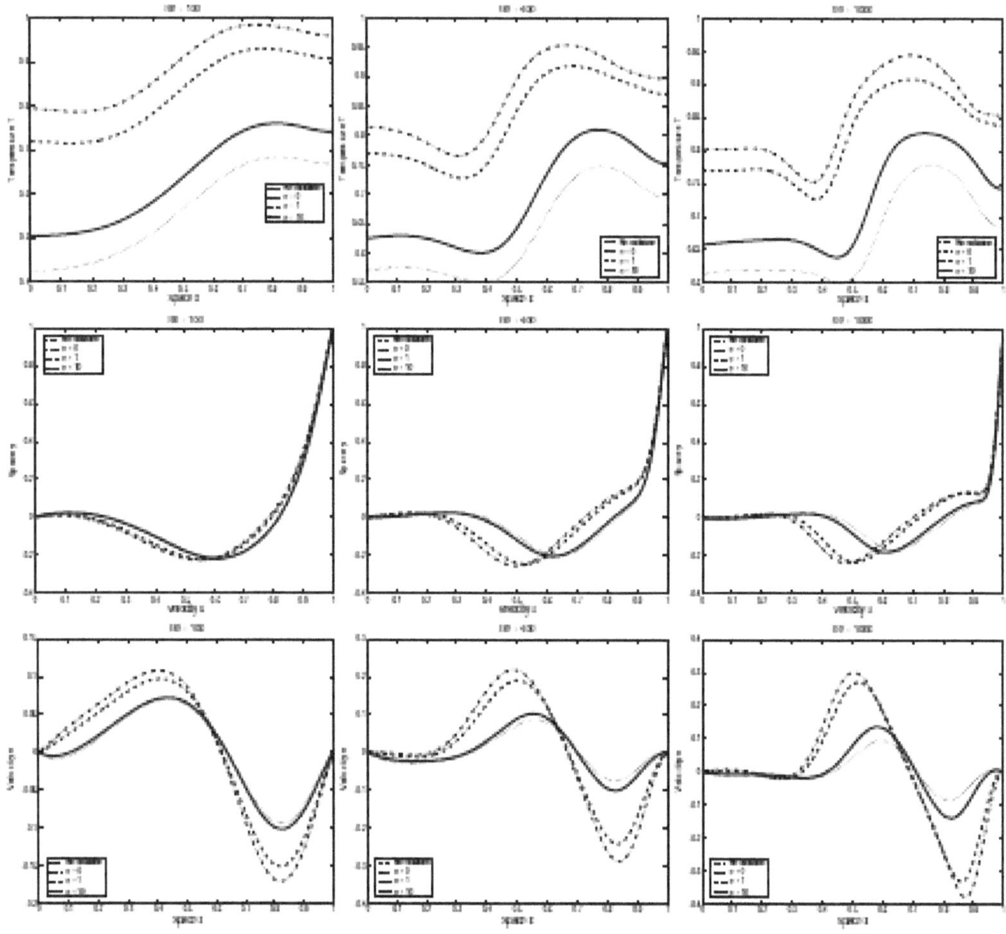

Fig. **11**: Comparison results for forced convection without and with radiation using different scattering coefficients and Reynolds numbers.

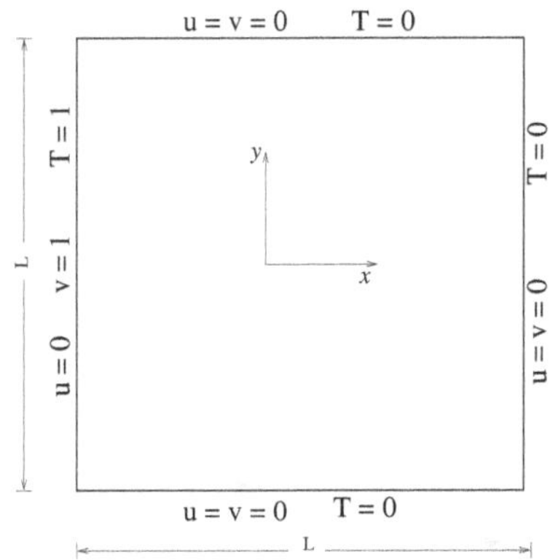

Fig. **12**: Schematic diagram of lid-driven cavity.

Table **3**: Average Nusselt number on the cavity walls at different Reynolds numbers.

	$Re = 5 \times 10^4$		$Re = 2 \times 10^5$		$Re = 4 \times 10^5$	
Wall	Present	Ref. [50]	Present	Ref. [50]	Present	Ref. [50]
Left	66.993	68.560	108.102	109.696	127.733	128.550
Right	41.108	39.422	81.913	80.558	105.310	99.412
Top	16.621	15.426	56.950	56.562	76.004	75.416
Bottom	17.855	18.854	57.887	59.133	78.151	78.844

2×10^5 and 4×10^5. The spatial domain is discretized into 128×128 gridpoints and numerical results are presented at steady-state time.

In Fig. **13**, the streamlines, velocity vectors, and isotherms obtained for different Reynolds numbers are displayed. It can be observed that the flow develops recirculating regions in the cavity corners and the structure of these recirculating regions is strongly influenced by the values of Reynolds numbers. In addition, the vertex centre is very close to the geometrical centre of the cavity for all Reynolds numbers. The convective heat transfer in recirculating flow is also affected by Reynolds numbers. There is excellent agreement between the presented numerical results and the numerical predictions [50] and the experimental results [48] available in the literature. As can be demonstrated the numerical models remain stable and highly accurate even for relatively coarse grids without solving Riemann problems or requiring special characteristics decomposition.

The profiles of the u-velocity component along the vertical centre line and the v-velocity component along the horizontal centre line are shown in Fig. **14** for $Re = 2 \times 10^5$. In this figure, the numerical predictions from [50] and experimental data from [48] have been included. The boundary layers for the velocity components can be clearly observed. There is excellent agreement between the computed results and those published in [50, 48]. Furthermore, computed vortex strengths and temperature distribution on the cavity walls are found to be consistent with those obtained for lid-driven squared cavity flow at corresponding Reynolds numbers.

In order to quantitatively assess the accuracy of the proposed models, Table **3** summarizes a comparison between published results for average Nusselt number on the cavity walls and results obtained with the relaxation method presented in this chapter. The obtained results mostly compare favorably with all the model results from references [50, 48]. As is obvious from the table, the small differences in comparison with other methods can be attributed to the turbulence model used in the computations. It should be noted that the authors in [50] used the κ-ε model to compute the effective viscosity and the thermal conductivity.

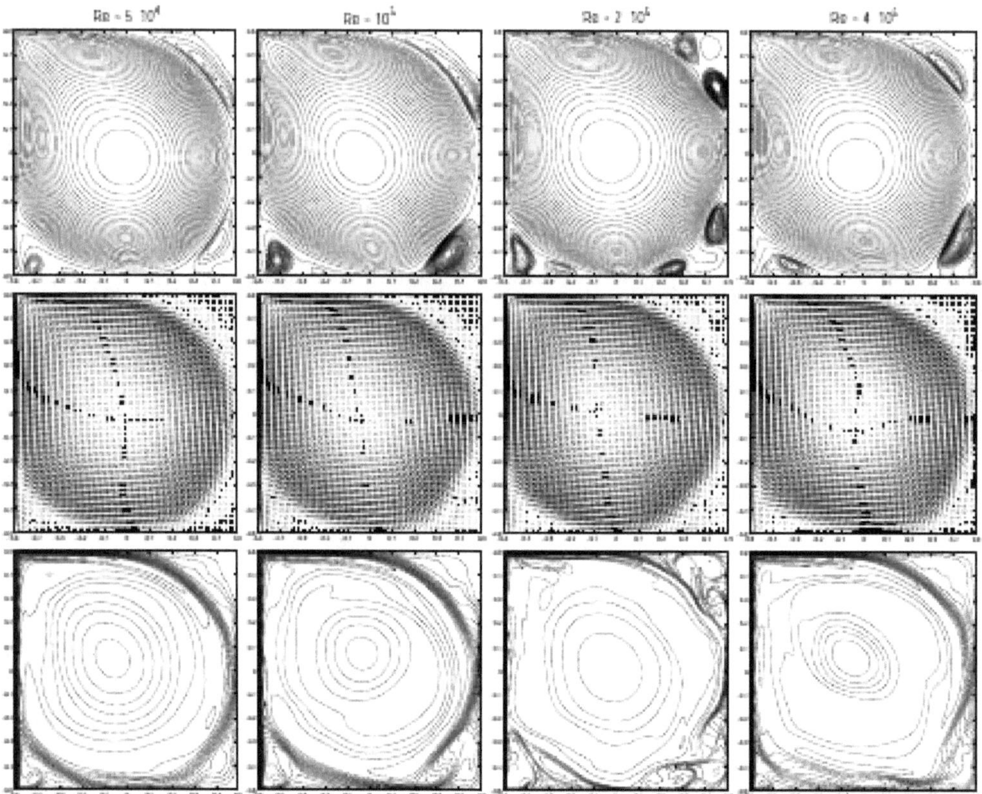

Fig. **13**: Streamlines (top row), velocity field (middle row) and isotherms (bottom row) for lid-driven cavity flow. From left to right: Re $= 5 \times 10^4$, Re $= 10^5$, Re $= 2 \times 10^5$ and Re $= 4 \times 10^5$.

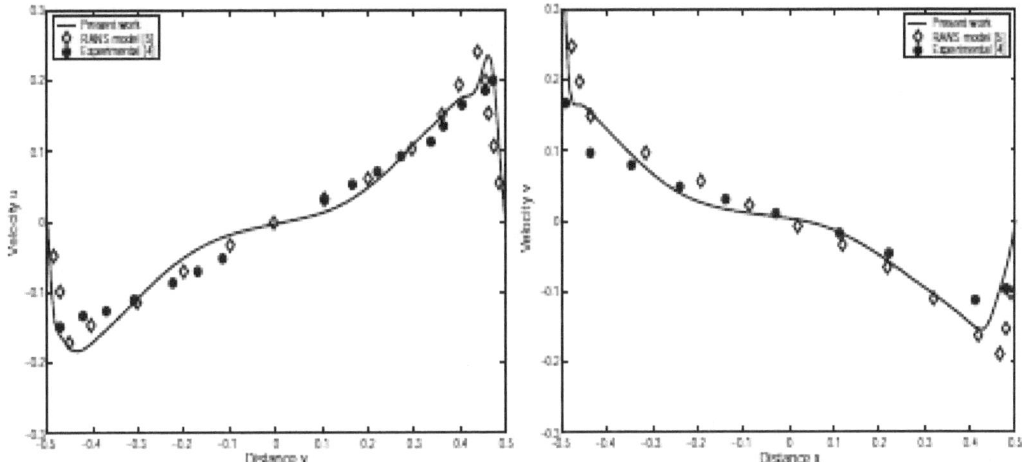

Fig. **14**: Variation of u-velocity at $x = 0$ (left) and v-velocity at $y = 0$ (right) for lid-driven cavity flow at $Re = 2 \times 10^5$.

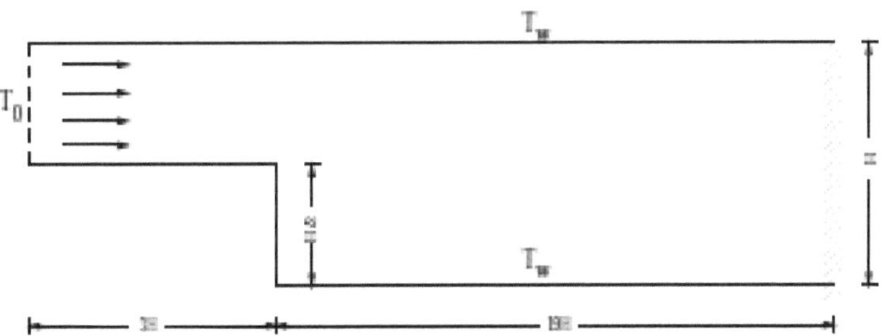

Fig. **15**: Schematic diagram of backward facing step.

[illegible]

The final example is the large-eddy simulation of thermal flow over a backward facing step. The turbulent flow past a backward facing step produces distinctly different flow regimes such as boundary layers, reattachment, flow separation reversal and recovery. The major flow feature is the recirculation zone appearing behind the backward facing wall. The size and the position of the reattachment point are often used for method validation. The computational domain has height $H = 1$, length before the step 3, height of the step 0.5, and the length after the step 19, see Fig. **15**. The velocity boundary conditions for the step geometry include the no-slip velocity for all solid walls, a parabolic velocity profile with maximum inflow $u = 1$ at the inlet, and zero normal stress at the outlet. For the temperature variable, a dimensionless temperature of $T_0 = -0.5$ is imposed at the channel inlet and $T_w = 0.5$ on the fixed walls. Adiabatic boundary conditions are used for the temperature on the channel exit.

Considered values of Reynolds numbers, based on the step length and the free-stream velocity at the inlet, range from 400 to 1000. In all these computations, the Prandtl number $\mathrm{Pr} = 0.71$, the Smagorinsky constant $c_s = 0.15$, and the computational domain is discretized into 440×50 gridpoints [8]. As in the previous test example, only steady-state results are presented. Here, the time integration process is stopped when the difference in the L^∞-norm of the velocity components in two consecutive steps were less than 10^{-5}. This is achieved after about 3×10^4 time steps in both examples.

The obtained results for different Reynolds numbers are shown in Fig. **16**. The results indicate that a number

Table **4**: Reattachment length x_R/H for backward facing step flow at different Reynolds numbers.

	Re = 400	Re = 600	Re = 800	Re = 1000
Present work	8.651	10.135	11.007	12.325
Numerical results from [37]	8.629	10.481	11.814	—
Experimental results from [2]	8.507	11.148	14.334	16.962

of eddy structures, consisting of large-scale vortices in the recirculating region, and the location of the reattachment slightly varies due to the size of these eddies. A thermal boundary layer is also detected after reattachment on the heated horizontal wall. Near the reattachment location, a nonlinear trend is observed. This nonlinearity in the isotherms is moving downstream when Re is increased. It was also observed that the wall shear rate in the separation zone varied markedly with the Reynolds number and the wall heat transfer remained relatively constant. At Re = 400 the flow exhibits a small vortex in the recirculating region, and the heat transfer is dominated by conduction. At Re = 600, the heat transfer along the channel is dominant due to convection effects, but between the top and bottom boundaries the effect of diffusion is more dominant. At Re = 800, the effect of convection between the top and bottom boundaries increases, but it is still very weak. Finally, at Re = 1000, convection effects are dominant in all directions.

Results for reattachment lengths are listed in Table **4** together with the numerical results from [37] and the experimental data from [2], proving the excellent performance of our discrete-velocity relaxation modelling. However, although numerical results present good agreement, the differences with the experimental results grow as the Reynolds number increases. These discrepancies were explained as resulting from significant three-dimensionality in the experiments that arises when the second recirculating region extends in the upper channel wall at Re > 600.

As a final remark we would like to comment on the accuracy of the relaxation scheme. To this end we present in Fig. **17** the distribution of streamwise velocity along a number of axial distances in the flow channel at Re = 1000. In this figure we have also plotted numerical results obtained using the second-order relaxation scheme proposed in [35] and the third-order relaxation scheme developed in [7]. Comparisons for other Reynolds numbers, not reported here, have also been carried out.

The main conclusions drawn from this comparative study are:

(i) At low Reynolds numbers there is very little difference in the results obtained by all relaxation schemes, this is attributed to the fact that at low Reynolds numbers, the physical diffusion present in the mathematical models is higher than the numerical diffusion introduced by the relaxation methods such that all the obtained results are roughly the same.

(ii) At high Reynolds numbers, convective terms in the equations become dominant and for such flow regimes, numerical dissipation introduced by the second- and third-order schemes becomes more visible, thus the fifth-order scheme accurately resolves the flow problem with very little numerical diffusion.

(iii) Reattachment lengths obtained by the fifth-order scheme are smaller with respect to those obtained by the second- and third-order schemes, which is evidence for the dissipative character of the later schemes.

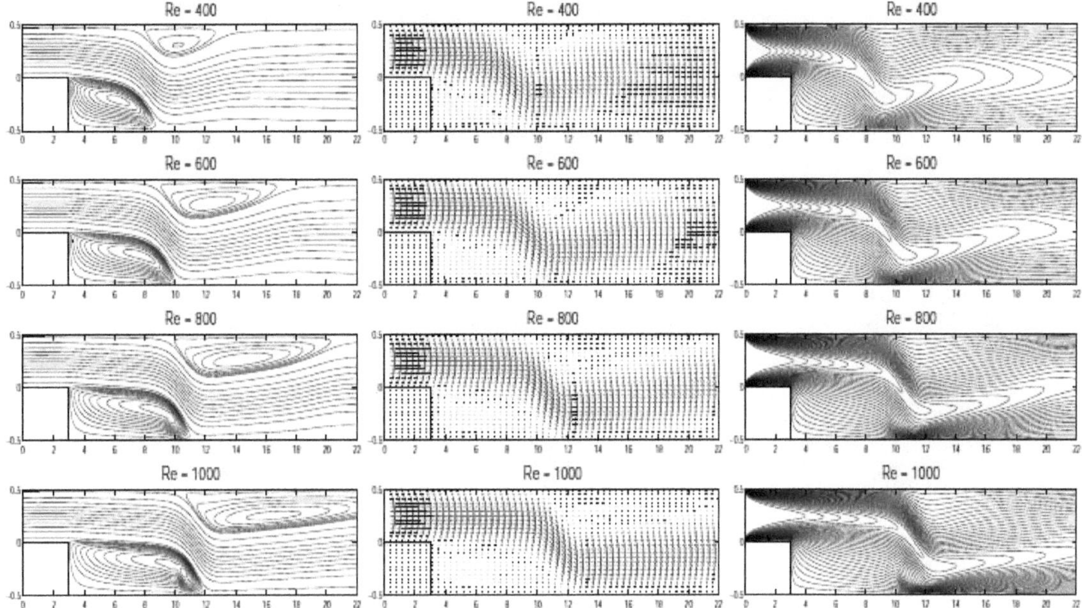

Fig. 16: Streamlines (left column), velocity field (middle column) and isotherms (right column) for backward facing step flow. From top to bottom: Re = 400, Re = 600, Re = 800 and Re = 1000.

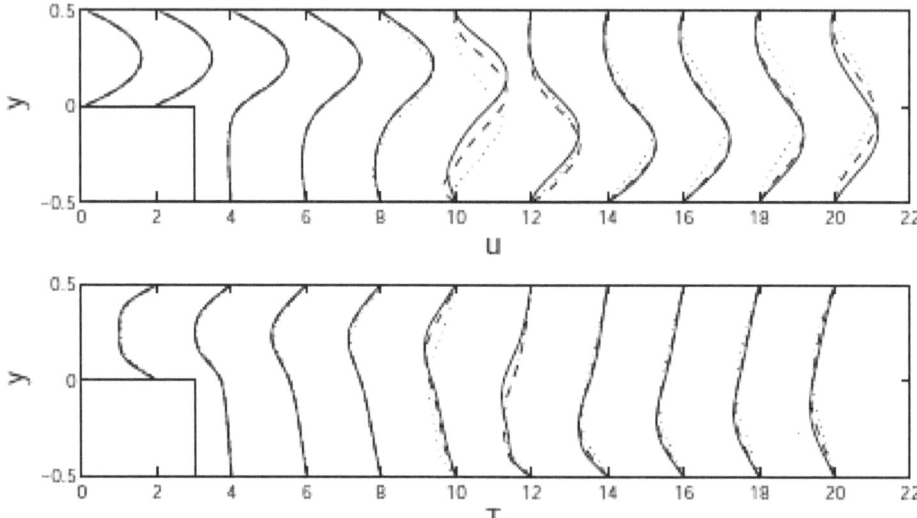

Fig. **17**: Profiles of the axial u-velocity (top) and the temperature (bottom) for backward facing step flow at Re = 1000. Numerical results obtained using second-order (dotted lines), third-order (dashed lines) and fifth-order (solid lines) relaxation schemes.

CONCLUSIONS

In this chapter, a class of lattice-Boltzmann discrete-velocity moment models for coupled convection and radiation systems have been presented including a third-order relaxation method for solving the systems. The mass, momentum and energy equations are obtained from the nine-velocity distributions of flow and temperature variables. The relaxation systems can also be simplified using suitable scaling in the discrete-velocity equations. The radiation coupling is accounted for in the energy equation by considering the S_N-discrete radiative transfer in participating media. Natural and forced convection-radiation problems have been recovered in the relaxation limit. Further a class of discrete-velocity models for large-eddy simulation of turbulent flows with heat transfer has also been presented. In the limit for small Mach and small Knudsen numbers the relaxation system reduces to the LES model of thermal incompressible Navier-Stokes equations. It has been shown that the lattice-Boltzmann discrete-velocity models reproduce all the known features of coupled convection-radiation phenomena in the relaxation limit.

A numerical approach to solve the relaxation systems has also been presented. A high-order relaxation scheme based on a fifth-order WENO reconstruction for the spatial discretization and a third-order IMEX method for the time integration is employed. Riemann problem solvers and hence nonlinear iterations are avoided in the solution procedure. The methods are stable, accurate and converge uniformly to the correct limit with an asymptotic-preserving property. Numerical examples illustrate the accuracy and ability of the methods to simulate well-established flow problems from computational fluid dynamics. The presented results agree well with those published in the literature for the considered problems.

ACKNOWLEDGEMENT

Part of this work was supported by the Deutsche Forschungsgemeinschaft in the collaborative research center SFB568 "Flow and Combustion in Future Gas Turbine Combustion Chambers". M.K. Banda is deeply grateful to the University of KwaZulu-Natal Competitive Grant 2006 "Flow Models with Source Terms, Scientific Computing, Optimization in Applications", the National Research Foundation FA2006020200012 and the career award for Y-rated researchers.

CONFLICT OF INTEREST

The authors confirm that this chapter content has no conflict of interest.

REFERENCES

[1] Aregba-Driollet D, Natalini R, Tang SQ. Diffusive kinetic explicit schemes for nonlinear degenerate parabolic systems. Math Comp 2004; 73: 63-94.

[2] Armaly BF, Durst F, Periera JCF, Schönung B. Experimental and theoretical investigation of backward-facing step flow. J Fluid Mech 1983; 127: 473-496.

[3] Balsara DS, Shu CW. Monotonicity Preserving Weighted Essentially Non-Oscillatory Schemes with Increasingly High Order of Accuracy. J Comput Phys 2000; 160: 405-452.

[4] Banda MK, Klar A, Seaïd M. A Lattice-Boltzman Relaxation Scheme for Coupled Convection-Radiation Systems. J Comput Phys 2007; 226: 1408-1431.

[5] Banda MK, Seaïd M, Klar A, Pareschi L. Compressible and Incompressible Limits for Hyperbolic Systems with Relaxation. J Comp Appl Math 2004; 168: 41-52.

[6] Banda MK, Seaïd M, Klar A, Pareschi L. Lattice-Boltzmann type Relaxation Systems and Higher Order Relaxation Schemes for the Incompressible Navier-Stokes Equation. Math Comp 2008; 77: 943-965.

[7] Banda MK, Seaïd M. Higher-Order Relaxation Schemes for Hyperbolic Systems of Conservation Laws. J Numer Math 2005; 13: 171-196.

[8] Banda MK, Seaid M, Teleaga I. Large-Eddy Simulation of Thermal Flows based on Discrete-Velocity Models. SIAM J Sci Comput 2008; 30: 1756-1777.

[9] Banda MK, Seaïd M, Teleaga I. Discrete-Velocity Relaxation Methods for Large-Eddy Simulation. Appl Math Comp 2006; 182: 739-753.

[10] Bardos C, Golse F, Levermore D. Fluid dynamic limits of kinetic equations: Formal derivations. J Stat Phys 1991; 63: 323-344.

[11] Benzi R, Succi S, Vergassola M. The Lattice-Boltzmann equation: Theory and applications. Phys Reports 1992; 222: 145-197.

[12] Berselli LC, Iliescu T, Layton WJ. Mathematics of Large Eddy Simulation of Turbulent Flows. Springer, 2006.

[13] Caflisch RE, Jin S, Russo G. Uniformly accurate schemes for hyperbolic systems with relaxation. SIAM J Numer Anal 1997; 34: 246-281.

[14] Cao N, Chen S, Jin S, Martinez D. Physical symmetry and lattice symmetry in lattice Boltzmann methods. Phys Rev E 1997; 55: R21-R24.

[15] Chatelain A, Ducros F, Metais O. LES of turbulent heat transfer: proper convection numerical schemes for temperature transport. Int J Numer Meth Fluids 2004; 44: 1017-1044.

[16] Chen H, Chen S, Matthaeus W. Recovery of the Navier-Stokes equations using a Lattice-Gas Boltzmann Method. Phys Rev A 1992, 45: 5339-5342.

[17] Chen S, Doolen GD. Lattice Boltzmann method for fluid flows. Ann Rev Fluid Mech 1998; 30: 329-364.

[18] Chorin AJ. Numerical solution of the Navier-Stokes equations. Math Comp 1968; 22: 745-762.

[19] De Masi A, Esposito R, Lebowitz JL. Incompressible Navier-Stokes and Euler limits of the Boltzmann equation. Comm Pure Appl Math 1989; 42: 1189-1214.

[20] De Valhl Davis D. Natural Convection of Air in a Square Cavity: A Benchmark Solution. Int J Numer Meth Fluids 1983; 3: 249-264.

[21] Fiveland W. Discrete-Ordinates Solutions of the Radiative Transport Equation for Rectangular Enclosures. J Heat Transfer 1984; 106: 699-706.

[22] Geurts BJ. Elements of direct and large-eddy simulation. Edwards, 2004.

[23] Ghia U, Ghia KN, Shin CT. High-Re Solutions for Incompressible Flow using the Navier-Stokes Equations and Multigrid Method. J Comput Phys 1982; 48: 387-411.

[24] Giraud L, d'Humieres D, Lallemand P. A lattice Boltzmann model for Jeffreys viscoelastic fluid. Europhys Lett 1998; 42: 625-630.

[25] Guo Z, Shi B, Zheng C. A coupled lattice BGK model for the Boussinesq equations. Int J Numer Meth Fluids 2002; 39: 325-342.

[26] He X, Luo L. On the Theory of Lattice Boltzmann Method: From the Boltzmann Equation to the Latice Boltzmann Equation. Phys Rev E 1997; 56: 6811-6893.

[27] He X, Chen S, Doolen GD. A novel thermal model for the lattice Boltzmann Method in Incompressible Limit. J Comput Phys 1998; 146: 282-300.

[28] He X, Luo LS. A priori derivation of the lattice Boltzmann equation. Phys Rev E 1997; 55: 6333-6336.

[29] He X, Luo LS. Lattice Boltzmann model for the incompressible Navier-Stokes equation. J Stat Phys 1997; 88: 927-944.

[30] Higuera FJ, Succi S, Benzi R. Lattice gas dynamics with enhanced collision. Europhys Lett 1989; 9: 345-349.

[31] d'Humières D. Generalized Lattice-Boltzmann Equations. in: AIAA Rarefied Gas Dynamics: Theory and Applications. Progress in Astronautics and Aeoronautics 1992; 159: 450-458.

[32] Inamuro T, Yoshino M, Ogino F. Accuracy of the lattice Boltzmann method for small Knudsen number with finite Reynolds number. Phys Fluids 1997; 9: 3535-3542.

[33] Iwatsu R, Hyun JM, Kuwahara K. Mixed convection in a driven cavity with a stable vertical temperature gradient. Int J Heat Mass Transfer 1993; 36: 1601-1608.

[34] Jiang GS, Shu C-W. Efficient Implementation of Weighted ENO Schemes. J Comput Phys 1996; 126: 202-212.

[35] Jin S, Xin Z. The relaxation schemes for systems of conservation laws in arbitrary space dimensions. Comm Pure Appl Math 1995; 48: 235-277.

[36] Junk M, Klar A, Luo LS. Asymptotic analysis of the lattice Boltzmann Equation. J Comput Phys 2005; 210: 676-704.

[37] Kim J, Moin JP. Application of a fractional-step method to incompressible Navier-Stokes equations. J Comput Phys 1985; 59: 308-323.

[38] Kim J, Moin P. Transport of passive scalars in a turbulent channel flow. Proceedings of the 6th Symposium on Turbulent Shear Flows 1987.

[39] Klar A, Pareschi L, Seaïd M. Uniformly Accurate Schemes for Relaxation Approximations to Fluid Dynamic Equations. Appl Math Lett 2003; 16: 1123-1127.

[40] Klar A. An asymptotic-induced scheme for nonstationary transport equations in the diffusive limit. SIAM J Numer Anal 1998; 35: 1073-1094.

[41] Klar A. Relaxation schemes for a Lattice-Boltzmann type discrete velocity model and numerical Navier-Stokes limit. J Comput Phys 1999; 148: 1-17.

[42] Kundu PK, Cohen IM. Fluid Mechanics. 3rd Ed., Elsevier 2004, Oxford.

[43] Landau LD, Lifschitz EM. Fluid Mechanics (Course of Theoretical Physics; vol. 6), 2nd Ed., Butterworth-Heinemann 1987, London.

[44] Liotta F, Romano V, Russo G. Central schemes for balance laws of relaxation type. SIAM J Numer Anal 2000; 38: 1337-1356.

[45] Lyons SL, Hanratty TJ, McLaughlin JB. Direct numerical simulation of passive heat transfer in a turbulent channel flow. Int J Heat Mass Transfer 1991; 34: 1149-1161.

[46] Massaioli F, Benzi R, Succi S. Exponential tails in two-dimensional Rayleigh-Bernard Convection. Europhys Lett 1993; 21: 305-310.

[47] Mihalas D, Mihalas BS. Foundations of Radiation Hydrodynamics. Oxford University Press 1983, New York.

[48] Mills RD. On the closed motion of fluid in a square cavity. J R Aeronaut Soc 1965; 69: 116-120.

[49] Modest MF. Radiative Heat Transfer. McGraw-Hill 1993, New York.

[50] Nonino C, Del Giudice S. Finite element analysis of turbulent forced convection in lid-driven rectangular cavities. Int J Numer Meth Fluids 1988; 25: 313-329.

[51] Pareschi L, Russo G. Implicit-Explicit Runge-Kutta Schemes and Applications to Hyperbolic Systems with Relaxation. J Sci Comput 2005; 25: 129-155.

[52] Pareschi L, Russo G. Implicit-Explicit Runge-Kutta schemes for stiff systems of differential equations. In: Recent trends in numerical analysis. Nova Science Publishers, Inc Commack, NY, USA 2000: 269-288.

[53] Puppo G, Bianco F, Russo G. High order central schemes for hyperbolic system of conservation laws. SIAM J Sci Comput 1999; 21: 294-322.
[54] Qian YH, d'Humieres D, Lallemand P. Lattice BGK models for the Navier-Stokes equation. Europhys Lett 1992; 17: 479-484.
[55] Sagaut P. Large Eddy Simulation for Incompressible Flows. Springer, 2001.
[56] Seaïd M. Non-oscillatory relaxation methods for the shallow water equations in one and two space dimensions. Int J Numer Meth Fluids 2004; 46: 457-484.
[57] Seaïd M, Klar A. Efficient Preconditioning of Linear Systems Arising from the Discretization of Radiative Transfer Equation. Lecture Notes Comput Sci Engrg 2003; 35: 211-236.
[58] Seaïd M. High-Resolution Relaxation Scheme for the Two-Dimensional Riemann Problems in Gas Dynamics. Numer Meth Part Diff Eqs 2006; 22: 397-413.
[59] Seaïd M, Frank M, Klar A, Pinnau R, Thömmes G. Efficient Numerical Methods for Radiation in Gas Turbines. J Comp Appl Math 2004; 170: 217-239.
[60] Seaïd M, Klar A. Asymptotic-Preserving Schemes for Unsteady Flow Simulations. Computers & Fluids 2006; 35: 872-878.
[61] Shi Y, Zhao TS, Guo ZL. Finite difference-based lattice Boltzmann simulation of natural convection heat transfer in a horizontal concentric annulus. Comput & Fluids 2006; 35: 1-15.
[62] Shi Y, Zhao TS, Guo ZL. Thermal Lattice Bhatnagar-Gross-Krook model for flows with viscous heat dissipation in the incompressible limit. Phys Rev E 2004; 70: 066310.
[63] Shan X, He X. Discretization of the Velocity Space in the Solution of the Boltzmann Equation. Phys Rev Lett 1998; 80: 65-68.
[64] Smagorinsky JS. General Circulation Experiments with the Primitive Equations. Monthly Weather Rev 1963; 91: 99-164.
[65] Sone Y. Asymptotic theory of a steady flow of a rarefied gas past bodies for small Knudsen numbers. In: Gatignol R, Soubbaramayer (eds). Advances in Kinetic Theory and Continuum Mechanics. Proceedings of a Symposium Held in Honour of Henri Cabannes (Paris), Springer New York 1990, 19-31.
[66] Struchtrup H, Weiss W. Maximum of local entropy production becomes minimal stationary process. Phys Rev Lett 1998; 80: 5048-5051.
[67] Versteeg HK, Malalasekera W. An introduction to Computational Fluid Dynamics. Prentice Hall, Englewood Cliffs, NJ, 1995.

Progress in Computational Physics, Vol. *3*, 2013, 91-126

CHAPTER 4

Asymptotic Analysis of Lattice Boltzmann Methods for Flow-Rigid Body Interaction

Alfonso Caiazzo [1,*], **Michael Junk**[2]

[1] *Weierstrass Institute for Applied Analysis and Stochastics, Berlin, Germany*, [2] *University of Konstanz, Konstanz, Germany*

Abstract: In this chapter we perform a detailed asymptotic analysis of different numerical schemes for the interaction between an incompressible fluid and a rigid structure within a lattice Boltzmann (LB) framework. After introducing the basic ideas and the main tools for asymptotic analysis of bulk LBM and boundary conditions [19, 21], we concentrate on moving boundary LB schemes. In particular, we investigate in detail the initialization of new fluid nodes created by the variations of the computational fluid domain, when a solid objects moves through a fixed computational grid. We discuss and analyze the equilibrium-non equilibrium (EnE) refill algorithm [6], reporting comparisons with other methods, based on numerical and theoretical considerations. Secondly, we focus on force computation through the Momentum Exchange Algorithm (MEA). Starting from the original scheme (as proposed in [30]), we introduce a correction which, motivated by the analysis, improves Galilean invariance properties of the force computation [5, 7, 32]. Moreover, precise accuracy estimates for the force computation are derived. Our analysis yields first order accuracy of the global force computation, while it shows that the classical MEA is not suitable for accurate local forces evaluation. This problem is fixed with the proposed modification, providing a detailed proof of the accuracy results.

Keywords: Asymptotic analysis, fluid-structure interaction, momentum-exchange algorithm, moving boundary problems, lattice Boltzmann node re-initialization.

1. INTRODUCTION

Back to its origin, the lattice Boltzmann method (LBM) was developed [1, 9, 37] as an evolution of the former *Lattice Gas Cellular Automata* (LGCA) [13, 43]. As a numerical scheme, the LBM can be considered as a *simple*, *low order accuracy* (first order in time, second order in space), *explicit integrator* of the Navier-Stokes equations (see e.g. [19]). However, the low accuracy properties are compensated by a favorable algorithmic formulation, which allows efficient implementations, and by the ability of LBM in modeling flows through complex geometries.

Among others, the treatment of fluid-structure interaction within the LBM has become a very relevant issue in recent years. This problem can be approached from different points of view. We consider the interaction with an incompressible fluid with rigid bodies, described by subsets of computational domain moving through the (fixed) LB cartesian grid, according to the force exerted by the fluid field. Such problem requires additional algorithms to handle the fluid-solid interaction, which should be able to preserve the properties of the original scheme, without spoil efficiency and accuracy, in order to keep the LBM numerically competitive.

We use the asymptotic analysis to investigate the problem and the properties of the considered schemes. Besides providing rigorous accuracy results for the LBM with periodic boundary conditions [19, 26, 27, 28], this approach has been successfully used to investigate initial [4, 20] and boundary conditions [21, 23], including a variety of open boundary treatments [24, 25].

In section 4, we deal with the *refill*, i.e. the reinitialization of LB nodes created by the movement of solid objects through the fixed lattice. As a key result, we show formally that reinitialization limited to equilibrium distribution spoils the accuracy of the method near the boundaries [35], which might result in instabilities of

Address correspondence to: Alfonso Caiazzo, WIAS Berlin, Mohrenstrasse 39, 10117 Berlin, Germany; Tel: ++49 30 20372 332; Fax: ++49 30 20372 317; E-mail: Alfonso.Caiazzo@wias-berlin.de

Matthias Ehrhardt (Ed.)

fluid-solid coupling, especially when dealing with colliding particles [32]. As observed in [35], the problem can be solved including an approximation of the non-equilibrium contribution. We describe and analyze the algorithm proposed in [6], which preserve the LB accuracy with a minimal computational effort.
Force computation is analyzed in detail in section 5. First, we compare different *extrapolation approaches*, where fluid stresses on the interface are approximated from the LB variables at neighboring nodes. Then, we focus on the Momentum Exchange Algorithm [30, 40, 45], which can be efficiently formulated in the framework of LBM, and only uses LB variables. Applying the results of the asymptotic analysis, we show that the original algorithm might yield relevant losses in Galilean invariance [32], and that it is not suitable for computation of local stresses [5, 7]. After proposing a corrected version, the main result consists in a precise accuracy statement, which shows that this corrected MEA is able to compute global forces with first order accuracy, while it only achieves a sub-optimal convergence for local interface stresses.

2. PRELIMINARIES

2.1. Fluid-solid model

Let us consider a domain $\Omega \subset \mathbb{R}^d$ $(d = 2,3)$ decomposed into an incompressible fluid and a rigid body part, separated by an interface Γ:

$$\Omega = \Omega_F(t) \cup \Gamma(t) \cup \Omega_S(t).$$

Given the state of the system at the initial time $t = 0$, the dynamics of the fluid can be described by an initial boundary value incompressible Navier-Stokes problem,

$$\begin{aligned} &\begin{cases} \nabla \cdot \mathbf{u} = 0 \\ \partial_t \mathbf{u} + \nabla p + \mathbf{u} \cdot \nabla \mathbf{u} = \nu \nabla^2 \mathbf{u} + \mathbf{G} \end{cases} \quad t > 0,\ \mathbf{x} \in \Omega_F(t) \\ &\mathbf{u}(t,\mathbf{x}) = \mathbf{u}_B(t,\mathbf{x}),\ \mathbf{x} \in \Gamma(t) \\ &\mathbf{u}(0,\mathbf{x}) = \mathbf{u}_0(\mathbf{x}), \quad \mathbf{x} \in \Omega_F(0). \end{aligned} \tag{1}$$

where $\mathbf{u}(t,\mathbf{x})$ represents the velocity of the flow field at time t and position $\mathbf{x}$, and $p(t,\mathbf{x})$ is the kinematic pressure, i.e. the dynamic pressure divided by the incompressible density, which is assumed to be equal to one. Moreover, $\mathbf{u}_B(t,\mathbf{x})$ denotes the velocity of the interface point $\mathbf{x} \in \Gamma(t)$ at time t and $\mathbf{G}$ denotes a given volume force acting on the fluid.
For the solid domain, we consider the motion of a rigid body (or of a set of rigid bodies), described by the motion of the center of mass $\mathbf{x}_{CM}$ and the angular velocity ω according to Newton's equations

$$\begin{cases} \ddot{\mathbf{x}}_{CM} = \overline{\mathbf{F}}_S = \dfrac{\mathbf{F}_S}{m} \\ \dot{\omega} = \overline{\mathbf{T}}_S = M^{-1}\mathbf{T}_S. \end{cases} \tag{2}$$

In (2), m and M denote respectively the mass and the moment of inertia tensor of the rigid body.
In absence of external source, the total force and torque acting on the solid are given by the integration over the interface Γ of the hydrodynamical forces:

$$\begin{aligned} \mathbf{F}_S(t) &= \int_{\Gamma(t)} (-p(t,\mathbf{x})\mathbf{I} + \mathbf{S}(t,\mathbf{x})) \cdot \mathbf{n}(\mathbf{x}) d\gamma(\mathbf{x}), \\ \mathbf{T}_S &= \int_{\Gamma(t)} (\mathbf{x} - \mathbf{x}_{CM}(t)) \wedge [(-p(t,\mathbf{x})\mathbf{I} + \mathbf{S}(t,\mathbf{x})) \cdot \mathbf{n}]\, d\gamma(\mathbf{x}) \end{aligned} \tag{3}$$

where

$$\mathbf{S} = \mathbf{S}[\mathbf{u}] = \nu \left(\nabla \mathbf{u} + \nabla \mathbf{u}^T\right) \tag{4}$$

is the viscous stress tensor and $\mathbf{n}$ the normal vector to Γ, pointing out of the solid domain. In what follows, we will refer to $\mathbf{F}_S$ and $\mathbf{T}_S$ as total boundary force and total boundary torque. Moreover, let us denote with $\mathbf{t}_1, \ldots, \mathbf{t}_{d-1}$ the tangential vectors to Γ. We define the **local stresses** (normal and tangential) as

$$f_{\mathbf{n}} := -p + (\mathbf{S} \cdot \mathbf{n}) \cdot \mathbf{n}, \quad f_{\mathbf{t}_\alpha} := (\mathbf{S} \cdot \mathbf{n}) \cdot \mathbf{t}_\alpha,\ \alpha = 1,, \ldots, d-1. \tag{5}$$

Our fluid-structure interaction model is defined by the set of equations (1)-(2), coupled through the interface quantities $\mathbf{u}_B$, $\mathbf{F}_S$ and $\mathbf{T}_S$.
In our numerical approach, without loss of generality we assume that equations (2) are solved with a simple *explicit* integration in time (first order Euler method):

$$\begin{aligned}
&\hat{\mathbf{u}}_{CM}^{n+1} - \hat{\mathbf{u}}_{CM}^{n} = \Delta t \overline{\mathbf{F}}_S^n \\
&\hat{\omega}^{n+1} - \hat{\omega}^{n} = \Delta t \overline{\mathbf{T}}_S^n \\
&\hat{\mathbf{x}}_{CM}^{n+1} - \hat{\mathbf{x}}_{CM}^{n} = \Delta t \left(\hat{\mathbf{u}}_{CM}^{n+1} + \hat{\mathbf{u}}_{CM}^{n} \right),
\end{aligned} \tag{6}$$

where Δt denotes the time step.

2.2. Lattice Boltzmann Method

The NSE (1) are discretized in time using a lattice Boltzmann method. In this section, we shortly recall the derivation of the numerical method, presenting the notations used in the rest of the chapter. For simplicity, we focus on a two dimensional case. For more detailed discussions, overviews and classifications of the lattice Boltzmann schemes, we refer, for example, to [1, 9, 10, 11, 41, 42, 43].

2.2.1. *Space and time discretizations*

Let us introduce an uniform partition of the time interval $(0,T]$ with time step Δt and a discretization of Ω with a *regular* (Cartesian) spatial grid

$$\mathcal{G}(\Delta x) \subset \mathbb{Z}^d, \tag{7}$$

whose nodes are defined by

$$\mathbf{j} \in \mathcal{G}(\Delta x) \stackrel{\text{def}}{\Longleftrightarrow} \mathbf{x_j} = \mathbf{j}\Delta x \in \Omega. \tag{8}$$

For the LBM, we consider the case when the two discretizations are related by the *diffusive* scaling

$$\Delta t = \beta \Delta x^2, \tag{9}$$

for a certain parameter $\beta > 0$, which can be used to tune the time discretization size. In what follows, we assume $\beta = 1$, describing the implementation of the algorithm using integer indices, n for time and $\mathbf{j}$ for space.
Finally, let us denote with $\mathbb{V} = \{\mathbf{c}_i\}_{i=0,\ldots,b}$ a lattice Boltzmann finite velocity set. For simplicity, we focus on the D2Q9 model in two dimensions (figure **1**, left), with nine vectors (in lattice units)

$$\begin{aligned}
&\mathbf{c}_0 = (0,0), \\
&\mathbf{c}_1 = (1,0), \quad \mathbf{c}_2 = (1,1), \quad \mathbf{c}_3 = (0,1), \quad \mathbf{c}_4 = (-1,1) \\
&\mathbf{c}_5 = (-1,0), \quad \mathbf{c}_6 = (-1,-1), \quad \mathbf{c}_7 = (0,-1), \quad \mathbf{c}_8 = (1,-1).
\end{aligned} \tag{10}$$

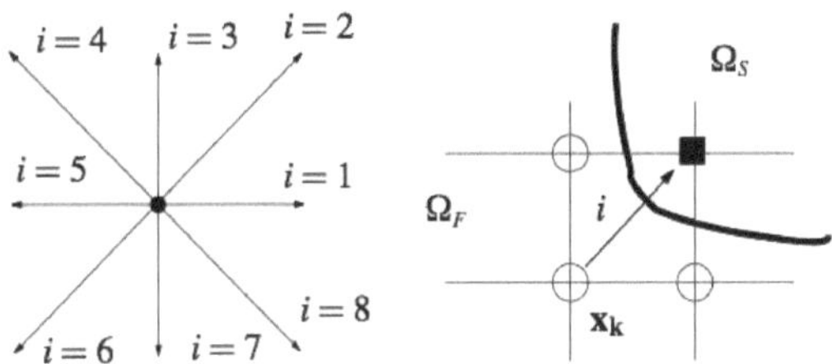

Fig. **1**: **Left.** D2Q9 discrete velocity model. **Right.** A fluid node $\mathbf{x_k}$ is a *boundary node* if it has at least a solid neighbor $\mathbf{x_k} + h\mathbf{c}_i$. The pair $(\mathbf{k}, i)$ is then a *boundary couple.*

Fluid and solid nodes From the computational point of view, we define the *fluid nodes* as the grid points which belong to the fluid domain Ω_F or to the interface Γ, and as *solid nodes* the grid points belonging to Ω_S. Moreover, we introduce the set of (fluid-solid) *boundary nodes*, containing the fluid nodes with at least one solid neighbor:

$$\mathcal{B} = \left\{ \mathbf{j} \in \mathcal{G} \mid \mathbf{x_j} \in \Omega_F \cup \Gamma, \mathbf{x_j} + h\mathbf{c}_i \in \Omega_S, \text{ for some } i \in \{1,\ldots,b\} \right\}$$

and the set of *boundary couples*

$$\mathcal{K} = \{(\mathbf{k},i) \in \mathcal{G} \times \{1,\ldots,b\} \mid \mathbf{x_k} \in \Omega_F \cup \Gamma,\ \mathbf{x_k} + h\mathbf{c}_i \in \Omega_S\}$$

(see Fig. **1**, right). Note that, in case of moving boundary problems, the sets $\mathcal{B}$ and $\mathcal{K}$ are also time-dependent.

2.2.2. *Bulk algorithm and BGK approximation*

The time iteration of the lattice Boltzmann method is governed by the lattce Boltzmann equation (LBE)

$$\hat{f}_i(n+1,\mathbf{j}+\mathbf{c}_i) = \hat{f}_i(n,\mathbf{j}) + J_i(\hat{f}(n,\mathbf{j})), \tag{11}$$

describing the evolution of the particle distribution

$$\hat{f}^n : \mathcal{G}(\Delta x) \to \mathbb{R}^{b+1}, \quad \hat{f}^n(\mathbf{j}) := \left(\hat{f}_i(n,\mathbf{j})\right)_{i=0,\ldots,b}, \tag{12}$$

where each component $\hat{f}_i(n,\mathbf{j})$ denotes the density of particles moving in direction $\mathbf{c}_i$, at time step t_n and at position $\mathbf{x_j}$. On the right hand side of (11), the *collision* operator $J_i(f)$ models the effects of the collisions between particles, producing variations in the distributions.
Let us introduce the arrays

$$\mathbf{1} := (1,\ldots,1) \in \mathbb{R}^{b+1},$$
$$\mathbf{c}_\alpha := (\mathbf{c}_{0\alpha},\ldots,\mathbf{c}_{b\alpha}) \in \mathbb{R}^{b+1}, \quad \alpha \in \{x,y\}$$

and the scalar product

$$\langle f,g \rangle = \sum_{i=0}^{b} f_i g_i, \quad f,g \in \mathbb{R}^{b+1}. \tag{13}$$

Using this compact notation, the macroscopic **density** ρ and **velocity** $\mathbf{u}$, computed as moments of the distribution $\hat{f}$ with respect to the velocity vectors $\mathbf{c}_i$ can be denoted as:

$$\rho(f) := \sum_{i=0}^{b} f_i = \langle f,\mathbf{1} \rangle, \quad \mathbf{u}_\alpha(f) := \sum_{i=0}^{b} \mathbf{c}_{i\alpha} f_i = \langle f,\mathbf{c}_\alpha \rangle. \tag{14}$$

Different numerical methods are defined, specifying the form of J_i in equation (11). We choose an operator in a *linear relaxation* form,

$$J(f) = A\,(f^{eq} - f) + \mathbf{g}, \tag{15}$$

where A is a $b \times b$ positive semi definite matrix, with eigenvalues ω_i, which satisfy for stability reasons [19, 27, 34]

$$0 < \omega_i < 2, \tag{16}$$

and the additional vector $\mathbf{g} \in \mathbb{R}^{b+1}$ is used to include volume forces.
In (15) f_i^{eq} is the *equilibrium distribution*, whose form depends on the discrete velocity space. In general, it is a function of f through the density ρ and the velocity $\mathbf{u}$, defined in equation (14):

$$f^{eq} = f^{eq}(f) = H_i^{eq}(\rho(f),\mathbf{u}(f)). \tag{17}$$

For the D2Q9 model (and for the D3Q15 model in three dimensions) we have

$$H_i^{eq}(\rho,\mathbf{u}) = f_i^* \left(\rho + c_s^{-2}\mathbf{c}_i \cdot \mathbf{u} + \frac{c_s^{-4}}{2} \left(|\mathbf{c}_i \cdot \mathbf{u}|^2 - c_s^2 |\mathbf{u}|^2 \right) \right) \tag{18}$$

where $c_s = \frac{1}{\sqrt{3}}$ is the *lattice sound speed*, and

$$\begin{aligned} f_i^* &= \tfrac{4}{9}, \quad \text{for } i = 0 \\ f_i^* &= \tfrac{1}{9}, \quad \text{for } i = 1,3,5,7 \\ f_i^* &= \tfrac{1}{36}, \quad \text{for } i = 2,4,6,8. \end{aligned} \tag{19}$$

In general, the function f_i^{eq} must be chosen is such a way to satisfy a set of algebraic requirements and symmetries, in relation with the Maxwellian equilibrium of kinetic theory [19, 42]. In the equilibrium H_i^{eq} we can identify a linear and a quadratic part:

$$H_i^{eq}(\rho,\mathbf{u}) = H_i^{L(eq)}(\rho,\mathbf{u}) + H_i^{Q(eq)}(\mathbf{u},\mathbf{u}). \tag{20}$$

Using the notation

$$(\mathbf{v} \otimes \mathbf{w})_{\alpha\beta} = \frac{1}{2}\left(v_\alpha w_\beta + v_\beta w_\alpha\right), \tag{21}$$

the following relations hold:

$$\begin{aligned} &\sum_{i=1}^{b} H_i^{eq}(\rho,\mathbf{u}) = \sum_{i=1}^{b} H_i^{L(eq)}(\rho,\mathbf{u}) = \rho \\ &\sum_{i=1}^{b} \mathbf{c}_i H_i^{eq}(\rho,\mathbf{u}) = \sum_{i=1}^{b} \mathbf{c}_i H_i^{L(eq)}(\rho,\mathbf{u}) = \mathbf{u} \\ &\sum_{i=1}^{b} \mathbf{c}_i \otimes \mathbf{c}_i H_i^{L(eq)}(\rho,\mathbf{u}) = c_s^2 \rho \mathbf{I} \\ &\sum_{i=1}^{b} \mathbf{c}_i \otimes \mathbf{c}_i H_i^{Q(eq)}(\mathbf{u},\mathbf{u}) = \mathbf{u} \otimes \mathbf{u}. \end{aligned} \tag{22}$$

Remark (Mass and momentum conservation). The first two identities in (22) are related to the conservation of mass and momentum. If the matrix A satisfies

$$\mathbf{1}^T A = 0, \; \mathbf{c}_\alpha^T A = 0$$

also the collision operator J (in the form (15)) conserves locally mass and momentum. In general, due to the properties of the equilibrium distribution, to have a momentum conserving collision operator it suffices that the vector $\mathbf{1}$ and $\mathbf{c}_\alpha$ are eigenvectors of the matrix A [19].

The most general form of the matrix A leads to the *Multiple Relaxation Time* model (MRT) (see e.g. [12, 34]), where different relaxation parameters are defined for the f-moments of different orders. The simplest collision operator (*Single relaxation time*) is known as the *BGK approximation* (based on the homonym original model for the Boltzmann equation [2]). Formally, it is obtained by setting

$$A = \frac{1}{\tau}\mathbf{I}, \tag{23}$$

with one relaxation time $\tau > \frac{1}{2}$. We restrict to this case. An analysis of a wide class of linear collision operators, including also the MRT model, has been presented in [19]. The BGK-lattice Boltzmann method reads

$$\hat{f}_i(n+1,\mathbf{j}+\mathbf{c}_i) = \hat{f}_i(n,\mathbf{j}) + \frac{1}{\tau}(f_i^{eq}(\hat{f}) - \hat{f}_i)(n,\mathbf{j}) + g_i(n,\mathbf{j}). \tag{24}$$

The relaxation time τ in equation (24) is related to a dimensionless viscosity through $\nu = c_s^2(\tau - \frac{1}{2})$. The term g_i depends on the additional source terms in (1) through

$$g_i(n,\mathbf{j}) = c_s^{-2} h^3 f_i^* \mathbf{c}_i \cdot \mathbf{G}(t_n,\mathbf{x}_\mathbf{j}). \tag{25}$$

The time advancing (11) is usually split into two sub-steps, a **local collision**:

$$\hat{f}_i^c(n,\mathbf{j}) = \hat{f}_i(n,\mathbf{j}) + J_i(\hat{f}(n,\mathbf{j})) \tag{26}$$

and an **advection**:

$$\hat{f}_i(n+1,\mathbf{j}+\mathbf{c}_i) = \hat{f}_i^c(n,\mathbf{j}). \tag{27}$$

In what follows, we will refer to $\hat{f}_i^c$ as the *post-collision* distribution.

2.2.3. *Implementation of boundary conditions*

To include the Dirichlet boundary conditions for the average velocity, an additional boundary algorithm has to be coupled to (24). In particular, the propagation step (27) has to be substituted with an appropriate boundary algorithm defining the *incoming* distribution at the boundary nodes, i.e. these distributions which enter the fluid domain. Among the available approaches, we consider the *BFL* rule [3] (also called *Interpolated Bounce-Back*).

Let us consider a *boundary node* $\mathbf{k} \in \mathcal{B}$, such that $\mathbf{x_k} = h\mathbf{k} \in \Omega_F \cup \Gamma$, $\mathbf{x_k} + h\mathbf{c}_i \in \Omega_S$, and let i^* be such that $\mathbf{c}_{i^*} = -\mathbf{c}_i$. For the incoming distribution $\hat{f}_{i^*}$ at $\mathbf{k}$, we define

$$\hat{f}_{i^*}(n+1,\mathbf{k}) = \begin{cases} 2q\hat{f}_i^c(n,\mathbf{k}) + (1-2q)\hat{f}_i^c(n,\mathbf{k}-\mathbf{c}_i) + 2c_s^{-2} f_i^* \mathbf{c}_i \cdot \mathbf{u}_B & q \le \frac{1}{2} \\ \frac{1}{2q}\hat{f}_i^c(n,\mathbf{k}) + (1-\frac{1}{2q})\hat{f}_{i^*}^c(n,\mathbf{k}) + \frac{1}{q}c_s^{-2} f_i^* \mathbf{c}_i \cdot \mathbf{u}_B & q > \frac{1}{2} \end{cases} \tag{28}$$

where $q \in [0,1)$ is the node-boundary distance along the link $\mathbf{c}_i$ and $\mathbf{u}_B$ is the boundary velocity evaluated at

$$\mathbf{b}_i(\mathbf{k}) = \mathbf{x_k} + qh\mathbf{c}_i \in \Gamma. \tag{29}$$

Note that the variable q can assume the value zero, since the interface Γ has been considered as part of the computational fluid domain. It should also be noted that, at certain exceptional nodes, a neighboring fluid node may not be available. In such situations, the BFL rule has to be replaced by a suitable one-point rule of sufficient accuracy [23].

Neumann boundary conditions Algorithm (28) imposes Dirichlet boundary conditions on the velocity. To impose other types of conditions, commonly used in CFD (no stress, do-nothing conditions, pressure conditions), different boundary rules must be used. In the analysis (section 3.3.1.) we will focus on (28). However, the same analysis technique can be used for general boundary rules, and theoretical results can be also extended to Neumann boundary conditions [23, 24, 25, 44].

Initial conditions We do not discuss here the initialization lattice Boltzmann schemes, given initial velocity and pressure fields. For the analysis and implementations of consistent initialization algorithms, we refer to [4, 20, 38].

2.2.4. *Summary*

In conclusion, the lattice Boltzmann algorithm on the set of variables $\hat{f}^n$ in the fluid domain is defined by equations (24) and (28), together with the update of the macroscopic variables (14) and the evaluation of the equilibrium distribution (18):

Algorithm 1 (Basic LBM).

Given initial pressure and velocity fields, initialize the set of discrete variables $\hat{f}^0$
for $n = 0, \ldots, N_T$
  evaluate $\mathbf{u}$, ρ as in (14) for each node $\mathbf{j} \in \mathcal{G}(h)$
  compute equilibrium distribution (18)
  compute local collision through (24) for each node $\mathbf{j} \in \mathcal{G}(h)$
  update inner nodes using (24)
  update incoming populations at boundary nodes (e.g. using (28) for Dirichlet nodes)
end

3. ASYMPTOTIC ANALYSIS OF LATTICE BOLTZMANN METHOD

The asymptotic analysis is a widely used tool, to investigate *equations depending on a small parameter* $h \ll 1$ [29]. In this section, we describe the basic ideas of the asymptotic analysis for the LB algorithm 1 (section 2.2.), considering as small parameters the size of the spatial discretization. As a result, accuracy of the numerical solutions for pressure and velocity will be rigorously stated.

3.1. Preliminary definitions

Let be given a set of small parameters $H \subset (0,1]$ with *an accumulation point at* 0[1]. For each $h \in H$ (a particular discretization), the lattice Boltzmann method is defined by a set of equations, at each time step $n = 1, \dots, N_T$, for the particle distributions $\hat{f}^n$. Let us denote the time iteration (including both collision and propagation steps (26)-(27) and the boundary rule (28)) as

$$\hat{f}_h^{n+1} - L_h(n, h, \hat{f}_h^n) = 0, \tag{30}$$

Remark (Definition of **numerical solution**). Note that equation (30) implicitly defines the lattice Boltzmann solution as the *solution of a particular set of discrete equations*, for each discretization $h \in H$.

Asymptotic expansion The scope of the asymptotic analysis is to *understand* the behavior of the solution $\hat{f}_h$ of (30) for small values of h, by approximating it with a **prediction**.
Namely, we look for an approximation of $\hat{f}_h$ in the form of a truncated asymptotic expansion in series of h

$$\hat{f}_h(n, \mathbf{j}) \approx F_h(t_n, \mathbf{x_j}), \quad \text{with} \quad F_h = \sum_{k \leq K} h^k f^{(k)}, \quad \text{for } K \in \mathbb{N}. \tag{31}$$

We further assume that the coefficients $f^{(k)}$ are smooth and h-independent. The function F_h will be called *regular prediction.* Actually, regularity assumptions could be replaced with a less restrictive requirement. However, this assumption allows to simplify the investigation, whilst the regularity we need to perform the analysis can be fixed *a posteriori.*

Precision of asymptotic prediction The meaning of the symbol "$\approx$" in (31) has to be specified. To this aim, we consider the following definitions [29]. For two real functions $f, g : H \to \mathbb{R}_+$, we define

$$f \in O(g), \text{ for } h \to 0 \overset{\text{def}}{\Longleftrightarrow} \exists K : |f| \leq K|g| \text{ eventually (evt.)} \tag{32}$$

(f is *of the same order of* g). Additionally,

$$f \in o(g), \text{ for } h \to 0 \overset{\text{def}}{\Longleftrightarrow} \forall \delta > 0 : |f| < \delta |g| \text{ evt.} \tag{33}$$

(f is *of order higher than* g). The above conditions are standard notations in performing asymptotic investigations.

Definition 1 (Residue)**.** For an asymptotic prediction F_h of the form (31), we define the residue $r(F,\cdot) : H \to \mathbb{R}_+$ as

$$r(F,h) := \sum_{n=1}^{N_T} \|r_i^n(F,h)\|^2, \quad \text{with } r_i^{n+1}(F,h) := F_{i_h}^{n+1} - L_{i_h}(n,h,F_h^n) \tag{34}$$

Definition 2 (Precision order)**.** Let F_h, G_h be two asymptotic predictions for the LBM in the form (31). We say that F is more precise than G iff.

$$r_F \in o(r_G).$$

[1] The accumulation point is needed for practical reasons, if we aim at the limit $h \to 0$.

3.2. Expansion coefficients of the regular predictions

Inserting the expansion (31) into the algorithm (30) one gets

$$r_i^{n+1}(F,h)(\mathbf{j}) =$$
$$\sum_{k\in K} h^k f_i^{(k)}(t_{n+1},\mathbf{x}_{\mathbf{j}+\mathbf{c}_i}) - \sum_{k\in K} h^k f_i^{(k)}(t_n,\mathbf{x}_\mathbf{j}) - \frac{1}{\tau}\left(f_i^{eq}\left(\sum_{k\in K} h^k f^{(k)}(t_n,\mathbf{x}_\mathbf{j})\right) - \sum_{k\in K} h^k f_i^{(k)}(t_n,\mathbf{x}_\mathbf{j})\right) - g_i(n,\mathbf{j}). \tag{35}$$

According to the definition of the forcing term (25), we expand $g_i(n,\mathbf{j})$ in a power series as

$$g_i(n,\mathbf{j}) = \sum_{k\in K} h^k g_i^{(k)}(t_n,\mathbf{x}_\mathbf{j}), \text{ with } g_i^{(3)} = f_i^* c_s^{-2}\mathbf{c}_i\cdot\mathbf{G}, \quad g_i^{(k)} = 0, \quad \text{for } k\neq 3.$$

Using a Taylor expansion around the point $(t_n,\mathbf{x}_\mathbf{j})$, we get

$$r_i^{n+1}(F,h)(\mathbf{j}) = \sum_{k\in K} h^k \left(h^2\partial_t + h\mathbf{c}_i\cdot\nabla + \ldots\right) f_i^{(k)}(t_n,\mathbf{x}_\mathbf{j}) - \\ \sum_{k\in K} h^k\left[\frac{1}{\tau}\left(f_i^{eq,(k)}(F_h(n,\mathbf{j})) - f_i^{(k)}(t_n,\mathbf{x}_\mathbf{j})\right) - g_i^{(k)}(t_n,\mathbf{x}_\mathbf{j})\right] + \ldots, \tag{36}$$

where $f^{eq,(k)}$ groups all the terms of order h^k of the equilibrium function. The dots at the end of the equation allude to a remainder, which contains all the terms of the expansion with indices $k\notin K$. We do not deal with this technical problem, assuming that the remainder will be asymptotically ignorable.

Remark (Definition of coefficients). The idea of the analysis is to define iteratively the coefficients $f_i^{(k)}$, for $k = 0,1,\ldots$ in such a way that every newly introduced order improves the precision of the truncated expansion.
In view of (36), the residue can be expressed as a power series in h. Hence, the precision of the prediction can be improved choosing the coefficients which cancel higher and higher orders in h. In other words, we will try to construct, order by order, the coefficients $f^{(k)}$ such that

$$r^{n+1,(k)}(F,h) = 0. \tag{37}$$

This will result in a series of PDEs, whose solution will define our prediction F_h.

Equilibrium of order k To derive an explicit expression of $f^{eq,(k)}$ we consider the equilibrium as a function of the moments of the particle distribution (equation (17)):

$$f_i^{eq}(F_h) = H_i^{eq}\left(\rho(F_h),\mathbf{u}(F_h)\right),$$

where ρ and $\mathbf{u}$ are defined as (equation (14))

$$\rho(F_h) = \sum_i F_{hi}, \quad \mathbf{u}(F_h) = \sum_i F_{hi}\mathbf{c}_i. \tag{38}$$

Considering F_h as a power series (31), analogous expansions are inherited from the moments:

$$\rho(F_h) = \sum_{k\in K} h^k\rho^{(k)}(F_h), \quad \mathbf{u}(F_h) = \sum_{k\in K} h^k\mathbf{u}^{(k)}(F_h), \tag{39}$$

where the coefficients of order k (h-independent) are obtained taking the moments of the coefficients of F of the corresponding order:

$$\rho^{(k)}(F_h) = \sum_i f_i^{(k)}, \quad \mathbf{u}^{(k)}(F_h) = \sum_i f_i^{(k)}\mathbf{c}_i. \tag{40}$$

Introducing the expression (18) for H_i^{eq}, the *equilibrium of order* k is defined isolating (splitting H_i^{eq} as in equation (20))

$$f_i^{eq,(k)}(F_h) = H_i^{L(eq)}(\rho^{(k)}(F_h),\mathbf{u}^{(k)}(F_h)) + \sum_{m+l=k} H_i^{Q(eq)}(\mathbf{u}^{(m)}(F_h),\mathbf{u}^{(l)}(F_h)). \tag{41}$$

Observe that $f_i^{eq,(k)}$ depends only on the coefficients of the prediction F, through equations (40).
In the leading orders we have

$$f^{eq,(0)} = f_i^* \rho^{(0)} + c_s^{-2} f_i^* \mathbf{c}_i \cdot \mathbf{u}^{(0)} + H_i^{Q(eq)}(\mathbf{u}^{(0)}, \mathbf{u}^{(0)}), \tag{42}$$

$$f^{eq,(1)} = f_i^* \rho^{(1)} + c_s^{-2} f_i^* \mathbf{c}_i \cdot \mathbf{u}^{(1)} + 2H_i^{Q(eq)}(\mathbf{u}^{(0)}, \mathbf{u}^{(1)}), \tag{43}$$

$$f^{eq,(2)} = f_i^* \rho^{(2)} + c_s^{-2} f_i^* \mathbf{c}_i \cdot \mathbf{u}^{(2)} + H_i^{Q(eq)}(\mathbf{u}^{(1)}, \mathbf{u}^{(1)}) + 2H_i^{Q(eq)}(\mathbf{u}^{(0)}, \mathbf{u}^{(2)}). \tag{44}$$

3.3. Hydrodynamic equations

Without describing the computation in detail (see [19] for general results), we give explicitely only the equations for the leading order moments.
In what follows we define the *pressure coefficients*

$$p^{(k)}(F) = c_s^2 \rho^{(k)}(F). \tag{45}$$

Additionally, we fix the relaxation time in order to have $\nu = c_s^2(\tau - \frac{1}{2})$ (ν being the viscosity in the original Navier-Stokes problem).

$$\begin{aligned}
& F \in \mathcal{A} \mid \forall t > 0,\ \forall \mathbf{x} \in \Omega : \\
& f_i^{(0)}(t,\mathbf{x}) = f_i^*, \\
& f_i^{(1)}(t,\mathbf{x}) = f_i^* c_s^{-2} \mathbf{c}_i \cdot \mathbf{u}(t,\mathbf{x}), \\
& f_i^{(2)}(t,\mathbf{x}) = f_i^* c_s^{-2} p(t,\mathbf{x}) + f_i^* c_s^{-2} \mathbf{c}_i \cdot \mathbf{v}^{(2)}(t,\mathbf{x}) + H_i^{Q(eq)}(\mathbf{u},\mathbf{u})(t,\mathbf{x}) \\
& \qquad\qquad - \tau f_i^* c_s^{-2} (\mathbf{c}_i \cdot \nabla) \mathbf{c}_i \cdot \mathbf{u}(t,\mathbf{x}), \\
& f_i^{(3)}(t,\mathbf{x}) = f_i^* c_s^{-2} q^{(3)}(t,\mathbf{x}) + f_i^* c_s^{-2} \mathbf{c}_i \cdot \mathbf{v}^{(3)}(t,\mathbf{x}) \\
& \qquad\qquad - \tau \left[(\mathbf{c}_i \cdot \nabla) f_i^{(2)}(t,\mathbf{x}) + \left(\frac{(\mathbf{c}_i \cdot \nabla)^2}{2} + \partial_t \right) f_i^{(1)}(t,\mathbf{x}) - g_i^{(3)}(t,\mathbf{x}) \right] \\
& f_i^{(k)}(t,\mathbf{x}) = 0, \quad \text{for } k \geq 4,
\end{aligned} \tag{46}$$

where the functions $\mathbf{u}$, p are a smooth solution to the Navier-Stokes problem

$$\begin{cases} \nabla \cdot \mathbf{u} = 0 \\ \partial_t \mathbf{u} + \nabla p + \mathbf{u} \cdot \nabla \mathbf{u} = \nu \nabla^2 \mathbf{u} + \mathbf{G}, \end{cases} \tag{47}$$

the couple $(\mathbf{v}^{(2)}, q^{(3)})$ solves the system

$$\begin{cases} \nabla \cdot \mathbf{v}^{(2)} = 0 \\ \partial_t \mathbf{v}^{(2)} + \nabla q^{(3)} - \nu \nabla^2 \mathbf{v}^{(2)} = 0, \end{cases} \tag{48}$$

and $(\mathbf{v}^{(3)}, q^{(4)})$ satisfies

$$\begin{cases} \nabla \cdot \mathbf{v}^{(3)} = -c_s^{-2} \partial_t p + \dfrac{1}{2} \nabla \cdot \mathbf{G} \\ \partial_t \mathbf{v}^{(3)} + \nabla q^{(4)} + 2\nabla \cdot \left(\mathbf{u} \otimes \mathbf{v}^{(3)} \right) = \nu \nabla^2 \mathbf{v}^{(3)} \end{cases}. \tag{49}$$

Remark (Initial data). The lack of initial data in the previous PDE systems is due to the fact that we are not considering the initial conditions for the algorithm.

3.3.1. *oundary conditions*

The prediction coefficients obtained in (46) are valid for the inner (fluid) nodes. For the boundary nodes, inserting the inner prediction into the boundary algorithm (28), we obtain additional conditions, i.e. boundary conditions for the systems (47),(48), (49). In particular, the BFL rule (28) yields the relation

$$\mathbf{u}(t,\mathbf{x}) = \mathbf{u}_B(t,\mathbf{x}) \qquad \mathbf{x} \in \partial\Omega_F(t). \tag{50}$$

Remark (Bounce-back rule). Note that, using a standard bounce-back algorithm, the definition of the prediction coefficients as in (46) is inconsistent with the assumed smoothness. In particular, this inconsistency indicates the formation of boundary layers which can be observed in practice. See e.g. [21] for details.

3.3.2. *Accuracy results*

The idea is to use the truncated expansion as a *prediction* (which is meant now, as a function which *can predict*) of the lattice Boltzmann *solution* $\hat{f}$.
Let us assume that the numerical solution is stable in time. Since suitable moments of $\hat{F}_i$ yield the Navier-Stokes solution $\mathbf{u}_{NS}$, p_{NS}, we conclude that the corresponding moments of $\hat{f}_i$ approximate these fields:

$$\frac{1}{h}\hat{\mathbf{u}} = \frac{1}{h}\sum_i \hat{f}_i \mathbf{c}_i = \mathbf{u}_{NS} + O(h^2) \tag{51}$$

$$\hat{p} := c_s^2 \frac{\sum_i \hat{f}_i - 1}{h^2} = p_{NS} + O(h). \tag{52}$$

Since we have stopped at the second order coefficients, the uncertainty on the third order coefficient appears on the remainders of equations (51)-(52).
As additional result, we can compute

$$\hat{\mathbf{S}}[\mathbf{u}] = -\frac{1}{\tau c_s^2 h^2}\sum_i \mathbf{c}_i \otimes \mathbf{c}_i \left(\hat{f}_i - f_i^{eq}(\hat{f})\right) \tag{53}$$

as a first order approximation of the tensor $\mathbf{S}[\mathbf{u}] \equiv \nabla\mathbf{u} + \nabla\mathbf{u}^T$.

4. NODE INITIALIZATIONS IN MOVING BOUNDARY PROBLEMS

The LBM defined in algorithm 1 is based on a fixed grid. Dealing with *moving boundaries*, the standard scheme has to be completed with an additional routines to initialize the variables on the nodes created by the variations of the computational domain.
Recently, LB algorithms to deal with moving boundaries have been proposed and numerically tested, pointing out the main difficulties of the task [5, 6, 16, 35]. In this section, we will investigate the problem and possible solutions using the asymptotic expansion approach.

4.1. Refill step

Let us consider the fluid problem (1), assuming that the solid domain $\Omega_S(t)$ moves with a given rigid body velocity field. Let $\mathbf{k} \in \mathcal{G}$ be a node such that $\mathbf{x_k} \in \Omega_S(t_n)$ and $\mathbf{x_k} \in \Omega_F(t_{n+1}) \cup \Gamma(t_{n+1})$, i.e., $\mathbf{x_k}$ enters the fluid domain at time t_{n+1}. We need to refill the new fluid point, defining the distributions $\hat{f}_i(n+1,\mathbf{k})$. We will call the process dealing with *initialization* of a new fluid node a **refill step**.
We want to approach the problem from the point of view of asymptotic analysis. In particular, we aim at a refill algorithm which can approximate the asymptotic prediction up to the order h^2, in such a way that the inner expansion, and therefore the approximation property for $\mathbf{u}$ and p, holds also for the moving boundary algorithm.

4.2. Benchmark: Disk in incompressible Stokes flow

We construct an analytical solution to (1) (restricting to a Stokes problem), considering $\Omega = [0,L] \times [0,L]$, and taking Ω_S as a circle of radius R. Let us assume that the circle moves with constant horizontal speed U. We defined the velocity field (see Fig. **2**) [15]

$$\begin{aligned} u(t,x,y) &= U\left(\log\left(\xi(x,t)^2+\eta(y)^2\right)+\frac{2\eta(y)^2}{\xi(t,x)^2+\eta(y)^2}+\frac{\xi(t,x)^2-\eta(y)^2}{\left(\xi(t,x)^2+\eta(y)^2\right)^2}\right) \\ v(t,x,y) &= -U\frac{2\xi(t,x)\eta(y)}{\xi(t,x)^2+\eta(y)^2}\left(1-\frac{1}{\xi(x,t)^2+\eta(y)^2}\right) \end{aligned} \tag{54}$$

and the pressure

$$p(t,x,y) = -\frac{4U\nu}{R}\frac{\xi(t,x)}{\xi(t,x)^2+\eta(y)^2}, \tag{55}$$

where

$$\xi(t,x) = \frac{x-(x_C^0+Ut)}{R}, \quad \eta(y) = \frac{y-y_C^0}{R} \tag{56}$$

are reference coordinates with respect to the center of the disk.

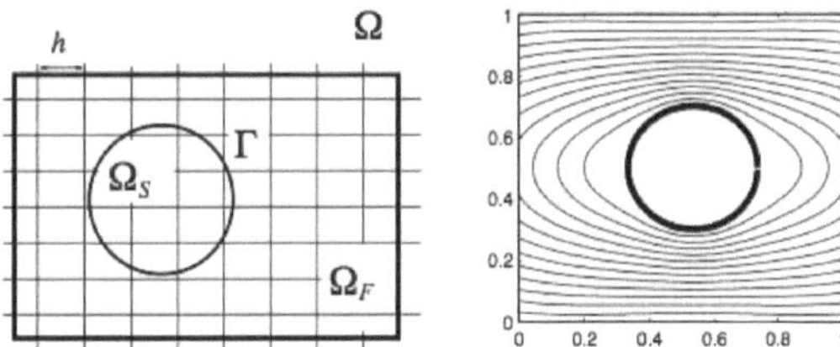

Fig. 2: **Left.** Moving boundary model problem: sketch of the fluid and solid domains, discretized by a regular cartesian grid. **Right.** Disk in flow benchmark. Streamlines of the velocity field (54). In the numerical simulation, we use a circle of radius $R = 0.6$ in a square box of edge length $L = 3$. Viscosity is $\nu = 0.03$.

Including the volume force

$$\mathbf{G}(t,x,y) = \partial_t \mathbf{u}(t,x,y) \tag{57}$$

and assigning the fields (54) as Dirichlet boundary conditions on the edge of Ω, u, v and p as defined in (54)-(55) solve the Stokes problem in Ω, where the solid disk $\Omega_S(t)$ moves along x with velocity U. We refer to this test problem as *Disk-in-Flow* (DiF).
Moreover, by adding the constant horizontal speed $-U$ to (54), we define a *fixed boundary* variant of the problem, called DiF^0, whose exact horizontal velocity $u-U$ is now zero along the disk boundary, while v and p are the same as in (54)-(55).

4.3. Equilibrium-non Equilibrium (EnE) Refill

Equilibrium refill Let us begin analyzing a simple approach [35], which initializes the populations on the new fluid nodes using the equilibrium distribution for approximate density and velocity.
Let $\mathbf{k} \in \mathcal{G}(h)$ the node to be refilled at time step $n+1$. We compute first $\tilde{\rho}_{n+1,\mathbf{k}}$ and $\tilde{\mathbf{u}}_{n+1,\mathbf{k}}$, approximations of density and velocity at $\mathbf{k}$, defining then the *equilibrium refill*:

$$\hat{f}_i(n+1,\mathbf{k}) = H_i^{eq}(\tilde{\rho}_{n+1,\mathbf{k}}, \tilde{\mathbf{u}}_{n+1,\mathbf{k}}). \tag{58}$$

The particular choice of the extrapolation rule for $\tilde{\rho}$ and $\tilde{\mathbf{u}}$ might depend on the considered flow and motion of the boundary. For example, we can use a backward extrapolation according to the boundary velocity $\mathbf{u}_B$ at a point of interface close to the new node [35].

To analyze the algorithm, we insert the inner expansion (46) into (58), used at a new node **k**. We find that the coefficient $f^{(2)}$ cannot be defined consistently for the whole domain. In particular using the equilibrium refill, the *non-equilibrium part* (in the second order)

$$f_i^{neq,(2)} = -\tau f_i^* c_s^{-2} \mathbf{c}_i \cdot \nabla \mathbf{u} \cdot \mathbf{c}_i \tag{59}$$

will be missing at the new refilled point. As a consequence, the prediction corresponding to (46) and the related accuracy results are no longer justified nor assured.

Non-equilibrium correction According to the analysis, all we need to keep the scheme as accurate as the interior LBM is a refill step able to reconstruct *equilibrium and non equilibrium* with enough accuracy. In practice, we complete the initialization of the populations (58) including an approximation of the non equilibrium part by a first order extrapolation, i.e. simply *copying* the non equilibrium part from a neighbor of the new fluid node. For completeness, it must be mentioned that a similar idea was used in [17], to implement the Dirichlet boundary condition.

Remark (Accuracy of the extrapolation). Due to the first order accuracy in pressure of the LBM, using a higher order extrapolation only removes some error terms while other terms at the same order remain. So the error may be smaller but not of higher order.

In practice, we propose the following algorithm to refill the node **k**.

Algorithm 2 (EQ+non EQ (EnE) Refill).

1 choose a (flow depending) extrapolation direction $\mathbf{c}_i^{ex}$ (not outgoing at **k**)

2 compute approximations $\tilde{\rho}_{n+1,\mathbf{k}}$, $\tilde{\mathbf{u}}_{n+1,\mathbf{k}}$

3 *equilibrium* + *non equilibrium* refill:

$$\hat{f}_j(n+1,\mathbf{k}) = H_j^{eq}(\tilde{\rho}_{n+1,\mathbf{k}}, \tilde{\mathbf{u}}_{n+1,\mathbf{k}}) + f_j^{neq}(n+1,\mathbf{k}+\mathbf{c}_i^{ex}), \quad j = 1,\ldots,b. \tag{60}$$

Asymptotic Analysis By construction, the new populations contain a second order correct equilibrium part, and a first order correct non-equilibrium part (approximation from a neighboring node). In conclusion, with an algorithm which requires little additional work, compared to the EQ-refill, we achieve the same accuracy in pressure and velocity as the interior LBM.

4.4. Numerical tests

The equilibrium refill is now applied to problem DiF. Figure **3** shows the maximum error in pressure over $\Omega_F(t)$ versus time. The peaks appearing in correspondence to the refill steps do not decrease on the finer grid. In the order plot in fig. **5** (left) the errors for different grid sizes are compared. We observe a first order accuracy for the velocity but an inconsistent pressure (the error does not decrease using a finer discretization).

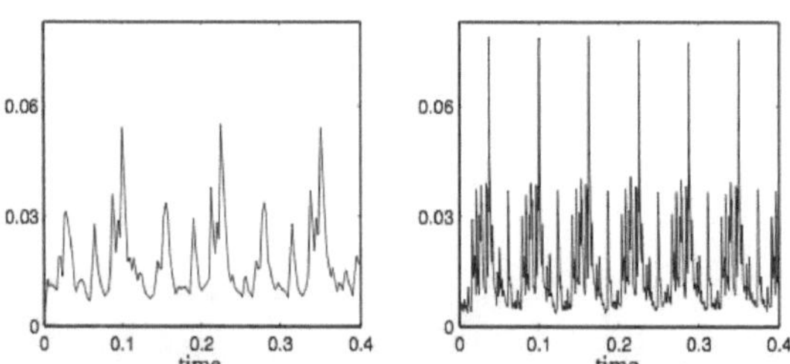

Fig. **3**: DiF Benchmark (with moving boundary). Error in pressure (maximum over $\Omega_F(t)$) using the algorithm (58) to initialize the new nodes. Grid sizes $h = 0.05$ (left) and $h = 0.025$ (right) are shown.

Simulating the problem DiF, we observe (fig. **4** and fig.**5**, right) improvements in accuracy, which is now comparable with the accuracy of the standard LBM.

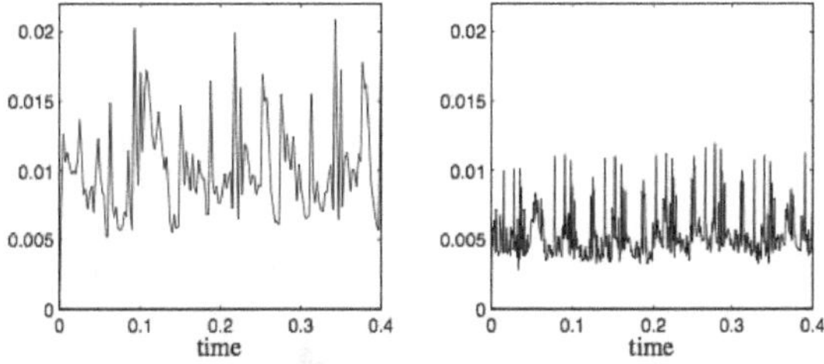

Fig. **4**: Results for the benchmark DiF employing the non-equilibrium corrected refill. Maximum error in pressure over $\Omega_F(t)$, $h = 0.05$ (left) and $h = 0.025$ (right). Unlike in the previous test case, now the results improve on the finer grid.

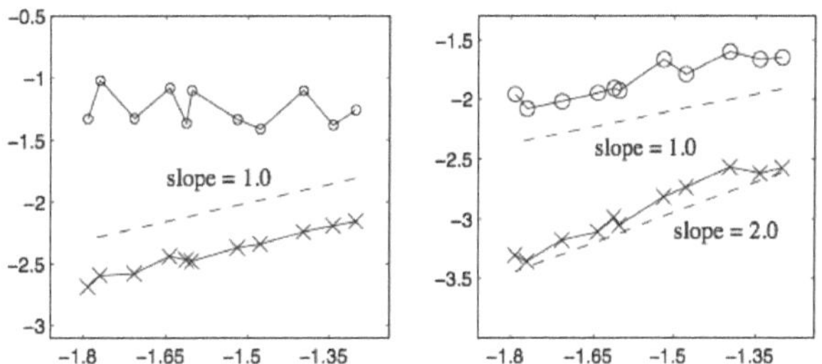

Fig. **5**: DiF Benchmark. Double logarithmic plot of maximum errors (in space and time) in pressure ($\circ$) and velocity ($\times$), versus grid size. The dashed curve shows a reference line of slope one. **Left.** For the *equilibrium refill* (left), we get a first order accurate velocity and an inconsistent pressure, while the EQ+nonEQ refill(right) allows to recover the original accuracy.

5. FORCE EVALUATIONS

In this section, we complete the discussion of the coupled fluid-solid problem (1)-(2) investigating additional routines for the force evaluation in the lattice Boltzmann framework. In particular, we aim at an efficient routine to evaluate fluid force and torque acting on the solid object:

$$\mathbf{F}_S = \int_\Gamma (-p\mathbf{I}+\mathbf{S})\cdot\mathbf{n}\,d\gamma, \qquad \mathbf{T}_S = \int_\Gamma \mathbf{r}(\gamma)\wedge[(-p\mathbf{I}+\mathbf{S})\cdot\mathbf{n}]\,d\gamma \tag{61}$$

(where $\mathbf{r}(\gamma)$ denotes the vector connecting a point on the interface with the rigid body center of mass). Since the best characteristics of the LBM lie in its *efficiency*, these additional algorithms should be able to preserve this property, in order to keep the LBM numerically competitive. After describing general extrapolation approaches, we will consider in detail the *Momentum Exchange algorithm* (originally proposed in [30]), which models the fluid-boundary interaction based on simple particle dynamics, and which requires only a low additional computational effort. We will discuss an *asymptotic analysis* of the algorithm, tackling in general the *consistency* and the *accuracy* of the method.

5.1. Benchmark

In order to set up a lattice Boltzmann method including the boundary force evaluation, we consider a variant of the *disk-in-channel* problem defined in section 3.3.2.. Let $\Omega = [0,L]\times[0,1]$, let $\Omega_S(t)$ be a disk with radius R and let $\Omega_F(t) = \Omega\backslash\Omega_S(t)$.
The fluid obeys the incompressible NS equations (1), while the disk moves with translational velocity $\mathbf{u}^D_{CM}$ and rotational velocity ω^D defined by

$$\mathbf{u}(t,\mathbf{x}) = \mathbf{u}^D_{CM}(t) + \omega^D(t)\times(\mathbf{x}_{CM}(t)-\mathbf{x}), \quad t>0,\ \mathbf{x}\in\Gamma(t), \tag{62}$$

For simplicity, to analyze the numerical scheme the motion of the disk is assumed to be known.
We will consider periodic boundary conditions as well as a given motion in a channel, including

$$\begin{aligned} \mathbf{u}(t,x,1) &= \mathbf{u}_N(t),\ t>0 \\ \mathbf{u}(t,x,0) &= \mathbf{u}_S(t),\ t>0 \end{aligned} \tag{63}$$

Physically, this situation models a cross section of a flow around a periodic array of long cylinders. To test the algorithm, we use two simple exact solutions of (1) around a disk of radius $R = 0.2$ in the unit square $\Omega = [0,1)^2$. We denote these benchmarks as DiF_0 and DiF_1.
In the case of DiF_0, we move the disk with constant velocity $\mathbf{u}_0$ in a flow with the same constant velocity $\mathbf{u}(t,\mathbf{x}) = \mathbf{u}_0$, zero pressure, and vanishing body force $\mathbf{G} = 0$. As a consequence, the local stresses vanish and the total boundary force is zero.
In the case DiF_1, we again move the body with a constant velocity $\mathbf{u}_0$ in the constant flow field $\mathbf{u}(t,\mathbf{x}) = \mathbf{u}_0$. However, to obtain a non-trivial local force, we choose a periodic function p_0 and define the body force $\mathbf{G} = \nabla p_0$ which generates a pressure $p(t,\mathbf{x}) = p_0(\mathbf{x})$. For the particular choice

$$p_0(x,y) = \sin(2\pi x)\cos(2\pi y) \tag{64}$$

we obtain

$$\begin{aligned} f_{\mathbf{t}}(t,\gamma) &= 0, \\ f_{\mathbf{n}}(t,\gamma) &= \underbrace{-\sin(2\pi(x_C(t)+R\cos\gamma))\cos(2\pi(y_C(t)+R\sin\gamma))}_{-p_0(x(t,\gamma),y(t,\gamma))}, \end{aligned} \tag{65}$$

where $\gamma\in[0,2\pi]$ denotes a one dimensional parametrization of the interface.
In the future, to quickly identify the parameters of a benchmark of type DiF discussed in this section, we will use a notation as in Fig. **6**.

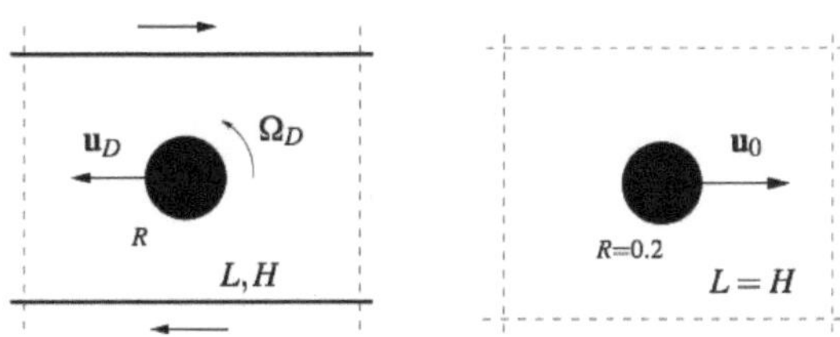

Fig. **6**: Benchmark CiF. Dashed lines represent periodic boundaries, bold lines are solid walls. On the right, the sketch for CiF_0 and CiF_1 is depicted.

5.2. Extrapolation approaches

Before describing in detail the Momentum Exchange algorithm, we present a short overview of other possible approaches to evaluate the interaction (61). They will serve for future comparison.
In general, the force (3), as well as the local interaction (5), could be evaluated using a quadrature on the interface:

$$\hat{\mathbf{F}}_S = \sum_{r=1}^{G} w_i(-p(\gamma_i)\mathbf{n}(\gamma_i) + \mathbf{S}(\gamma_i)\cdot\mathbf{n}(\gamma_i))(\gamma_{i+1} - \gamma_i). \tag{66}$$

The previous formula is based on a partition

$$P_\Gamma = \{\gamma_1, \ldots, \gamma_G\} \subset [0, 2\pi),$$

of the interface, and uses the values of p and $\mathbf{S}$ on those points. Unfortunately, approximations of pressure and stress tensor, $\hat{p}$ and $\hat{\mathbf{S}}$, are only available at the lattice nodes. To have an operational formula within a LBM, (66) has to be combined with additional routines, to *extrapolate* the values on the interface.
Several choices are available at this stage. Without too many details, we describe some examples, particularly fitting in the LB-framework.

5.2.1. *Boundary Node approximations*

Let us recall the definition of the set of *boundary couples* at a certain time step n:

$$\mathcal{K}^n := \left\{(\mathbf{k}, i) \in \mathcal{G}(h) \times \{1, \ldots, b\} \mid \mathbf{x_k} \in \overline{\Omega}_F(t_n),\ \mathbf{x}_{\mathbf{k}+\mathbf{c}_i} \in \Omega_S(t_n)\right\}. \tag{67}$$

Each boundary couple $(\mathbf{k}, i) \in \mathcal{K}^n$ defines a point $\mathbf{b}_i(\mathbf{k}) \in \Gamma$, when the link crosses the interface.
Hence, we can equivalently construct the set of the intersections grid-boundary at time t_n

$$\mathcal{N}^n(\Gamma) = \overline{\mathcal{G}(h)} \cap \Gamma(t_n) = \{\mathbf{b}_i(\mathbf{k}) \mid (\mathbf{k}, i) \in \mathcal{K}^n\} \tag{68}$$

Identifying each point of Γ with its coordinate in $[0, 2\pi)$, the set $\mathcal{N}^n$ can be ordered

$$\mathcal{N}^n(\Gamma) = \{\mathbf{b}(\gamma_r) = \mathbf{b}^n_{i_r}(\mathbf{k}_r), \mid r = 1, \ldots, |\mathcal{N}^n|\}$$

and used as discretization of the interface Γ.
The simplest approximation for pressure and stress tensor, is done using the values at the corresponding **boundary node**:

$$\tilde{p}_{BN}(\mathbf{b}_i(\mathbf{k})) = \hat{p}(n, \mathbf{k}) \tag{69}$$

(and similarly for the components of $\mathbf{S}$).

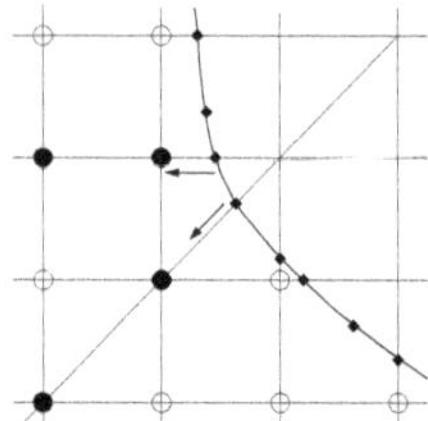

Fig. 7: Lattice link extrapolation. Using the partition defined by the intersections grid-interface, the pressure on the boundary point (⋄) can be extrapolated with the nodes (•) encountered following, $\mathbf{c}_i$-backward, the intersecting link.

5.2.2. *Lattice-Link extrapolation*

Otherwise, with the same partition $\mathcal{N}^n$, approximated value can be obtained via a linear (or quadratic) **extrapolation following the link $\mathbf{c}_i$** (see Fig. 7):

$$\tilde{p}_{LL}(\mathbf{b}_i(\mathbf{k})) = (1+q)\hat{p}(n,\mathbf{k}) - q\hat{p}(n,\mathbf{k}-\mathbf{c}_i) \tag{70}$$

being $q = h^{-1}\|\mathbf{x_k} - \mathbf{b}_i(\mathbf{k})\|$ (in lattice units). A disadvantage of these approaches, is that the partition is unstructured. Of course, simplified variants are possible, which only consider intersection with horizontal, or vertical, or diagonal links.

5.2.3. *Bilinear extrapolation*

Differently, we can define a LB-independent discretization $P_\Gamma = \{\mathbf{b}(\gamma_r)\} \subset \Gamma$, and take the pressure and stress tensor at the **closest lattice node**. Calling

$$\mathbf{k}^{\gamma_r} := \underset{\mathbf{j}\in I_F(n)}{\operatorname{argmin}}\{\|\mathbf{x_j} - \mathbf{b}(\gamma_r)\|\}, \tag{71}$$

we can use the approximation

$$\tilde{p}(\mathbf{b}(\gamma_r)) = \hat{p}(n,\mathbf{k}^{\gamma_r}). \tag{72}$$

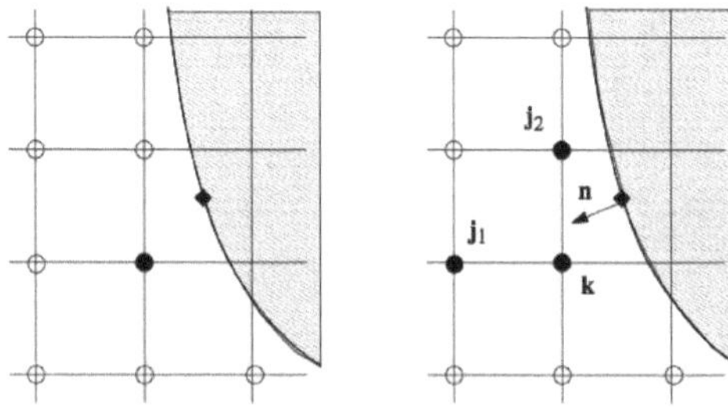

Fig. 8: Closest node extrapolation. Taking the values at the closest lattice nodes, approximations of p and $\mathbf{S}$ on a boundary point (⋄) can be constructed. **Left:** A simple way, is to take the value on the closest node (•). **Right:** More complicated, is a bilinear extrapolation, based on three not aligned nodes (•).

A bit more complicated is a **three-points bilinear** extrapolation (an example is drawn in Fig. 8), obtained by searching, for any γ in the partition, three nodes

$$\mathbf{k}^\gamma \in \mathcal{G}(h) : \|\mathbf{x}_{\mathbf{k}^\gamma} - \mathbf{b}(\gamma)\| = \min_{\mathbf{j}\in I_F(n)}\{\|\mathbf{x_j} - \mathbf{b}(\gamma)\|\},$$

$$\mathbf{j}^1,\mathbf{j}^2 : \|\mathbf{k}^\gamma - \mathbf{j}^r\| \text{ is minimal}, \quad \mathbf{k}^\gamma - \mathbf{j}^1,\ \mathbf{k}^\gamma - \mathbf{j}^2 \text{ linearly independent}$$

and by approximating bilinearly

$$\tilde{p}_{BIL}(\mathbf{b}(\gamma)) = \frac{\hat{p}(n,\mathbf{j}^1)-\hat{p}(n,\mathbf{k}^\gamma)}{\mathbf{j}_x^1-\mathbf{k}_x^\gamma}(x(\gamma)-\mathbf{k}_x^\gamma)+\frac{\hat{p}(n,\mathbf{j}^2)-\hat{p}(n,\mathbf{k}^\gamma)}{\mathbf{j}_y^2-\mathbf{k}_y^\gamma}(y(\gamma)-\mathbf{k}_y^\gamma)+\hat{p}(n,\mathbf{k}^\gamma) \tag{73}$$

(assuming $\mathbf{j}_y^1=\mathbf{k}_y^\gamma$, $\mathbf{j}_x^2=\mathbf{k}_x^\gamma$, as in figure **8**).

5.2.4. *LLFQ algorithm*

The LLFQ algorithm (from the authors *H.Li, X.Lu, H.Fang, Y.Qian.*) has been proposed and tested in [36]. The partition P_Γ is pre-defined on the interface, and for each $\mathbf{b}(\gamma_r)\in P_\Gamma$, we proceed as follows. According to the notation in Fig. **9**, we consider the lattice node $\mathbf{D}_r$, close to $\mathbf{b}(\gamma_r)$, the point $\mathbf{C}_r$, intersection between the grid and the line $\mathbf{bD}_r$ and the nodes $\mathbf{A}_r$ and $\mathbf{B}_r$, located in opposite sides of $\mathbf{C}_r$. From the populations $\hat{f}_i(\mathbf{A}_r)$, $\hat{f}_i(\mathbf{B}_r)$, approximations for $\hat{f}_i(\mathbf{C}_r)$ are constructed:

$$\tilde{f}_i(\mathbf{C}_r)=\hat{f}_i(\mathbf{A}_r)\overline{\mathbf{B}_r\mathbf{C}_r}+\hat{f}_i(\mathbf{B}_r)\overline{\mathbf{A}_r\mathbf{C}_r},$$

then, using linear extrapolation,

$$\tilde{f}_i^{LLFQ}(\mathbf{b}(\gamma_r))=\frac{1}{\overline{\mathbf{D}_r\mathbf{C}_r}}\left(\hat{f}_i(\mathbf{D}_r)\overline{\mathbf{b}_r\mathbf{C}_r}-\tilde{f}_i(\mathbf{C}_r)\overline{\mathbf{b}_r\mathbf{D}_r}\right) \tag{74}$$

is defined. From the resulting distributions, $\tilde{p}$ and $\tilde{\mathbf{S}}$ are extracted.

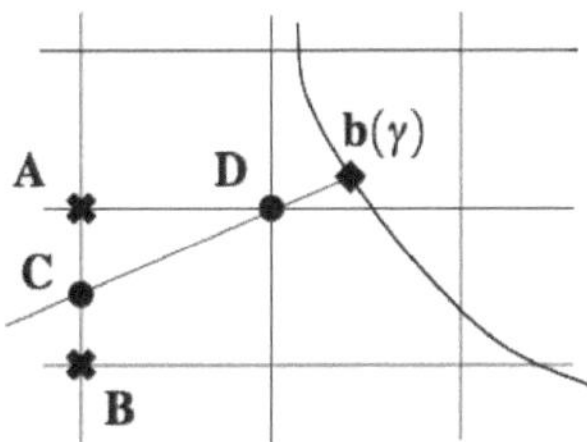

Fig. **9**: The approximation in [36] is constructed by extrapolating the LB population on a pre-selected boundary point $\mathbf{b}(\gamma)$ (diamond) from a close node **D** and a point **C** (black circles). In turn, the populations on **C** have to be interpolated from two nodes **A** and **B** (crosses). Furthermore, an average over the possible choices of **D** is considered.

Many choices of the node D_r are possible. In fact, an arithmetic average between more points is performed. For more details, see [36].

5.2.5. *Numerical tests*

As a preliminary test for the defined algorithms, we perform the computation of the stresses in the problem CiF_1 defined in section 5.1.. Only normal stresses (due to the pressure) are present. Figure **10** shows a comparison between the different approaches. In this simple case, all of them produce similar and regular results.

About the accuracy, assuming to get a first order accurate pressure running the LBM, we cannot expect to go beyond that precision. It is experimentally confirmed in the order plot in Fig. **11**.

The LB-grid defined discretization, consisting of a bigger amount of nodes, shows noisy and irregular behaviors. More smooth are the bilinear and LLFQ approximations.

Remark (Complexity of extrapolation). In a general situation, the implementation of the previous method can be awkward, especially dealing with *irregular boundaries* and *complex geometries*. In fact, the nodes containing the information used in equations (69)-(74) have to be searched, case by case, around the boundary point. The resulting routines reduce the efficiency of the LBM. The property of **locality** of operations is lost.

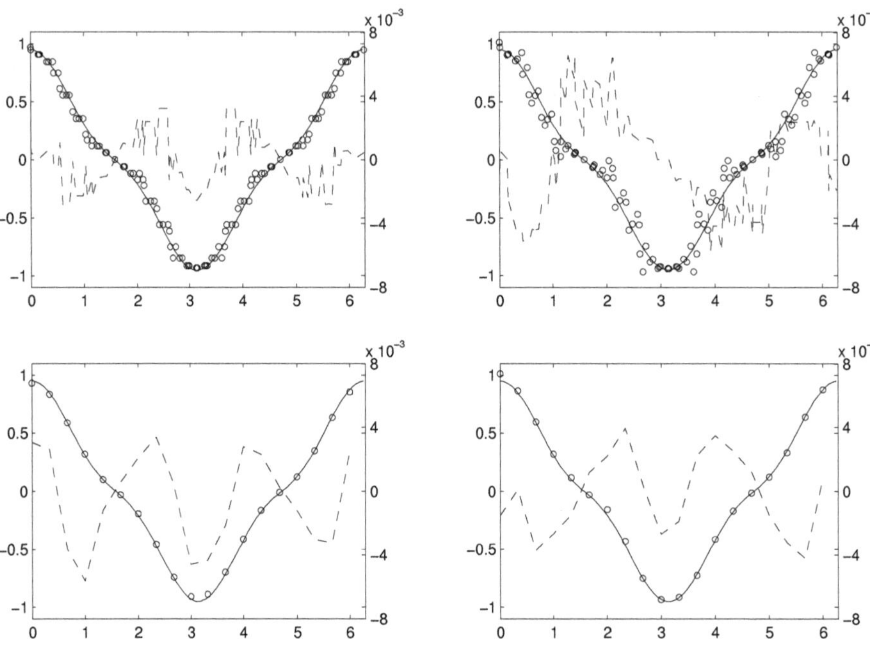

Fig. **10**: The presented extrapolation methods are compared, applied to the problem CiF_1. We sampled with small circles (ordinate on the left y-axis) the pressure, while the exact solution (64) is superimposed as a bold line. The dashed line (right y-axis) is the error in the result for the stress $\mathbf{S}_{xx}$ (exact solution $\mathbf{S}_{xx} = 0$). **Top-left:** *boundary node* approximation. **Top-right:** *lattice-link* extrapolation. **Bottom-left:** *bilinear* approximation. **Bottom-right:** LLFQ algorithm.

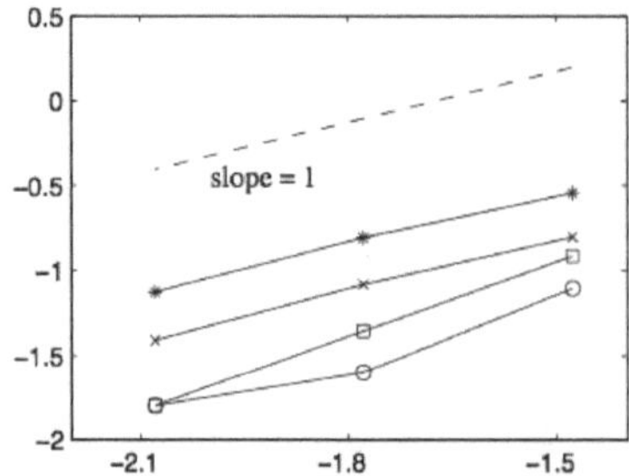

Fig. **11**: Investigation of the pressure accuracy order (double logarithmic plot), for the approximation methods (pressure) described above: *boundary node* ($\times$), *bilinear* ($*$), *lattice-link* ($\circ$), LLFQ ($\square$). The dashed reference lines indicates first order accuracy.

Concerning the accuracy, it has to be remarked that the LBM provides only *low order* pressure and stress tensor (first order in h). Therefore, for general problems, increasing the complexity and the order of the approximations (for example going from linear to quadratic, etc.), does not assure any gain in precision.
In view of these considerations, a good solution consists of a rather simple scheme, well-fitting in the lattice framework, which does not require too much exchange of information and, as a consequence, is not supposed to produce high accuracy results.

5.3. Momentum exchange algorithm

We investigate the features of the *Momentum Exchange Algorithm* (MEA), proposed in its original form by A. Ladd in [30], which allows to evaluate the interaction between fluid and boundary using directly the variables of LBM. For easiness of notation, in what follows we will assume to deal with an interface whose shape is constant in time. Thus, the dependence on the time step will be omitted from the discrete sets describing the lattice-interface relationships.

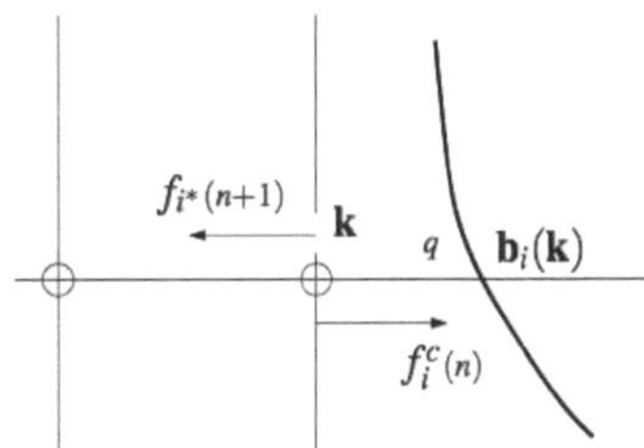

Fig. **12**: Applying the boundary rule on a boundary couple $(\mathbf{k}, i)$, the momentum exchanged in the point $\mathbf{b}_i(\mathbf{k})$ is defined as the difference between the incoming $f_{i^*}(n+1, \mathbf{k})$ and the population *after collision* $f_i^c(n, \mathbf{k})$.

The idea is to consider the momentum transferred to the solid from each *boundary fluid node* $\mathbf{x_k}$ interacting with the boundary along a link $\mathbf{c}_i$ (see Fig. **12**). The net momentum is given by the sum of momentum due to the particles moving with opposite velocities $\mathbf{c}_i$ and $\mathbf{c}_{i^*}$.

$$\phi_i(n, \mathbf{k}) \equiv \mathbf{c}_i \hat{f}_i^c(n, \mathbf{k}) - \mathbf{c}_{i^*} \hat{f}_{i^*}(n+1, \mathbf{k}) = \mathbf{c}_i \left(\hat{f}_{i^*}(n+1, \mathbf{k}) + \hat{f}_i^c(n, \mathbf{k})\right). \tag{75}$$

(f_i^c is the distribution *after collision*). Denoting with B^n the set of boundary couples (equation (67)) the force is approximated by

$$\hat{\mathbf{F}}(n) = \sum_{(\mathbf{k},i)\in\mathcal{K}} \phi_i(n, \mathbf{k}). \tag{76}$$

In practice, the algorithm can be summarized as

Algorithm 3 (Momentum Exchange Algorithm).
Given the boundary couple set $\mathcal{K} = \{(\mathbf{k}_r, i_r),\ r = 1, \dots |\mathcal{K}|\}$
`initialize` $\hat{\mathbf{F}} = 0$
`DO for` $r = 1, \dots, |\mathcal{K}|$
  `LB-collision:` $\rightarrow \hat{f}_i^c(n, \mathbf{k})$
  `boundary condition:` $\rightarrow \hat{f}_{i^*}(n+1, \mathbf{k})$
  `momentum exchanged:` $\phi_i(n, \mathbf{k}) = \mathbf{c}_i \left(\hat{f}_{i^*}(n+1, \mathbf{k}) + \hat{f}_i^c(n, \mathbf{k})\right)$
  `update:` $\hat{\mathbf{F}} = \hat{\mathbf{F}} + \phi_i(n, \mathbf{k})$
`end`

Numerical tests Algorithm 3 is now tested on the problem CiF_0. In absence of pressure, we compare the results for the local stresses, by evaluating the momentum exchanged *point by point* along the boundary, when the flow and the cylinder are fixed ($\mathbf{u}_0 = 0$), or both moving with the same velocity $\mathbf{u}_0 = (5, 0)$ (Fig. **13**). Despite the trivial exact solution $f_\mathbf{t} = f_\mathbf{n} = 0$, we observe the presence of local forces in relevant orders, *different* in the two cases and *highly irregular*, even if both pressure and velocity are exact in the domain.

Remark (Quadrature properties of MEA). Since the boundary forces are represented by an integral over the interface, our primary interest is to investigate the properties of the Momentum Exchange Algorithm intended as *integration routine*. Thus, in order to decouple the errors coming from the approximation of the fluid fields from the one due to the approximate quadrature along the boundary, we use the simple test cases CiF_0 and CiF_1, where the flow is easily solved correctly. In other words, the fields on the boundary are correctly reproduced by the LBM and the arising errors in the force computation are due exclusively to the integration properties of the MEA[2].

Even in the first simple benchmark, when the boundary stresses vanish, we do not obtain correct results. Also, it has to be observed that the observed problem appears in the leading orders of the results independently on the size of the velocity $\mathbf{u}_0$. In fact, the relatively high value $\mathbf{u}_0 = (5,0)$ (in physical units) has been employed only to amplify the error, in order to make the inconsistency better visible.

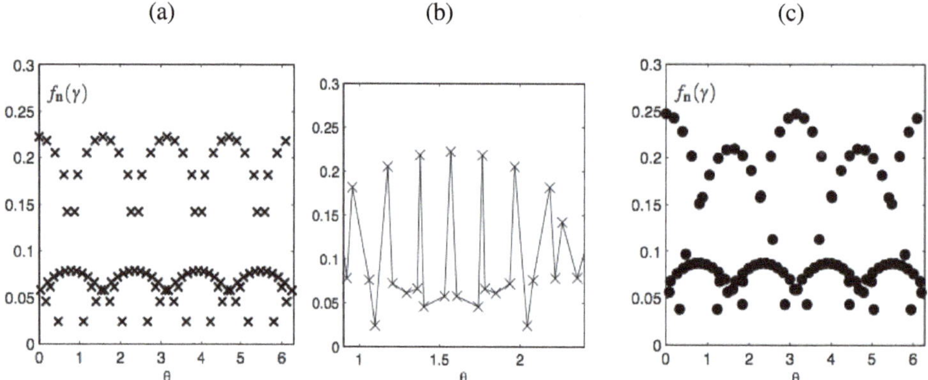

Fig. **13**: Results of the MEA for the normal stress $f_{\mathbf{n}}$, simulating CiF_0 on a 25×25 grid. For each boundary point $\mathbf{b}_i(\mathbf{k})$ the value computed with (75) at the corresponding node $\mathbf{x}_{\mathbf{k}}$ is drawn. The exact solution is $f_{\mathbf{n}} = 0$. **(a)** Flow at rest, $\mathbf{u}_0 = (0,0)$. The results show strong oscillations. Note that consecutive points ($\times$) are not connected by lines for clarity. The discrete data appear to be $\frac{\pi}{2}$-periodic. **(b)** Zoom on a small part of the boundary of amplitude $\frac{\pi}{2}$ around the north pole $\gamma = \frac{\pi}{2}$ now with connecting lines to demonstrate the oscillation. **(c)** Same model as in (a), but with $\mathbf{u}_0 = (5,0)$. The MEA breaks the Galilean invariance (in the relevant order).

5.4. Asymptotic Analysis

To investigate the properties of the algorithm, we consider the approximation of global and local forces as a function of the LB-output:

$$\hat{\mathbf{F}}_h = \sum_{(\mathbf{k},i)\in\mathcal{K}} \phi_i(n,\mathbf{k}) = \mathbf{F}(\hat{f}_h),\ \hat{\mathbf{f}}_h = \sum_{(\mathbf{k},i)\in\mathcal{K}'\subset\mathcal{K}} \phi_i(n,\mathbf{k}) = \mathbf{f}(\hat{f}_h)$$

$\mathcal{K}'$ being a properly chosen subset of $\mathcal{K}$.
In this way, we can expand the results of MEA using the LB prediction

$$F_h^{LBM} = \sum_{k=0}^{2} h^k f^{LBM,(k)},$$

with coefficients (46) (reported below):

[2]It must be excluded also the dependence of the results on the choice between single and multiple relaxation time (MRT) collision operator, as well as on the particular LB-model.

$$
\begin{aligned}
f_i^{LBM,(0)} &= f_i^*, \\
f_i^{LBM,(1)} &= f_i^* c_s^{-2} \mathbf{c}_i \cdot \mathbf{u}, \\
f_i^{LBM,(2)} &= f_i^* c_s^{-2} p + H_i^{Q(eq)}(\mathbf{u},\mathbf{u}) - \tau f_i^* c_s^{-2} (\mathbf{c}_i \cdot \nabla) \mathbf{c}_i \cdot \mathbf{u},
\end{aligned}
$$

where $\mathbf{u}$ and p solve the Navier-Stokes problem (1).

By definition, an asymptotic expansion for the force is the sum over $\mathcal{K}$ of the asymptotic expansions for the momentum exchange $\phi_i(\mathbf{k})$.

At the same time, a prediction for $\phi_i(\mathbf{k})$ is derived inserting (46) into (75), and using the relation (28) for the population updated with the boundary conditions. We have (dropping the time dependence for brevity):

$$
\phi_i(\mathbf{k}) = \phi_i^{(0)}(\mathbf{b}_i(\mathbf{k})) + h^2 \phi_i^{(2)}(\mathbf{b}_i(\mathbf{k})) + O(h^3), \tag{77}
$$

with

$$
\begin{aligned}
\phi_i^{(0)} &= 2 f_i^* \mathbf{c}_i \\
\phi_i^{(2)} &= 2 f_i^* c_s^{-2} \left(p + \frac{c_s^{-2}}{2} \left(|\mathbf{c}_i \cdot \mathbf{u}_B|^2 - c_s^2 \mathbf{u}_B^2 \right) - c_s^{-2} \nu \mathbf{c}_i \cdot \nabla \mathbf{u}_B \cdot \mathbf{c}_i \right) \mathbf{c}_i .
\end{aligned} \tag{78}
$$

All the quantities in equations (77)-(78) are evaluated at the boundary point $\mathbf{b}_i(\mathbf{k})$.

Plane horizontal boundary To better understand equation (78), we look at a simple example, with a horizontal boundary on the top of the fluid flow (Fig. **14** left).

In our convention for the discrete velocities of the model D2Q9 (Fig. **1**), the last row of fluid nodes interacts with the solid along the directions $i = 2,3,4$. Computing explicitly the sum $\Phi(\mathbf{k}) = \phi_2(\mathbf{k}) + \phi_3(\mathbf{k}) + \phi_4(\mathbf{k})$ with equation (78), for a particular boundary node $\mathbf{x_k}$, we have (suppressing the argument $\mathbf{k}$)

$$
\Phi = \frac{1}{3} \begin{pmatrix} 0 \\ 1 \end{pmatrix} + h^2 \begin{pmatrix} -\frac{1}{2}[\mathbf{S}_{xy}(\mathbf{b}_2) + \mathbf{S}_{xy}(\mathbf{b}_4)] \\ \frac{1}{6}[p(\mathbf{b}_2) + 4p(\mathbf{b}_3) + p(\mathbf{b}_4)] - \mathbf{S}_{yy}(\mathbf{b}_3) \end{pmatrix} + \\
+ h^2 \begin{pmatrix} \frac{1}{2}[u_B(\mathbf{b}_2) v_B(\mathbf{b}_2) + u_B(\mathbf{b}_4) v_B(\mathbf{b}_4)] \\ \frac{1}{6}[v_B(\mathbf{b}_2) + 4 v_B(\mathbf{b}_3)^2 + v_B(\mathbf{b}_4)^2] \end{pmatrix} + O(h^3) \tag{79}
$$

The zero order term predicts a surplus of pressure[3], and it is not related to the integral (3). The second order is a combination of quadrature formulas over a small interval on the boundary for the functions p, $\mathbf{S}$, plus a quadratic function of velocity, which breaks the *Galilean invariance* (as happened in the test problem CiF_0, Fig. **13**).

5.4.1. *Corrected and averaged momentum*

After discovering the unwanted terms in expression (78), we can use the expansion to define a *corrected* momentum exchange algorithm, based on the values

$$
\overline{\phi}_i(\mathbf{k}) = \phi_i(\mathbf{k}) - 2 f_i^* \mathbf{c}_i - h^2 f_i^* c_s^{-4} \left(|\mathbf{c}_i \cdot \mathbf{u}_B(\mathbf{b}_i(\mathbf{k}))|^2 - c_s^2 \mathbf{u}_B(\mathbf{b}_i(\mathbf{k}))^2 \right) \mathbf{c}_i . \tag{80}
$$

Using this modification, the simple test problem CiF_0 with zero boundary stresses is now solved correctly. We continue our analysis with problem CiF_1 where a prescribed pressure distribution appears on the boundary. Results obtained with the modified MEA are shown in Fig. **15**.

Obviously, the approximation of the local stresses is still unsatisfactory[4]. In the special case of horizontal boundary the sum of momentum exchange in a boundary node (79) had a clear relation with an approximate

---

[3]Unless (as pointed out in [40]) we define the pressure using

$$
\hat{p} := c_s^2 \hat{\rho} .
$$

The term of order zero in equation (78) would be than encompassed into the pressure. This, however, besides not having the meaning of a physical fluid pressure, does not suffice to correct the original algorithm in a general case.

[4]It should be remarked that the MEA in its original form (algorithm 3) has not been designed for local stress evaluation. However, for our purpose it represents just an initial step, to be further improved.

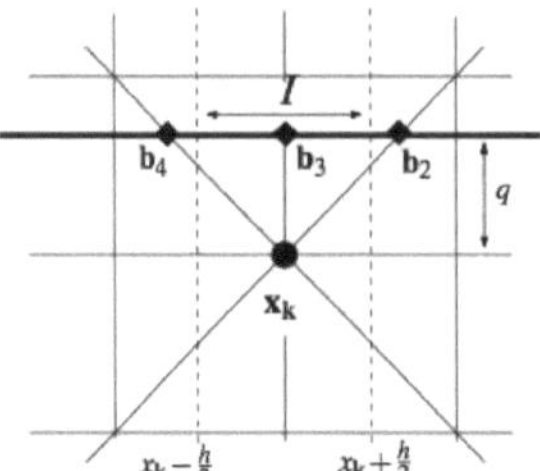

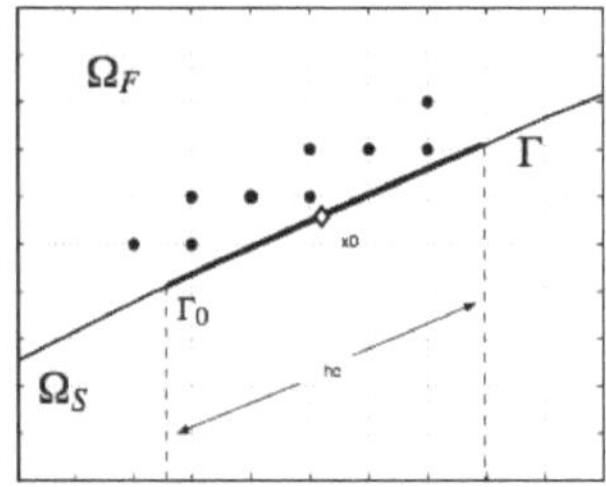

Fig. **14**: **Left:** Formula (78) for horizontal boundary. The points on the boundary ($\diamond$) where the functions are evaluated can be interpreted as nodes of a quadrature rule for the integral (3) over an interval I of length h. The location of such nodes depends on the distance q. **Right:** Meaning of the *coarser grid* h_c introduced in theorem 5. For a point $\mathbf{g}_0 \in \Gamma$ ($\diamond$), the local boundary $\Gamma_0(\mathbf{g}_0, h_c)$ (bold line inside the circle) can be identified as a ball centered in $\mathbf{g}_0$ and diameter h_c, intersected with the interface. The momentum exchange is evaluated at the points interacting with Γ_0 ($\bullet$).

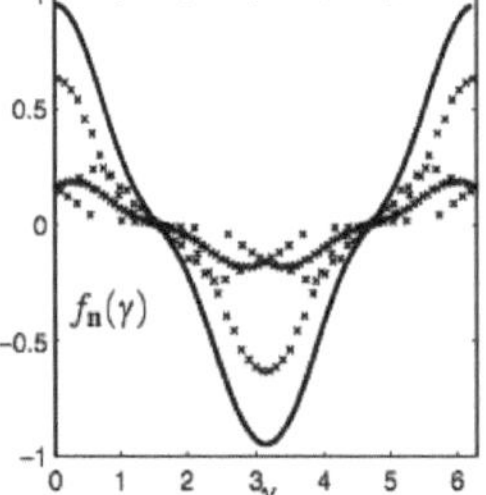

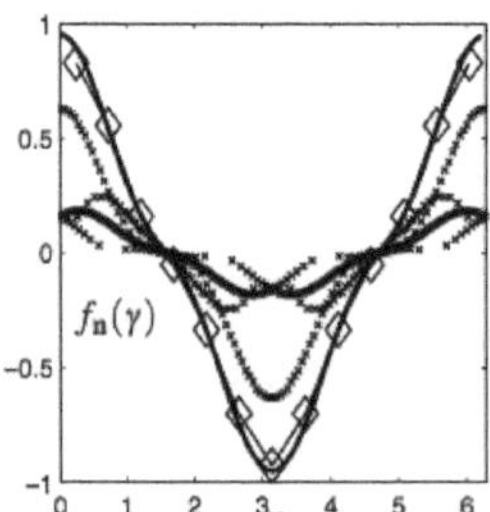

Fig. **15**: Problem CiF_1 (pressure different from 0). **Left:** The symbols ($\times$) denote the values (80) for each boundary couple $(\mathbf{k}, i)$ versus the related boundary point $\mathbf{b}_i(\mathbf{k}) \in \Gamma$, identified by $\gamma \in [0, 2\pi)$, for a 50×50 grid. The solid line is the exact solution (65). **Right:** Results using a 100×100 grid. The approximation on the fine scale (crosses) is more noisy, but does not improve the approximation of the local stresses. Averaged values (as in algorithm 4 and Corollary 6), computed grouping the points according to a grid $h_c = h^{0.5}$ ($\diamond$), are indicated by diamonds.

integration rule. However, for general curved boundary the distribution of these points and of the outgoing directions along the interface is extremely *irregular*, and the momentum exchange in a single boundary node might not be directly related to an approximation of the stresses on the interface. Moreover, the ME-interaction is *discrete*, i.e. using directly the momentum exchange $\overline{\phi}_i(\mathbf{k})$ as approximation of the stress in the point $\mathbf{b}_i(\mathbf{k})$ allows only to define the boundary interaction in special points (the intersections between grid and lattice). In other words, the MEA does not allow to define the force acting on an arbitrary $\mathbf{b} \in \Gamma$. To overcome these problems, we have analyzed an *averaged* value of the momentum exchanged along small intervals on the boundary. In practice, we choose a partition $\{\mathbf{b}_m\} \subset \Gamma$ based on a coarse grid size $h_c > h$. The approximation of the local force in $\mathbf{b}_m$ is computed summing all the momentum exchange contributions (with a proper weight relating h and h_c) of the couples $(\mathbf{k}, i) \in \mathcal{K}$ such that the corresponding $\mathbf{b}_i(\mathbf{k})$ belongs to an h_c-neighborhood of $\mathbf{b}_m$.

Algorithm 4 (Coarsening procedure).

let be given:
 the set $\mathcal{B} = \{\gamma_r = \gamma(\mathbf{b}_{i_r}(\mathbf{k}_r)) \mid (\mathbf{k}_r, i_r) \in \mathcal{K}\}$ (intersections grid-boundary)
 the values $\mathcal{M}_{MEA} := \{\phi_i(\mathbf{k}) \mid (\mathbf{k}, i) \in \mathcal{K}\}$
set the coarse grid parameter h_c
define partition of the interface:
 $\{\mathbf{b}_m = \mathbf{b}(\gamma_m^c) \mid \gamma_m^c = mh_c,\ m = 1, \ldots, M\}$
`for` $l = 1, M$
 initialize the averaged forces $\mathbf{F}_m^c = 0$
 `for` $j = 1, |\mathcal{B}|$
  `if` $(m - \frac{1}{2})h_c < \gamma_r < (m + \frac{1}{2})h_c$
   $\mathbf{F}_m^c = \mathbf{F}_m^c + \phi(\mathbf{k}_r)$
 `end`
normalize $\mathbf{F}_m^c = \mathbf{F}_m^c \dfrac{h}{h_c}$
`end`

This algorithm leads to better results (Fig. **15**, right). From a theoretical point of view, the improvements can be rigorously stated using the asymptotic prediction for the momentum exchanged.

Theorem 5 (Accuracy property of 2D MEA). *Let* $\mathbf{b}_0 \in \Gamma$ *be a point on a smooth one-dimensional interface of finite length. Given the LB-grid size h, we consider a* coarser *grid* h_c, *such that* $h \in o(h_c)$, *and the related interval* $\Gamma_0(\mathbf{b}_0, h_c) = \left\{\mathbf{b} \in \Gamma : \|\mathbf{b} - \mathbf{b}_0\| < \dfrac{h_c}{2}\right\}$. *Defining local averages of the exact and approximate normal stress in* Γ_0

$$\mathcal{I}(\mathbf{b}_0, h_c) = \frac{1}{h_c} \int_{\Gamma_0} (-p\mathbf{I} + \mathbf{S}) \cdot \mathbf{n}\, d\sigma, \quad \overline{\Phi}(\mathbf{b}_0, h_c) = \frac{h}{h_c} \sum_{(\mathbf{k},i):\mathbf{b}_i(\mathbf{k}) \in \Gamma_0} \frac{\overline{\phi}_i(\mathbf{k})}{h^2},$$

the following estimate holds

$$\left|\mathcal{I}(\mathbf{b}_0, h_c) - \overline{\Phi}(\mathbf{b}_0, h_c)\right| \in O\left(h_c + \frac{h}{h_c}\right). \tag{81}$$

A detailed proof for the D2Q9 case is given in section 5.5. Shortly, it is based on writing the sum $\bar{\Phi}(\mathbf{b}_0, h_c)$ in terms of the functions p and $\mathbf{S}$, using equation (78) combined with a Taylor expansion around the node $\mathbf{b}_0$. The resulting expressions can be viewed as approximate integration rules on the interface. Unfortunately, the weights of the arising quadrature formulas do not sum up exactly to one at every node, which rules out first order accuracy. However, using some arithmetical properties of the weights, it can be shown that the deviation from one goes to zero, if the weights are summed over subsets of the interface which are large compared to the grid size h of the regular grid. On the other hand, the Taylor approximation is less accurate if it is used on a coarse mesh of typical distance h_c. Hence, a balance between fine and coarse grid arises in equation (81), and to obtain an optimal error bound, a good compromise is required.

Corollary 6 (Optimal coarsening). The optimal coarsening is obtained choosing $h_c = \sqrt{h}$. Equation (81) gives

$$\left|\mathcal{I}(\mathbf{b}_0, \sqrt{h}) - \bar{\Phi}\left(\mathbf{b}_0, \sqrt{h}\right)\right| \in O\left(\sqrt{h}\right). \tag{82}$$

To illustrate the result of theorem 5, we compare the momentum exchange evaluated along the boundary point in problem CiF_1, averaged according to several coarser grids of type $h_c = h^\alpha$. The order plot in Fig. **16** confirms the predicted result. The best rate is obtained when $h_c = \sqrt{h}$.

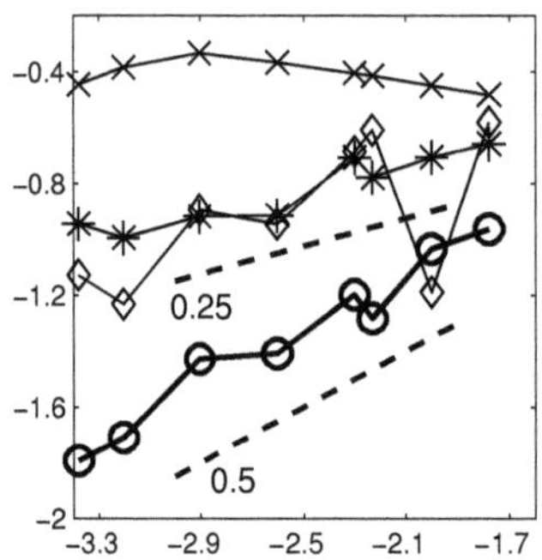

Fig. **16**: Double logarithmic plot of the error in the local forces versus the grid size h. Comparisons of different coarser grids $h_c = h^\alpha$, with $\alpha = 0.25(\diamond), 0.5(\circ), 0.75(*), 0.9(\times)$. The dashed lines represent reference slopes.

It should be noted that the averaging does not affect the accuracy of the global force evaluation.

Test on the cylinder in flow In Fig. **17** we show the result (double logarithmic plot) for the MEA used to compute the total force on the disk in the problem CiF_1. The value for the exact solution is obtained using a high accuracy numerical quadrature for the pressure (64). The plot confirms the expected first order accuracy.

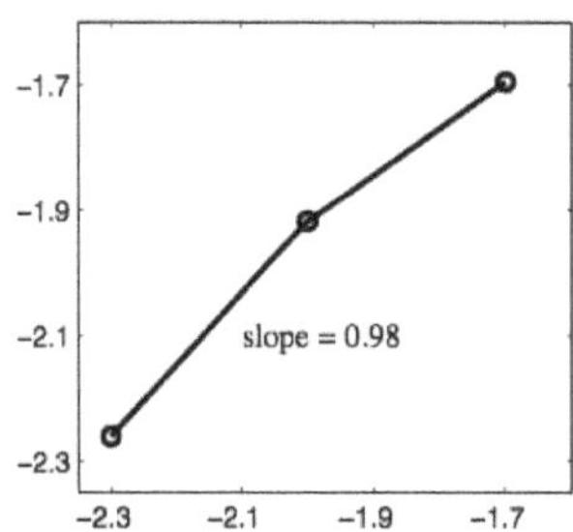

Fig. **17**: Double logarithmic plot (logarithm of error *vs* logarithm of the grid size) for the MEA, used to compute the force in the problem CiF_1. The value used as exact solution is a high order numerical approximation of the integral (3).

Remark (Lees-Edwards boundary conditions). The importance of the corrected MEA has been demonstrated in [32, 33] in the context of suspension flows with Lees-Edwards boundary conditions. In this case, suspensions move in a periodic channel, but a difference in velocity is imposed at the two sides of the periodic boundary. Hence, Galilean invariance of the force computation becomes highly necessary.

5.5. Accuracy properties of MEA

Theorem 5 (Accuracy property of 2D MEA) *Let* $\mathbf{b}_0 \in \Gamma$ *be a point on a smooth one-dimensional interface of finite length. Given the LB-grid size h, we consider a* coarser *grid* h_c*, such that* $h \in o(h_c)$*, and the related interval* $\Gamma_0(\mathbf{b}_0, h_c) = \left\{ \mathbf{b} \in \Gamma : \|\mathbf{b} - \mathbf{b}_0\| < \frac{h_c}{2} \right\}$. *Defining local averages of the exact and approximate normal stress in* Γ_0

$$\mathcal{I}(\mathbf{b}_0, h_c) = \frac{1}{h_c} \int_{\Gamma_0} (-p\mathbf{I} + \mathbf{S}) \cdot \mathbf{n} \, d\sigma, \quad \overline{\Phi}(\mathbf{b}_0, h_c) = \frac{h}{h_c} \sum_{(\mathbf{k},i):\mathbf{b}_i(\mathbf{k}) \in \Gamma_0} \frac{\overline{\phi}_i(\mathbf{k})}{h^2},$$

the following estimate holds

$$\left| \mathcal{I}(\mathbf{b}_0, h_c) - \overline{\Phi}(\mathbf{b}_0, h_c) \right| \in O\left(h_c + \frac{h}{h_c} \right). \tag{83}$$

The proof proceeds as follows. In section 5.5.1. we introduce notations and quantities which will be useful during the demonstration. Then, we analyze in section 5.5.2. the effect of averaging the momenta exchanged along the interface, proving the convergence toward the local integral for straight boundaries. In section 5.5.3. we extend the proof to the total force evaluation.
We perform a detailed computation for $d = 1$, i.e. with a one-dimensional interface in a domain $\Omega \subset \mathbb{R}^2$. The case $d = 2$ is a straightforward extension.

5.5.1. *Preliminary definitions*

Let us recall the definition of the set of boundary couples

$$\mathcal{K} = \left\{ (\mathbf{k}, i) \in \mathcal{G}(h) \times \{1, \ldots, 8\} \mid \mathbf{x_k} \in \overline{\Omega}_F, \ \mathbf{x_k} + h\mathbf{c}_i \in \Omega_S \right\}, \tag{84}$$

which depends on the domain Ω, the grid $\mathcal{G}(h)$ and the shape of the boundary Γ. As previously observed, each boundary couple in $\mathcal{K}$, identifies uniquely a point on the interface:

$$\begin{array}{ccc} \mathcal{K} & \rightarrow & \mathcal{N}(\Gamma) \subset \Gamma \\ (\mathbf{k}, i) & \mapsto & \mathbf{b}_i(\mathbf{k}) \end{array}$$

i.e. the intersection between the interface Γ and the link $\mathbf{c}_i$ starting from the node $\mathbf{k}$ (Fig. **18**, left). We will also refer to $\mathcal{N}(\Gamma)$ as the set of *lattice-interface intersections.*
The base of the proof of theorem 5 is the expansion derived at $\mathbf{b}_i(\mathbf{k})$ for the corrected momentum exchange $\bar{\phi}$ (equations (78) and (80)):

$$\bar{\phi}_i(\mathbf{k}) = h^2 \phi_i^{(2)}(\mathbf{b}_i(\mathbf{k})) + O(h^3), \tag{85}$$

where

$$\phi_i^{(2)}(\mathbf{b}_i(\mathbf{k})) = \left[2 f_i^* c_s^{-2} \left(p(\mathbf{b}_i(\mathbf{k})) - c_s^{-2} \nu \mathbf{c}_i \cdot \nabla \mathbf{u}_B(\mathbf{b}_i(\mathbf{k})) \cdot \mathbf{c}_i \right) \right] \mathbf{c}_i. \tag{86}$$

Those expressions, for different i, are composed by evaluating a *macroscopic* function ζ (like p or $\mathbf{S}_{\alpha,\beta}$) at the boundary point $\mathbf{b}_i(\mathbf{k})$ related to the couple $(\mathbf{k}, i) \in \mathcal{K}$, with particular weights. To generalize the notations, we introduce the vector $\mathbf{a}^{\zeta,i} = (a_x^{\zeta,i}, a_y^{\zeta,i})$ which indicates the two-dimensional weight multiplying the function ζ in ϕ_i:

$$\phi_i^{(2)}(\mathbf{b}) = \sum_{\zeta} \mathbf{a}^{\zeta,i} \zeta(\mathbf{b}). \tag{87}$$

For instance, for $\phi_1^{(2)} = \left(\frac{2}{3} p - \mathbf{S}_{xx}, 0 \right)$, we have $\mathbf{a}^{p,1} = \left(\frac{2}{3}, 0 \right)$, $\mathbf{a}^{\mathbf{S}_{xx},1} = (-1, 0)$.
As next, we consider the following projection operators defined on $\mathcal{K}$

$$\Pi_{\mathcal{G}} : \mathcal{K} \rightarrow \mathcal{G}(h), \quad \Pi_{\mathbb{V}} : \mathcal{K} \rightarrow \{1, \ldots, 8\}. \tag{88}$$

We indicate with $X(\Gamma) = \Pi_{\mathcal{G}}(\mathcal{K})$ the set of boundary *grid points.*

For ease of notation, we also define a map on the set $X(\Gamma)$, which gives all the outgoing directions at a boundary node $\mathbf{k} \in X$:

$$\begin{array}{rcl} \mathbb{P}:\ X(\Gamma) & \rightarrow & \mathcal{P}(\{1,\ldots,8\}) \\ \mathbf{k} & \mapsto & \Pi_{\mathbb{V}}\left(\Pi_{\mathcal{G}}^{-1}(\mathbf{k})\right). \end{array} \tag{89}$$

This map is obviously strictly related to the grid and the shape of the boundary, as the set $\mathcal{K}$ is.

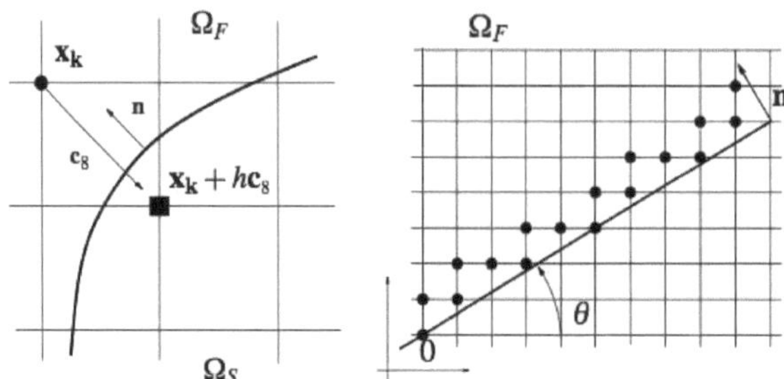

Fig. **18**: **Left.** Sketch of a boundary node (circle) in the fluid domain Ω_F, close to the interface Γ. The couple $(\mathbf{k},8) \in B$, being the point $\mathbf{x_k} + h\mathbf{c}_8 \in \Omega_S$.The vector $\mathbf{n}$ is the outgoing (with respect to the solid) normal to the interface. **Right.** Boundary points for a straight boundary with inclination θ. Here $\tan\theta = \frac{3}{5}$. According to (90), the fluid region is on the top. $\mathbf{n}$ is the solid-outgoing normal.

5.5.2. *Straight boundaries*

First, we consider theorem 5 in the special case of *straight boundaries*. Performing the proof will help in understanding the way to proceed in more general cases.
We assume the interface Γ to be a straight line, with constant inclination $\theta \in [0,2\pi)$. Without loss of generality, we consider (see Fig. **18**, right) $\Omega = [0,1] \times [0,\tan\theta]$ and the solid domain

$$\Omega_S = \{(x,y) \in \Omega \mid y < x\tan\theta\}. \tag{90}$$

In other words, the boundary (named θ-*boundary* in the future) is inclined by θ with respect to the x axis. In this particular situation, the outgoing normal to the interface is $\mathbf{n} = (-\sin\theta, \cos\theta)$.
Actually, we also assume $0 \leq \theta \leq \frac{\pi}{4}$. Otherwise, a similar proof can be repeated rotating the reference system, i.e. exchanging the role of x and y or of the discrete links in the following formulas.

Frequency of appearance The structure of $\mathbb{P}(\mathbf{k})$ (outgoing direction in the node $\mathbf{k}$) can be described through some algebraic relations. Let us consider a node $\mathbf{k} = (k_x,k_y)$. From

$$\mathbf{x_k} \in \Omega_F \cup \Gamma \iff k_y \geq \tan\theta k_x$$

and

$$\begin{array}{rcl} i \in \mathbb{P}(\mathbf{k}) & \Leftrightarrow & \mathbf{x_k} + h\mathbf{c}_i \in \Omega_S \\ & \Leftrightarrow & hk_y + hc_{iy} < \tan\theta(hk_x + hc_{ix}) \\ & \Leftrightarrow & hk_y < \tan\theta hk_x + h(\tan\theta c_{ix} - c_{iy}), \end{array}$$

we conclude

$$i \in \mathbb{P}(\mathbf{k}) \Leftrightarrow 0 \leq k_y - \tan\theta k_x < \tan\theta c_{ix} - c_{iy}. \tag{91}$$

Inequality (91) has solutions only if the right hand side is greater than zero. We call **boundary crossing link** any discrete velocity $\mathbf{c}_i$ such that $\tan\theta c_{ix} - c_{iy} > 0$. To fix the ideas, we look at the two extreme situations.

If $\theta = 0$, relation (91) reduces to

$$i \in \mathbb{P}(\mathbf{k}) \Leftrightarrow 0 \leq k_y < -c_{iy} \; (k_y \geq 0). \tag{92}$$

In this case the boundary points (see Fig. **19**, left) are the nodes **k** such that $k_y = 0$. From the condition (92), we have $c_{iy} < 0$, i.e. $i = 6,7,8$. If $\theta = \frac{\pi}{4}$ (Fig. **19**, right), the condition is $0 < k_y - k_x < c_{ix} - c_{iy}$. It is satisfied by $i = 1,7$ and $k_x = k_y$, or by $i = 8$ and $k_y = k_x, k_x + 1$. Notice that there are two couples with the same k_x, containing $i = 8$, and $\tan\theta c_{8x} - c_{8y} = 2$, while only one with $i = 1$ or $i = 7$ and $\tan\theta c_{1x} - c_{1y} = \tan\theta c_{7x} - c_{7y} = 1$. In these examples the quantity $c_{ix}\tan\theta - c_{iy}$ estimates how many k_y satisfy the inequality

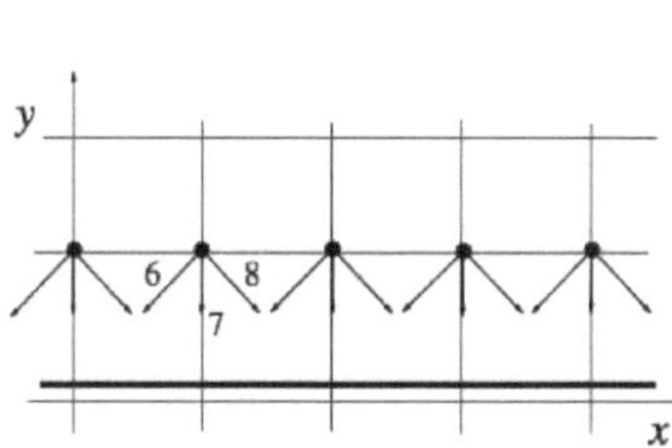

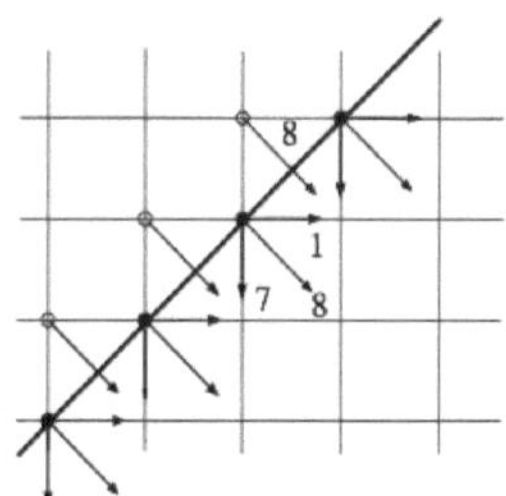

Fig. **19**: Sketch of the boundary direction for the cases $\theta = 0$, $\theta = \frac{\pi}{4}$. In the second example, for a fixed k_x, the link $i = 8$ appears twice (nodes $\circ$ and $\bullet$), with respect to the directions $i = 1,7$ (nodes $\bullet$).

for a fixed k_x. This suggests a further characterization of the θ-boundary.

Definition 3 (Frequency of appearance)**.** If the interface is a θ-boundary, we define the *frequency of appearance* of the link i in the set $\mathbb{P}(\mathbf{k})$ as

$$m_{\theta,i} \equiv \max\left(\tan\theta c_{ix} - c_{iy}, 0\right). \tag{93}$$

As remarked in (91),

$$i \in \mathbb{P}(\mathbf{k}) \Leftrightarrow m_{\theta,i} > 0. \tag{94}$$

Within the restriction $0 \leq \theta \leq \frac{\pi}{4}$, we have

$$\tan\theta c_{ix} - c_{iy} > 0 \Leftrightarrow i \in \{1,6,7,8\}.$$

Remark. Since $m_{\theta,i} \propto (\mathbf{c}_i \cdot \mathbf{n})$, $\mathbf{c}_i$ has to be an *outgoing* direction. Such condition has been already announced. However, the function $m_{\theta,i}$ contains more information. In some sense, $m_{\theta,i}$ *counts* the boundary couples. This concept will be formalized in general in the following part. Through $m_{\theta,i}$ a set of properties relating the macroscopical weights $\mathbf{a}^{\zeta,i}$ to the integral (3) can be proven.

Lemma 7. *Let us define*

$$\mathbf{W}^{\theta}(\zeta) = \sum_i m_{\theta,i} \begin{pmatrix} a_x^{i,\zeta} \\ a_y^{i,\zeta} \end{pmatrix}. \tag{95}$$

It holds

$$\mathbf{W}^{\theta}(p) = \begin{pmatrix} \tan\theta \\ -1 \end{pmatrix}, \mathbf{W}^{\theta}(\mathbf{S}_{xx}) = \begin{pmatrix} -\tan\theta \\ 0 \end{pmatrix}, \mathbf{W}^{\theta}(\mathbf{S}_{xy}) = \begin{pmatrix} 1 \\ -\tan\theta \end{pmatrix}, \mathbf{W}^{\theta}(\mathbf{S}_{yy}) = \begin{pmatrix} 0 \\ 1 \end{pmatrix}. \tag{96}$$

Proof. We will proof only the relation for $W_x^{\theta}(p)$ (others are similar). Direction by direction, we have

$$\mathbf{a}^{p,1} = \left(\frac{2}{3}, 0\right) = -\mathbf{a}^{p,5}, \; \mathbf{a}^{p,2} = \left(\frac{1}{6}, \frac{1}{6}\right) = -\mathbf{a}^{p,6},$$

$$\mathbf{a}^{p,3} = \left(0, \frac{2}{3}\right) = -\mathbf{a}^{p,7}, \; \mathbf{a}^{p,4} = \left(-\frac{1}{6}, \frac{1}{6}\right) = -\mathbf{a}^{p,8}.$$

Hence,

$$\sum_i a_x^{p,i} \max\left(\tan\theta c_{ix} - c_{iy}, 0\right) =$$
$$= \frac{2}{3}\tan\theta + \frac{1}{6}(\tan\theta - 1) + \frac{1}{6}(\tan\theta + 1) = \tan\theta.$$

□

A direct consequence is the following

Lemma 8. *Let* $\mathbf{b} \in \Gamma$. *It holds*

$$\sum_\zeta \mathbf{W}^\theta(\zeta)\zeta(\mathbf{b}) = \frac{1}{\cos\theta}\mathbf{n}(\mathbf{b}) \cdot (-p(\mathbf{b}) + \mathbf{S}(\mathbf{b})). \tag{97}$$

The last relation expresses a connection between the *discrete sum* of the ME ϕ and the *continuous integral* of the hydrodynamical fields. Furthermore, it serves to identify the origin of the integral measure of the momentum exchange algorithm.

Remark. The particular dependence on the abscissae comes from the restriction $0 \leq \theta < \frac{\pi}{4}$. In general, the axis which spans the angle θ with the interface is involved.

Averaged momentum exchanged For a point $\mathbf{b}_0 \in \Gamma$ and a parameter h_c, we define the coarse interval

$$I(\mathbf{b}_0, h_c) = \left\{(x,y) \in \Gamma \mid |x - \mathbf{b}_{0x}| \leq \frac{h_c}{2}\right\} \tag{98}$$

as the subset of Γ whose projection on the x-axis is an $\frac{h_c}{2}$-neighborhood of $\mathbf{b}_{0x}$,

$$B(\mathbf{b}_0, h_c) := \{(\mathbf{k}, i) \in \mathcal{K} \mid \mathbf{b}_i(\mathbf{k}) \in I(\mathbf{b}_0, h_c)\} \subset \mathcal{K},$$

as the set of the lattice-interface intersections contained in the subset $I(\mathbf{b}_0, h_c)$, and

$$X(\mathbf{b}_0, h_c) = \Pi_{\mathcal{G}}\left(B(\mathbf{b}_0, h_c)\right),$$

of the boundary grid nodes of the elements $B(\mathbf{b}_0, h_c)$.

Remark. Note that, in the context of MEA, the set $X(\mathbf{b}_0, h_c)$ contains all the grid nodes exchanging momentum with the subset $I(\mathbf{b}_0, h_c)$.

According to the statement of theorem 5, we need a coarse parameter $h_c(h)$ such that $h \in o(h_c)$.
For instance, a class of grids satisfying this requirement is defined by

$$h_c = h^\alpha, \quad 0 \leq \alpha < 1.$$

In general, in what follows, we define a coarse grid as

$$h_c(h) = L(h)h, \tag{99}$$

$L(h)$ being the number of finer grid nodes belonging to a coarse grid step. Since $h \in o(h_c)$, we have

$$\lim_{h \to 0} L(h) = +\infty. \tag{100}$$

The idea is to prove a set of estimates regarding the boundary set and the momentum exchange which leads to the continuous integral after taking the average over several grid nodes. To do this, we relate the boundary set to the shape of the boundary, through the frequency of appearance $m_{\theta,i}$.

Lemma 9. *Let us denote with $\chi_A(x)$ the characteristic function of a set A. We have*

$$\frac{1}{L}\sum_{\mathbf{k}\in X(\mathbf{b}_0,Lh)} \chi_{\mathbb{P}(\mathbf{k})}(i) = m_{\theta,i} + O\left(\frac{1}{L}\right).$$

Thus, $m_{\theta,i}$ represents *asymptotically* the *number of times per unit of length* a link i appears as outgoing direction in a set of points of length L on the x-axis.

Proof. The sum

$$\sum_{\mathbf{k}\in X(\mathbf{b}_0,Lh)} \chi_{\mathbb{P}(\mathbf{k})}(i)$$

expresses the number of times the boundary is crossed by a link $\mathbf{c}_i$ outgoing from nodes $\mathbf{k} \in X(\mathbf{b}_0,Lh)$.
We restrict to $0 \le \tan\theta \le 1$. Moreover, for simplicity we can shift the axis in order to have $\mathbf{b}_0 = h\frac{L}{2}$, i.e.

$$\forall \mathbf{k} \in X(\mathbf{b}_0,h_c):\ 0 \le k_x \le L-1. \tag{101}$$

Of course, the links $i = 2,3,4,5$ satisfy trivially the property, holding $m_{\theta,i} = 0$.
Now, let us consider the directions such that $m_{\theta,i} > 0$. Calling $\mathbf{k} = (k_x,k_y)$, the condition $\chi_{\mathbb{P}(\mathbf{k})}(i) = 1$ means

$$k_x \tan\theta \le k_y < (k_x + c_{ix})\tan\theta - c_{iy}. \tag{102}$$

Since $k_x,k_y \in \mathbb{Z}$, we can count the integer solutions of inequality (102). Namely, for fixed $k_x \in \{0,\dots,L-1\}$ and $\mathbf{c}_i \in \mathbb{V}$, we have

$$\lceil (k_x + c_{ix})\tan\theta - c_{iy}\rceil - \lceil k_x \tan\theta \rceil \tag{103}$$

possible choices for k_y, such that $\chi_{\mathbb{P}(\mathbf{k})}(i) = 1$ ($\mathbf{k}$ is a boundary node and $\mathbf{c}_i$ is an outgoing direction). In equation (103), $\lceil\ \rceil$ denotes the *upper integer part*. For $\alpha \in \mathbb{R}$:

$$\lceil \alpha \rceil = \min_{n\in\mathbb{Z}}\{n \mid n \ge \alpha\}. \tag{104}$$

So that, summing up for $k_x \le L-1$,

$$\sum_{\mathbf{k}\in X(\mathbf{b}_0,Lh)} \chi_{\mathbb{P}(\mathbf{k})}(i) = \sum_{k_x=0}^{L-1}\left(\lceil (k_x + c_{ix})\tan\theta - c_{iy}\rceil - \lceil k_x \tan\theta \rceil\right). \tag{105}$$

Using simple arithmetic relations, one can show that [5], for $c_{ix} \in \{-1,0,1\}$ it holds

$$\sum_{\mathbf{b}\in X(\mathbf{b}_0,Lh)} \chi_{\mathbb{P}(\mathbf{k})}(i) = Lm_{\theta,i} + O(1). \tag{106}$$

Dividing by L, the proof of lemma 9 is complete. □

As a consequence, we obtain an estimate for the number of points in the boundary set.

Corollary 10.

$$\frac{|X(\mathbf{b}_0,Lh)|}{L} = 1 + \tan\theta + O\left(\frac{1}{L}\right).$$

Coarse grids Lemmas 8 and 9 motivate the origin of the *coarser* grid. In fact, considering a discretization parameter $h_c(h) = L(h)h$ such that (equation (100))

$$\lim_{h\to 0} L(h) = +\infty,$$

for small h we can translate the previous properties in asymptotic estimates.
Let $\mathbf{b}_0 \in \Gamma$ be given. Considering the coarse interval $I(\mathbf{b}_0,h_c)$, as defined in (98), we focus on the subintegral

$$\mathcal{I}(\mathbf{b}_0,h_c) = \int_{I(\mathbf{b}_0,h_c)} \mathbf{n}\cdot(-p+\mathbf{S})\,d\gamma := \int_{I(\mathbf{b}_0,h_c)} \mathbf{Z}\,d\gamma. \tag{107}$$

Using a Taylor expansion we have

$$\mathcal{I}(\mathbf{b}_0,h_c) = \frac{h_c}{\cos\theta}\mathbf{Z}(\mathbf{b}_0) + O(h_c^2). \tag{108}$$

At the same time,

$$\Phi(\mathbf{b}_0,h_c) = \sum_{(\mathbf{k},i)\in B(\mathbf{b}_0,h_c)} \bar{\phi}_i(\mathbf{k}) = \sum_{i=1}^{8}\left(\sum_{\mathbf{k}\in X(\mathbf{b}_0,h_c)} \chi_{\mathbb{P}(\mathbf{k})}(i)\bar{\phi}_i(\mathbf{k})\right). \tag{109}$$

Inserting the prediction (78) for (the corrected) $\bar{\phi}_i$ and its Taylor expansion around $\mathbf{b}_0$, the previous formula becomes

$$\Phi(\mathbf{b}_0,h_c) = h^2\left(\sum_{i=1}^{8}\bar{\phi}_i^{(2)}(\mathbf{b}_0)\sum_{\mathbf{k}\in X(\mathbf{b}_0,h_c)} \chi_{\mathbb{P}(\mathbf{k})}(i) + O(h_c)\right). \tag{110}$$

Remark. We average over a coarse grid cell, which contains L nodes of the original grid along the x-axis. Moreover, since the quantities in which we are interested are of order h^2 in $\overline{\Phi}(\mathbf{b}_0,h_c)$, we consider an averaged and rescaled sum.

$$\overline{\Phi}(\mathbf{b}_0,h_c) = \frac{1}{L}\frac{\Phi(\mathbf{b}_0,h_c)}{h^2}. \tag{111}$$

We have

$$\begin{aligned}\overline{\Phi}(\mathbf{b}_0,h_c) &= \sum_{i=1}^{8}\phi_i^{(2)}(\mathbf{b}_0)\sum_{\mathbf{k}\in X(\mathbf{b}_0,h_c)}\left(\frac{\chi_{\mathbb{P}(\mathbf{k})}(i)}{L} + O(L^{-1}h_c)\right) = \\ &= \sum_{i=1}^{8} m_{\theta,i}\phi_i^{(2)}(\mathbf{b}_0) + O(|X(\mathbf{b}_0,h_c)|L^{-1}h_c).\end{aligned}$$

Estimating the right hand side using lemma 9 and corollary 10 we find

$$\overline{\Phi}(\mathbf{b}_0,h_c) = \frac{1}{\cos\theta}\mathbf{Z}(\mathbf{b}_0) + O\left(\frac{h}{h_c}\right) + O(h_c). \tag{112}$$

Being $L = \frac{h_c}{h}$, the relation between (108) and (110) is now

$$\left|\frac{\mathcal{I}(\mathbf{b}_0,h_c)}{h_c} - \overline{\Phi}(\mathbf{b}_0,h_c)\right| \in O\left(h_c + \frac{h}{h_c}\right). \tag{113}$$

In the enunciate of theorem 5, we deal with the coarse set

$$\Gamma_0(\mathbf{b}_0,h_c) = \left\{\mathbf{b}\in\Gamma \mid \|\mathbf{b}-\mathbf{b}_0\| < \frac{h_c}{2}\right\}$$

which is slightly different from the $I(\mathbf{b}_0,h_c)$ considered in equation (113). However, for straight interfaces these two sets are asymptotically equal, in the sense that their lengths are of the same order, and the number of nodes of difference (belonging to one but not to another) goes to zero as L goes to infinity. Hence, the validity of the argument is not affected, i.e. estimate (113) proves theorem 5 for straight boundaries.

5.5.3. *Generalization for curved boundaries*

To generalize the proof, we follow the path which yielded equation (113), investigating the modification we need in a more general case.
Let us assume that the interface Γ is described by the graph of a *smooth* function:

$$\exists g \text{ smooth } : \Gamma = \{(x,y) \in \Omega \mid y = g(x)\}, \tag{114}$$

In this case, the fluid and solid domains are identified by

$$\Omega_F = \{(x,y) \in \Omega \mid y > g(x)\}, \; \Omega_S = \{(x,y) \in \Omega \mid y < g(x)\} \tag{115}$$

(compare with equation (90)). Note that without loss of generality, we assume the boundary to be described as an implicit function $y = g(x)$. However, the following arguments can be generalized also to the case where the coordinates are swapped.

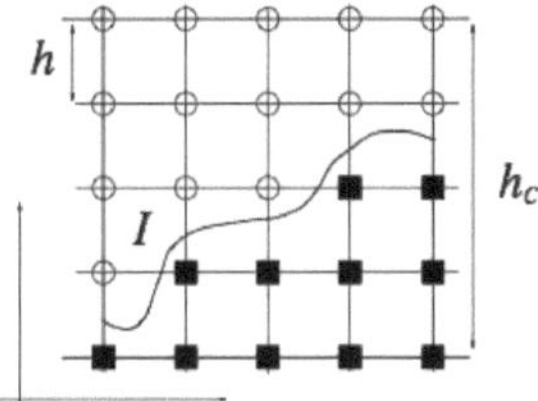

Fig. **20**: Example of a coarse cell $h_c \times h_c$ in a lattice of size h. The squares represent the solid node, the circles the fluid ones. The set $I = I(\mathbf{b}_0, h_c)$ (bold line) is the portion of interface contained in the cell.

Assuming to have chosen a point $\mathbf{b}_0 \in \Gamma$ and a coarse parameter h_c, defining the related coarse interval $I(\mathbf{b}_0, h_c)$ as in equation (98), we focus on a single cell $h_c \times h_c$ (Fig. **20**).

Averaged inclination We look for a geometric condition on a couple $(\mathbf{k}, i)$ such that $\mathbf{b}_i(\mathbf{k})$ belongs to the set $B(\mathbf{b}_0, h_c)$. According to (115),

$$(\mathbf{k}, i) \in B(\mathbf{b}_0, h_c) \iff \begin{array}{l} hk_y \geq g(hk_x) \\ h(k_y + c_{iy}) < g(hk_x + hc_{ix}). \end{array} \tag{116}$$

If $g(x)$ is smooth, we have

$$g(hk_x + hc_{ix}) = g(hk_x) + \int_{hk_x}^{hk_x + hc_{ix}} g'(s)ds, \tag{117}$$

and using $s = h(k_x + c_{ix}\sigma)$ yields

$$g(hk_x + hc_{ix}) = g(hk_x) + hc_{ix} \int_0^1 g'(hk_x + hc_{ix}\sigma)d\sigma\,. \tag{118}$$

This allows to rewrite (116) as

$$(\mathbf{k}, i) \in B(\mathbf{b}_0, h_c) \iff 0 \leq k_y - \frac{g(hk_x)}{h} < c_{ix} \int_0^1 g'(hk_x + hc_{ix}\sigma)d\sigma - c_{iy}. \tag{119}$$

In view of (119), for a given node $\mathbf{k}$, the direction $\mathbf{c}_i$ intersects the interface iff

$$c_{ix} \int_0^1 g'(hk_x + hc_{ix}\sigma)d\sigma - c_{iy} > 0. \tag{120}$$

Different interpretations of (119) are possible. In words, equation (120) expresses the fact that the vector $\mathbf{c}_i$ from the node $\mathbf{k}$ is a **boundary crossing link** in terms of the average of the function $g'(x)$ in the range $h[k_x, k_x + c_{ix}]$.

Let us consider a boundary couple $(\mathbf{k},i) \in B(\mathbf{b}_0,h_c)$. We can denote with

$$\langle g' \rangle(\mathbf{b}_i(\mathbf{k})) = \int_0^1 g'(hk_x + hc_{ix}\sigma)d\sigma. \tag{121}$$

the average of $g'(x)$ in the x-cell containing the intersection $\mathbf{b}_i(\mathbf{k})$ (Fig. **21**). As well, we can introduce the angle $\langle\theta\rangle \in \left[-\frac{\pi}{2},\frac{\pi}{2}\right]$ such that

$$\tan\langle\theta\rangle(\mathbf{b}_i(\mathbf{k})) = \langle g' \rangle(\mathbf{b}_i(\mathbf{k})), \tag{122}$$

i.e. the *average inclination* of the interface in $h[k_x, k_x + c_{ix}]$.
The definition of $\langle\theta\rangle$ can be extended to any point of the interface[5], i.e.

$$\forall \mathbf{b} \in \Gamma: \ \langle\theta\rangle(\mathbf{b}) = \text{ average inclination in the } x\text{-cell containing } \mathbf{b}.$$

In other words, we are looking at the interface Γ as a piecewise straight line, with inclinations equal to the average on the corresponding cell (Fig. **21**).

Remark. Condition (119) is purely geometric, and related to a property of the grid and the interface. The averaged inclination (123) will be used to transport the results proven for straight boundaries also in the case of a curved interface.

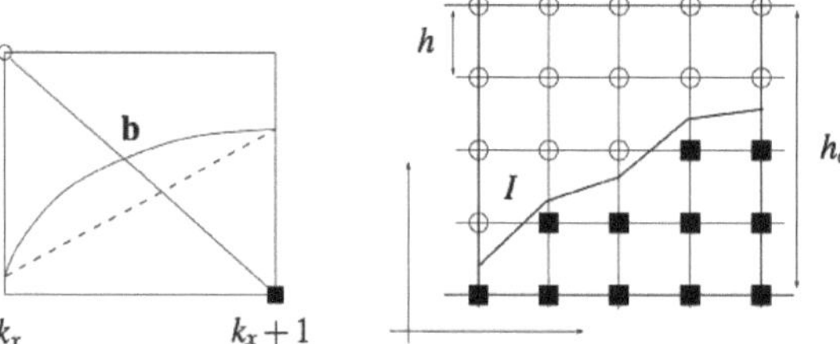

Fig. **21**: **Left:** Averaged inclination of the interface between the abscissae k_x and $k_x + 1$. Formula (121) expresses the averaged inclination of the interface, over the interval of size h containing the intersection lattice-boundary $\mathbf{b}$. The dashed line is a straight boundary with inclination equal to the averaged one. **Right:** In a $h_c \times h_c$ cell (Fig. **20**), using (121) we approximate the interface with a piecewise straight approximation.

Generalized frequency of appearance Relation (119) can be rewritten as

$$\forall (\mathbf{k},i) \in B(\mathbf{b}_0,h_c): \ c_{ix}\tan\langle\theta\rangle(\mathbf{b}_i(\mathbf{k})) - c_{iy} > 0. \tag{124}$$

It generalizes the previous relation (91) in case of curved boundaries, where the inclination is no longer constant.
In analogy with the previous results, we introduce the *generalized frequency of appearance*

$$\langle m \rangle_{(\mathbf{k},i)} = \max\left\{ c_{ix}\int_0^1 g'(hk_x + hc_{ix}\sigma)d\sigma - c_{iy}, 0 \right\}. \tag{125}$$

In case of straight boundary with constant inclination θ, $g'(x) = \tan\theta$, $\langle m \rangle$ does not depend on $\mathbf{k}$ and it coincides with the frequency of appearance previously defined.

[5]Formally, any $\mathbf{b} \in \Gamma$ is contained in one and only one x-cell:

$$\exists!\mathbf{k}(\mathbf{b}) \in \mathcal{G}(h): \ \mathbf{b} \in h[k_x(\mathbf{b}), k_x(\mathbf{b})+1) \times h[k_y(\mathbf{b}), k_y(\mathbf{b})+1),$$

and we can define

$$\langle\theta\rangle(\mathbf{b}) = \tan^{-1}\left(\int_0^1 g'(hk(\mathbf{b})_x + h\sigma)d\sigma\right). \tag{123}$$

Lemma 11.

$$\forall \mathbf{k} \in X(\mathbf{b}_0, h_c) : \langle m \rangle_{(\mathbf{k},i)} = m_{\langle \theta \rangle(\mathbf{b}_0),i} + O(h_c).$$

Proof. Let us take $(\mathbf{k},i) \in \mathcal{G}(h) \times \{1,\ldots,8\}$. We have

$$\langle m \rangle_{(\mathbf{k},i)} = \begin{cases} c_{ix} \tan\langle \theta \rangle(\mathbf{b}_i(\mathbf{k})) - c_{iy} & (\mathbf{k},i) \in B(\Gamma) \\ 0 & (\mathbf{k},i) \notin B(\Gamma). \end{cases} \tag{126}$$

Taylor expanding g around $\mathbf{b}_{0x}$, for the points $\mathbf{b}_i(\mathbf{k})$ in the considered coarse cell, it holds

$$\tan\langle \theta \rangle(\mathbf{b}_i(\mathbf{k})) = \tan\langle \theta \rangle(\mathbf{b}_0) + O(h_c).$$

Hence,

$$\langle m \rangle_{(\mathbf{k},i)} = \begin{cases} c_{ix} \tan\langle \theta \rangle(\mathbf{b}_0) - c_{iy} + O(h_c) & (\mathbf{k},i) \in B(\Gamma) \\ 0 & (\mathbf{k},i) \notin B(\Gamma). \end{cases} \tag{127}$$

The straight frequency of appearance reads

$$m_{\langle \theta \rangle(\mathbf{b}_0),i} = \max\{c_{ix} \tan\langle \theta \rangle(\mathbf{b}_0) - c_{iy}, 0\}.$$

It might happen, that for $(\mathbf{k},i) \in B(\mathbf{b}_0, h_c)$ we have $\langle m \rangle_{(\mathbf{k},i)} > 0$ but $m_{\langle \theta \rangle(\mathbf{b}_0),i} = 0$, or vice versa. In words, $(\mathbf{k},i)$ is a boundary couple for Γ, but not for a straight approximation with inclination $\langle \theta \rangle(\mathbf{b}_0)$. However, in this case

$$c_{ix} \tan\langle \theta \rangle(\mathbf{b}_0) \le 0$$
$$c_{ix} \tan\langle \theta \rangle(\mathbf{b}_0) + O(h_c) > 0,$$

which imply

$$\langle m \rangle_{(\mathbf{k},i)},\ m_{\langle \theta \rangle(\mathbf{b}_0),i} \in O(h_c).$$

Hence, $\langle m \rangle_{(\mathbf{k},i)} = m_{\langle \theta \rangle(\mathbf{b}_0),i} + O(h_c)$. □

Therefore, in the considered $h_c \times h_c$ cell, up to an error $O(h_c)$ we can consider the frequency of appearance of a straight interface with inclination equal to the average in the coarse cell (Fig. **22**).
Obviously, the boundary couples of a straight approximation of the interface do not coincide with the boundary couples of the original one. However, the lemma tells that this happens only when the frequency of appearance is of order $O(h_c)$. In other words the difference between the boundary set of the interface Γ and of its straight approximation is *small*, for small h_c. Formally,

Lemma 12.

$$\frac{1}{L} \sum_{\mathbf{k} \in X(\mathbf{b}_0, Lh)} \chi_{\mathbb{P}(\mathbf{k})}(i) = m_{\langle \theta \rangle(\mathbf{b}_0),i} + O\left(\frac{1}{L}\right) + O(h_c).$$

Proof. Let $\mathbf{k} \in X(\mathbf{b}_0, Lh)$ and $\mathbf{c}_i \in \mathbb{V}$ be fixed. As before (lemma 9), we estimate the sum counting, for each k_x, how many choices for k_y we have, such that $\mathbf{k} = (k_x, k_y) \in X(\mathbf{b}_0, Lh)$ and $i \in \mathbb{P}(\mathbf{k})$.
According to condition (116), it has to hold

$$\frac{g(hk_x)}{h} \le k_y < \frac{g(hk_x + hc_{ix})}{h} - c_{iy}.$$

Hence, there are

$$\left\lceil \frac{g(hk_x + hc_{ix})}{h} - c_{iy} \right\rceil - \left\lceil \frac{g(hk_x)}{h} \right\rceil$$

possible solutions $k_y \in \mathbb{Z}$. For simplicity, since $h_c = Lh$, we assume (see lemma 9)

$$\forall \mathbf{k} \in X(\mathbf{b}_0, Lh) : 0 \le k_x \le L - 1$$

(it can be done shifting the interface along the x-axis).

We have

$$\sum_{\mathbf{k}\in X(\mathbf{b}_0,Lh)} \chi_{\mathbb{P}(\mathbf{k})}(i) = \sum_{k_x=0}^{L-1} \left\lceil \frac{g(hk_x+hc_{ix})}{h} - c_{iy} \right\rceil - \left\lceil \frac{g(hk_x)}{h} \right\rceil = \\ = -c_{iy}L + \sum_{k_x=0}^{L-1} \left\lceil \frac{g(hk_x+hc_{ix})}{h} \right\rceil - \left\lceil \frac{g(hk_x)}{h} \right\rceil. \tag{128}$$

Now, we split the cases $c_{ix} = 0, 1, -1$.
If $c_{ix} = 0$, equation (128) reduces to

$$\sum_{\mathbf{k}\in X(\mathbf{b}_0,Lh)} \chi_{\mathbb{P}(\mathbf{k})}(i) = -c_{iy}L. \tag{129}$$

If $c_{ix} = 1$ we have a telescopic sum, which yields

$$\sum_{\mathbf{k}\in X(\mathbf{b}_0,Lh)} \chi_{\mathbb{P}(\mathbf{k})}(i) = -c_{iy}L + \left\lceil \frac{g(Lh)}{h} \right\rceil - \left\lceil \frac{g(0)}{h} \right\rceil = \\ = -c_{iy}L + \frac{g(Lh)-g(0)}{h} + O(1). \tag{130}$$

From

$$g(Lh) - g(0) = \int_0^{Lh} g'(x)dx = Lh \int_0^1 g'(Lh\sigma)d\sigma,$$

using the Taylor expansion for $g'(x)$ we can estimate

$$\frac{g(Lh)-g(0)}{h} = L\left(\langle g'\rangle(\mathbf{b}_0) + O(h_c)\right),$$

which yields

$$\sum_{\mathbf{k}\in X(\mathbf{b}_0,Lh)} \chi_{\mathbb{P}(\mathbf{k})}(i) = Lm_{\langle\theta\rangle(\mathbf{b}_0),i} + O(Lh_c) + O(1). \tag{131}$$

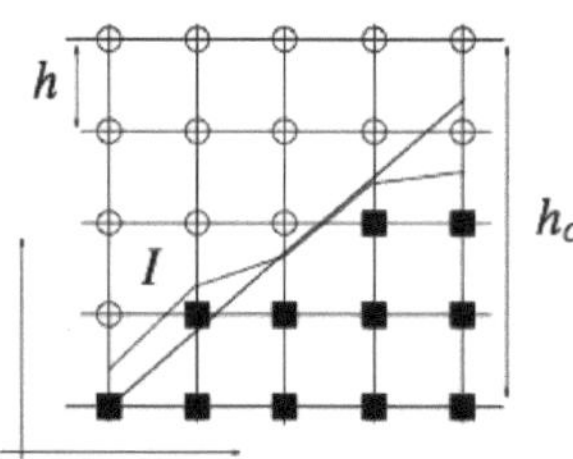

Fig. **22**: From the subset of interface $I(\mathbf{b}_0, h_c) \subset \Gamma$ of Fig. **20**, we reduce to a straight interface with inclination equal to the average in the LB cell (of size h) containing $\mathbf{b}_0$ (bold line). Differences in the structure of the boundary node set of Γ and of the straight approximation appear only if the frequency of appearance $\langle m\rangle_{(\mathbf{k},i)}$ is small.

The case $c_{ix} = -1$ is analogous, so that we can conclude

$$\forall c_{ix} \in \{0,1,-1\} : \frac{1}{L} \sum_{\mathbf{k}\in X(\mathbf{b}_0,Lh)} \chi_{\mathbb{P}(\mathbf{k})}(i) = m_{\langle\theta\rangle(\mathbf{b}_0),i} + O(h_c) + O\left(\frac{1}{L}\right).$$

□

The last result is analogous to lemma 9, only using $\langle\theta\rangle(\mathbf{b}_0)$ instead of θ, and with an additional term $O(h_c)$. Proceeding in a way analogous to the straight boundary case, the statement of theorem 5 can be proved for general shape boundaries.

6. SUMMARY

In this chapter, we investigated in detail the interaction of a lattice Boltzmann fluid with a moving rigid body. First, we described the basic ideas of the asymptotic analysis for the lattice Boltzmann method. As next, we discussed consistent node initialization for moving boundary problems based on an approximation of the lattice Boltzmann asymptotic expansion for the inner domain. Finally, we investigated the Momentum Exchange Algorithm for the evaluation of the fluid forces on the solid body. We show that the original formulation might yield inconsistencies in relevant orders, while the algorithm is not suitable for local forces computation. Introducing a correction based on the asymptotic expansion results, we described a rigorous proof of the accuracy properties of the algorithm. In particular, we showed that global force evaluation is first order accurate, while local interface efforts can only be approximated up to $O(\sqrt{h})$. Throughout the chapter, theoretical conclusions have been validated with several numerical benchmarks.

ACKNOWLEDGEMENT

The Authors are grateful to Martin Rheinländer (Ruprecht-Karls-University Heidelberg) for the collaboration on the asymptotic analysis, and to Zhaoxia Yang (University of Konstanz) for the useful discussions on boundary conditions. Moreover, Alfonso Caiazzo wishes to thank Eric Lorenz (University of Amsterdam) for the useful discussion and collaboration about the application of Momentum Exchange for suspension flows modeling.

CONFLICT OF INTEREST

The authors confirm that this chapter content has no conflict of interest.

REFERENCES

[1] R. Benzi, S. Succi, M. Vergassola, "The lattice Boltzmann equation: Theory and applications", *Phys. Rep.* vol. 222, pp. 147–197, 1992.
[2] P.L. Bhatnagar, E.P. Gross, M. Krook, "A model for collision processes in gases I: Small amplitude processes in charged and neutral one-component systems", *Phys. Rev, E*, vol. 94, pp. 511–525, 1954.
[3] M. Bouzidi, M. Firdaouss, P. Lallemand, "Momentum transfer on a Boltzmann-lattice fluid with boundaries", *Phys. Fluids*, vol. 13, pp. 3452–3459, 2001.
[4] A. Caiazzo, "Analysis of lattice Boltzmann initialization routines", *J. Stat. Phys.* vol. 121, pp. 37–48, 2005.
[5] A. Caiazzo, "Asymptotic Analysis of lattice Boltzmann method for Fluid-Structure Interaction problems", PhD thesis, Technische Universität, Kaiserslautern and Scuola Normale Superiore, Pisa, 2007.
[6] A. Caiazzo, "Analysis of lattice Boltzmann nodes initialization in Moving Boundary problems", *Prog. Comp. Fluid Dyn.*, vol. 8, pp. 3–10, 2008.
[7] A. Caiazzo, M. Junk, "Boundary Forces in lattice Boltzmann: analysis of Momentum Exchange algorithm", *Comp. Math. w. Appl.*, vol. 55, pp. 1415–1423, 2008.
[8] A. Caiazzo, M. Junk, M. Rheinländer, "Comparison of analysis techniques for the lattice Boltzmann method", *Comp. Math. w. Appl.*, vol. 58, pp. 883–897, 2009.
[9] S. Chen, G.D. Doolen, "Lattice Boltzmann method for fluid flows". *Ann. Rev. Fluid Mech.*, vol. 30, pp. 329–364, 1992.
[10] P.J. Dellar, "Incompressible limits of lattice Boltzmann equations using multiple relaxation times", *J. Comput. Phys.*, vol. 190, pp. 351–370, 2003.
[11] D. d'Humières, M. Bouzidi, P. Lallemand, "Thirteen-velocity three-dimensional lattice Boltzmann model", *Phys. Rev. E*, vol. 63, pp. 1–7, 2001.
[12] D. d'Humières, I. Ginzburg, M. Krafczyk, P. Lallemand, L.-S. Luo, "Multiple-relaxation-time lattice Boltzmann models in three dimensions", *Philos. Trans. R. Soc. London A*, vol. 360, pp. 437–451, 2002.
[13] U. Frisch, D. d'Humières, B. Hasslacher, P. Lallemand, Y. Pomeau, J. Rivet, "Lattice-gas hydrodynamics in two and three dimensions", *Complex Systems*, vol. 1, pp. 649–707, 1987.
[14] U. Frisch, B. Hasslacher, Y. Pomeau, "Lattice-gas automata for the Navier-Stokes equation", *Phys. Rev. Lett.*, vol. 56, pp. 1505–1508, 1986.
[15] G. Galdi, *An Introduction to the Mathematical Theory of the Navier-Stokes Equation. Volume 1*, Springer-Verlag, Berlin, 1998.
[16] I. Ginzbourg, D. d'Humiéres, "The multireflection boundary conditions for lattice Boltzmann models", *Phys. Rev. E*, vol. 68, pp. 066614, 2003.
[17] Z. Guo, C. Zheng, B. Shi, "An extrapolation method for boundary conditions in lattice Boltzmann method", *Phys. Fluids*, vol. 14, pp. 2007–2010, 2002.
[18] M. Junk, "A finite difference interpretation of the lattice Boltzmann method", *Num. Meth. PDE*, vol. 17, pp. 383–402, 2001.
[19] M. Junk, A. Klar, L.-S. Luo, "Asymptotic analysis of the lattice Boltzmann Equation", *J. Comput. Phys.*, vol. 210, pp. 676–704, 2005.
[20] M. Junk, M. Rheinländer, P. Van Leemput, "Smooth initialization of lattice Boltzmann schemes", *Comp. Math. w. Appl.*, vol. 58, pp. 867–882, 2009.
[21] M. Junk, Z. Yang, "Analysis of lattice Boltzmann Boundary Conditions", *Proc. Appl. Math. Mech.*, vol. 3, pp. 76–79, 2003.
[22] M. Junk, Z. Yang, "Asymptotic analysis of finite difference method", *Appl. Math. Comput.*, vol. 158, pp. 267–301, 2004.
[23] M. Junk, Z. Yang, "A one-point boundary condition for the lattice Boltzmann method", *Phys. Rev. E*, vol. 72, pp. 066701, 2005.
[24] M. Junk, Z. Yang, "Outflow boundary conditions for the lattice boltzmann method", *Prog. Comp. Fluid. Dyn.*, vol. 8, pp. 38–48, 2008.

[25] M. Junk, Z. Yang, "Pressure boundary condition for the lattice Boltzmann method", *Comp. Math. Appl.*, vol. 58, pp. 922–929, 2009.
[26] M. Junk, Z. Yang, "Convergence of Lattice Boltzmann Methods for Stokes Flows in Periodic and Bounded Domains", *Numer. Math.*, vol. 112, pp. 65–87, 2009.
[27] M. Junk, W.A. Yong, "Rigorous Navier-Stokes limit of the lattice Boltzmann equation", *Asymptot. Anal.*, vol. 35, pp. 165–185, 2003.
[28] M. Junk, W.A. Yong, "Weighted L^2-stability of the lattice Boltzmann method", *SIAM J. Numer. Anal.*, vol. 47, pp. 1651–1665, 2009.
[29] J. Kevorkian, J.D. Cole, *Perturbation Methods in Applied Mathematics*, Springer-Verlag, Berlin, 1981.
[30] A.J.C. Ladd, "Numerical simulations of particular suspensions via a discretized Boltzmann equation. Part 1 (Theory)", *J. Fluid Mech.*, vol. 271, pp. 285–310, 1994.
[31] A.J.C. Ladd, "Numerical simulations of particular suspensions via a discretized Boltzmann equation. Part 2 (Numerical results)", *J. Fluid Mech.*, vol. 271, pp. 311–339, 1994.
[32] E. Lorenz, A. Caiazzo, A.G. Hoekstra, "Corrected momentum exchange method for lattice Boltzmann simulations of suspension flow", *Phys. Rev. E*, vol. 79, pp. 036705, 2009.
[33] E. Lorenz, A.G. Hoekstra, A. Caiazzo, "Lees-Edwards boundary conditions for lattice Boltzmann suspension simulations", *Phys. Rev. E*, vol. 79, pp. 036706, 2009.
[34] P. Lallemand, L.-S. Luo, "Theory of the lattice Boltzmann method: Dispersion, dissipation, isotropy, Galileian invariance and stability", *Phys. Rev. E*, vol. 61, pp. 6546–6562, 2000.
[35] P. Lallemand, L.-S. Luo, "Lattice Boltzmann method for moving boundaries", *J. Comput. Phys.*, vol. 184, pp 406–421, 2003.
[36] H. Li, X. Lu, H. Fang, Y. Qian, "Force evaluations in lattice Boltzmann simulations with moving boundaries in two dimensions", *Phys. Rev. E*, vol. 70, pp. 026701, 2004.
[37] G.R. McNamara, G. Zanetti, "Use of the Boltzmann equation to simulate lattice-gas automata", *Phys. Rev. Lett.*, vol. 61, pp. 2332–2335, 1998.
[38] R. Mei, L.-S. Luo, P. Lallemand, D. d'Humières, "Consistent initial conditions for lattice Boltzmann simulations", *Comput. Fluids*, vol. 35, pp. 855–862, 2006.
[39] R. Mei, D. Yu, W. Shyy, L.-S. Luo, "An accurate curved boundary treatment in the lattice Boltzmann method", *J. Comput. Phys.*, vol. 155, pp. 307–330, 1999.
[40] R. Mei, D. Yu, W. Shyy, L.-S. Luo, "Force evaluation in the lattice Boltzmann method involving curved geometry", *Phys. Rev. E*, vol. 65, pp. 041203, 2002.
[41] R.R. Nourgaliev, T.N. Dinh, T.G. Theofanous, D. Joseph, "The lattice Boltzmann equation: Theoretical Interpretation, Numerics and Implications", *Int. J. Multiphase Flow*, vol. 29, pp. 117–169, 2003.
[42] S. Succi, *The Lattice Boltzmann Equation for Fluid Dynamics and Beyond*, Oxford University Press, Oxford, 2001.
[43] D.A. Wolf-Gladrow, *Lattice Gas Cellular Automata and Lattice Boltzmann Models*, Springer, Berlin, 2000.
[44] Z. Yang, "Analysis of Lattice Boltzmann Boundary Conditions", PhD thesis, University of Konstanz, 2007.
[45] D. Yu, R. Mei R, L.-S. Luo, W. Shyy, "Viscous flow computation with the method of the lattice Boltzmann equation", *Prog. in Aerosp. Sci.*, vol. 39, pp. 329–367, 2003.

Progress in Computational Physics, Vol. 3, 2013, 127-154

CHAPTER 5

Numerical Lifting for Lattice Boltzmann Models

Ynte Vanderhoydonc[1,*], Wim Vanroose[1], Christophe Vandekerckhove[2], Pieter Van Leemput[3], Dirk Roose [4]

[1] *Department of Mathematics and Computer Science, Universiteit Antwerpen, Belgium,* [2] *Alfons Van Zandyckestraat 9, 9800 Meigem, Belgium,* [3] *Priester Daensstraat 12, 2840 Rumst, Belgium,* [4] *Department of Computer Science, Katholieke Universiteit Leuven, Belgium*

Abstract: In this chapter we give an overview of various lifting strategies for lattice Boltzmann models (LBMs). A lifting operator finds for a given macroscopic variable the corresponding distribution functions, mesoscopic variables of the lattice Boltzmann model. This is, for example, useful in coupled LBM and partial differential equation (PDE) models, where one part of the domain is described by a macroscopic PDE while another part is modeled by a LBM. Such a hybrid coupling results in missing data at the interfaces between the different models. The lifting operator provides the correct boundary conditions for the LBM domain at the interfaces. We discuss the accuracy, computational cost and convergence rate of some analytical and numerical lifting procedures.

Keywords: Chapman-Enskog expansion, computational cost, Constrained Runs, hybrid models, initialization, lattice Boltzmann models, lifting operator, macroscopic partial differential equations, numerical Chapman-Enskog expansion, spatial coupling.

1. INTRODUCTION

This chapter discusses lifting operators. A lifting operator is an important tool in a multiscale method. It is used to find for a given macroscopic state the corresponding state of the lattice Boltzmann model (LBM) with an accuracy that goes beyond the equilibrium distribution.

The Boltzmann equation is a kinetic equation that describes the evolution of distribution function $f(x,v,t)$ that counts the number of particles in an infinitesimal volume around a certain point x in space with a certain velocity v at some time $t \geq 0$. The lattice Boltzmann model is a special discretization of this Boltzmann equation that considers only a limited number of velocities from a discrete set defined by the geometry of a grid. Such a LBM uses simple collision and propagation rules that update the distribution functions $f(x_i,v_j,t)$ of idealized particles with time step Δt on a grid defined by grid points x_i with limited velocities v_j such that $v_j\Delta t$ connect neighboring grid points corresponding to space step Δx.

A LBM is a mesoscopic model, with a level of detail that lies between the macro- and microscopic models. They are developed to avoid some of the computational cost of a full kinetic description. More and more modeling is currently done at the mesoscopic level with the help of lattice Boltzmann models. For example, LBMs are used to simulate complex fluid systems such as droplet breakup in turbulence [26], blood cell dynamics [7] and the reproduction of the effects of rough walls, shear thinning and granular flow [9].

Macroscopic models, in contrast, are typically described by a few low order moments and their evolution is simulated by using a macroscopic partial differential equation (PDE). Advection-diffusion-reaction PDEs are macroscopic equivalent descriptions for the LBMs considered in this chapter. For these PDEs very efficient numerical methods have been developed to simulate them with appropriate discretizations. The spatial differential operators are approximated by, for example, finite differences.

There are several applications where these macroscopic moments need to be mapped to the underlying distribution functions of the LBM. For example, starting a LBM from given macroscopic initial conditions

Address correspondence to: Ynte Vanderhoydonc, Universiteit Antwerpen, Department of Mathematics and Computer Science, Campus Middelheim, Middelheimlaan 1, 2020 Antwerpen, Belgium; Tel: ++32 3 265 3859; Fax: ++32 3 265 3777; E-mail: ynte.vanderhoydonc@ua.ac.be

Matthias Ehrhardt (Ed.)

includes some arbitrariness. Typically, there are a large set of microscopic variables and only a few macroscopic variables are given. When the initialization is not consistent it leads to solutions with steep initial layers. Therefore, a lifting operator is necessary to initialize the LBM.

Another important application of a lifting operator is found in hybrid models that combine lattice Boltzmann and macroscopic PDE models. In many cases LBM simulations can be confined to small subdomains where the system needs a detailed description. Elsewhere in the system, where the evolution is slow — for example, away from reaction fronts — macroscopic models can be used. Coupling these different models into a hybrid model is an attractive feature since this can increase the efficiency compared to simulating the full mesoscopic model. Hybrid LBM and PDE models are found in several situations. The spatial coupling was considered by Latt et al. [16]. They couple the lattice Boltzmann model with a finite difference Navier-Stokes solver. An application of this type of coupling can be found in the PDE framework Peano [17]. However, this requires at each interface the construction of a one-to-many map that maps the variables of the PDE to those of the LBM. A lifting operator can offer such a one-to-many map.

A lifting operator can also be found in a hybrid solution method to compute steady states of lattice Boltzmann models [23]. It is inspired by multigrid, where a fine scale model is solved by a correction scheme based on an error equation on a coarse level that corrects the fine scale model. The resulting error needs to be transferred back to the finer grid by using a lifting operator.

This chapter summarizes the work on numerical lifting operators for LBMs that has been done in a series of papers [27, 28, 29, 30, 31, 32, 33, 34, 35, 36, 37]. These lifting operators have been used both for the initialization of lattice Boltzmann models, for coupling of LBM and PDE models and in hybrid solution methods.

This chapter is organized as follows. Section 2. defines the model problem including LBMs and their macroscopic equivalent PDEs. Section 3. includes some lifting strategies to deal with the initialization of the LBM and the missing data at the interfaces of the hybrid model. An analytical method that can be used as a lifting operator is the Chapman-Enskog expansion [2, 27] which writes the distribution functions of the LBM as a function of the density, a given macroscopic variable, and its spatial derivatives. This method is outlined in Section 3.2. The Constrained Runs (CR) algorithm is a numerical method that is analyzed in Section 3.3. It is based on the separation of time scales between the fast detailed microscopic or mesoscopic processes and the slowly varying macroscopic variables. Time trajectories of such systems are quickly attracted toward a slow manifold described by the macroscopic variables [13]. The Constrained Runs scheme [31] performs a series of lattice Boltzmann simulations and resets the macroscopic variable, i.e. the lowest moment, to its initial condition after each run while leaving the higher order moments unchanged. The main drawbacks of these methods are the need to derive analytical expressions and the computational cost. A numerical Chapman-Enskog method [37] is proposed in Section 3.5. to deal with these drawbacks. It combines the ideas of Constrained Runs and the Chapman-Enskog expansion to make the lifting procedure computationally less expensive. Section 4. presents the application of the hybrid spatial coupling. A conclusion can be found at the end of this chapter.

2. MODEL PROBLEM

2.1. Lattice Boltzmann Models

Here we repeat the basic properties of lattice Boltzmann models in order to have a consistent notation. Lattice Boltzmann models (LBMs) [38, 25] describe the evolution of particle distribution functions $f_i(x,t) = f(x,v_i,t)$ defined on a space-time grid with lattice spacing Δx in space and Δt in time. We limit ourselves in this introduction to the D1Q3 model, i.e. only three values are considered for the velocity v_i on the one-dimensional domain

$$v_i = c_i \frac{\Delta x}{\Delta t}, \quad c_i = i \in \{-1,0,1\},$$

with c_i the dimensionless lattice velocity. This corresponds to particles moving to the left (with $c_{-1} = -1$), particles moving to the right ($c_1 = 1$) and particles that do not stream ($c_0 = 0$).

2.1.1. *The Lattice Boltzmann Equation*

The lattice Boltzmann equation (LBE) describing the evolution of the distribution functions is

$$\begin{aligned} f_i(x+c_i\Delta x, t+\Delta t) &= f_i^\star(x,t^\star) \\ &= f_i(x,t)+\Omega_i(x,t), \end{aligned} \tag{1}$$

for $i \in \{-1,0,1\}$. As equation (1) shows, a LBM time step is executed in two stages. First, the term Ω_i, the relaxation to the local equilibrium is evaluated and used to update the value of $f_i(x,t)$ to post-collision values $f_i^\star(x,t^\star)$. Afterwards, these values stream to their neighboring nodes according to their velocity. The LBE evolution (1) for one time step Δt is illustrated in Fig. **1**.

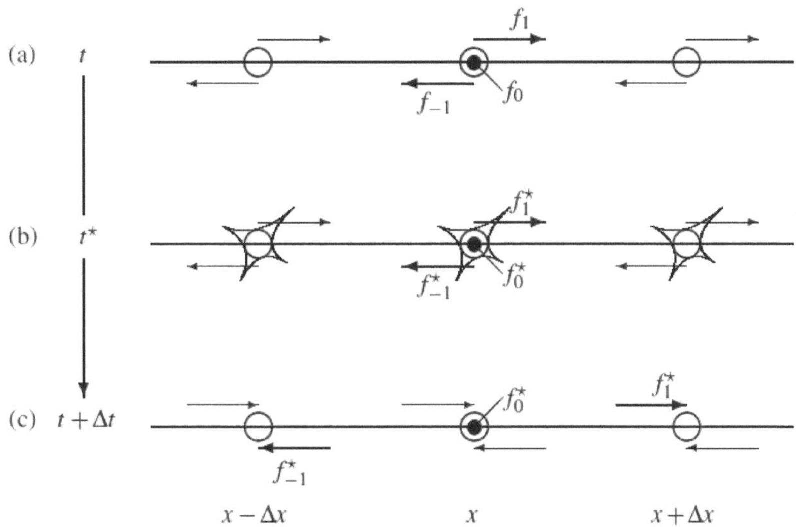

Fig. **1**: Evolution of the D1Q3 LBM for one time step Δt at a lattice site x on a one-dimensional domain [27]. (a) The distributions $f_i(x,t)$, $i \in \{-1,0,1\}$. (b) These variables undergo local collisions. As a result, we obtain post-collision values $f_i^\star(x,t^\star)$. (c) Subsequently, these values propagate to adjacent lattice sites according to their velocity. We obtain the values $f_i(x+c_i\Delta x, t+\Delta t) = f_i^\star(x,t^\star)$.

The particle density $\rho(x,t)$ is defined as the zeroth order velocity moment of the distribution functions

$$\rho(x,t) = \sum_{i=-1}^{1} f_i(x,t) = \sum_{i=-1}^{1} f_i^{eq}(x,t).$$

The second equality is explained below by equation (3).
The diffusive collisions are modeled by the Bhatnagar-Gross-Krook (BGK) collision model [1]. The collision operator Ω_i is defined as [3, 4, 15, 19]

$$\Omega_i(x,t) = -\omega\left(f_i(x,t) - f_i^{eq}(x,t)\right), \tag{2}$$

and models the collisions as a relaxation to a local diffusive equilibrium distribution $f_i^{eq}(x,t)$ which is defined below in (4). Since it is assumed that no particles are created or destroyed during diffusive collisions, we have

$$\begin{aligned} &\sum_{i=-1}^{1} \Omega_i(x,t) = 0 \quad \Leftrightarrow \\ &\sum_{i=-1}^{1} f_i^{eq}(x,t) = \sum_{i=-1}^{1} f_i(x,t) = \rho(x,t). \end{aligned} \tag{3}$$

The latter expresses that the density ρ is a locally conserved quantity during the BGK diffusive collisions. In turn, the equilibrium distribution $f_i^{eq}(x,t)$ is a functional of the conserved variable ρ and is defined as [5, 20, 38]

$$f_i^{eq}(x,t) = \nu_i \rho(x,t), \tag{4}$$

with ν_i satisfying the constraints

$$\sum_{i=-1}^{1} \nu_i = 1 \quad \text{and} \quad \nu_{-1} = \nu_1.$$

This leaves one degree of freedom for the choice of ν_i. For reaction-diffusion systems, all equilibrium weights are chosen equal [20, 6], i.e.

$$\nu_i = \frac{1}{3}.$$

The BGK relaxation coefficient ω in (2) is defined as

$$\omega = \frac{2}{1 + \frac{2D}{\sum_i c_i^2 \nu_i} \frac{\Delta t}{\Delta x^2}}, \tag{5}$$

with D the macroscopic diffusion coefficient. The BGK approximation represents a relaxation to the equilibrium with associated time scale $\tau = \frac{1}{\omega}$. The link between ω and D is explained in Section 2.3.
For later use, we rewrite the LBE (1) with the BGK collision model (2) as

$$f_i(x + c_i\Delta x, t + \Delta t) = (1-\omega) f_i(x,t) + \omega f_i^{eq}(x,t). \tag{6}$$

2.1.2. *Initialization*

In general, starting a microscopic simulator from given macroscopic initial conditions includes some arbitrariness because there are, typically, a large set of microscopic variables and only a few macroscopic variables are given. Because the macroscopic initial state does not contain enough information to initialize the microscopic simulator, the missing information has to be filled in. In molecular dynamics, for example, the initial positions and velocities of all the particles have to be computed from the given temperature and density profile. Typically, the particles are positioned on a square lattice while the velocities are drawn from a Maxwellian distribution.
In a similar way, LBM simulations are traditionally bootstrapped from the initial macroscopic fields by setting the distribution functions in each point to the local BGK equilibrium. For steady state calculations, one then relies on the subsequent time simulation to damp any initialization errors likely to occur. However, careless initialization can lead to significant initialization errors with considerable effects on the computed solution [18, 24].
The same initialization problem occurs in the lifting procedure in the equation-free framework. Using a multiple time scales argument, Kevrekidis et al. argued in [14] that the effect of the errors from the initializations will disappear very fast (compared to the macroscopic time step) as the higher order moments of the microscopic distribution quickly become slaved by the lower order ones (corresponding to the macroscopic variables). I.e. it is assumed that the long-term dynamics of the microscopic model take place on or very near to a lower-dimensional slow manifold, which can be parameterized adequately by the macroscopic variables, and that any orbit started away from this manifold is very quickly attracted to it (at a time scale much smaller than the typical macroscopic time scales). However, this fast slaving process does not imply that all influences of an inaccurately reconstructed initial state disappear quickly. As such, the initialization can have a significant impact on the accuracy of the solution. A discussion of the difficulties associated with an inaccurate initial state can be found in [28] and [35]. The considered macroscopic model in [28] is the FitzHugh-Nagumo PDE system, while the mesoscopic model is an equivalent lattice Boltzmann model. The steady states and periodic solutions of this LBM can be analyzed with techniques of numerical bifurcation analysis. It is the ideal benchmark to compare the coarse-grained (macroscopic level) and mesoscopic bifurcation results. Indeed if the mesoscopic model has a steady state, the best one can hope for is to compute the same steady state with the coarse-grained time stepper (and similarly for periodic solutions). The considered LBM is fully deterministic which makes the initialization of the LBM at each coarse-grained

time step also fully deterministic. Therefore, this model is an ideal example to study the errors caused by the inaccurate initial mesoscopic state in the coarse-grained time stepper. Different possible initialization schemes are tested by considering a bifurcation analysis. Accuracy and stability results of the coarse time stepper for the diffusion problem are considered in [35].
Hence the focus of this chapter is the construction of appropriate lifting operators to deal with this initialization problem for LBMs.

2.2. Moments of the One-Particle Distribution Functions

Besides the density, one can define the dimensionless first and second order velocity moments of the distribution functions as

$$\begin{aligned}\phi(x,t) &= \sum_{i=-1}^{1} c_i f_i(x,t),\\ \xi(x,t) &= \frac{1}{2}\sum_{i=-1}^{1} c_i^2 f_i(x,t).\end{aligned}$$

These are called "momentum" ϕ and "energy" ξ. The tranformation between the distribution functions and the moments is straightforward since the matrix M below is invertible.

$$\begin{aligned}\begin{bmatrix}\rho\\ \phi\\ \xi\end{bmatrix} &= \begin{bmatrix}1 & 1 & 1\\ 1 & 0 & -1\\ \frac{1}{2} & 0 & \frac{1}{2}\end{bmatrix}\begin{bmatrix}f_1\\ f_0\\ f_{-1}\end{bmatrix},\\ &= M\begin{bmatrix}f_1\\ f_0\\ f_{-1}\end{bmatrix}.\end{aligned} \tag{7}$$

It is therefore equivalent to describe the state of the system via $\{f_1, f_0, f_{-1}\}$ or $\{\rho, \phi, \xi\}$.

2.3. Chapman-Enskog Expansion

When the system's solution varies slowly on a macroscopic length and time scale, both the LBM and PDE can be used to describe its evolution. Under this condition, the LBM can be reduced to the PDE using a multiscale Chapman-Enskog expansion [10, 2]. In [27] a detailed derivation for the LBM is given by use of this Chapman-Enskog expansion. It assumes diffusive scaling $x_\varepsilon = \varepsilon x$, $t_\varepsilon = \varepsilon^2 t$, with ε a small tracer parameter, such that the space and time derivatives are scaled as

$$\frac{\partial}{\partial x} = \varepsilon\frac{\partial}{\partial x_\varepsilon} \quad \text{and} \quad \frac{\partial}{\partial t} = \varepsilon^2\frac{\partial}{\partial t_\varepsilon}.$$

The expansion is given by

$$f_i(x,t) = f_i^{[0]}(x,t) + f_i^{[1]}(x,t) + f_i^{[2]}(x,t) + \mathcal{O}(\Delta x^3), \tag{8}$$

with the zeroth and first order contribution, $f_i^{[0]}$ and $f_i^{[1]}$, given by

$$f_i^{[0]} = f_i^{eq} = \frac{1}{3}\rho, \quad f_i^{[1]} = -\frac{c_i \Delta x}{3\omega}\frac{\partial \rho}{\partial x},$$

while the second order contribution $f_i^{[2]}$ is derived as

$$\begin{aligned}f_i^{[2]} &= \frac{\Delta t}{6\omega}(3c_i^2 - 2)\frac{\partial \rho}{\partial t},\\ &= -\frac{\Delta x^2}{18\omega^2}(\omega - 2)(3c_i^2 - 2)\frac{\partial^2 \rho}{\partial x^2}.\end{aligned}$$

The PDE that results by summing over the velocities in expansion (8) is

$$\frac{\partial \rho}{\partial t} = -\frac{\Delta x^2}{3\omega\Delta t}(\omega-2)\frac{\partial^2 \rho}{\partial x^2} = D\frac{\partial^2 \rho}{\partial x^2}, \tag{9}$$

with

$$D = -\frac{\Delta x^2}{3\omega\Delta t}(\omega-2), \tag{10}$$

the diffusion coefficient. According to how many detail one wants for the description of the system, the mesoscopic LBE (6) or the macroscopic PDE (9) can be used for it. Example 1 is included to clarify the main differences between these different mathematical models. With an appropriate initialization, a modeling error occurs between them. It is the mesoscopic difference between the mathematically different LBM and diffusion PDE model.

Example 1. Initialize a LBM with initial condition $f_i(x,0)$, $i \in \{-1,0,1\}$ — determination of $f_i(x,0)$ is considered in the remaining of this chapter, at this point we just focus on the difference between the different types of description, namely the LBM and the macroscopic equivalent PDE — and perform some LBM time steps. The corresponding density can be obtained in every time step by applying $\rho(x,t) = \sum_i f_i(x,t)$. Simulate a full lattice Boltzmann model on a one-dimensional domain with parameters given below in (11).

$$\begin{aligned} &L=10, \quad n=200, \quad \Delta x=\frac{L}{n}=0.05, \quad \Delta t=0.001, \quad \rho(x,0)=e^{-(x-\frac{L}{2})^2}, \\ &D=1 \quad \text{with corresponding} \quad \omega=0.9091. \end{aligned} \tag{11}$$

L is the length of the one-dimensional domain, n the number of spatial grid points, Δx and Δt the space and time steps, $\rho(x,0)$ the given initial density and D the diffusion coefficient. At the boundaries, we impose periodic boundary conditions on the one-dimensional domain.

This LBM simulation is compared in Fig. **2** with the simulation of the macroscopic equivalent diffusion PDE (9) build up with the same parameters.

Fig. **2** contains the simulation of the density of a full PDE model (top left), the density of the full LBM (top right) and the absolute difference $|\rho_{\text{PDE}} - \rho_{\text{LBM}}|$ between them (bottom). The resulting absolute difference $|\rho_{\text{PDE}} - \rho_{\text{LBM}}|$ is the modeling error. This is the mesoscopic difference between the mathematically different LBM and PDE model.

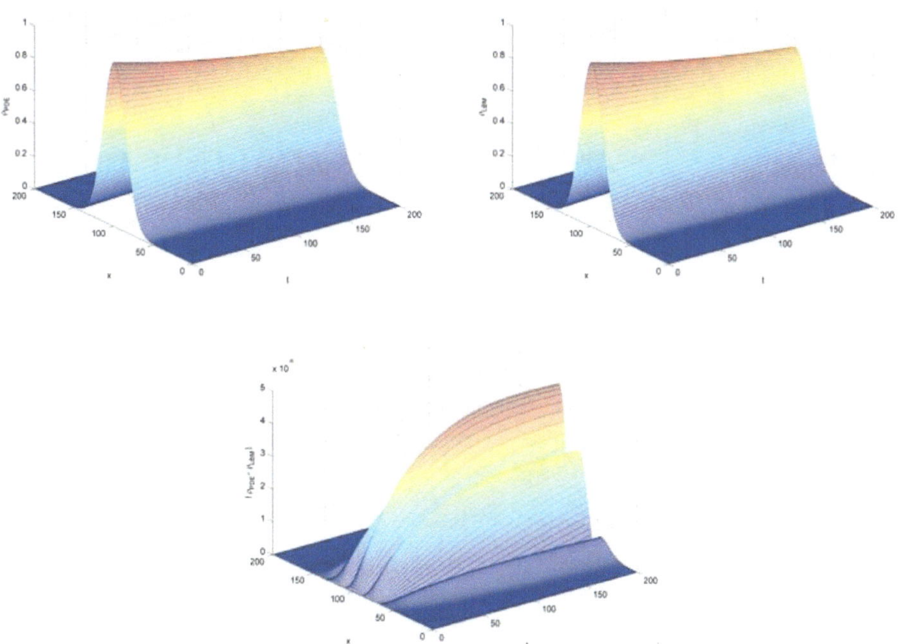

Fig. **2**: Simulation of a full PDE model (top left) compared with the simulation of a full LBM (top right). The absolute difference $|\rho_{\text{PDE}} - \rho_{\text{LBM}}|$ is represented in the bottom figure. The parameters of the simulation are listed in (11) for the one-dimensional spatial domain. The horizontal axes represent the spatial grid points in x and the time steps in t while the vertical axes show the density of respectively the PDE model ρ_{PDE} at the top left, the LBM ρ_{LBM} at the top right and the absolute difference between them at the bottom. The bottom figure represents the modeling error, the mesoscopic difference between the LBM and the PDE model.

Note the diffusive effect of the density as time proceeds, the density decays by spreading out over the spatial domain. The modeling error, shown at the bottom of Fig. **2**, is largest when there are steep gradients. This is further explained in Section 3.2. The macroscopic PDE model is not accurate enough to describe such small scale steep effects. This is what happens, for example, in steep reaction fronts where a LBM or a full kinetic description is necessary.

3. LIFTING OPERATORS

3.1. Introduction

The problem for initializing the LBM is described in Section 2.1.2. Starting the LBM from a given density, for example, is not straightforward. It requires a lifting operator to create this mapping from density to distribution functions. Possible lifting operators constructed for this mapping are discussed in this section. A traditional approach is the Chapman-Enskog expansion (Section 3.2.). It analytically writes the distribution functions as a series of the density and its spatial derivatives. Numerical approaches are the Constrained Runs algorithm (Section 3.3.), the smooth initialization procedure (Section 3.4.) and the numerical Chapman-Enskog expansion (Section 3.5.). These methods are outlined in the remaining of this section.

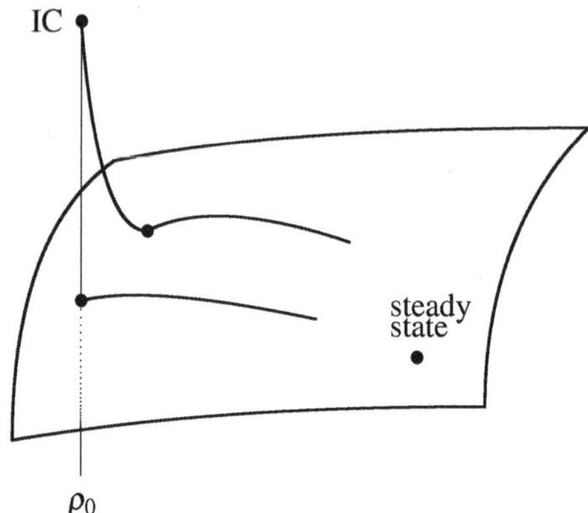

Fig. **3**: Sketch of a slow attracting manifold in phase space [27]. Starting from an initial condition (IC) away from the manifold, the system's evolution can be described in terms of the macroscopic variables $\rho(t)$ on the manifold after a short healing phase. Here we assume that a stable steady state exists.

3.2. Chapman-Enskog Expansion

A traditional approach to construct a $\rho(x,t)$ to $f(x,v,t)$ map is based on the Chapman-Enskog expansion which is only readily available for a few analytic solvable problems. As derived in Section 2.3., the expansion for diffusive systems with diffusion coefficient (10) is

$$f_i(x,t) = \frac{1}{3}\rho(x,t) - \frac{c_i\Delta x}{3\omega}\frac{\partial\rho(x,t)}{\partial x} - \frac{\Delta x^2}{18\omega^2}(\omega-2)(3c_i^2-2)\frac{\partial^2\rho(x,t)}{\partial x^2} + \mathcal{O}(\Delta x^3). \quad (12)$$

The derivatives can be approximated by, for example, finite differences. The distribution functions are then expressed, in an explicit way, as a function of the density in three successive grid points.

From this expansion it can be seen that a traditional bootstrap for LBM simulations from the initial macroscopic fields by setting the distribution functions in each point to the local BGK equilibrium is not an accurate method when the system is not in equilibrium. The higher order derivatives of the density are necessary to obtain a better lifting procedure. This suggests why in Example 1 the modeling error is largest when there are steep gradients.

3.3. Constrained Runs Scheme

It is well known that in phase space the dynamics are quickly attracted to a slow manifold [13] as represented in Fig. **3**. On this slow manifold, the dynamics are parametrized by macroscopic variables. Gear et al. [11] proposed the Constrained Runs (CR) scheme in the context of stiff singularly perturbed ordinary differential equations (ODEs) to consistently map slow initial data to the full system state. For LBM simulations, this corresponds to mapping macroscopic variables, like the density, to the distribution functions of the LBM. The CR-scheme then performs a series of short microscopic simulations and resets the lowest moments of the microscopic variables to the macroscopic initial condition after each run while leaving the higher order moments unchanged. In this way, the scheme computes an approximation of the slaved state corresponding to the initial macroscopic data. As a result, the microscopic variables are initialized both consistent with the initial macroscopic state and close to the slow manifold.

The origin of this algorithm is outlined in Section 3.3.1. while the application to LBMs is discussed in Sections 3.3.2. and 3.3.3.

3.3.1. *Origin*

The CR-algorithm finds its origin in systems of ODEs [11]. Given system

$$\begin{aligned} \frac{\partial u(t)}{\partial t} &= p(u(t), v(t)), \\ \frac{\partial v(t)}{\partial t} &= q(u(t), v(t)), \end{aligned} \tag{13}$$

where only the initial condition for u, namely $u(0) = u_0$, is given. The aim is to find $v(0) = v_0$ such that the initial condition (u_0, v_0) lies on (or close to) the slow manifold. The latter can be formulated by the function $v = v(u)$.
Gear et al. [11] proposed to obtain the v-value from equation

$$\frac{\mathrm{d}^{m+1} v(t=0)}{\mathrm{d}t^{m+1}} = 0, \tag{14}$$

that can be approximated by a forward difference

$$\Delta^{m+1} v(t) \approx \Delta t^{m+1} \frac{\mathrm{d}^{m+1} v(t)}{\mathrm{d}t^{m+1}}. \tag{15}$$

It can be shown that this difference approximation used in the CR-algorithm can be interpreted as a backward extrapolation [32]. It corresponds with a backward extrapolation in time based on a polynomial of degree m that passes through the values v_l $(l = 1, \ldots, m+1)$ while the known variable u is reset to its original initial value u_0. The used coefficients of the forward finite difference formulas at time t are listed in Table **1** for different degrees of m.

Table **1**: Coefficients of the forward finite difference formulas at time t for different degrees of m in equation (15). For example, for $m = 1$, this formula corresponds to $\Delta t^2 \mathrm{d}^2 v(t)/\mathrm{d}t^2 \approx v(t) - 2v(t+\Delta t) + v(t+2\Delta t)$. Using equation (14), this leads to $v(t) = 2v(t+\Delta t) - v(t+2\Delta t)$.

m	$t+\Delta t$	$t+2\Delta t$	$t+3\Delta t$	$t+4\Delta t$
0	1	0	0	0
1	2	-1	0	0
2	3	-3	1	0
3	4	-6	4	-1

The general algorithm for a constant extrapolation, $m = 0$, is given in Algorithm 1.

Algorithm 1 Constrained Runs for a constant extrapolation in time in the system of ODEs (13)

Require: Initial condition $u(0) = u_0$
  Choose v_0, norm $||.||$ and a tolerance θ
  repeat
    Advance the model with one time step Δt:
      u_1 and v_1 at time $t = \Delta t$
    Difference approximation $\Delta v_0 = v_1 - v_0$
    $v_0 \rightarrow v_0 + \Delta v_0$
    Reset u to u_0, the given initial condition
  until $||\Delta v_0|| < \theta$

Similar algorithms can be constructed for higher degrees of m by advancing the model during more time steps and using (14) for different values of m.
We now describe the application of Constrained Runs to LBMs.

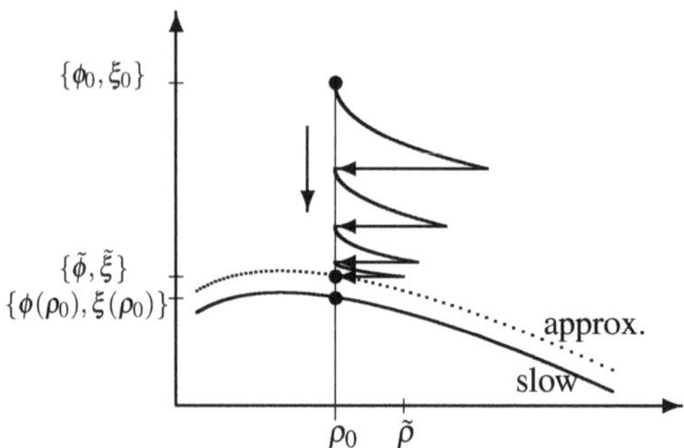

Fig. **4**: Schematic representation of the Constrained Runs algorithm for the LBM with a constant backward extrapolation in time (Algorithm 2) [27]. This algorithm converges toward the slow manifold. ρ is expressed on the horizontal axis and the missing moments ϕ and ξ on the vertical axis. We start iterating with ρ_0, the known density and initial guesses ϕ_0 and ξ_0 for the missing moments. After each step of the LBM, the density is reset to its initial value. This algorithm gives an approximation for the missing values ϕ and ξ on the slow manifold.

3.3.2. *CR-scheme for LBMs with a Constant Extrapolation in Time*

The CR-scheme [12, 11] is a fixed point iteration scheme that computes the full state of a microscopic time simulator on (or close to) the slow manifold corresponding to the given macroscopic variables. A similar idea can be applied for LBMs. For initializing the LBM scheme via the CR-algorithm, some LBM steps are performed before resetting the density. The number of steps determines the accuracy of the scheme. Doing only one LBM step (constant extrapolation in time) results in Algorithm 2 and is visualized in Fig. **4**. Higher order versions can be considered (Section 3.3.3.) although most analytical and theoretical results are available for the constant extrapolation in time.
The CR-procedure for LBMs iterates upon the higher order moments ϕ and ξ, given $\rho_0 = \rho(x,0)$ since it is equivalent to determine $\{f_1, f_0, f_{-1}\}$ or $\{\rho, \phi, \xi\}$. The missing moments are denoted as

$$v = \begin{pmatrix} \phi \\ \xi \end{pmatrix},$$

a long vector $v \in \mathbb{R}^{2n}$, the variable $u = \rho \in \mathbb{R}^n$ denotes the known initial condition $u(0) = u_0 \in \mathbb{R}^n$. Here n is the number of spatial grid points.
The vector v^k denotes the k-th iterate of the CR-algorithm and the iterations are related by

$$v^{k+1} = \mathscr{C}_m(u_0, v^k), \tag{16}$$

where $\mathscr{C}_m$ denotes one step of the CR-algorithm and m is related to the order of the time derivative that is set to zero in equation (14).
Fig. **4** sketches the procedure's evolution. We denote the fixed point of the CR-scheme by $\{\tilde{\phi}, \tilde{\xi}\}$ and the corresponding density from the LBM simulation by $\tilde{\rho}$. Due to the LBM simulation, $\tilde{\rho}$ will be different from the ρ_0 to reset to (that is, unless the initial condition is a steady state of the microscopic model itself and $\tilde{\rho} = \rho_0$). The corresponding higher order moments $\{\tilde{\phi}, \tilde{\xi}\}$ are only approximations to the missing higher order moments $\{\phi(\rho_0), \xi(\rho_0)\}$ of the slaved state. The values $\{\rho_0, \tilde{\phi}, \tilde{\xi}\}$ correspond to a point close to the slow manifold.
The details for a one-dimensional diffusive LBM are given in Algorithm 2. For a given density profile ρ_0, a good initial guess for the initial distribution functions is the BGK equilibrium distribution (4) with $v_i = 1/3$. The LBM is then repeatedly used to evolve the state for a short time Δt. After each such simulation, the lowest moment of the distribution functions is reset to the initial density profile. A straightforward choice

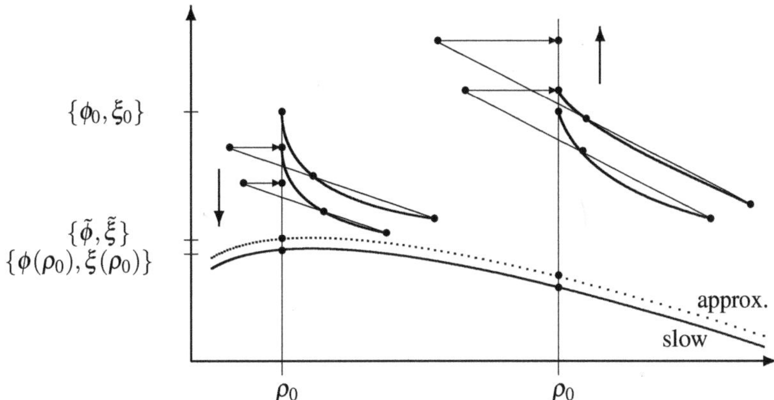

Fig. **5**: Sketch of the evolution of the Constrained Runs initialization scheme with a linear extrapolation in time (Algorithm 3) [27]. The higher order moments ϕ and ξ are plotted with respect to the macroscopic variable ρ. We perform a backward linear extrapolation using the ϕ and ξ values obtained from two LBM simulations and reset ρ to the given ρ_0 afterwards. The left iteration is stable, the right one unstable.

for the stopping criterion is

$$\begin{aligned} ||\phi^{k+1}(x) - \phi^k(x)|| < \theta \quad \text{and} \\ ||\xi^{k+1}(x) - \xi^k(x)|| < \theta, \end{aligned} \tag{17}$$

with tolerance θ chosen by the user.

Algorithm 2 Constrained Runs scheme for diffusive LBM (constant extrapolation in time)

Require: $\rho_0 = \rho(x,0)$ and a tolerance θ
  Choose initial distribution f^0, e.g. equilibrium distribution with $v_i = \frac{1}{3}$
  $\phi^0 = (v_1 - v_{-1})\rho_0$
  $\xi^0 = \frac{1}{2}(v_1 + v_{-1})\rho_0$
  repeat
    LBM simulation on f^k (6) over time Δt: this leads to f^{k+1}
    Transform f^{k+1} via (7) into the moments, ρ^{k+1}, ϕ^{k+1} and ξ^{k+1}
    Reset the density ρ^{k+1} to ρ_0, the given initial density
    Transform the moments $(\rho_0, \phi^{k+1}, \xi^{k+1})$ back to distribution functions f^{k+1} via (7)
  until stopping criterion (17)

Theoretical results are obtained in [30] and [31]. An extensive analysis of the stability and convergence of the CR initialization scheme for the LBM with BGK collisions for one-dimensional reaction-diffusion problems is performed. The CR-scheme is unconditionally stable and convergent and the resulting higher order moments correspond to the Chapman-Enskog slaving relations up to first order. The asymptotic convergence rate equals $|1-\omega|$ with ω the BGK relaxation parameter.

3.3.3. *Higher Order CR-scheme for LBMs*

Higher order variants of the Constrained Runs algorithm use a higher order m in (14). A sketch of the evolution of such a higher order Constrained Runs scheme is shown in Fig. **5** and Algorithm 3 for a linear extrapolation in time ($m = 1$).

Both the constant and linear extrapolation schemes can be written as stationary iterations in the unknown coefficients. The first is unconditionally stable if the underlying LBM is stable.

The higher order versions (higher order extrapolation, including linear extrapolation) of this Constrained Runs algorithm can be unstable. In [33] the instability is circumvented by reformulating the fixed point

Algorithm 3 Constrained Runs scheme for diffusive LBM (linear extrapolation in time)

Require: $\rho_0 = \rho(x,0)$ and a user-defined tolerance θ
Choose initial distribution functions, e.g. equilibrium distribution with $v_i = \frac{1}{3}$
$\phi^0 = (v_1 - v_{-1})\rho_0$
$\xi^0 = \frac{1}{2}(v_1 + v_{-1})\rho_0$
repeat
First LBM simulation on f^k (6) over time Δt: this leads to f^{k+1}
Calculate ϕ_1^{k+1} and ξ_1^{k+1} of the resulting distribution functions
Second LBM simulation on f^{k+1} (6) over time Δt
Calculate ϕ_2^{k+1} and ξ_2^{k+1} of the resulting distribution functions
Extrapolation of higher order moments
$\phi^{k+1} \leftarrow 2\phi_1^{k+1} - \phi_2^{k+1}$
$\xi^{k+1} \leftarrow 2\xi_1^{k+1} - \xi_2^{k+1}$
Reset the density to ρ_0
Transform $(\rho_0, \phi^{k+1}, \xi^{k+1})$ back to distribution functions f^{k+1} via (7)
until stopping criterion (17)

problem.
In general, equation (16) is nonlinear and the fixed point can be found by a Newton iteration. This means solving

$$g_m(u_0, v) := v - \mathscr{C}_m(u_0, v) = 0,$$

for a macroscopic value u_0. Newton's method gives an update to the guesses as

$$v^{k+1} = v^k + \delta v^k,$$

where the corrections δv^k are found by solving the linear system

$$\begin{aligned} A(u_0, v^k) \cdot \delta v^k &= \frac{\partial g_m}{\partial v}(u_0, v^k) \cdot \delta v^k, \\ &= \left(I - \frac{\partial \mathscr{C}_m}{\partial v}(u_0, v^k) \right) \cdot \delta v^k, \\ &= -g_m(u_0, v^k), \end{aligned}$$

with $A = \frac{\partial g_m}{\partial v}$ the linearization (Jacobian matrix) of g_m and $\frac{\partial \mathscr{C}_m}{\partial v}$ the linearization of $\mathscr{C}_m$. The linearization can be estimated with the help of the approximation

$$Ae_i \approx \frac{g_m(u_0, v + \varepsilon e_i) - g_m(u_0, v)}{\varepsilon},$$

with e_i the unit vector, $i = 1, \ldots, 2n$ and ε small.
This was then used to create a lifting operator for the full state. A serious drawback is that this method requires the construction and solution of large Jacobian systems, which can make it prohibitive for realistic applications. Many evaluations of the underlying Boltzmann model are then required to construct the Jacobian matrix.
Similarly, matrix-free methods like the generalized minimal residual method (GMRES) that combine the solution of the system with the estimation of the Jacobian require many matrix-vector products. This is because the spectrum is unfavorable for fast convergence [33, 32]. GMRES approximates the exact solution $x^* = A^{-1}b$ by the vector x_n from the Krylov subspace $\mathscr{K}_n =< r_0, Ar_0, \ldots, A^{n-1}r_0 >$ generated by A and r_0, such that the two-norm of the residual $r_n = b - Ax_n$ is minimized. An example of the convergence history of applying GMRES is presented in Fig. **6**. This figure is produced in [33] for various values of ω, $m = 0$ and 128 grid points. The theoretical upper bounds computed in [33] are also depicted in the figure. The convergence can be accelerated with an appropriate preconditioner [33]. After applying the preconditioner, the spectral properties are more favorable such that GMRES converges faster.

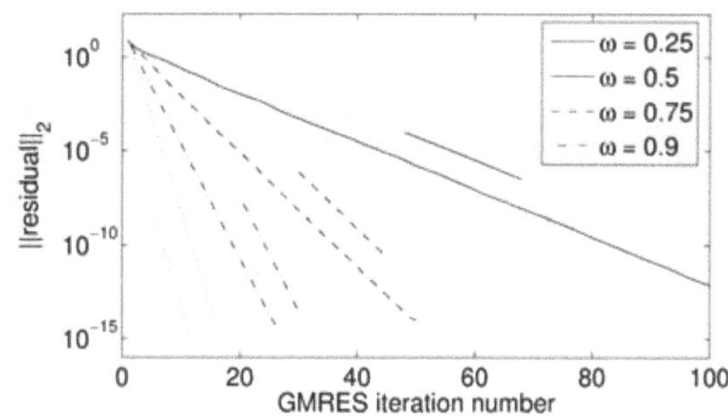

Fig. **6**: GMRES convergence history [33] for the LBM with 128 grid points and various values of ω when $m = 0$. The theoretical upper bounds are also depicted (short lines).

3.3.4. *Consistent Initialization on a Slow Manifold*

A recent method proposed to initialize on a slow manifold is discussed in [34]. The technique is presented for the computation of coarse-scale steady states of dynamical systems with time scale separation by using a fine-scale simulator but can be modified to provide initial conditions on the slow manifold. The approach uses short bursts of the fine-scale simulator to track changes in the coarse variables of interest. Constrained Runs is based on the moments of the full state space, which is typically much bigger than the space of macroscopic variables. The method proposed in [34] performs most of the computations only in the space of the observables. As an extra surplus, this method remains stable even if Constrained Runs has instability problems.

3.4. Smooth Initialization

A procedure similar to the Constrained Runs algorithm is the smooth initialization principle. The missing initial data are defined in such a way that the output behaves smoothly, without initial errors. This procedure will not be outlined here but can be found in [29]. [36] continued this work with a focus on the specific interfaces for hybrid coupling problems.

3.5. Numerical Chapman-Enskog Expansion

The methods discussed in Sections 3.2. and 3.3. are well known methods to construct a lifting operator for LBMs. However, each has some drawbacks.

A drawback of the use of the Chapman-Enskog expansion (Section 3.2.) is the necessity to construct the expressions analytically.

The Constrained Runs scheme (Section 3.3.) can be used to approximate these expressions numerically. However, higher order Constrained Runs can be unstable and requires a Newton-Krylov method to avoid stability problems. This makes the lifting method computationally expensive since it requires many evaluations of the underlying lattice Boltzmann model to construct and solve the Jacobian matrix.

In the context of a hybrid model local updates around the ghost points can bring the computational cost down [36]. However, it still remains computationally expensive to use in practice, especially in higher dimensional problems.

In [37] a numerical Chapman-Enskog expansion is proposed that alleviates the drawbacks. It combines the ideas of Constrained Runs and the Chapman-Enskog expansion. Instead of using Constrained Runs to find for each grid point the missing moments ϕ and ξ of the distribution functions, it uses Constrained Runs to find the unknown coefficients of the Chapman-Enskog expansion. This has several advantages. First, it significantly reduces the number of unknowns in the lifting since we only need to find the coefficients rather than the full state $f_i(x,t)$, $i \in \{-1,0,1\}$. And secondly, it can be done off-line before the calculations.

Indeed, once the coefficients are found they can be reused in every time step to realize the lifting since they remain constant. A summary of the use and advantages of this method is presented below.
The solution $f_i(x,t)$, $i \in \{-1,0,1\}$ of a LBM with an infinite domain and parameters Δx, Δt and ω can be written as a series of $\rho(x,t)$, the macroscopic density. This is presented in equation (18) below.

$$\begin{aligned} f_i(x,t) = f_i^{eq}(x,t) + \alpha_i \frac{\partial \rho}{\partial x} + \beta_i \frac{\partial^2 \rho}{\partial x^2} + \delta_i \frac{\partial^3 \rho}{\partial x^3} + \varepsilon_i \frac{\partial^4 \rho}{\partial x^4} + \dots \\ + \gamma_i \frac{\partial \rho}{\partial t} + \zeta_i \frac{\partial^2 \rho}{\partial t^2} + \dots + \eta_i \frac{\partial^2 \rho}{\partial x \partial t} + \dots, \end{aligned} \tag{18}$$

where

$$i \in \{-1,0,1\} \quad \text{and} \quad \alpha = \begin{pmatrix} \alpha_1 \\ \alpha_0 \\ \alpha_{-1} \end{pmatrix} \in \mathbb{R}^3, \quad \beta = \begin{pmatrix} \beta_1 \\ \beta_0 \\ \beta_{-1} \end{pmatrix} \in \mathbb{R}^3, \dots, \tag{19}$$

are fixed constants that only depend on ω, Δx and Δt. The justification of this expansion of the distribution functions in the form (18) is outlined in [37].
With the help of expansion (18) it is possible to build a lifting operator that constructs the distribution functions for a given density. The determination of the vectors of constants (19), the coefficients of such a lifting operator (18), is discussed below. To simplify the discussion and notation we limit ourselves to a truncated series

$$f(x,t) = f^{eq}(x,t) + \alpha \frac{\partial \rho}{\partial x} + \beta \frac{\partial^2 \rho}{\partial x^2} + \gamma \frac{\partial \rho}{\partial t}, \tag{20}$$

where α, β and γ are the vectors containing the constants. The method is easily generalized to include higher order terms which is considered later on in this section.
Using the fact that equation (20) is valid for every possible grid point, we can consider three grid points x_j, x_k and x_l and set up a linear system for the nine unknowns (three vectors each containing three constants). Where j, k and l are certain indices determined in a later stage of this section.

$$\left(\begin{array}{ccc|ccc|ccc}
\frac{\partial\rho(x_j)}{\partial x} & & & \frac{\partial^2\rho(x_j)}{\partial x^2} & & & \frac{\partial\rho(x_j)}{\partial t} & & \\
 & \frac{\partial\rho(x_j)}{\partial x} & & & \frac{\partial^2\rho(x_j)}{\partial x^2} & & & \frac{\partial\rho(x_j)}{\partial t} & \\
 & & \frac{\partial\rho(x_j)}{\partial x} & & & \frac{\partial^2\rho(x_j)}{\partial x^2} & & & \frac{\partial\rho(x_j)}{\partial t} \\ \hline
\frac{\partial\rho(x_k)}{\partial x} & & & \frac{\partial^2\rho(x_k)}{\partial x^2} & & & \frac{\partial\rho(x_k)}{\partial t} & & \\
 & \frac{\partial\rho(x_k)}{\partial x} & & & \frac{\partial^2\rho(x_k)}{\partial x^2} & & & \frac{\partial\rho(x_k)}{\partial t} & \\
 & & \frac{\partial\rho(x_k)}{\partial x} & & & \frac{\partial^2\rho(x_k)}{\partial x^2} & & & \frac{\partial\rho(x_k)}{\partial t} \\ \hline
\frac{\partial\rho(x_l)}{\partial x} & & & \frac{\partial^2\rho(x_l)}{\partial x^2} & & & \frac{\partial\rho(x_l)}{\partial t} & & \\
 & \frac{\partial\rho(x_l)}{\partial x} & & & \frac{\partial^2\rho(x_l)}{\partial x^2} & & & \frac{\partial\rho(x_l)}{\partial t} & \\
 & & \frac{\partial\rho(x_l)}{\partial x} & & & \frac{\partial^2\rho(x_l)}{\partial x^2} & & & \frac{\partial\rho(x_l)}{\partial t}
\end{array}\right)
\left(\begin{array}{c} \alpha_1 \\ \alpha_0 \\ \alpha_{-1} \\ \hline \beta_1 \\ \beta_0 \\ \beta_{-1} \\ \hline \gamma_1 \\ \gamma_0 \\ \gamma_{-1} \end{array}\right)
= \left(\begin{array}{c}
f_1(x_j,t) - f_1^{eq}(x_j,t) \\
f_0(x_j,t) - f_0^{eq}(x_j,t) \\
f_{-1}(x_j,t) - f_{-1}^{eq}(x_j,t) \\ \hline
f_1(x_k,t) - f_1^{eq}(x_k,t) \\
f_0(x_k,t) - f_0^{eq}(x_k,t) \\
f_{-1}(x_k,t) - f_{-1}^{eq}(x_k,t) \\ \hline
f_1(x_l,t) - f_1^{eq}(x_l,t) \\
f_0(x_l,t) - f_0^{eq}(x_l,t) \\
f_{-1}(x_l,t) - f_{-1}^{eq}(x_l,t)
\end{array}\right). \tag{21}$$

For a given $f_i(x,t)$ where $i \in \{-1,0,1\}$, the linear system (21) will give the coefficients α, β and γ.

Remark 1. If a PDE in closed form exists that describes the evolution of ρ in the form of $\rho_t + a\rho_x = D\rho_{xx}$, then the linear system (21) will be singular. Indeed, the PDE will give a relation between ρ_t, ρ_x and ρ_{xx} in each of the grid points x_j, x_k and x_l. As a result, every element in the last three columns of the linear system (21) can be written as a linear combination of the first six columns. In practice, however, the PDE is only an approximation and the system will be close to singular.

However, linear system (21) only delivers the correct coefficients if f_i is smooth enough such that f_i^{eq} satisfies the smoothness condition. This is the case when f_i lies on the slow manifold. The Constrained Runs algorithm offers a way to reach the slow manifold in an iterative way.
We combine the ideas of the CR-algorithm to reach the slow manifold and the Chapman-Enskog expansion to find the unknown constants on this slow manifold. The numerical procedure to do so is given in Section 3.5.1.

Remark 2. The choice of the grid points x_j, x_k and x_l for equation (21) should be such that the condition number of the matrix is optimal.

Remark 1 clarifies why we do not solve the linear system in (21) and first focus on one that is not close to singular in Section 3.5.1.

3.5.1. *Numerical Procedure to Construct the Lifting Operator*

From the previous discussion it is clear that the coefficients of the lifting operator can be extracted from a linear system once f approaches the slow manifold. But it is desirable to avoid many LBM steps that are required to obtain smooth enough distribution functions on the slow manifold.
We need a way to circumvent these LBM time steps and at the same time avoid the singularity in the matrix that determines the coefficients. This can be done by formulating a fixed point for the unknown coefficients as considered in the next paragraph.

Distribution Functions that Approach the Slow Manifold This discussion is limited to the first few terms of the expansion. The singular system can be avoided by taking a series that only contains spatial derivatives. Such a series can represent the same state since the time derivative $\partial_t \rho$ is often related to the spatial derivatives through the macroscopic PDE. For example, suppose that a PDE of the form $\rho_t + a\rho_x = D\rho_{xx}$ describes the behavior of ρ. It is then possible to eliminate $\partial_t \rho$ from the expansion. The coefficients are then $\tilde{\alpha} = \alpha - \gamma a$ and $\tilde{\beta} = \beta + \gamma D$. The distribution functions can now be represented as a series with only spatial derivatives.

$$f(x,t) = f^{eq}(x,t) + \tilde{\alpha}\frac{\partial \rho}{\partial x} + \tilde{\beta}\frac{\partial^2 \rho}{\partial x^2}. \tag{22}$$

Rewrite $\tilde{\alpha}$ and $\tilde{\beta}$ as α and β but bear in mind that the coefficients in equation (20) and equation (22) are different.
Again, once the distribution functions are close to the slow manifold, we can extract the coefficients α and β from the linear system

$$\left(\begin{array}{ccc|ccc}
\frac{\partial\rho(x_j)}{\partial x} & & & \frac{\partial^2\rho(x_j)}{\partial x^2} & & \\
 & \frac{\partial\rho(x_j)}{\partial x} & & & \frac{\partial^2\rho(x_j)}{\partial x^2} & \\
 & & \frac{\partial\rho(x_j)}{\partial x} & & & \frac{\partial^2\rho(x_j)}{\partial x^2} \\
\hline
\frac{\partial\rho(x_k)}{\partial x} & & & \frac{\partial^2\rho(x_k)}{\partial x^2} & & \\
 & \frac{\partial\rho(x_k)}{\partial x} & & & \frac{\partial^2\rho(x_k)}{\partial x^2} & \\
 & & \frac{\partial\rho(x_k)}{\partial x} & & & \frac{\partial^2\rho(x_k)}{\partial x^2}
\end{array}\right)
\left(\begin{array}{c}
\alpha_1 \\ \alpha_0 \\ \alpha_{-1} \\ \hline \beta_1 \\ \beta_0 \\ \beta_{-1}
\end{array}\right)
= \left(\begin{array}{c}
f_1(x_j,t) - f_1^{eq}(x_j,t) \\
f_0(x_j,t) - f_0^{eq}(x_j,t) \\
f_{-1}(x_j,t) - f_{-1}^{eq}(x_j,t) \\ \hline
f_1(x_k,t) - f_1^{eq}(x_k,t) \\
f_0(x_k,t) - f_0^{eq}(x_k,t) \\
f_{-1}(x_k,t) - f_{-1}^{eq}(x_k,t)
\end{array}\right), \tag{23}$$

which is constructed such that it avoids problems with a possible singularity.
However, we are still faced with the problem that f should lie close to the slow manifold which requires many LBM steps to eliminate any possible initialization errors.
An alternative is to combine Constrained Runs with the extraction of the coefficients. Consider a numerical function $h(\alpha,\beta;\rho,m)$ as described in Function 4. This function takes as input α, β, a fixed density ρ and an integer m — m is related to the claim of m-th order smoothness discussed below — as parameters. It first constructs, with this input, a state f with the help of series (22). These distribution functions are then used to perform multiple LBM steps. For each of these steps we can find the corresponding moments ϕ and ξ. On the moments, we can use the CR-algorithm (Section 3.3.) to find the moments that are closer to the slow manifold by considering the finite difference approximations of the m-th order smoothness condition (24).

$$\begin{aligned}
\frac{d^{m+1}\phi}{dt^{m+1}} &= 0, \\
\frac{d^{m+1}\xi}{dt^{m+1}} &= 0.
\end{aligned} \tag{24}$$

This leads to a new guess for the moments ϕ and ξ. These moments results in new coefficients α and β, by applying the linear system, equation (23), on the distribution functions f_i corresponding to the new ϕ and ξ and the given ρ.
The idea is now to determine α and β such that they are invariant under this numerical function $h(\alpha,\beta;\rho,m)$. Indeed, if the initial and final state can be described by the same α and β then the lifted f is close to the slow manifold since the underlying iterations are based on the CR-algorithm for which the fixed point is found by performing LBM steps.
Also in this case, a Newton iteration can be used instead of performing a fixed point iteration with $h(\alpha,\beta;\rho,m)$. Instead of constructing the expensive Jacobian on the moments in every grid point and every time step — as is done in the original Constrained Runs algorithm in combination with Newton's method in Section 3.3. — we do so on the coefficients α and β only once before the actual hybrid simulation. This reduces the computational cost significantly because the size of the system is much smaller.

Function 4 $h(\alpha,\beta;\rho,m)$

Require: Guess on coefficients α, β, given density ρ, order m of the smoothness in equation (24).
1: Construct lifting operator: $f = f^{eq} + \alpha\frac{\partial\rho}{\partial x} + \beta\frac{\partial^2\rho}{\partial x^2}$.
2: Compute corresponding moments ϕ and ξ by applying equation (7).
3: Perform $m+1$ LBM time steps to compute $\frac{d^{m+1}\phi}{dt^{m+1}} = 0$ and $\frac{d^{m+1}\xi}{dt^{m+1}} = 0$ by a forward finite difference formula This results in new moments ϕ and ξ. {find moments closer to slow manifold}
4: Revert back to distribution functions f by applying equation (7).
5: Select grid points x_j and x_k to construct linear system (23).
6: Solve the system for α and β.
7: **return** α, β.

Macroscopic PDE Once the fixed point is found, we have α and β that lifts ρ to the distribution functions close to the slow manifold.

By performing two more LBM steps, $\partial_t\rho$ can be calculated by using a forward finite difference formula. System (21) can be applied to find the vectors of constants of this larger system that includes time derivatives. There are now two possibilities: either the system is non-singular and it can be solved for the coefficients and only an approximate PDE can be found. Or it is too singular to be solved accurately but then the PDE can be extracted from the nullspace of the system.

Let us first discuss the situation where the matrix in (21) is non-singular. The system can then be solved for α, β and γ. The approximate PDE can be determined by summing over the obtained coefficients.

$$\frac{\partial\rho}{\partial t} = -\frac{\sum_i\alpha_i}{\sum_i\gamma_i}\frac{\partial\rho}{\partial x} - \frac{\sum_i\beta_i}{\sum_i\gamma_i}\frac{\partial^2\rho}{\partial x^2}.$$

This PDE is only approximate. Otherwise, if it would hold exactly, the system would be singular as expected. For a singular system, we know that one or more of the eigenvalues will be zero with a corresponding null eigenvector. Focusing on the null eigenvector $v = \{v_1, v_2, \ldots, v_9\}$, we know that $Av = 0$ with A the matrix in system (21). Using this, we obtain the PDE

$$\frac{\partial\rho(x_j)}{\partial t} = -\frac{v_1}{v_7}\frac{\partial\rho(x_j)}{\partial x} - \frac{v_4}{v_7}\frac{\partial^2\rho(x_j)}{\partial x^2}.$$

Numerical Procedure: Choice for Indices in Equation (23) The numerical function $h(\alpha,\beta;\rho,m)$ has a density $\rho(x,0)$ as a parameter and the solution for the coefficients should be independent of its choice of ρ. The coefficients only depend on constants ω, the spatial grid size Δx, time step Δt and the weights of f^{eq} of the underlying LBM. Since the coefficients does not depend on time, we can choose an arbitrary density $\rho(x,0)$ such that the matrix from (23) is away from singular and easily solvable. The spatial derivatives that are considered needs to exist and should not come close to zero during the LBM evolution since otherwise we would end up with hard to solve linear systems. For this reason, we consider $\rho(x,0) = \rho_{\text{test}}$. For example,

$$\begin{aligned}\rho_{\text{test}}(x,0) &= x + \frac{1}{2}x^2 \quad \text{for unknowns} \quad \alpha,\beta,\\ &= x + \frac{1}{2}x^2 + \frac{1}{6}x^3 \quad \text{for unknowns} \quad \alpha,\beta,\delta.\end{aligned}$$

Furthermore, the test domain used in the LBM inside the function $h(\alpha,\beta;\rho,m)$ can be much smaller than the domain on which we want to solve the LBM. A smaller test domain will not affect the constant coefficients of the lifting operator. However, the test domain should use the same Δx and Δt as the LBM of interest since the constants do depend on the chosen spacings in space and time.

Consider, for example, a domain for the complex system of length $L = 10$ build up with $n = 200$ grid points (the domain on which we want to solve the LBM). The space step is $\Delta x = L/n = 0.05$ and as time step $\Delta t = 0.001$ is considered further in Example 2. Since we know that the constants are only affected by these space and time steps, we should consider — together with $\rho(x,0) = \rho_{\text{test}}$ — a test domain with the same step sizes since the vectors of constants (19) are affected by these choices. The test domain is of length $L_{\text{test}} = 3$ such that $n_{\text{test}} = 60$ since $\Delta x = 0.05$. This number of grid points will make it possible to choose the indices x_j, x_k, ... such that system (23) is not close to singular. We can now return to the question which indices should be used in system (23). Focus on the fact that we do not want an effect of wrongly chosen boundary conditions in the smaller test domain. The grid points should be taken far enough from the edges and in points such that the system (23) does not become singular. The indices can be, for example, 10, 20, 30, ... spread over the test domain of 60 grid points.

Remark 3. Note that one can limit the LBM simulation to small intervals around the considered grid points x_j and x_k similar as in the gap-tooth scheme [22]. When the number of iterations needed in Newton's method are known, one knows how many LBM steps will be performed to find the new coefficients α and β. Then the size of the test domain can be shrunk to a smaller subdomain around x_j and x_k. This is the same idea as used in [36] to perform local updates for the CR-algorithm.

Higher Order Versions There are two ways to increase the accuracy. First, more terms in the expansion can be considered such that more derivatives of the density are taken into account. Second, we can require higher order smoothness in the CR-algorithm. This requires more LBM steps and uses a higher order finite difference formula to estimate the derivatives in time in equation (24). Both methods are outlined below.

More Terms in the Expansion So far, the lifting operator was derived for a truncated series (22). The proposed method can easily be extended by considering higher order derivatives, which will give better results. Consider the expansion below

$$f = f^{eq} + \alpha\frac{\partial\rho}{\partial x} + \beta\frac{\partial^2\rho}{\partial x^2} + \delta\frac{\partial^3\rho}{\partial x^3} + \varepsilon\frac{\partial^4\rho}{\partial x^4}. \tag{25}$$

This expansion requires the determination of more unknown coefficients. This can be done by considering — in addition to x_j and x_k — extra grid points x_l and x_m, which leads to a larger system of unknowns.

Remark 4. Since the lifting operator is an infinite series, taking more derivatives into account results in a better output. For example, when $\partial_t^2\rho$ is taken into account as an extra term in the expansion of the larger system (20) instead of just considering $\partial_t\rho$, it can be seen that the effect of the coefficient before $\partial_t^2\rho$ comes from the corresponding spatial derivatives that were considered. However, it will give a less comfortable PDE to work with.

Higher Order Smoothness Higher order smoothness can be performed on the moments ϕ and ξ as in the CR-algorithm by considering a higher order m in equation (24).
For the further conclusions, codes and results higher order derivatives and higher order smoothness is considered.

3.5.2. *Algorithm for Lifting Operator and Macroscopic PDE*

When we combine all the ideas discussed in the previous sections, we end up with an algorithm for a lifting operator that also delivers a macroscopic PDE — for example used to construct a hybrid model. The pseudocode is presented in Algorithm 5 while the complete algorithm is presented below. The algorithm starts by searching for the lifting operator based on the spatial derivatives after which it inserts time derivatives and creates the macroscopic PDE from it.
Start with an initial guess for $\{\alpha,\beta,\delta,\varepsilon\}$ in equation (25). Apply Function 4 $h(\alpha,\beta,\delta,\varepsilon;\rho,m)$ for a given ρ and a certain order m for the order of smoothness until the coefficients $\{\alpha,\beta,\delta,\varepsilon\}$ are invariant under this function. This results in coefficients $\{\alpha,\beta,\delta,\varepsilon\}$ that represent distribution functions closer to the slow manifold. The lifting operator is constructed at this point.
When these distribution functions are found based on the spatial derivatives only, we still need to determine the corresponding PDE by considering the null eigenvector or by a summation of the coefficients. By

performing two extra LBM steps — to estimate the time derivative with a forward finite difference formula — the coefficient γ belonging to the time derivative of the expansion below can be numerically calculated.

$$f = f^{eq} + \alpha\frac{\partial\rho}{\partial x} + \beta\frac{\partial^2\rho}{\partial x^2} + \delta\frac{\partial^3\rho}{\partial x^3} + \varepsilon\frac{\partial^4\rho}{\partial x^4} + \gamma\frac{\partial\rho}{\partial t}.$$

Since the PDE can be obtained from the numerically constructed distribution functions, the exact PDE obtained through the Chapman-Enskog expansion does not need to be obtained analytically.

Remark 5. Note that one can also consider $f^{eq}(x,t) = \kappa\rho(x,t)$ and determine the constants of vector κ in a similar setting by using an extra grid point to obtain a larger system of unknowns.

Remark 6. The LBM simulation is stable in the 2-norm when $0 \leq \omega \leq 2$. However, it is not necessary that the macroscopic PDE, when it is discretized with the same Δx and Δt and forward Euler, is also stable. Indeed, when $\omega \to 0$ for a fixed Δx and Δt the resulting diffusion coefficient D grows, see (5).

The pseudocode to apply the idea of the numerical Chapman-Enskog expansion as a lifting operator is given in Algorithm 5 together with the idea how to determine the PDE to construct a hybrid model from it.

Algorithm 5 Pseudocode numerical Chapman-Enskog expansion with m-th order smoothness

Require: Test domain that defines $\rho(x,0)$, initial guess $a^0 = \{\alpha,\beta,\delta,\varepsilon\}$ = zeros(12,1) and a user-defined tolerance tol, parameter m, the order of the smoothness condition.
repeat
  $a^{k+1} = a^k - \left(J(a^k)\right)^{-1} h(a^k;\rho,m)$ with h defined in Function 4.
until $\|a^{k+1} - a^k\| <$ tol.
Solution $a^* = \{\alpha,\beta,\delta,\varepsilon\}$ gives $f = f^{eq} + \alpha\frac{\partial\rho}{\partial x} + \beta\frac{\partial^2\rho}{\partial x^2} + \delta\frac{\partial^3\rho}{\partial x^3} + \varepsilon\frac{\partial^4\rho}{\partial x^4}$.
Perform 2 more LBM steps to determine $\frac{\partial\rho}{\partial t}$ numerically by a forward finite difference formula.
Construct system (21) — including higher order spatial derivatives — to determine the coefficients of the PDE by using the nullspace of the system.
return Lifting operator $f = f^{eq} + \alpha\frac{\partial\rho}{\partial x} + \beta\frac{\partial^2\rho}{\partial x^2} + \delta\frac{\partial^3\rho}{\partial x^3} + \varepsilon\frac{\partial^4\rho}{\partial x^4}$ and macroscopic PDE.

3.5.3. *Advantages of the Numerical Chapman-Enskog Expansion*

By constructing the lifting operator in this way, we avoid the need to construct the Chapman-Enskog expansion analytically. We only need to find the coefficients (vectors of constants) rather than the full state of the distribution functions as in the Constrained Runs algorithm. And it can be done off-line before the calculations. Indeed, once the coefficients are found they can be reused every time step to realize the lifting. This summarizes the idea of the numerical Chapman-Enskog expansion. As an extra surplus, the macroscopic PDE can be found to construct hybrid models.

4. NUMERICAL APPLICATION: HYBRID LBM AND PDE SIMULATION

The level of detail required to model the system changes from region to region and different models have to be used on different parts of the domain. At the interface between the regions (sometimes called the 'handshaking region'), there will be a mismatch in the kind (and number) of variables used by the different models. There, the variables have to be mapped to one another. Of particular interest are systems where the higher-level description breaks down on a localized part of the domain because there the particle interactions have become so complex that variables absent in the coarse higher-level model become important. Examples are the occurrence of cracks and defects in solids and the viscous fingering phenomena observed in ionized gases [8].

The hybrid model problem with periodic boundary conditions is discussed in this section. The model problem is shown in Fig. 7. It spatially couples a LBM and the finite difference discretization of a PDE.

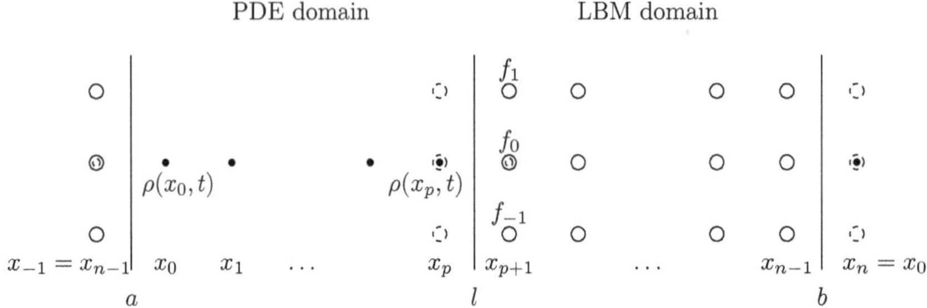

Fig. 7: Application of the lifting operator in a hybrid PDE and LBM model [36]. The domain $[a,b]$ in the hybrid model is split into $[a,l[$ on which we solve the PDE model and $[l,b]$ on which we solve the LBM. The solid points (•) represent the grid for the density ρ of the discrete PDE, the circles (○) represent the LBM variables $(f_1,f_0,f_{-1})^T$. The periodic boundary conditions and the coupling are implemented with ghostcells which are drawn by dashed circles. The density in the ghostcells of the PDE domain, in x_{-1} and x_{p+1}, are found by taking $\sum_i f_i$ in x_{n-1} and x_{p+1}, respectively. However, the ghostcells for the LBM domain, in x_p and x_n, require a lifting operator that lifts ρ to $(f_1,f_0,f_{-1})^T$ in these points. Details are described in Section 4.

A one-dimensional domain $[a,b]$ is considered that couples the diffusion PDE (9) on $[a,l[$ with the LBE (6) on $[l,b]$. Since we focus on the error caused by the coupling, we use the same grid spacings in space (Δx) and in time (Δt) for both regions. Using a different space-time grid is of particular interest for future work since the grid of the PDE domain can be further coarsened when ρ is smooth. Then some interpolation operator needs to be considered at the coupling point. [21] proposes a method to couple grids with different mesh sizes for classical lattice Boltzmann schemes. Numerical convergence studies are performed there on a simple lattice Boltzmann algorithm solving the advection-diffusion equation.

On the domain $[a,l[$, we discretize the diffusion PDE with cell centered central differences and forward Euler time discretization. The grid points x_j with $j\in\{0,1,\ldots,p\}$ cover this domain and for these points it holds that

$$\rho(x_j,t+\Delta t)=\rho(x_j,t)+\frac{D\Delta t}{\Delta x^2}\left(\rho(x_{j-1},t)-2\rho(x_j,t)+\rho(x_{j+1},t)\right).$$

Note that this discretization typically entails a stability condition of the form $\Delta t\leq C\Delta x^2$, e.g. for pure diffusion, the constant C equals $1/(2D)$.

On the domain $[l,b]$ we have the LBM for the grid points x_j with $j\in\{p+1,\ldots,n-1\}$

$$f_i(x_j+c_i\Delta x,t+\Delta t)=(1-\omega)f_i(x_j,t)+\omega f_i^{eq}(x_j,t).$$

The periodic boundary conditions lead to the following boundary conditions for the PDE domain

$$\forall t:\rho(x_{-1},t)=\textstyle\sum_i f_i(x_{n-1},t),$$
$$\forall t:\rho(x_{p+1},t)=\textstyle\sum_i f_i(x_{p+1},t).$$

The full domain has an initial condition $\forall j\in\{0,\ldots,n-1\}$

$$\rho(x_j,0)=q(x_j).$$

The aim is to construct the boundary conditions of the LBM domain in such a way that $\forall t>0$ and $\forall j\in\{0,\ldots,n-1\}$ the macroscopic density defined as $\rho(x_j,t)=\rho(x_j,t)$ if $j\in\{0,\ldots,p\}$ and $\rho(x_j,t)=\sum_i f_i(x_j,t)$ if $j\in\{p+1,\ldots,n-1\}$ behaves as the density of a LBM solved on the full domain.

To formulate these boundary conditions, a lifting operator is required that maps the density $\rho(x,t)$ in the ghost points x_0 and x_p, the unknown of the PDE, to the distribution functions $f_i(x,t)$, $i\in\{-1,0,1\}$ of the LBM.

Example 2. The considered model problem has the following parameters for a one-dimensional domain of length L.

$$L=10,\quad N=200,\quad \Delta x=\frac{L}{N}=0.05,\quad \Delta t=0.001,\quad \rho(x,0)=e^{-(x-\frac{L}{2})^2},\quad \omega=0.9091.$$

For these parameters the classical Chapman-Enskog expansion predicts a diffusion coefficient $D = 1$ (equation (10)).

The discussed lifting operators from Section 3. need to be used to solve the mapping problem at the interfaces of the hybrid domain. The first, analytical method is the Chapman-Enskog expansion (Section 3.2.).
The second method is the Constrained Runs algorithm (Section 3.3.). The numerical interfacing method is useful when the application of the analytical method is not feasible. However, it is computationally demanding because of the constrained iterations that are needed to compute the value of the unknown distribution functions at the interface.
As an advantage of the Constrained Runs algorithm combined with Newton's method, we should note that the lifting error can be smaller than the modeling error by using the Constrained Runs algorithm in the hybrid model [30]. This error is the mesoscopic difference between the LBM and the PDE when the hybrid density is compared with the density obtained from considering a full lattice Boltzmann model (reference solution).
Note that both coupling schemes must be applied in an intermediate region where the Chapman-Enskog relations are valid. This will affect the size of the LBM subdomain and may confine the overall efficiency in realistic applications.
Similar results can be obtained by applying a local update at the interface by using smooth initialization. This is considered in [36].
The next method combines the ideas of Constrained Runs and the Chapman-Enskog expansion. Instead of using Constrained Runs to find for each grid point the missing moments ϕ and ξ of $f(x,v,t)$, the numerical Chapman-Enskog expansion [37] uses Constrained Runs to find the unknown coefficients of the Chapman-Enskog expansion.

4.1. Chapman-Enskog Expansion in a Hybrid Model

A comparison figure up to zeroth order in Δx of (12) is given in Fig. **8** (top left). It plots the absolute difference $|\rho_{\text{hybrid}}(x,t) - \rho_{\text{LBM}}(x,t)|$, which consists of ρ_{hybrid}, the density of the hybrid model with assumption $f_i(x,t) = \frac{1}{3}\rho(x,t)$, $i \in \{-1,0,1\}$ — in both the ghost points and the initial distribution functions of the LBM — and ρ_{LBM}, the density of a full LBM. Full LBM considers a LBM on the whole spatial domain $[a,b]$. The used parameters are presented in Example 2.
This comparison shows two clear errors: an initial error and an error in the points where the different models are coupled. These errors diffuse over the domain. Both errors occur because of the assumption $f_i(x,t) = \frac{1}{3}\rho(x,t)$, $i \in \{-1,0,1\}$, the first term of the Chapman-Enskog expansion. It is obvious that this lifting operator is not sufficient.

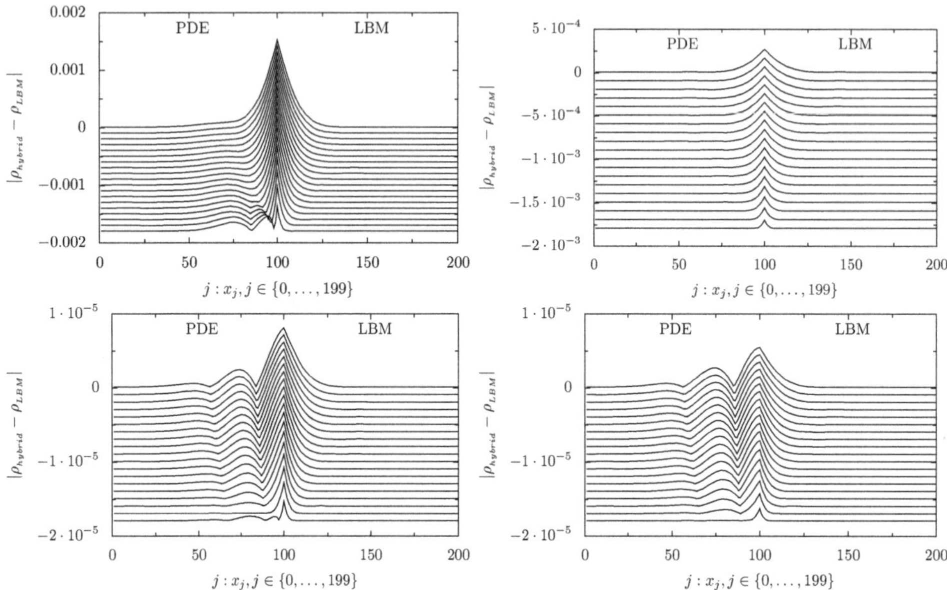

Fig. **8**: The absolute value of the difference between the density of the hybrid model and the full LBM after 200 time steps [36]. The difference is also shown at earlier time slots, but shifted down for clarity. The lines represent time steps between one and 200. The top line corresponds to time step 191 while the bottom line represents time step 11. The lines in between correspond to jumps with 10 time steps from 11 to 21,...,181 to 191. The domain is the interval [0,10] with cell centered grid points on the PDE domain [0,5[and LBM grid points on [5,10] as shown in Fig. **7**. The lifting operator is $f_i(x,t) = 1/3\,\rho(x,t)$ (top left), first order (top right), second order (bottom left) and third order Chapman-Enskog (bottom right) respectively. The lifting operator is used both to find the ghost points of the LBM domain and for the creation of the initial state for the LBM region. The model parameters are presented in Example 2.

The error decreases when higher order terms in the expansion are used in the lifting. The remaining figures of Fig. **8** show the absolute differences with the Chapman-Enskog expansion up to first, second, respectively third order as a lifting operator. However, the Chapman-Enskog expansion is hard to derive for general Boltzmann models, especially if there are velocity dependent collision rates in the model.

4.2. Constrained Runs Algorithm in a Hybrid Model

In hybrid models, the Jacobian matrices are significantly smaller than $2n \times 2n$ since we only lift in the ghost points of the LBM region. This requires only local updates which results in a smaller Jacobian [36].

4.2.1. *Constrained Runs with Constant Extrapolation*

We use the Constrained Runs algorithm with constant extrapolation in time. The function we solve with Newton's method is then

$$0 = g_0(u_0, \phi, \xi) = \begin{cases} \phi - \phi_1 \\ \xi - \xi_1, \end{cases}$$

where ϕ_1 and ξ_1 are the moments corresponding to the distribution functions after one step of LBM. The initial condition of the LBM is obtained by extracting the distribution functions from equation (7), starting from an initial guess for the moments ϕ and ξ.

This is used to reduce both the initial error and the error in the ghost points of the LBM. The resulting error is shown in Fig. **9** (top left) which shows $|\rho_{\text{hybrid}}(x,t) - \rho_{\text{LBM}}(x,t)|$. The error is of order 10^{-4} with the same

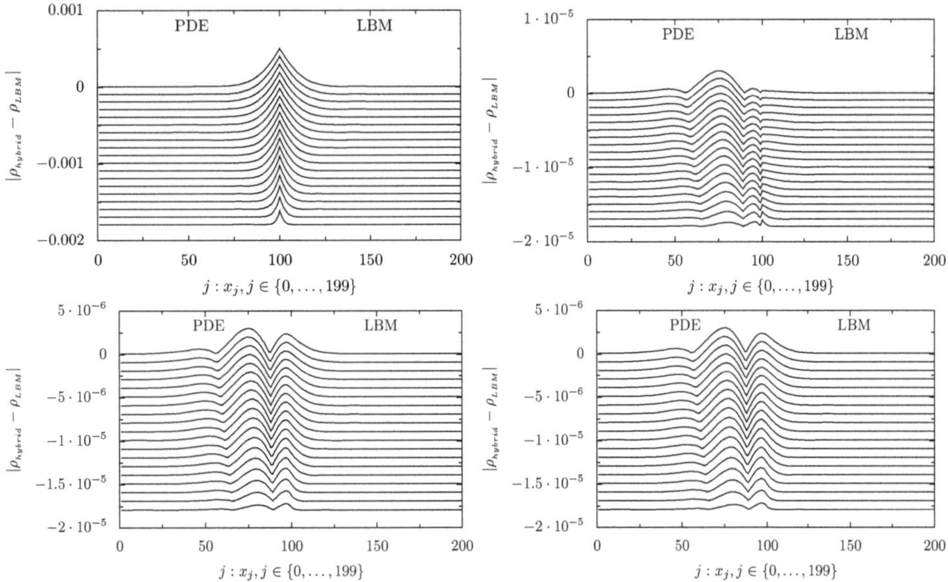

Fig. **9**: $|\rho_{\text{hybrid}} - \rho_{\text{LBM}}|$ after 200 time steps where the lifting operator based on the CR-algorithm is used in combination with the method of Newton [36]. We show results for constant (top left), linear (top right), quadratic (bottom left) and cubic (bottom right) backward extrapolation, respectively. The difference is also shown at earlier time slots, but shifted down for clarity. The domain that is considered, is shown in Fig. **7**. Model parameters are presented in Example 2.

declarations as in Fig. **8**. The initial distribution functions are based on the Constrained Runs algorithm combined with Newton's method for a constant extrapolation in time.

4.2.2. *Constrained Runs with Linear Extrapolation*

When linear extrapolation in time is used the function that we solve with Newton's method is

$$0 = g_1(u_0, \phi, \xi) = \begin{cases} \phi - 2\phi_1 + \phi_2 \\ \xi - 2\xi_1 + \xi_2, \end{cases}$$

where ϕ_1, ξ_1 and ϕ_2, ξ_2 are the moments corresponding to the distribution functions after one and two steps, respectively, of LBM. The initial condition of the LBM is obtained by extracting the distribution functions from equation (7), starting from an initial guess for the moments ϕ and ξ.
The results for dealing with the initial error and the error in the ghost points of the LBM by using a linear extrapolation in the CR-algorithm in combination with the method of Newton are given in Fig. **9** (top right). It shows a better absolute difference $|\rho_{\text{hybrid}} - \rho_{\text{LBM}}|$ than the difference based on a constant extrapolation.

4.2.3. *Constrained Runs with Higher Order Extrapolation*

In the bottom figures of Fig. **9**, higher order extrapolations are considered. The error that remains in these figures is the modeling error. This error is not caused by the coupling of both models but by the difference in both models. We compare a PDE model on $[a, l[$ with a full LBM. So it is obvious that a certain error is created on the densities.

4.3. Numerical Chapman-Enskog Expansion in a Hybrid Model

When the numerical Chapman-Enskog expansion (up to the sixth spatial derivative) is used in our one-dimensional hybrid model problem, Fig. **10** is obtained. Here, we have two possibilities. First, act as if we know the PDE (9) obtained from the exact Chapman-Enskog expansion. $|\rho_{\text{hybrid}} - \rho_{\text{LBM}}|$ is given in the left Fig. **10** for which the hybrid domain is shown in Fig. **7** and the PDE is the one given in (9). Second, use the PDE that is obtained from the proposed lifting operator through summing the proposed lifting operator or considering the nullspace as explained in Section 3.5.1. With this PDE, the result for $|\rho_{\text{hybrid}} - \rho_{\text{LBM}}|$ is shown in the right Fig. **10**.

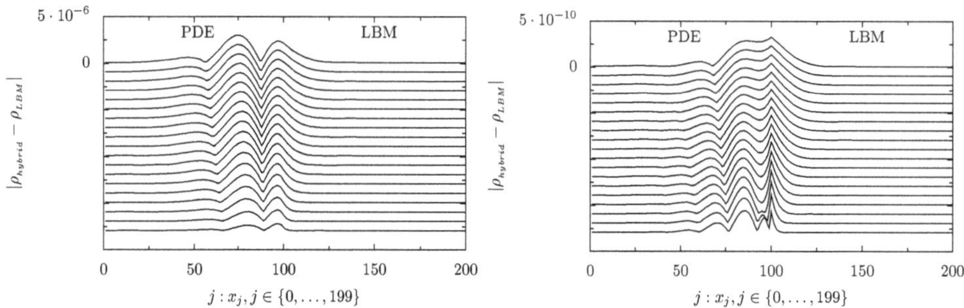

Fig. **10**: $|\rho_{\text{hybrid}} - \rho_{\text{LBM}}|$ after 200 time steps in the model problem of Example 2 [37]. The domain that is considered, is shown in Fig. **7**. To deal with the initial error and the error in the ghost points of the LBM domain the numerical Chapman-Enskog expansion (order spatial expansion 6) is used. The considered PDE in the hybrid domain is the analytically known PDE (9) in the left figure and the one that is obtained from the numerical Chapman-Enskog expansion in the right figure.

As can be seen in Fig. **10**, a change in the PDE — by considering the PDE obtained through the numerical Chapman-Enskog expansion — results in an even smaller modeling error compared to the one obtained via the classical Chapman-Enskog expansion.

4.4. Two Spatial Dimensions in a Hybrid Model

Two spatial dimensions are considered in this section. Two-dimensional problems can take different discrete sets of velocities into account. To give one example, we include D2Q5 (two spatial dimensions and 5 possible velocity directions).

The hybrid test domain for D2Q5 is represented in Fig. **11**. Again, the domain is split into subdomains. One part of the domain is described by the LBM while another part is described by a macroscopic PDE. Example 3 describes the parameters for the model problem in this two-dimensional setting.

Example 3. The considered model problem has the following parameters for a two-dimensional domain — described by 5 possible velocity directions (D2Q5) — of length $L \times L$ (with n^2 the number of grid points).

$$L = 10, \quad n = 200, \quad \Delta x = \Delta y = \frac{L}{n}, \quad \Delta t = 0.0001, \quad \omega = 1.6129.$$

For these parameters the classical Chapman-Enskog expansion for two spatial dimensions predicts a diffusion coefficient $D = 1$.

The comparison of $|\rho_{\text{hybrid}} - \rho_{\text{LBM}}|$ is represented in Fig. **12** for Example 3. The different lifting operators are used to obtain distribution functions from a given density. The used lifting operators are the equilibrium distribution function in the top left figure, the first order Chapman-Enskog expansion in the top right, the second order Chapman-Enskog expansion in the middle left, the numerical Chapman-Enskog expansion of spatial order expansion 4 in the middle right and the bottom — depending on the used PDE in the hybrid model.

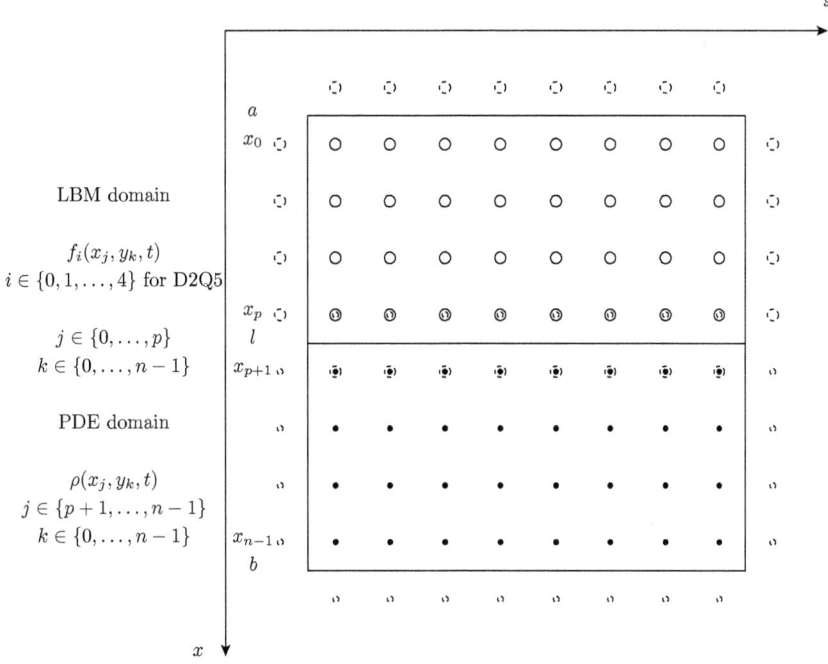

Fig. **11**: The two-dimensional spatial domain $[a,b] \times [a,b] \subset \mathbb{R}^2$ in the hybrid model is split into $[a,l[\times[a,b]$ on which we solve the LBM and $[l,b] \times [a,b]$ on which we solve the PDE model [37]. The solid points (•) represent the grid for the density ρ of the discrete PDE, the circles (∘) represent the LBM variables $f_i(x,y,t)$, $i \in \{0,\dots,4\}$ for D2Q5. The periodic boundaries and the coupling are implemented with ghostcells which are drawn by dashed circles. The density in the ghostcells of the PDE domain, in (x_p, y_k), $k \in \{0,\dots,n-1\}$ and (x_n, y_k), are found by taking $\sum_i f_i$ in (x_p, y_k) and (x_0, y_k), respectively. However, the ghostcells for the LBM domain, in (x_{-1}, y_k) and (x_{p+1}, y_k), require a lifting operator that lifts ρ to the distribution functions in these points.

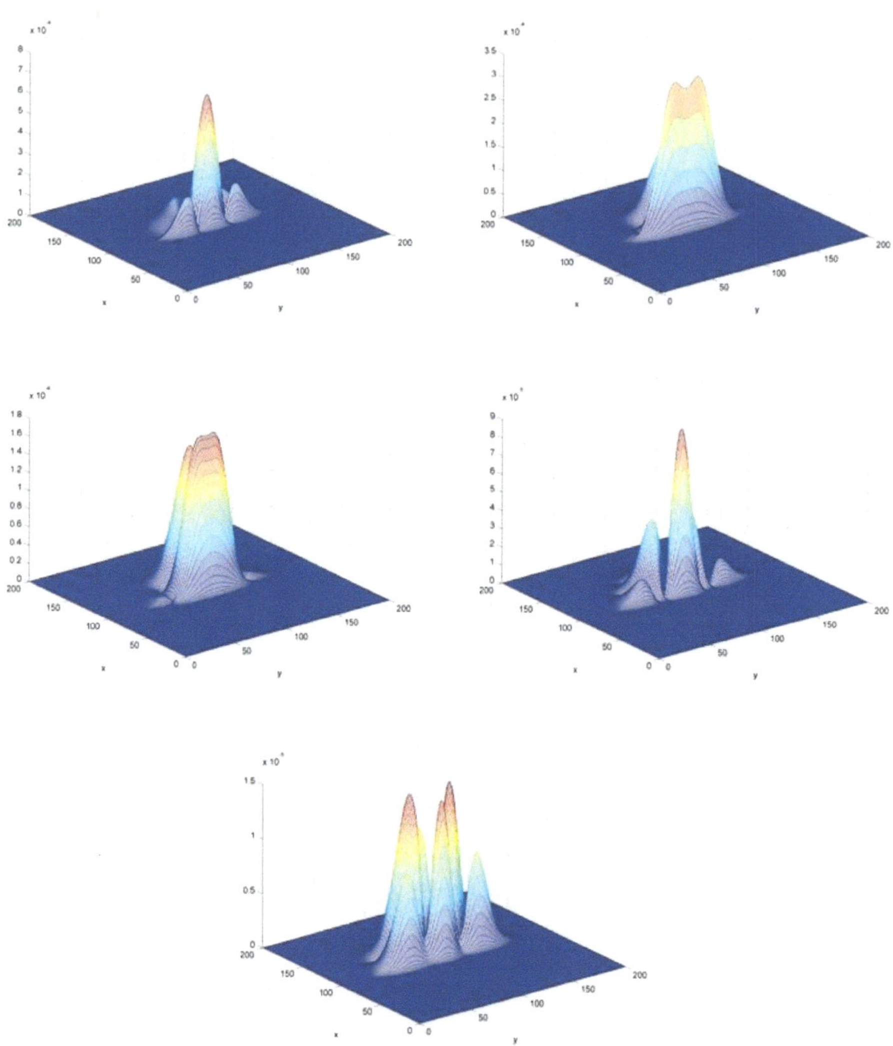

Fig. **12**: Comparison between different lifting operators for the model problem presented in Example 3 [37]. The absolute difference $|\rho_{\text{hybrid}} - \rho_{\text{LBM}}|$ is plotted at time step 200. To deal with the initial error and the error in the ghost points of the LBM we use: the equilibrium distribution function (top left), the first order Chapman-Enskog expansion (top right), the second order Chapman-Enskog expansion (middle left), the numerical Chapman-Enskog expansion (order expansion 4) where the PDE in the hybrid domain is the analytically known PDE (middle right) and the numerical Chapman-Enskog expansion where the considered PDE in the hybrid domain is the one that is obtained from the numerical Chapman-Enskog expansion (bottom).

CONCLUSIONS

We focused in this chapter on multiscale interfacing with the help of a numerical lifting operator. Given the macroscopic quantities, meaningful values for the LBM variables have to be derived. This one-to-many mapping problem occurs in various applications.

When a LBM is spatially coupled to a discretized PDE in a hybrid model, this mapping problem occurs at the interfaces. To compute the LBM variables from the PDE variables at the interfaces in each time step,

one needs a lifting operator. Either the Chapman-Enskog based slaving relations, the Constrained Runs (CR) scheme or the numerical Chapman-Enskog expansion can be applied.
The Chapman-Enskog expansion is an analytically derived expansion that expresses the distribution functions as a function of the density and its spatial derivatives.
The CR-scheme is a numerical scheme that can be used when the Chapman-Enskog relations are unavailable or difficult to obtain analytically. The basic Constrained Runs scheme performs a series of short microscopic simulations and resets the lowest moments of the microscopic variables to the macroscopic initial condition after each run. Mathematically, this scheme is a fixed point iteration. For diffusive problems, it is known that the convergence rate is $|1-\omega|$, with ω the BGK relaxation parameter. The state converged upon is a first order accurate approximation of the slaved state defined by the Chapman-Enskog relations, i.e. the microscopic state corresponding to the macroscopic initial condition. Higher order Constrained Runs schemes for the LBM use interpolation techniques to improve accuracy. However, these schemes can become unstable depending on the LBM parameters. The fixed point iteration can be replaced by a Newton iteration to avoid the instability. The linear systems can be solved by a preconditioned Krylov subspace method. This approach is computationally expensive. As an advantage of the higher order Constrained Runs algorithm combined with Newton-Krylov methods, we should note that the error made by the lifting procedure can be smaller than the modeling error in the considered hybrid model. This error compares the density of the hybrid model with the density of a full LBM simulation (reference solution).
The numerical Chapman-Enskog method is constructed to reduce the computational cost of a numerical lifting operator. It combines the ideas of Constrained Runs and the Chapman-Enskog expansion. Instead of using Constrained Runs to find for each grid point the missing moments ϕ and ξ of $f(x,v,t)$, it uses Constrained Runs to find the unknown coefficients of the Chapman-Enskog expansion. This numerical lifting method has several advantages. First, it significantly reduces the number of unknowns in the lifting steps: we only need to find the coefficients rather than the full state $f(x,v,t)$. And secondly, it can be done off-line before the calculations. Indeed, once the coefficients are found they can be reused in every time step and at every grid point to realize the lifting, at no significant additional cost. A third advantage is that the expansion gives, as a spin-off, the transport coefficients of the macroscopic PDE. Furthermore, increasing the order m of the smoothness condition improves the accuracy of the lifting operator and the PDE. This is illustrated in [37].

ACKNOWLEDGMENT

This work is supported by research project *Hybrid macroscopic and microscopic modelling of laser evaporation and expansion*, G.017008N, funded by 'Fonds Wetenschappelijk Onderzoek' together with an 'ID-beurs' of the University of Antwerp. It presents research results of the Belgian Network DYSCO (Dynamical Systems, Control, and Optimization), funded by the Interuniversity Attraction Poles Programme, initiated by the Belgian State, Science Policy Office.

CONFLICT OF INTEREST

The authors confirm that this chapter content has no conflict of interest.

REFERENCES

[1] Bhatnagar PL, Gross EP, Krook M. A Model for Collision Processes in Gases. Small I. Amplitude Processes in Charged and Neutral One-Component Systems. Phys Rev 1954; 94: 511-525.
[2] Chapman S, Cowling TG. The mathematical theory of non-uniform gases. Cambridge University Press 1953.
[3] Chen H, Chen S, Matthaeus WH. Recovery of the Navier-Stokes equations using a lattice-gas Boltzmann method. Phys Rev A 1992; 45: R5339-R5342.
[4] Chen S, Chen H, Martinez D, Matthaeus W. Lattice Boltzmann model for simulation of magnetohydrodynamics. Phys Rev Lett 1991; 67: 3776-3779.
[5] Chopard B, Dupuis A, Masselot A, Luthi P. Cellular Automata and Lattice Boltzmann techniques: An approach to model and simulate complex systems. Adv Compl Syst 2002; 5: 103-246.
[6] Dawson SP, Chen S, Doolen GD. Lattice Boltzmann computations for reaction-diffusion equations. J Chem Phys 1993; 98: 1514-1523.
[7] Dupin MM, Halliday I, Care CM, Munn LL. Lattice Boltzmann modelling of blood cell dynamics. Int J Comput Fluid Dyn 2008; 22: 481-492.
[8] Ebert U, van Saarloos W, Caroli C. Propagation and structure of planar streamer fronts. Phys Rev E 1997; 55: 1530-1549.
[9] Flekkoy EG, Herrmann HJ. Lattice Boltzmann models for complex fluids. Physica A: Stat Mech Appl 1993; 199: 1-11.
[10] Frisch U, d'Humières D, Hasslacher B, Lallemand P, Pomeau Y, Rivet J. Lattice Gas Hydrodynamics in Two and Three Dimensions. Compl Syst 1987; 1: 649-707.

[11] Gear CW, Kaper TJ, Kevrekidis IG, Zagaris A. Projecting to a slow manifold: singularly perturbed systems and legacy codes. SIAM J Appl Dyn Syst 2005; 4: 711-732.
[12] Gear CW, Kevrekidis IG. Constraint-defined Manifolds: A Legacy Code Approach to Low-Dimensional Computation. J Sci Comput 2005; 25: 17-28.
[13] Gorban AN, Karlin IV. Invariant Manifolds for Physical and Chemical Kinetics. Springer Lecture Notes in Physics 2005.
[14] Kevrekidis IG, Gear CW, Hyman JM, et al. Equation-Free, Coarse-Grained Multiscale Computation: Enabling Microscopic Simulators to Perform System-Level Analysis. Commun Math Sci 2003; 1: 715-762.
[15] Koelman JMVA. A simple Lattice Boltzmann scheme for Navier-Stokes fluid flow. Europhys Lett 1991; 15: 603-607.
[16] Latt J, Chopard B, Albuquerque P. Spatial coupling of a Lattice Boltzmann fluid model with a Finite Difference Navier-Stokes solver arXiv:physics/0511243v1.
[17] Mehl M, Neckel T, Neumann Ph. Navier-Stokes and Lattice-Boltzmann on octree-like grids in the Peano framework. Int J Numer Meth Fluids 2011; 65: 67-86.
[18] Mei R, Luo L, Lallemand P, d'Humières D. Consistent initial conditions for lattice Boltzmann simulations. Comput & Fluids 2006; 35: 855-862.
[19] Qian YH, d'Humières D, Lallemand P. Lattice BGK models for Navier-Stokes Equation. Europhys Lett 1992; 17: 479-484.
[20] Qian YH, Orszag SA. Scalings in Diffusion-Driven Reaction $A + B \to C$: Numerical Simulations by Lattice BGK Models. J Stat Phys 1995; 81: 237-253.
[21] Rheinländer M. A consistent grid coupling method for lattice-Boltzmann schemes. J Stat Phys 2005; 121: 49-74.
[22] Samaey G, Roose D, Kevrekidis I. The Gap-Tooth Scheme for Homogenization Problems, Multiscale Model & Simul 2005; 4: 278-306.
[23] Samaey G, Vandekerckhove C, Vanroose W. A multilevel algorithm to compute steady states of lattice Boltzmann models. Springer-Verlag. Coping with Complexity: Model Reduction and Data Analysis 2011; 75: 151-167.
[24] Skordos PA. Initial and boundary conditions for the lattice Boltzmann method. Phys Rev E 1993; 48: 4823-4842.
[25] Succi S. The lattice Boltzmann equation for fluid dynamics and beyond. Oxford University Press 2001.
[26] Toschi F, Perlekar P, Biferale L, Sbragaglia M. Droplet breakup in homogeneous and isotropic turbulence. arXiv:1010.1795v1 [physics.flu-dyn] 2010.
[27] Van Leemput P. Multiscale and equation-free computing for lattice Boltzmann models. PhD thesis K.U. Leuven 2007.
[28] Van Leemput P, Lust KWA, Kevrekidis I. Coarse-grained numerical bifurcation analysis of lattice Boltzmann models. Physica D 2005; 210: 58-76.
[29] Van Leemput P, Rheinländer M, Junk M. Smooth initialization of lattice Boltzmann schemes. Comput & Math Appl 2009; 58: 867-882.
[30] Van Leemput P, Vandekerckhove C, Vanroose W, Roose D. Accuracy of hybrid lattice Boltzmann/finite difference schemes for reaction-diffusion systems. Multiscale Model & Simul 2007; 6: 838-857.
[31] Van Leemput P, Vanroose W, Roose D, Mesoscale analysis of the equation-free Constrained Runs initialization scheme. Multiscale Model & Simul 2007; 6: 1234-1255.
[32] Vandekerckhove C. Macroscopic simulation of multiscale systems within the equation-free framework. PhD thesis K.U. Leuven 2008.
[33] Vandekerckhove C, Kevrekidis I, Roose D. An efficient Newton-Krylov implementation of the Constrained Runs scheme for initializing on a slow manifold. J Sci Comput 2009; 39: 167-188.
[34] Vandekerckhove C, Sonday B, Makeev A, Roose D, Kevrekidis I. A common approach to the computation of coarse-scale steady states and to consistent initialization on a slow manifold. Comput Chem Eng 2011; 35: 1949-1958.
[35] Vandekerckhove C, Van Leemput P, Roose D. Accuracy and stability of the coarse time-stepper for a lattice Boltzmann model. J Algo & Comput Techn 2008; 2: 249-273.
[36] Vanderhoydonc Y, Vanroose W. Lifting in hybrid lattice Boltzmann and PDE models. Comput Vis Sci 2011; 14: 67-78.
[37] Vanderhoydonc Y, Vanroose W. Numerical extraction of a macroscopic PDE and a lifting operator from a lattice Boltzmann model. Multiscale Model & Simul 2012; 10: 766-791.
[38] Wolf-Gladrow DA. Lattice-gas cellular automata and lattice Boltzmann models. Springer 2000.

Progress in Computational Physics, Vol. *3*, 2013, 155-183

CHAPTER 6

A Multiscale Lattice Boltzmann Method for Reaction-Diffusion Processes in Chemically and Physically Heterogeneous Environments

Davide Alemani*

EPFL, Lausanne, Switzerland

Abstract: A recent Lattice Boltzmann (LB) method applied to diffusion-reaction processes for chemical and environmental applications is shown, in particular to physicochemical processes that take place in environmental systems, such as aquatic systems, porous media, sediment, soils and biofilm layers on inert substrates. The inherently multi-scale problem in space and time of such types of applications has been investigated via the combination of two techniques, the time splitting and the grid refinement methods. We will describe in detail how the two techniques have been combined to produce a new numerical method that has been implemented in the computer program MHEDYN. Recent results on metal flux (biouptake) and dynamic speciation at bio-interface in ligand mixtures will be presented, specifically on the roles of simple, fulvic and aggregate complexes on metal of particular interest for their ecotoxicology in freshwater ligand mixtures at planar consuming interfaces. Furthermore, the role of Michaelis-Menten boundary condition at consuming interface will be pointed out under specific conditions. Further extension of the model to biofilm with the inclusion of convection driven by chemical gradient will be discussed.

Keywords: Chemical gradient, computer program MHEDYN, diffusion-reaction process, ecotoxicology, environmental systems, grid refinement, ligand mixtures, metal flux (biouptake), Michaelis-Menten boundary condition, multiscale problem, physicochemical process, time splitting.

1. INTRODUCTION

The work presented here is inspired from the wide complexity of the physical systems and consequently by the necessity to simplify their complexity into fundamental processes. It deals with a wide variety of physicochemical processes that take place in environmental systems, such as aquatic systems, porous media, sediment, soils and biofilm layer on inert substrate. In particular the attention is focused on metal complexes in aquatic systems (Fig. **1**). In these systems, the values of the physicochemical parameters linked to the metal species, such as rate and equilibrium constants, or diffusion coefficients, may vary over orders of magnitudes depending on the nature of the chemical ligands and the physical structure of the medium. The model introduced here consider two processes coupled together: diffusion and chemical reaction. The general problem studied is the set of reaction-diffusion equations for a metal M in a chemical solution with a collection of ligands and complexes. The specific purpose is to compute the flux of the metal M at a consuming surface, as bioanalogical sensors or microorganisms, and investigate the impact of complexation with ligands in environmental systems. The general framework of application of the work presented here deals with the uptake, by a consuming surface, of metal ions complexed by environmental ligands, as described in Fig. **1**. It shows schematically the most important physicochemical processes that take place in aquatic systems, near a consuming surface, represented by a bioanalogical sensor or a microorganism. Many biophysicochemical processes in aquatic systems are dynamic [13, 36, 46]. For instance the biouptake of metals by microorganisms depends on hydrodynamics, metal transfer through the plasma membrane and metal transport in solution by diffusion, as well as chemical kinetics of complex formation/dissociation in solution [49, 50].

Natural complexants include various types of compounds [14], often significantly more complicated than "simple ligands" such as OH^-, CO_3^{2-}, aminoacids, oxalate, because both electrostatic and covalent interactions with the metals need to be considered. In general they can be classified as follows [14]:

Address correspondence to: Davide Alemani, EPFL SV IBI1 UPDALPE, Office AAB019, Station 15, CH-1015, Lausanne, Switzerland; Tel: ++41 21 693 0963; E-mail: davide.alemani@epfl.ch, davide.alemani75@gmail.com

Matthias Ehrhardt (Ed.)

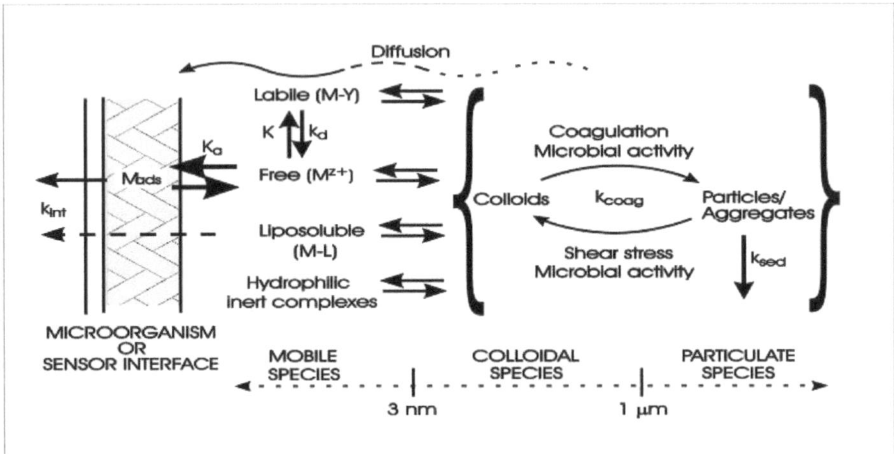

Fig. **1**: Schematic diagram of the physicochemical processes that take place near a consuming surface, electrode or microorganism.

1. Simple organic and inorganic ligands, which are often found in large excess compared to transition and b metals
2. Organic biopolymers, the most important of which are humic/fulvic compounds
3. Particles and aggregates in the size range 1-1000 nm, largely composed of inorganic solids such as clays, iron oxide etc.

Each type of complexant has its own specific properties which should be considered properly for correct computation of dynamic fluxes. These aspects are discussed in detail in [18]. Key aspects to consider are briefly summarized below:

- simple complexants are small sized, forming quickly diffusing compounds which are complexes, often labile or semi-labile, with weak to intermediate stability. Thus, when present, these complexes can be expected to contribute to metal bioavailability. But this contribution is limited by their stability.
- Humics and fulvics are "small polyelectrolytes" (1-3 nm) with intermediate diffusion coefficients, i.e. intermediate mobility. In addition they include a large number of different site types, forming metal complexes with widely varying stability and formation/dissociation kinetics. Thus the corresponding contribution to the flux is expected to depend largely on this chemical heterogeneity through the metal/ligand ratio under the given conditions.
- Particulate complexants are often aggregates of various particles and polymers. Thus they may be also chemically heterogeneous, even though relatively chemically homogeneous particles may also be found. The important sites of particles (e.g. -FeOOH sites on iron oxide) form complexes with intermediate to strong stability and intermediate to slow chemical kinetics. The key property of these particles is that their size distribution is often very wide, i.e. their diffusion coefficient may vary from intermediate to very low values. So it is expected that their contribution to bioavailability will be largely dependent on the size class.

The computation of metal flux, at consuming interfaces, in complicated environmental systems including many ligands, is a difficult task due to the many coupled dynamic physical and chemical processes. Theoretical concepts have been developed long time ago [33, 37] to compute a metal flux regulated by reaction-diffusion processes at consuming voltammetric electrodes, in solution containing a single ligand. Such theories and concepts have been applied to bioanalogical sensors and biouptake [16, 48]. Theories have also been extended recently to the case of solutions containing many ligands [29, 45, 39]. However, most papers refer to 1/1 ML complexes with simple ligands, with exceptions of a few ones [42] dealing with successive complexes. In addition, the ligand, in most cases, is considered as being in excess compared to the total metal concentration.

As far as computation codes are concerned, the situation of metal flux dynamic computation is at odds with the case of thermodynamic distribution of metal complexes for which a wealth of codes have been developed [7, 41]. To our knowledge only two numerical codes have been published [17, 39] for metal flux computation in presence of large mixtures of ligands, which considers a wide range of chemical kinetics and diffusion coefficients, as it is usually the case in natural waters. However, the numerical code published in [17] is applicable only in excess of ligands compared to metal, while the algorithm developed in [39] investigates specifically reaction-diffusion processes in gel and resin layers and is limited to the specific analysis of the dynamic features of diffusion gradients in thin film devices. The processes illustrated in Fig. **1** belong to the wide class of multiscale processes, because their physicochemical parameters vary in a wide range of values. In order to deal with these types of processes, we recently proposed a numerical method based on the Lattice Boltzmann approach that can be applied to compute metal fluxes in presence of such ligands and their mixture for any ligand concentration values, and to estimate the relative impact of each type of complex on the overall metal flux at a consuming surface (e.g. organism or bioanalogical dynamic sensor) [5, 6].

In particular, we have developed a procedure that couples the Lattice Boltzmann approach with two standard techniques:

- The time splitting method, to discriminate fast from slow processes [34]
- The grid refinement method, to localise and resolve large variations of gradient concentrations [27]

The procedure has been implemented in the numerical code MHEDYN which can be download free of charge from the web `http://www.mhedyn-project.com/`. Here we will present interesting results obtained with this numerical code on metal flux (biouptake) and dynamic speciation at bio-interface in ligand mixtures specifically on the roles of simple, fulvic and aggregate complexes on metal of particular interest for their ecotoxicology (for example Pb) in freshwater ligand mixtures at planar consuming interfaces.

2. THE PHYSICAL PROBLEM

Reaction-diffusion processes are common in environmental chemistry and biological systems. They can be highly non-linear, involve many species and often take place in irregular geometries. As a consequence, several time and spatial scales characterize the processes and accurate numerical solutions are difficult to obtain.

The general environmental reaction-diffusion problem involves the solution of a set of complexation reactions for a metal M in a heterogeneous system with several ligands of different nature. For instance a metal M can react simultaneously with a first ligand $^{1}\mathrm{L}$ and a second ligand $^{2}\mathrm{L}$:

$$\begin{aligned} \mathrm{M} + {}^{1}\mathrm{L} &\rightleftharpoons \mathrm{M}^{1}\mathrm{L} \\ \mathrm{M} + {}^{2}\mathrm{L} &\rightleftharpoons \mathrm{M}^{2}\mathrm{L} \end{aligned} \tag{1}$$

The ligands $^{1}\mathrm{L}$ and $^{2}\mathrm{L}$ may have completely different chemical properties, different diffusion coefficients and may or may not be in large excess with respect to M. The reaction of M with different ligands is called parallel complexation, because the metal M in solution can bind with two or more ligands at the same time. Moreover, each complex can react with the same ligand to generate a new complex and so on, via a set of successive reactions. For instance, considering the above mentioned reactions, $\mathrm{M}^{1}\mathrm{L}$ may bind with $^{1}\mathrm{L}$ and $\mathrm{M}^{2}\mathrm{L}$ may bind with $^{2}\mathrm{L}$:

$$\begin{aligned} \mathrm{M}^{1}\mathrm{L} + {}^{1}\mathrm{L} &\leftrightharpoons \mathrm{M}^{1}\mathrm{L}_2 \\ \mathrm{M}^{2}\mathrm{L} + {}^{2}\mathrm{L} &\leftrightharpoons \mathrm{M}^{2}\mathrm{L}_2 \end{aligned} \tag{2}$$

The subscript of L refers to the stoichiometry of L in the complex. The type of reactions (2) is called successive or sequential complexation reactions.

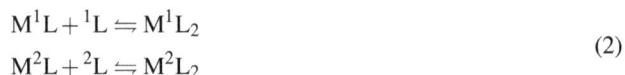

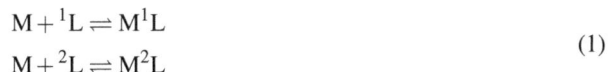

Parallel and successive complexation reactions are very typical in environmental chemical solutions. Such

Table **1**: Ranges of the typical concentration values of the more important metal ion M (page 2, from [13])

Element	**Open sea waters** (mol m^{-3})	**Fresh waters** (mol m^{-3})
Mn	$10^{-7} - 10^{-5}$	$10^{-6} - 10^{-2}$
Fe	$10^{-7} - 10^{-5}$	$10^{-4} - 10^{-2}$
Ni	$10^{-6} - 10^{-3}$	$10^{-6} - 10^{-3}$
Cu	$10^{-6} - 10^{-3}$	$10^{-6} - 10^{-3}$
Zn	$10^{-8} - 10^{-3}$	$10^{-6} - 10^{-3}$
Cd	$10^{-9} - 10^{-4}$	$10^{-7} - 10^{-5}$
Pb	$10^{-8} - 10^{-4}$	$10^{-7} - 10^{-3}$

reactions are a simplification of the real environmental processes that occur in nature, nevertheless until now, no dynamic numerical simulation that takes into account both types of reactions (1) and (2) at the same time has been developed at our present knowledge.

2.1. The Prototype Problem

In open sea waters and fresh waters the concentration of inorganic elements varies on a very wide range over orders of magnitude [13]. Table **1** shows that the concentrations of important trace metal ions range from 10^{-9} mol m^{-3} up to 10^{-2} mol m^{-3}. In environmental systems, trace metals are found in different forms, including free hydrated ions, and complexes with well-known inorganic ligands, with poorly defined natural ligands or as adsorbed species on the surfaces of particles and colloids [14]. Their chemical reactions in the external medium greatly influence their biological effects [50].
The basic process of adding a ligand to a free metal or a complex is the same for parallel and successive reactions and can be reduced to the simple 1:1 reaction:

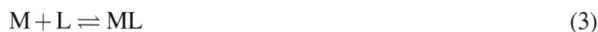

$$\mathrm{M} + \mathrm{L} \rightleftharpoons \mathrm{ML} \tag{3}$$

It is important, therefore, to understand the basics of this simple process in order to fully understand the behaviour of more complicated systems.
Thus, as a first step, the discussion below is focused on the prototype problem under planar diffusion. Most properties and considerations made for a planar geometry are valid also for spherical geometry. Moreover, planar diffusion is also adequate to describe spherical diffusion, provided the sphere radius is large enough and the time domain of interest is small enough. For instance, for a sphere of radius r_0, the planar diffusion is accurate within $a\%$ if $\delta/r_0 \leq a/100$. Here δ denotes the diffusion layer of the metal in solution. It is defined by equation (11).
The prototype problem is shown in Fig. **2** which depicts the concentration profiles of M and ML at the surface of a consuming sensor or organism. One of the most interesting and important physicochemical and biological tasks is to understand the role played by chemical complexations and physical transport of M and ML in the surrounding environment of the sensor or organism with regards to their uptake. As shown in Fig. **2**, the metal ion M in solution can form a complex ML with a ligand L via reaction (3), with equilibrium constant K and association and dissociation rate constants k_a and k_d. M, ML and L diffuse in solution with diffusion coefficients D_M, D_{ML} and D_L. The plane $x = 0$ contains a surface which consumes M but not ML or L. If the consuming surface is a Hg voltammetric electrode, M can be reduced into the metal species M^0 via the redox reaction $M + ne^- \leftrightharpoons M^0$, when a sufficiently negative potential E is applied. Then M^0 diffuses in the amalgam (extension to diffusion in the same solution is straightforward) with diffusion coefficient D_{M^0}. On the other hand, if the consuming surface is a microorganism, the metal M first binds with a complexing site at the surface of the membrane and is then internalized inside the microorganism. This process is the so-called Michaelis-Menten mechanism [50].

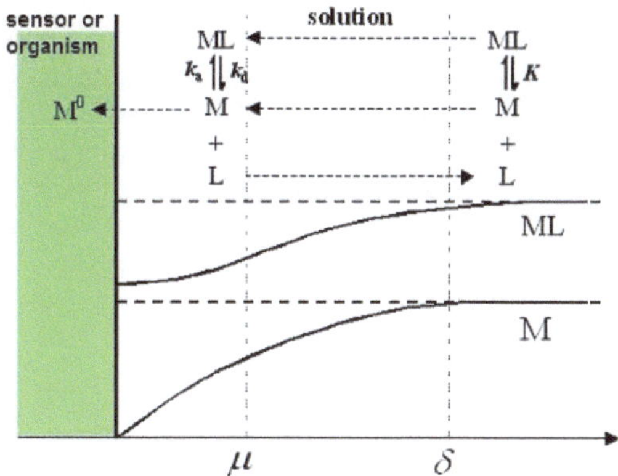

Fig. 2: Schematic representation of the physicochemical problem. The metal ion M can form a complex ML with a ligand L, having stability constant K, an association rate constant k_a and a dissociation rate constant k_d. Each of the three species diffuse in solution. M can also be consumed at the interface through various reactions (see text). The diffusion layer, δ, is the region in the vicinity of the consuming surface where the concentration is significantly different from the bulk value. The reaction layer μ is such that any M dissociated from ML is supposed to be consumed at the interface more quickly than recombined to L.

2.1.1. *The Governing Equations in Planar Geometry*

The equilibrium constant of the reaction (3), $K = k_a/k_d$ expresses the relation between M, L and ML in the bulk solution

$$K = \frac{[\mathrm{ML}]^*}{[\mathrm{M}]^*[\mathrm{L}]^*}$$

where $[\mathrm{X}]^*$ are the bulk concentrations of the species X=M, L and ML respectively. Relevant environmental cases are those where $[\mathrm{ML}]^* \geq [\mathrm{M}]^*$, i.e. $K[\mathrm{L}]^* \geq 1$, and where $[\mathrm{L}]^*_{\mathrm{tot}} \geq [\mathrm{M}]^*_{\mathrm{tot}}$. In order to compact the notation, the functions $[\mathrm{X}]=[\mathrm{X}](x,t)$, which represent the values of the concentrations of the species involved in the processes, are introduced (X=M, L, ML and M^0).

The planar semi-infinite diffusion-reaction problem for the species M, L and ML, is described by the following system of partial differential equations in the x-axis, for all $t > 0$:

$$\frac{\partial[\mathrm{M}]}{\partial t} = D_{\mathrm{M}} \frac{\partial^2[\mathrm{M}]}{\partial x^2} + R_{\mathrm{M}}, \tag{4}$$

$$\frac{\partial[\mathrm{ML}]}{\partial t} = D_{\mathrm{ML}} \frac{\partial^2[\mathrm{ML}]}{\partial x^2} + R_{\mathrm{ML}}, \tag{5}$$

$$\frac{\partial[\mathrm{L}]}{\partial t} = D_{\mathrm{L}} \frac{\partial^2[\mathrm{L}]}{\partial x^2} + R_{\mathrm{L}}, \tag{6}$$

$$\frac{\partial[\mathrm{M}]^0}{\partial t} = D_{\mathrm{M}^0} \frac{\partial^2[\mathrm{M}]^0}{\partial x^2}, \tag{7}$$

where the R_{X}'s with X=M, L and ML are the rates of formation of M, L and ML respectively:

$$R_{\mathrm{M}} = k_d[\mathrm{ML}] - k_a[\mathrm{M}][\mathrm{L}] \tag{8}$$

$$R_{\mathrm{L}} = R_{\mathrm{M}} \tag{9}$$

$$R_{\mathrm{ML}} = -R_{\mathrm{M}} \tag{10}$$

Equations (4)–(6) are defined for all $x \in (0, +\infty)$, while equation (7) is defined for all $x \in (-\infty, 0)$.

2.2. Space Scales: Diffusion and Reaction Layer Thicknesses

It is important to introduce here two crucial space scale parameters, connected with the physicochemical properties, which describe the spatial behaviour of the system: the diffusion layer thickness δ_M and the reaction layer thickness μ_{ML}.
As schematically depicted in Fig. **2**, the diffusion layer can be understood for each species as the region in the vicinity of an electrode where the concentration is significantly different from its bulk value. The value of the diffusion layer thickness depends on the consumption of M at the surface, on its diffusion coefficient, on time and on hydrodynamic conditions. In many cases, in unstirred solutions, δ_M, can be expressed as [33]:

$$\delta_M = \sqrt{\pi D_M t} \tag{11}$$

where t is the total time in which diffusion occurs.
The reaction layer is associated with the formation rate of a complex ML. Its thickness, μ_{ML}, corresponds to the distance from the consuming surface beyond which the deviation from the chemical equilibrium is taken to be negligibly small. Outside this layer, when M dissociates from ML, it can be only recombined to L after some short time. Inside this layer, the dissociated M is more often consumed at the interface than recombined to L. The value of μ_{ML} depends on the ratio of the diffusion rate of M over its recombination rate with L [33]:

$$\mu_{ML} = \sqrt{\frac{D_M}{k_a[L]^*}} \tag{12}$$

where $[L]^*$ is the bulk concentration of L.
For fast reactions (k_a large), this distance is a very thin layer. For intermediate k_a values, the rate of the chemical reaction plays a key role on the flux of the metal ion M towards the interface at $x = 0$.
The thicknesses of δ_M and μ_{ML} influence the numerical simulation of the reaction-diffusion process by playing a crucial role in the choice of the value of the grid size. In general, it has to be less than the minimum value taken by either μ or δ in order to be able to accurately resolve the concentration gradients of all the species, close to the consuming surface

2.3. Diffusion and Reaction Time Scales

Other two important parameters are essential to describe the behaviour of the system: the reactive and the diffusive time scales. The time scales of reaction can be defined by the recombination rate of M with L

$$t_R = \frac{1}{k_a[L]^*} \tag{13}$$

On the other hand, the time scale of diffusion is described by combining the expression of the diffusion layer (11) with the diffusion coefficient, [14]

$$t_D = \frac{\delta_M^2}{D_M} \tag{14}$$

Relevant cases are those for which the time scale of reaction is smaller or comparable to the time scale of diffusion. Diffusion coefficients of metals and complexes range in between 10^{-12} m^2 s^{-1} and 10^{-9} m^2 s^{-1}, so that the corresponding time scale is $t_D = 10^{-5} - 100$s.
Kinetic rate constants k_a can range from very low to very high values, usually in between 10^{-6} and 10^9m^3 mol^{-1}s^{-1}, so that the time scale of complex formation, equation (13), ranges in between 10^{-8}s and days. If $t_R \gg t_D$ then the complex is inert and only diffusive processes are important, while for $t_R < t_D$ diffusion and reaction both influence the flux.
In order to understand the influence of the complexation reaction on the flux, the flux computed in the tested conditions will be compared to:

1. The equally mobile and labile flux, J_{max}:

$$J_{max} = \frac{D_M[M]_{tot}^*}{\delta_M} \tag{15}$$

Table **2**: Range values of the main physicochemical parameters for the typical reaction diffusion process (3)

Metal Concentrations
10^{-8} mol m^{-3} – 10^{-3} mol m^{-3}
Diffusion Coefficients
10^{-12} m^2 s^{-1} – 10^{-9} m^2 s^{-1}
Kinetic Rate Constants
10^{-6} s^{-1} – 10^{9} s^{-1}

2. The "inert" flux, J_{in}:

$$J_{\mathrm{in}} = \frac{D_{\mathrm{M}}[\mathrm{M}]^*}{\delta_{\mathrm{M}}} \tag{16}$$

3. The "labile" flux, J_{lab}:

$$J_{\mathrm{lab}} = \frac{\bar{D_{\mathrm{M}}}[\mathrm{M}]^*_{\mathrm{tot}}}{\delta_{\mathrm{M}}} \tag{17}$$

The mobile-labile flux, J_{max}, is the case corresponding to the labile flux and hypothetical equal diffusion coefficients, i.e. $D_{\mathrm{ML}} = D_{\mathrm{M}}$. The inert flux, J_{in}, is the flux which would be obtained if the complex was inert, i.e. does not dissociate at all. It is equal to the diffusive flux of M without L, at the bulk concentration $[\mathrm{M}]^*$. The labile flux, J_{lab}, is the flux which would be obtained if metal and complexes were fully labile. It is equal to its diffusive flux, with an average diffusion coefficient defined as [45]:

$$\bar{D_{\mathrm{M}}} = \frac{\sum_i D_{\mathrm{ML}i}[\mathrm{ML}]_i{}^*}{[\mathrm{M}]^*_{\mathrm{tot}}} \tag{18}$$

for a fixed ligand L. The computation of the fluxes introduced above, enables to determine the lability of a complex ML, i.e. how much it affects the total flux of M and to establish its bioavailability in the surrounding solution [13, 15, 14, 50].

3. A TYPICAL MULTI-SCALE PROBLEM

To complete the general description of the prototype problem, Table **2** gives a summary of the typical range of metal concentrations, diffusion coefficients and kinetic rate constants for an environmental problem. The trace metal concentrations vary on a wide range of values (see Table **1**), the diffusion coefficients are low and they vary on three orders of magnitude and the complexation kinetic rate constants vary significantly in a range of fifteen orders of magnitude.

The four parameters, δ_{M}, μ_{ML}, t_{R} and t_{D} (equations (11), (12), (13) and (14)) are essential to describe the space-time scales of the processes involved in the system. Their values influence the physicochemical properties of an environmental systems and they are useful to determine the rate-limiting processes of the system.

Let us consider a typical set of values wherein the bulk concentration of L, $[\mathrm{L}]^*$ is in excess compared with the bulk concentration of M, $[\mathrm{M}]^*$: $[\mathrm{M}]^* = 10^{-3}$mol m^{-3}, $[\mathrm{L}]^* = 1$mol m^{-3}, $D_{\mathrm{M}} = 10^{-9}$m^{2}s^{-1} and $k_{\mathrm{a}}[\mathrm{L}]^* = 10^8$s^{-1}. If consumption of M at the planar surface is very fast, a diffusion gradient is established close to the electrode surface. After one second, the four key parameters take the following values: $\mu \sim 3$nm, $\delta \sim 60\mu$m, $(k_{\mathrm{a}}[\mathrm{L}]^*)^{-1} = 0.01\mu$s and $\delta^2_{\mathrm{M}}/D_{\mathrm{M}} \sim 3$s. Thus, clearly, the reaction and the diffusion processes take place at very different scales. For this reason the prototype problem (3) is considered as an example of a typical multiscale process.

Table **3** gives the typical ranges of space and time scales which are met in environmental systems. Diffusive space scales range usually from submicrometers to mm, depending on the geometry and diffusion coefficient of the species. Reactive space scales take very different values depending on the complexation reaction rates. They can take values as small as 1-10nm, for fully labile complexes. Such very small values are the most important limiting factors in terms of computer memory. This is because the grid sizes have to be chosen sufficiently small to follow the large concentration variations of the species involved in that space scales.

Table **3**: Typical ranges of diffusion and reaction layers and diffusion and reaction times in environmental systems.

	Space	**Time**
Reaction	μ	$(k_a[\mathrm{L}]^*)^{-1}$
	10^{-9}m $\div$ 10^{-3} m	10^{-8}s $\div$ 100 s
Diffusion	δ	δ^2/D
	10^{-7}m $\div$ 10^{-3} m	10^{-4} $\div$ 100 s

In order to localize and compute accurate concentration profiles in a thin layer of solution close to the interface, the grid should be refined within the specific region. The corresponding numerical methods are known in literature as grid refinement methods.
Table **3** also shows typical time scales of reaction and diffusion under environmental conditions. Typical reaction time scales can vary between 10^{-8} and 100 s^{-1}. The smallest values, corresponding to fully labile complexes, are the limiting factors in terms of computational time, since the computational time step should be short enough to ensure a sufficient accuracy. For this reason, a suitable numerical method, enabling to discriminate slow and fast processes, is necessary.
Multiscale problems are often met in real systems and they always represent a big challenge for the numerical simulation community. For that reason, a simplification is needed which on the one hand reduces the computational cost and the computer memory usage and, on the other hand, maintains a sufficient accuracy of the solution.
In order to achieve such a task, we have proposed to introduce the time splitting method and the grid refinement techniques in the Lattice Boltzmann framework for solving reaction-diffusion systems, not only for environmental or electrochemical applications but in general for a larger community of scientists that are interested in simulating and understanding multiscale phenomena.

4. THE MATHEMATICAL FORMULATION OF THE PROBLEM FOR MULTI-LIGAND APPLICATIONS

4.1. Reaction-Diffusion Equations

Let us suppose that the system includes n_l ligands and ${}^jn$ successive complexation reactions for each type of ligand, with $j = 1,\ldots,n_l$. A set of parallel and successive chemical reactions of the following kind will be considered:

$$\mathrm{M} + {}^j\mathrm{L} \underset{{}^jk_{\mathrm{a},1}}{\overset{{}^jk_{\mathrm{d},1}}{\leftrightharpoons}} \mathrm{M}^j\mathrm{L} \tag{19}$$

$$\mathrm{M}^j\mathrm{L}_{i-1} + {}^j\mathrm{L} \underset{{}^jk_{\mathrm{a},i}}{\overset{{}^jk_{\mathrm{d},i}}{\leftrightharpoons}} \mathrm{M}^j\mathrm{L}_i \qquad i = 2,\ldots,{}^jn \tag{20}$$

Chemical reactions (19) and (20) take place within the solution domain. Index i represents the stoichiometric number of ${}^j\mathrm{L}$ in the complex and the superscript j is limited to the nature of the ligand. The chemical rate associated to each reaction is given by:

$${}^jr_i = -{}^jk_{\mathrm{a},i}[\mathrm{M}^j\mathrm{L}_{i-1}][{}^j\mathrm{L}_i] + {}^jk_{\mathrm{d},i}[\mathrm{M}^j\mathrm{L}_i] \tag{21}$$

where ${}^jk_{\mathrm{a},i}$ and ${}^jk_{\mathrm{d},i}$ are the association and dissociation rate constants respectively. The association and dissociation rate constants define the equilibrium constant for each reaction, ${}^jK_i$. It is defined as:

$${}^jK_i = \frac{{}^jk_{\mathrm{a},i}}{{}^jk_{\mathrm{d},i}} = \frac{[\mathrm{M}^j\mathrm{L}_i]^*}{[\mathrm{M}^j\mathrm{L}_{i-1}]^*[{}^j\mathrm{L}_i]^*} \qquad i = 2,\ldots,{}^jn \tag{22}$$

The first equilibrium constant ${}^{j}K_1$ is:

$$ {}^{j}K_1 = \frac{{}^{j}k_{\mathrm{a},1}}{{}^{j}k_{\mathrm{d},1}} = \frac{[\mathrm{M}^{j}\mathrm{L}]^*}{[\mathrm{M}]^*[{}^{j}\mathrm{L}]^*} \tag{23} $$

All the species diffuse within the solution domain following the usual set of reaction-diffusion equations:

$$ \frac{\partial[\mathrm{M}]}{\partial t} = D_{\mathrm{M}}\nabla^2[\mathrm{M}] + \sum_{j=1}^{n_l} {}^{j}r_1 \tag{24} $$

$$ \frac{\partial[{}^{j}\mathrm{L}]}{\partial t} = D_{{}^{j}\mathrm{L}}\nabla^2[{}^{j}\mathrm{L}] + \sum_{i=1}^{{}^{j}n} {}^{j}r_i \tag{25} $$

$$ \frac{\partial[\mathrm{M}^{j}\mathrm{L}_i]}{\partial t} = D_{\mathrm{M}^{j}\mathrm{L}_i}\nabla^2[\mathrm{M}^{j}\mathrm{L}_i] - {}^{j}r_i + {}^{j}r_{i+1} \qquad i = 1,\ldots,{}^{j}n-1 \tag{26} $$

$$ \frac{\partial[\mathrm{M}^{j}\mathrm{L}]_s}{\partial t} = D_{\mathrm{M}^{j}\mathrm{L}_s}\nabla^2[\mathrm{M}^{j}\mathrm{L}]_s - {}^{j}r_s \qquad s = {}^{j}n \tag{27} $$

For all the problems studied in this work, it is assumed that the ligands and the complexes are not consumed at the micro-organism or electrode interface, i.e. null flux condition are fixed at $x = 0$ for these species. Only M can be consumed. Depending on the surface reactions, M satisfies different boundary conditions. Two types of boundary conditions corresponding to two problems are considered: the Nernst boundary conditions at voltammetric electrodes and the Michaelis-Menten boundary conditions at micro-organism surfaces.

4.2. Initial Conditions

Two types of initial conditions may be considered. The first one, supposes to begin the simulations at the chemical equilibrium, therefore the initial conditions correspond to the bulk equilibrium values for each species X:

$$ [\mathrm{X}](\mathbf{x},t) = [\mathrm{X}]^*(\mathbf{x},t) \qquad t = 0 \tag{28} $$

The second one supposes that the system is initially "empty", i.e. the concentration of species X is null. Therefore the corresponding initial condition is:

$$ [\mathrm{X}](\mathbf{x},t) = 0 \qquad t = 0 \tag{29} $$

4.3. Boundary Conditions

Depending on the nature of the problem, either finite diffusion or semi-infinite diffusion condition is applied to species X. When the chemical solution is stirred, the bulk concentrations of the species are maintained constant at a certain distance d from the active surface. This condition corresponds to the finite diffusion condition, which states that:

$$ [\mathrm{X}](\mathbf{x},t) = [\mathrm{X}]^*(\mathbf{x},t) \quad |\mathbf{x}| = d \tag{30} $$

When no stirring occurs in the solution domain the bulk concentration is only reached at $x \to \infty$. This condition corresponds to semi-infinite diffusion and it is given by:

$$ [\mathrm{X}](\mathbf{x},t) \to [\mathrm{X}]^*(\mathbf{x},t), \qquad \text{as} \quad x \to \infty \tag{31} $$

At the consuming surface S, there is no flux of $\mathrm{M}^{j}\mathrm{L}_i$ and ${}^{j}\mathrm{L}_i$ crossing the interface. Therefore:

$$ \left(\frac{\partial[\mathrm{M}^{j}\mathrm{L}_i]}{\partial n}\right)_{x\in S} = 0 \tag{32} $$

$$ \left(\frac{\partial[{}^{j}\mathrm{L}_i]}{\partial n}\right)_{x\in S} = 0 \tag{33} $$

where n is the normal vector of the surface.

Two types of boundary conditions for M are considered at the consuming surface. They are described below.

4.3.1. *Interfacial Boundary Condition for M: Nernst Equation*

For the voltammetric sensor, the Nernst boundary condition is considered. The metal M can be reduced at the electrode interface into its neutral form M^0, via the following redox process:

$$M^0 \overset{n_e^-}{\rightleftharpoons} M \tag{34}$$

where n_e is the number of electrons involved in the redox reaction. If a constant potential is applied at the electrode and the redox process can be considered reversible, then the Nernst condition applies:

$$[M](t) = [M^0](t)e^{(E-E_0)n_e f} \quad \text{at } x = 0 \tag{35}$$

where E_0 is the standard redox potential for the couple M/M^0 and f is the Faraday reduced constant ($f = \frac{F}{RT} = 38.92V^{-1}$). In the above equation another species has been introduced M^0. Hence, another boundary expression involving M^0 and/or M is necessary in order to solve the set of reaction-diffusion equations. This additional boundary condition comes from the flux conservation at the electrode surface. It is given by:

$$D_M \frac{\partial [M]}{\partial n} = D_M^0 \frac{\partial [M^0]}{\partial n} \qquad x \in S \tag{36}$$

The reduced form M^0 is present only inside the electrode and its evolution is followed by solving an appropriate diffusion equation:

$$\frac{\partial [M^0]}{\partial t} = D_{M^0} \nabla^2 [M^0] \tag{37}$$

To solve equation (37), an additional boundary condition for M^0 is needed at either $x = -r_0$ (micro-electrode) or $x \to -\infty$ (macroscopic electrode). In the following, the potential satisfy the following condition: $\Delta E = E - E_0 \ll -0.3V$. Under this assumption the electrode surface acts as a perfect sink for M and equation (37) involving M^0 can be disregarded.

4.3.2. *Interfacial boundary condition for M: Michaelis-Menten equation*

If the consuming surface S is a micro-organism, the mechanism of site adsorption and internalization is described by the Michaelis-Menten equation. This equation gives the internalization flux of M as a function of its volume concentration near the surface.

The general form of the Michaelis-Menten equation for a metal M is given in [30]:

$$\{R\}_{\text{tot}} \frac{d}{dt} \frac{K_a[M]}{1+K_a[M]} = D_M \nabla_n [M] - k_{\text{int}} \{R\}_{\text{tot}} K_a \frac{[M]}{1+K_a[M]} \tag{38}$$

where k_{int} is the internalization rate constant (s^{-1}), K_a is the adsorption constant of M on the sites at the membrane surface ($m^3 mol^{-1}$), $\{R\}_{\text{tot}}$ is the surface concentration of the free sites for the binding/transport of M (mol m^{-2}). In most cases, the assumption of steady-state for the Michaelis-Menten equation is reasonable [4]. Therefore, its expression is given by:

$$J_{\text{int}} = \frac{1}{A}\frac{dN}{dt} = \frac{k_{\text{int}} K_a \{R\}_{\text{tot}} [M]}{1+K_a[M]} \qquad x \in S \tag{39}$$

where $J_{\text{int}} = J_M$ is the internalization flux, A is the surface area (m^2), N is the number of moles of M passing through the interface S, t is the time (s), and $[M]$ is the volume concentration of M (mol m^{-3}).

Equation (39) is a mixed type boundary condition. Indeed, equation (39) contains both the flux of M at the surface, J_{int} and the concentration of M, $[M]$. Therefore, the version of equation (39) in terms of mixed boundary condition takes the following form:

$$D_M \nabla_n [M] = \frac{K_a \{R\}_{\text{tot}} k_{\text{int}} [M]}{1+K_a[M]} \tag{40}$$

The above equation is a (nonlinear) combination of [M] and its normal derivative at the surface S of the micro-organism, $\nabla_n[\mathrm{M}]$.

5. THE LATTICE BOLTZMANN METHOD FOR REACTION-DIFFUSION PROCESSES

A LB model can be interpreted as a discretization of the Boltzmann transport equation on a regular lattice of spacing Δx along each lattice direction and with discrete time step Δt [35]. The possible velocities for the pseudo-particles are the vectors $\mathbf{v}_i$. They are chosen so as to match the lattice constraints: if $\mathbf{x}$ is a lattice site, $\mathbf{x}+\mathbf{v}_i\Delta t$ is also a lattice point. The dynamics involves $z+1$ possible velocities, where z is the coordination number and $\mathbf{v}_0 = 0$ describes the population of rest particles. The lattice is identified by its spatial dimension d and its coordination number z indicating how many neighbors each lattice point has. Traditionally, the lattice is then referred to as a DdQz lattice (D stands for Dimension and Q for Quantities). For isotropy reasons the lattice topology must at least satisfy the conditions [22, 51]:

$$\sum_i v_{i\alpha} = 0 \quad \text{and} \quad \sum_i v_{i\alpha} v_{i\beta} = v^2 C_2 \delta_{\alpha\beta} \tag{41}$$

where C_2 is a numerical coefficient which depends on the lattice topology. The Greek indices label the spatial dimension and $v = \Delta x/\Delta t$. The first condition follows from the fact that if $\mathbf{v}_i$ is a possible velocity, then so is $-\mathbf{v}_i$.

In the LB approach a physical system is described through density distribution functions $f_i(\mathbf{x},t)$. For hydrodynamics and reaction-diffusion processes, $f_i(\mathbf{x},t)$ represents the distribution of particles entering a site $\mathbf{x}$ at time t and moving in direction $\mathbf{v}_i$. Therefore, in a LB approach, the description is finer than e.g. in a finite difference scheme, as information on the particle microscopic velocity is included. As it can be shown, an important consequence of this fact is that the f_i's also contain information on the spatial derivatives of the macroscopic quantities. Physical quantities can be defined from moments of these distributions. For instance, the local density is obtained by

$$\rho = \sum_{i=0}^{z} f_i \tag{42}$$

A LB model is determined by specifying:

- A lattice
- A general kinetic equation
$$f_i(\mathbf{x}+\mathbf{v}_i\Delta t, t+\Delta t) - f_i(\mathbf{x},t) = \Omega_i$$
where Ω_i is the collision term that must preserve the conservation laws of the system. For instance, in a diffusion process, particle number is conserved and, in a fluid, momentum is also conserved. In its simplest form (BGK model), the dynamics can be written as a relaxation to a given local equilibrium
$$f_i(\mathbf{x}+\mathbf{v}_i\Delta t, t+\Delta t) - f_i(\mathbf{x},t) = \omega(f_i^{eq}(\mathbf{x},t) - f_i(\mathbf{x},t)) \tag{43}$$
where ω is a relaxation parameter, which is a free parameter of the model.
- An equilibrium distribution f_i^{eq}, that contains all the information concerning the physical process investigated. It depends only on the local values of the macroscopic quantities and it changes according to whether hydrodynamics, reaction-diffusion or wave propagation is considered. For reaction-diffusion processes it takes the form [51]
$$f_i^{eq}(\mathbf{x},t) = \frac{[X](\mathbf{x},t)}{2d} \tag{44}$$
where $[X](\mathbf{x},t)$ is the volume concentration of X.

Let's focus the discussion on reaction-diffusion systems. The LBGK model stated in equation (43) will be used. Note that in this work, we consider Δx and Δt as real time and space variables. Δx is expressed in meters and Δt is expressed in seconds. As a consequence, $f_{X,i}(\mathbf{x},t)$ is expressed in mol/m^3.

Such a method has already been used for solving reaction-diffusion problems (see for instance [51, 23, 38]), for two main reasons:

- The LBGK model for reaction-diffusion systems is very simple and easy to establish, even in the presence of a large number of species and complicated boundary geometries
- The time step is limited only by accuracy and not by stability requirements [51]. Moreover, the computer code is rather simple.

In the following, the LB method in its reaction-diffusion form will be extended to solve multiligand reactive-diffusive processes, eqns (24)–(27).
Here DdQ2d lattices are considered, which means a cubic-like lattice in dimension d in which each lattice site has $2d$ neighbours, that is the possibility of particles at rest is excluded. The exclusion of the rest particles is acceptable according to what is reported in [47]: "*it is well known that 90° rotational invariance is sufficient to yield full isotropy for diffusive phenomena*". Moreover, according to [47] it is sufficient to use a square or a cubic lattice in two or three dimensions, respectively.
In $3D$ ($d=3$), the lattice velocities are therefore: $\mathbf{v}_1=(v,0,0)$, $\mathbf{v}_2=(-v,0,0)$, $\mathbf{v}_3=(0,v,0)$, $\mathbf{v}_4=(0,-v,0)$, $\mathbf{v}_5=(0,0,v)$, $\mathbf{v}_6=(0,0,-v)$, where

$$v=\Delta x/\Delta t. \tag{45}$$

The chemical species X are described by density distribution functions $f_{X,i}(\mathbf{x},t)$. According to the general method, the macroscopic concentrations $[X](\mathbf{x},t)$ at points $(\mathbf{x},t)$ are then given by:

$$[X](\mathbf{x},t)=\sum_{i=1}^{2d} f_{X,i}(\mathbf{x},t) \tag{46}$$

Following the general procedure of the LB method, the prototype problem expressed in equations (4)–(7) and the multiligand problem expressed in equations (24) –(27), can be represented as follows:

$$\boxed{f_{X,i}(\mathbf{x}+\mathbf{v}_i\Delta t,t+\Delta t)=f_{X,i}(\mathbf{x},t)+\Omega^{NR}_{X,i}(\mathbf{x},t)+\Omega^{R}_{X,i}(\mathbf{x},t)} \tag{47}$$

where $\Omega^{NR}_{X,i}(\mathbf{x},t)$ contains the non-reactive part of the interaction (e.g. diffusion) whereas $\Omega^{R}_{X,i}(\mathbf{x},t)$ contains all chemical reactions affecting species X (see for instance [20]).
It can be shown [22, 51] that corresponding partial differential equations (PDE) for the prototype and for the multiligand problems are obeyed by $[X](\mathbf{x},t)=\sum_i f_{X,i}(\mathbf{x},t)$ provided that the collision operators and the equilibrium functions are adequately chosen. For the prototype and the multiligand problems the non reactive operator $\Omega^{NR}_{X,i}(\mathbf{x},t)$ is given by:

$$\Omega^{NR}_{X,i}(\mathbf{x},t)=\omega_X(f^{eq}_{X,i}(\mathbf{x},t)-f_{X,i}(\mathbf{x},t)) \tag{48}$$

The quantity ω_X is a free parameter that tunes the transport coefficients. In case of a purely diffusive phenomenon, the relaxation parameter ω_X is related to the diffusion coefficients as [38]:

$$\omega_X=\frac{2}{1+2d\frac{D_X\Delta t}{\Delta x^2}} \tag{49}$$

On the other hand, the reactive operator, $\Omega^{R}_{X,i}(\mathbf{x},t)$, is given by

$$\Omega^{R}_{X,i}(\mathbf{x},t)=\frac{\Delta t}{2d}R_X \tag{50}$$

where the expression for R_X depends on the type of problem investigated. For the prototype problem it takes the form stated in equations (8), (9) and (10) for the metal M, the ligand L and the complex ML, respectively. For the multiligand problem it takes the following form

- For the metal M:

$$R_M=\sum_{j=1}^{n_l} {}^{j}r_1 \tag{51}$$

- For the ligands ^{j}L with $j = 1, \dots, n_l$

$$R_{^j\mathrm{L}} = \sum_{i=1}^{^j n} {}^j r_i \tag{52}$$

- for the complexes M^{j}L$_i$ with $j = 1, \dots, n_l$ and $i = 1, \dots, {}^j n - 1$

$$R_{\mathrm{M}^j\mathrm{L}_i} = -{}^j r_i + {}^j r_{i+1} \tag{53}$$

- For the complex M^{j}L$_s$ with $s = {}^j n$ and $j = 1, \dots, n_l$

$$R_{\mathrm{M}^j\mathrm{L}_s} = -{}^j r_s \tag{54}$$

Note that the following notation is adopted: n_l is the number of ligands present in solution and ${}^j n$ is the number of successive complexations for the ligand ^{j}L.
In order to satisfy the mass conservation, the equilibrium function takes the following form [51, 1]:

$$f^{eq}_{\mathrm{X},i}(\mathbf{x},t) = \frac{[\mathrm{X}](\mathbf{x},t)}{2d} \tag{55}$$

To the first order in the Chapman-Enskøg expansion and in the limit $\Delta x \to 0$ and $\Delta t \to 0$, with $\Delta x^2/\Delta t \to$ const, the distribution functions in equation (47) are shown to obey [22, 1]:

$$f_{\mathrm{X},i}(\mathbf{x},t) = f^{eq}_{\mathrm{X},i}(\mathbf{x},t) + f^{neq}_{\mathrm{X},i} \tag{56}$$

where

$$f^{neq}_{\mathrm{X},i} = -\frac{\Delta t}{2d\omega_{\mathrm{X}}} v_i \cdot \nabla[\mathrm{X}](\mathbf{x},t) \tag{57}$$

The above two equations (56) and (57) establish the relationship between the macroscopic concentration $[\mathrm{X}](\mathbf{x},t)$ and the density distribution functions $f_{\mathrm{X},i}(\mathbf{x},t)$. Note that these expressions are valid only for pre-collision values.

5.1. A Way to Compute the Flux

The computation of the flux through a surface S with normal vector n_S is defined as:

$$J_{\mathrm{M}} = -D_{\mathrm{M}} \nabla_{n_S} [\mathrm{M}] \tag{58}$$

and can be related to the microscopic density distribution functions $f_{\mathrm{X},i}(\mathbf{x},t)$ as follows. Recall that the normal vector n_S of a surface S is defined in each point $\mathbf{x} \in S$ as the outgoing unity vector perpendicular to the tangent space at the point $\mathbf{x}$. The operator ∇_{n_S} is the normal derivative at the surface S along the normal vector n_S.
By multiplying equation (57) by v_i and summing over i one obtains

$$\sum_i f^{neq}_{\mathrm{M},i} v_i = -\Delta t \frac{1}{2d\omega_{\mathrm{M}}} 2v^2 \nabla[\mathrm{M}] \tag{59}$$

Here, for a DdQ2d lattice, $\sum_i v_i v_i = 2v^2 \mathbf{1}$,where $\mathbf{1}$ is the $d \times d$ identity matrix. Since, from equation (49)

$$D_{\mathrm{M}} = \frac{v^2 \Delta t}{d} \left(\frac{1}{\omega_{\mathrm{M}}} - \frac{1}{2} \right)$$

one finally obtains

$$J_{\mathrm{M}} = -D_{\mathrm{M}} \nabla[\mathrm{M}] = \left(1 - \frac{\omega_{\mathrm{M}}}{2}\right) \sum_i f^{neq}_{\mathrm{M},i} v_i = d \frac{\omega_{\mathrm{M}} D_{\mathrm{M}}}{\Delta x} \frac{1}{v} \sum_i f^{neq}_{\mathrm{M},i} v_i \tag{60}$$

This expression is purely local and can be computed without having to discretize the concentration gradient. This feature is an interesting advantage of the LB approach. Note also that in case of a diffusive system, $\sum_i f^{eq}_{\mathrm{M},i} v_i = 0$ and thus $\sum_i f^{neq}_{\mathrm{M},i} v_i = \sum_i f_{\mathrm{M},i} v_i$.

The above expression (60) for J_M is valid in the bulk solution. Some care is needed when computing the flux of particles exactly at the consuming surface, which corresponds to the boundary condition. Then, all the f_i are not known and some of them must be computed according to the desired behaviour at the boundary. In order to use equation (60), the missing f_i's must be set up consistently with the theory, that is the f_i's values at the boundary have to be updated. However, the amount of particles that is consumed at the surface can always be computed directly from the balance between the number of particles reaching the surface and those leaving it during one time step [2].

5.2. The Regularised LBGK Method for Reaction-Diffusion Problem

The regularised LBGK method relies on the assumption that $f_{\mathrm{X},i}(\mathbf{x},t)$ is separated into its equilibrium $f^{eq}_{\mathrm{X},i}(\mathbf{x},t)$ and non equilibrium $f^{neq}_{\mathrm{X},i}$ part, equation (56). It consists in determine the $f^{neq}_{\mathrm{X},i}$ part of $f_{\mathrm{X},i}(\mathbf{x},t)$ such that

$$f_{\mathrm{X},i}(\mathbf{x},t) = f^{eq}_{\mathrm{X},i}(\mathbf{x},t) + f^{(1)}_{\mathrm{X},i} \tag{61}$$

where $f^{(1)}_{\mathrm{X},i}$ is the first order approximation of $f_{\mathrm{X},i}(\mathbf{x},t)$ ($f_{\mathrm{X},i}(\mathbf{x},t) = f^{(1)}_{\mathrm{X},i} + f^{(2)}_{\mathrm{X},i} + \ldots$). By substituting equation (61) in the LBGK method for reaction-diffusion introduced in equation (47) one gets, after some algebra:

$$f_{\mathrm{X},i}(\mathbf{x}+\mathbf{v}_i\Delta t, t+\Delta t) = f^{eq}_{\mathrm{X},i}(\mathbf{x},t) + (1-\omega_\mathrm{X}) f^{(1)}_{\mathrm{X},i} + \Omega^{\mathrm{NR}}_{\mathrm{X},i}(\mathbf{x},t) \tag{62}$$

Notice that only the non reactive operator has changed. By applying the Chapman-Enskøg expansion, $f^{(1)}_{\mathrm{X},i}$ can be written as

$$f^{(1)}_{\mathrm{X},i} = -\frac{\Delta t}{2d\omega_\mathrm{X}} \mathbf{v}_i \cdot \nabla[\mathrm{X}](\mathbf{x},t)$$

Therefore, the regularised LBGK method is finally written as [12]:

$$\boxed{f_{\mathrm{X},i}(\mathbf{x}+\mathbf{v}_i\Delta t, t+\Delta t) = f^{eq}_{\mathrm{X},i}(\mathbf{x},t) + \frac{(1-\omega_\mathrm{X})}{2v^2} \sum_j f_{\mathrm{X},j}(\mathbf{x},t)\mathbf{v}_i \cdot \mathbf{v}_j + \Omega^{\mathrm{R}}_{\mathrm{X},i}(\mathbf{x},t)} \tag{63}$$

The regularized method applied only to diffusive phenomena, has been developed in [12] for the first time. Here its extension to multi-ligand reaction-diffusion problems is considered.

This approach has the advantage to be more accurate, because the non equilibrium part of $f_{\mathrm{X},i}(\mathbf{x},t)$ is set to the first order approximation $f^{(1)}_{\mathrm{X},i}$ before the collision process. Moreover, as shown in [12] the time convergence is faster than the standard method.

5.3. The Numerical Initial Conditions

The initial conditions, defined in equations (28) and (29), are rewritten in terms of $f_{\mathrm{X},i}(\mathbf{x},t)$ in the following form (for 3D):

$$f_{\mathrm{X},i}(\mathbf{x},t) = \frac{[\mathrm{X}](\mathbf{x},t)}{6} \qquad t = 0 \tag{64}$$

for the bulk initial condition, (equation (28)) and

$$f_{\mathrm{X},i}(\mathbf{x},t) = 0 \qquad t = 0 \tag{65}$$

for the null initial condition, (equation (29)). Thus, with condition (64), at each point of the lattice, the distribution functions of species X take the value of the concentration of X divided by the number of lattice directions.

5.4. The Numerical Boundary Conditions

The numerical boundary conditions are written by using either the formulas derived from the Chapman-Enskøg expansion

$$f_{X,i}(\mathbf{x},t) = f^{eq}_{X,i}(\mathbf{x},t) + f^{(1)}_{X,i} \tag{66}$$

if one wants to recover a Neumann boundary condition, i.e. the flux is known, or the simple formula

$$f_{X,i}(\mathbf{x},t) = [X](\mathbf{x},t) - \sum_{j \neq i} f_{X,j}(\mathbf{x},t) \tag{67}$$

which gives the $f_{X,i}(\mathbf{x},t)$ as a function of the $[X](\mathbf{x},t)$ and the $f_{X,j}$ with $j \neq i$. The numerical form of the boundary conditions are listed below.

5.4.1. *Solution Side: Finite Diffusion*

The boundary condition of finite diffusion stated in equation (30) is

$$f_{X,i}(\mathbf{x},t) = [X]^* - \sum_{j \neq i} f_{X,j}(\mathbf{x},t) \qquad |\mathbf{x}| = d \tag{68}$$

where d is the distance of diffusion, the lhs is the missing value and the right hand side is known. This equation is equivalent to have a stirred solution at distance d from the consuming surface.

5.4.2. *Solution Side: Semi-Infinite Diffusion*

The boundary condition of semi-infinite diffusion is similar to the previous boundary condition of finite diffusion with the difference that the finite diffusion length d should be at least 5 times of the diffusion layer thickness $\delta = \sqrt{(\pi D_M t)}$. This condition ensures that the diffusion process is not affected by the boundary at distance d. Therefore the semi-infinite diffusion condition is

$$f_{X,i}(\mathbf{x},t) = [X](\mathbf{x},t) - \sum_{j \neq i} f_{X,j}(\mathbf{x},t) \qquad |\mathbf{x}| = 5\delta_X \tag{69}$$

where $f_{X,i}(\mathbf{x},t)$ is the missing value and the right hand side is known.

5.4.3. *Flux Boundary Condition at the Interface*

A flux boundary condition means that the concentration gradient at the surface along the normal vector n must be defined or computed. When the surface is not regular and does not match the lattice, such computation is not trivial. However, a good approximation is to discretize the surface with lattice points with a sufficient small grid size and compute the flux at the boundary at the closest lattice points outside the surface. In this way the surface is approximated by a collection of step functions, which in the limit of small grid size, takes the same shape as the surface boundary. Therefore, the cubic nature of the lattice is preserved and equation (60) can be used. As an example of application let us consider the simple 1D planar problem. The boundary conditions for L and ML, at $x = 0$ are

$$f_{L,1} = f_{L,3} \tag{70}$$

$$f_{ML,1} = f_{ML,3} \tag{71}$$

5.4.4. *Nernst and Flux Boundary Condition at the Interface*

Nernst boundary condition apply when the consuming surface is a voltammetric electrode on which the metal ion M is reversibly reduced into M^0. Again, to obtain the numerical boundary condition at $x = 0$, for M and M^0, conditions (35) and (36) must be transformed in terms of density distribution functions. By combining expression (67) with the Nernst condition (35) and equation (67) with the flux conservation, equation (36) to write the gradient, it is possible to find two equations with two unknowns along each Cartesian axis. For instance, along the x axis, one gets:

$$f_{\mathrm{M},1} + f_{\mathrm{M},3} = (f_{\mathrm{M}^0,1} + f_{\mathrm{M}^0,3})e^{(E-E_0)nf} \tag{72}$$

$$\beta_{\mathrm{M}}(f_{\mathrm{M},1} - f_{\mathrm{M},3}) = \beta^0_{\mathrm{M}}(f_{\mathrm{M}^0,1} - f_{\mathrm{M}^0,3}) \tag{73}$$

where n is the number of electrons, f is the Faraday reduced constant, $\beta_{\mathrm{M}} = \frac{\omega_{\mathrm{M}} D_{\mathrm{M}}}{\Delta x}$ and $\beta_{\mathrm{M}^0} = \frac{\omega_{\mathrm{M}^0} D_{\mathrm{M}^0}}{\Delta x_0}$, Δx and Δx_0 are the mesh sizes of the space in solution and inside the electrode, respectively. After some algebraic manipulation one gets:

$$f_{\mathrm{M}^0,3} = \frac{f_{\mathrm{M}^0,1}(\beta^0_{\mathrm{M}} - \beta_{\mathrm{M}} e^{(E-E_0)nf}) + 2f_{\mathrm{M},3}\beta_{\mathrm{M}}}{\beta_{\mathrm{M}} e^{(E-E_0)nf} + \beta^0_{\mathrm{M}}} \tag{74}$$

$$f_{\mathrm{M},1} = (f_{\mathrm{M}^0,1} + f_{\mathrm{M}^0,3})e^{(E-E_0)nf} - f_{\mathrm{M},3} \tag{75}$$

In the application discussed here below, the condition $E \ll E_0$, is satisfied, so that the electrode surface can be treated as a perfect sink for M. Under such conditions, one does not need to compute the concentration of M^0 and the conditions stated by the previous equations (74) and (75) can be simplified to a condition only on M

$$f_{\mathrm{M},1} = -f_{\mathrm{M},3}$$

5.4.5. *The Michaelis-Menten Condition at the Interface*

The Michaelis-Menten condition in its the complete form is stated in equation (38). Under the steady-state approximation, equation (39):

$$J_{\mathrm{int}} = \frac{1}{A}\frac{dN}{dt} = \frac{k_{\mathrm{int}} K_a \{R\}_{\mathrm{tot}}[\mathrm{M}]}{1 + K_a[\mathrm{M}]} \qquad x \in S \tag{76}$$

can be used. In this approximation the Michaelis-Menten condition becomes a mixed boundary condition. Therefore one can apply the equation (66) to the equation (76) and the numerical condition is rewritten along the Cartesian directions as

$$\begin{aligned} J_x &= \frac{\omega d D}{\Delta x}(f^{(1)}_1 - f^{(1)}_2) \\ J_y &= \frac{\omega d D}{\Delta x}(f^{(1)}_3 - f^{(1)}_4) \\ J_z &= \frac{\omega d D}{\Delta x}(f^{(1)}_5 - f^{(1)}_6) \end{aligned} \tag{77}$$

where J_x, J_y and J_z are the x, y and z components of the internalization flux.
For instance, for a planar surface, if $K_a[\mathrm{M}] \ll 1$ as it is often the case for trace metal, the Michaelis-Menten numerical boundary condition, reduces to solve the following algebraic equation in the unknown $f_{\mathrm{M},1}$:

$$\frac{2\omega_{\mathrm{M}} D_{\mathrm{M}}}{\Delta x}\left(f_{\mathrm{M},1} - f_{\mathrm{M},2}\right) = k_{\mathrm{int}} K_a \{R\}_{\mathrm{tot}}\left(f_{\mathrm{M},1} + f_{\mathrm{M},2}\right)$$

from which, after simple computations, one get:

$$f_{\mathrm{M},1} = -\frac{1 + \frac{2\omega_{\mathrm{M}} D_{\mathrm{M}}}{k_{\mathrm{int}} K_a \{R\}_{\mathrm{tot}}}}{1 - \frac{2\omega_{\mathrm{M}} D_{\mathrm{M}}}{k_{\mathrm{int}} K_a \{R\}_{\mathrm{tot}}}} f_{\mathrm{M},2} \tag{78}$$

Following the same reasoning, but with more calculation, one can extend the above condition for 3D spherical surface, by discretizing the surface with elementary cubes of size Δx.

6. THE TIME SPLITTING METHOD

We will outline below a numerical scheme which makes use of the time splitting method within the framework of the regularized LBGK scheme (47). The idea of the splitting method is to solve separately the diffusion and the reaction parts. This will produce additional integration errors, but will allow us to deal efficiently with a large difference in the characteristic time scales. For simplicity, we shall illustrate the splitting technique directly at the level of the partial differential equations so as to build two coupled problems.
Let c be the vector of concentration functions. Following the standard mathematical notation, the vector function c is introduced. Each entry of it contains the concentration function of each species. For instance, considering the prototype problem, equation (3), X=M, L and ML, hence $c = (c_{\mathrm{M}}, c_{\mathrm{L}}, c_{\mathrm{ML}})^{\top}$. The original problem is

$$\frac{\partial c}{\partial t} = T_{\mathrm{D}}c + T_{\mathrm{R}}c \tag{79}$$

where T_{D} and T_{R} are the diffusion and the reaction operators, respectively. With appropriate choices of T_{D} and T_{R}, our problem stated in equations (4)–(7) can be easily recovered.
The objective is to compute the concentration c at time $t+\Delta t$, $c(t+\Delta t)$, by solving the equation (79) on the time domain $(t, t+\Delta t]$ with the initial condition at time t, $c(t) = c_t$. By using the standard time splitting method, equation (79) is decomposed into

$$\frac{\partial c'}{\partial t} = T_{\mathrm{D}}c' \qquad \text{on } (t, t+\Delta t] \qquad \text{with } c'(t) = c_t \tag{80}$$

$$\frac{\partial c''}{\partial t} = T_{\mathrm{R}}c'' \qquad \text{on } (t, t+\Delta t] \qquad \text{with } c''(t) = c'(t+\Delta t) \tag{81}$$

The final value is $c(t+\Delta t) = c''(t+\Delta t)$. This decomposition is also called RD scheme, because the diffusion operator T_{D} is solved at the first step and the reaction operator T_{R} is solved at the second step. It can be shown [34] that the time splitting scheme is exact if the reactive and the diffusive operators commute. If they do not commute, the time splitting error is accurate at the second order in space and at least first order in time.

6.1. The Time Splitting Method in the LBGK Framework

We will consider two LB dynamics to solve the coupled system of equations (80) and (81), one for each process. The first equation is solved by applying the pure diffusive dynamics:

$$f_{\mathrm{X},i}(\mathbf{x} + \mathbf{v}_i\Delta t, t+\Delta t) = f_{\mathrm{X},i}(\mathbf{x}, t) + \Omega^{\mathrm{NR}}_{\mathrm{X},i}(\mathbf{x}, t)$$

The second equation is solved by applying the pure reactive dynamics:

$$f_{\mathrm{X},i}(\mathbf{x}, t+\Delta t) = f_{\mathrm{X},i}(\mathbf{x}, t) + \Omega^{\mathrm{R}}_{\mathrm{X},i}(\mathbf{x}, t)$$

The pure diffusive scheme is non-local and it can be decomposed into its collision and propagation parts. Notice that the decomposition into collision and propagation steps of whatever Lattice Boltzmann dynamics, can be considered as an application of the time splitting scheme to the original Boltzmann equation. In fact, it can be shown that the Lattice Boltzmann scheme, in its LBGK approximation, is equivalent to the discretization of the Boltzmann Equation in the BGK approximation

$$\frac{\partial f}{\partial t} + \mathrm{v}\cdot\nabla f = \omega(f^{\mathrm{eq}} - f)$$

by applying the time splitting method to separate the streaming operator $\mathrm{v}\cdot\nabla f$ from the collision operator $\omega(f^{\mathrm{eq}} - f)$. On the other hand the second scheme is purely local and it reduces only to one collision step.
The idea is to perform the diffusive collision, followed by the reactive collision and, finally, by the propagation.

Therefore, by using the RD splitting scheme, the complete numerical scheme (47) is now split into its pure diffusive and reactive parts as:

$$f_{X,i}(\mathbf{x},t') = f_{X,i}(\mathbf{x},t) + \Omega^{NR}_{X,i}(\mathbf{x},t) \qquad (t,t'] \tag{82}$$

$$f_{X,i}(\mathbf{x},t'') = f_{X,i}(\mathbf{x},t') + \Omega^{R}_{X,i}(\mathbf{x},t') \qquad (t',t''] \tag{83}$$

$$f_{X,i}(\mathbf{x}+\mathbf{v}_i\Delta t, t+\Delta t) = f_{X,i}(\mathbf{x},t'') \qquad t'' = t+\Delta t \tag{84}$$

Here we have assumed that the time variable t' for the diffusion equation runs differently from the time variable t'' of the reaction equation. In fact, the difficult problems to handle are those where the reaction time run faster than the diffusion time, i.e. $t_r \ll t_d$. These class of problems, where reaction is fast, are also known in literature as stiff problems.
The sub-problem (82) is a numerical initial-boundary-value problem involving only diffusion process, therefore it will be called process D. On the other hand, sub-problem (83) is a numerical initial value problem involving only a chemical reaction, therefore it will be called process R. In such a way, one can define several splitting procedures.
Notice that process R, as expressed in equation (83) is an explicit scheme, because the values of $\Omega_{X,i}(\mathbf{x},t)$ are computed at the actual time level. However, the very advantage of the time splitting method, is the possibility of using different techniques for sub-problems (82) and (83) and to use different time integration steps. In fact, during the simulation the time step Δt has to be understood as the main integration time and when not explicitly specified it will be used for both schemes D and R. Furthermore, to take advantage of the splitting techniques, an implicit scheme, which warrants a larger stability region [34], should be used for the reaction process (83). In this case, equation (83), is replaced with:

$$f_{X,i}(\mathbf{x},t'') = f_{X,i}(\mathbf{x},t') + \Omega^{R}_{X,i}(\mathbf{x},t'') \tag{85}$$

Several tests has been made to check the capability and the accuracy of the various splitting modes for stiff reaction processes, i.e. fast chemical reaction [4]. The results of the tests allowed us to choose the RD splitting as the most convenient for stiff reactive processes.

7. THE GRID REFINEMENT METHODS

Grid refinement (GR) is a technique in which the computational mesh is made finer in the regions where small scale processes take place, while to save CPU time and memory, a coarser grid is used in less demanding regions. Many strategies have been developed in the implementation of different GR in the Lattice Boltzmann approach, see for instance [32, 27, 25]. Many theoretical studies, the so-called adaptative grid [8, 9, 10, 11], involves the investigation of the correct grid size to use in these layers when they become thin. There are clearly two opposite constraints. If the fine grid spans a large part of the computational domain, the numerical solution will be more accurate, but the CPU time and memory usage may increase beyond acceptable limits. The question is then to know how to divide in an efficient way the domain in coarse and fine grids. Note that, obviously, there can be more than two grid levels. Another question is to determine the most appropriate change of mesh size between two adjacent grids as well as the best way to couple the physical quantities on both side of the grid interface.

7.1. The Reason to Refine the Grid

Let's consider a simple diffusion problem with only one species, say M, and let's vary its reaction rate of formation, say k. A 1D problem is considered in a semi-infinite domain $(0,+\infty)$, with initial condition

$$[\mathrm{M}] = [\mathrm{M}]^*$$

boundary semi-infinite diffusion condition

$$[\mathrm{M}] \to [\mathrm{M}]^* \qquad x \to +\infty$$

and perfect sink condition

$$[\mathrm{M}] = 0 \qquad x = 0$$

The time splitting exact solution (equation (80) and (81)) in the time interval $[0,t]$ is [34]:

$$[\mathrm{M}] = [\mathrm{M}]^* e^{kt} \mathrm{erf}\Big(\frac{x}{2\sqrt{D_{\mathrm{M}}t}}\Big) \tag{86}$$

If the exact solution is considered in two points, say x and $x+\Delta x$, one can write:

$$[\mathrm{M}](x+\Delta x,t) - [\mathrm{M}](x,t) = [\mathrm{M}]^* e^{kt}\Big[\mathrm{erf}\Big(\frac{x+\Delta x}{2\sqrt{D_{\mathrm{M}}t}}\Big) - \mathrm{erf}\Big(\frac{x}{2\sqrt{D_{\mathrm{M}}t}}\Big)\Big] \tag{87}$$

The first order Taylor series expansion around x of an amount Δx, with $\Delta x \to 0$, allows us to write:

$$[\mathrm{M}](x+\Delta x,t) - [\mathrm{M}](x,t) = [\mathrm{M}]^* e^{kt} \Delta x \frac{e^{-\frac{x^2}{2\sqrt{D_{\mathrm{M}}t}}}}{\sqrt{\pi D_{\mathrm{M}}t}} \tag{88}$$

In general, one expects that $[\mathrm{M}]$ will vary continuously to ensure numerically accuracy. So, if

$$\frac{|[\mathrm{M}](x+\Delta x,t) - [\mathrm{M}](x,t)|}{[\mathrm{M}]^*} < \theta,$$

then the following condition is obtained:

$$\Delta x < e^{-kt} \sqrt{\pi D_{\mathrm{M}} t}\, e^{\frac{x^2}{2\sqrt{D_{\mathrm{M}}t}}} \cdot \theta \tag{89}$$

Inequality (89) says that: a) Δx has an upperbound and b) it two rates k_1 and k_2 are considered, such that $k_1 > k_2$, then $\Delta x_{k_1} < \Delta x_{k_2}$. Therefore, for very fast reaction dynamics, in order to get a variation between x and $x+\Delta x$ less than θ, the grid size Δx has to be chosen small enough, accordingly with inequality (89).
Furthermore, it is not necessary to use the inequality (89) in all the domain. It is advisable to restrict its usage only to region where this is really needed, i.e. close to the consuming surface and, more precisely, within the region comparable to the reaction layer thickness, μ, equation (12).

7.2. The Grid Refinement Scheme

Grid refinement amounts to using different grid spacing along the computational domain Ω. Such a choice is necessary when large variations of mass occur in Ω. This is the case, for instance, when k_{a} value is large, i.e. when the reaction layer thickness μ_{ML}, equation (12), is between 10^{-9}m and 10^{-6}m. In this region it is necessary to set grid sizes less than the reaction layer thickness.
However, it is not required to have small grid sizes in the whole domain. Hence, the domain Ω is divided into s sub-grids $\mathrm{G}_1,\ldots,\mathrm{G}_s$, such that $\Omega = \mathrm{G}_1 \cup \cdots \cup \mathrm{G}_s$. Each sub-grid is made by n_i points, such that the total number of points $n_x = \sum_{i=1}^{s}(n_i - 1) + 1$. The G_i are discretized in time with a time step Δt_i and in space with a grid size Δx_i, such that $\Delta x_{i+1} > \Delta x_i$. In order to make a simple but sufficiently general description, the domain is divided in three sub-grids, i.e. $s = 3$, even though for real computations, s can be larger.
The set up of the grid is illustrated in Fig. **3**. Points A and D are boundary points, while points B and C

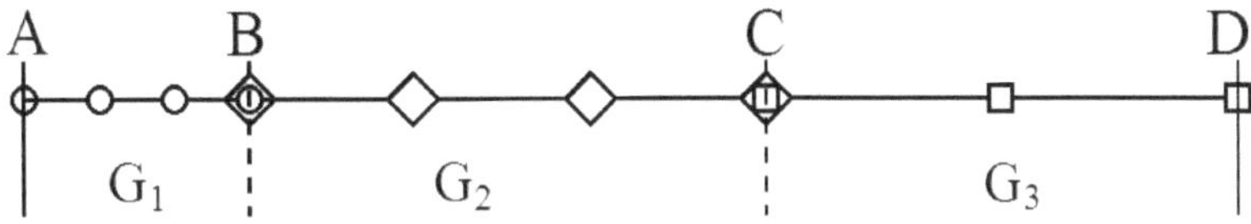

Fig. **3**: Set up of a 1D computational domain with three sub-grids G_i, $i = 1,2,3$. Circle, rhombus and square markers represent points belonging to G_1, G_2 and G_3 respectively.

are sub-grid interface points of G1-G2 and G2-G3 respectively. Notice that points B and C belong to both adjacent sub-grids.

At the first step, the smallest grid size, Δx_1 is chosen. As mentioned above, it should be smaller than the smallest reaction layer thickness. Secondly, the smallest time step, Δt_1 is chosen. To satisfy convergence conditions, Δt_1 must be smaller than the smallest time scale of the process, usually $\Delta t_1 < \mathrm{Min}_{ij}({}^j k_{a,i}[{}^j\mathrm{L}])^{-1}$. The total number of points, n_x, the number of point of G_1, n_1 and the length of the domain, l are chosen arbitrarily.

Let's define, here, the density distribution functions $f_{\mathrm{X},i}$ of the LB method at each sub-grid, which will be useful in the rest of the section to describe the grid refinement methods. Let $f_{\mathrm{X},i}^{(\mathrm{G_k})}(x_j \in \mathrm{G_k}, t)$ for $j = 1, \ldots, n_k$ be the density functions of each species X at the points of sub-grid G_k for $k = 1, 2, 3$.

Points B and C are critical, because, at these points, not all the density distribution functions are computed by the numerical scheme. Some of them remain unknowns and need to be estimated invoking either 1) time interpolation or 2) conservation of mass and flux.

Time interpolation is performed whenever one needs to compute a function which cannot be computed with the evolution equation. For instance, let us suppose that in G_2 $\Delta t_2 = 2\Delta t_1$. The function $f_{\mathrm{X},2}^{(\mathrm{G1})}(\mathrm{B}, t + \Delta t_1)$ is not known and to compute it, one interpolates from the corresponding functions already computed in the sub-grid G_2, for instance by using a first order interpolation in time, $f_{\mathrm{X},2}^{(\mathrm{G1})}(\mathrm{B}, t + \Delta t_1) = \frac{1}{2}(f_{\mathrm{X},2}^{(\mathrm{G2})}(\mathrm{B}, t - \Delta t_1) + f_{\mathrm{X},2}^{(\mathrm{G2})}(\mathrm{B}, t + 2\Delta t_1))$.

The condition of conservation of mass and flux only works for the special 1D case. A complete detailed explanation of it can be found in [3]. Such a condition, which will be called "grid interface condition", in terms of density distribution functions, e.g. at the point B, reads:

$$f_{\mathrm{X},1}^{(\mathrm{G1})}(\mathrm{B}, t) + f_{\mathrm{X},2}^{(\mathrm{G1})}(\mathrm{B}, t) = f_{\mathrm{X},1}^{(\mathrm{G2})}(\mathrm{B}, t) + f_{\mathrm{X},2}^{(\mathrm{G2})}(\mathrm{B}, t) \tag{90}$$

$$\frac{\omega_1}{\Delta x_1}(f_{\mathrm{X},1}^{(\mathrm{G1})}(\mathrm{B}, t) - f_{\mathrm{X},2}^{(\mathrm{G1})}(\mathrm{B}, t)) = \frac{\omega_2}{\Delta x_2}(f_{\mathrm{X},1}^{(\mathrm{G2})}(\mathrm{B}, t) - f_{\mathrm{X},2}^{(\mathrm{G2})}(\mathrm{B}, t)) \tag{91}$$

where $f_{\mathrm{X},1}^{(\mathrm{G2})}(\mathrm{B}, t)$ and $f_{\mathrm{X},2}^{(\mathrm{G1})}(\mathrm{B}, t)$ are the unknowns, while $f_{\mathrm{X},2}^{(\mathrm{G2})}(\mathrm{B}, t)$ and $f_{\mathrm{X},1}^{(\mathrm{G1})}(\mathrm{B}, t)$ are computed from the evolution scheme. Note that, the above "grid interface condition" works only for reactive-diffusive schemes in 1D. For hydrodynamic problems and for diffusive 2D or 3D, this approach is not directly applicable, but we believe it is "equivalent" to the grid refinement proposed by [25] based on the conservation of the equilibrium functions and on the rescaling of the non-equilibrium functions. The approach given in [25] is based on the fact that along the sub-grids, the conserved quantities does not depend on the lattice resolution but only on physical quantities. My approach follows the same line: with conditions (90) and (91), the same values to the physical quantities are imposed (mass concentration and flux) across the sub-grid interfaces. The equation (90) is the conservation of mass

$$[\mathrm{X}]^{(\mathrm{G_1})}(\mathrm{B}, t) = [\mathrm{X}]^{(\mathrm{G_2})}(\mathrm{B}, t)$$

while the equation (91) is the conservation of flux

$$J_{\mathrm{X}}^{(\mathrm{G_1})}(\mathrm{B}, t) = J_{\mathrm{X}}^{(\mathrm{G_2})}(\mathrm{B}, t)$$

In the grid refinement method proposed here the time step is maintained constant among the sub-grids:

$$\Delta t_i = \text{constant} \quad i = 1, 2, 3 \tag{92}$$

The big advantage of this method is the absence of constraints on the grid sizes Δx_i. This is a very important and useful feature, only valid for 1D problem, because it allows us to choose the grid spacing arbitrarily. Note that no time interpolation is required. Therefore this method has the great advantage of being simple and easy to write in a numerical code. The corresponding algorithm is shown in Algorithm 1.

Algorithm 1 Numerical scheme with the grid refinement method.

1. **Start at time t.**

2. Application of the evolution scheme to each sub-grid. We obtain the density distribution functions in all the points of G_3 at $t = t + \Delta t$, G_2 at $t = t + \Delta t$ and G_1 at $t = t + \Delta t$, except for $f_{X,1}^{(G_1)}(A, t + \Delta t)$, $f_{X,2}^{(G_1)}(B, t + \Delta t)$, $f_{X,1}^{(G_2)}(B, t + \Delta t)$, $f_{X,2}^{(G_2)}(C, t + \Delta t)$, $f_{X,1}^{(G_3)}(C, t + \Delta t)$ and $f_{X,2}^{(G_3)}(D, t + \Delta t)$.

3. Application of boundary conditions to compute $f_{X,1}^{(G_1)}(A, t + \Delta t)$ and $f_{X,2}^{(G_3)}(D, t + \Delta t)$. Application of the grid interface condition to compute $f_{X,2}^{(G_1)}(B, t + \Delta t)$ and $f_{X,1}^{(G_2)}(B, t + \Delta t)$. Again, application of the grid interface condition to compute $f_{X,2}^{(G_2)}(C, t + \Delta t)$ and $f_{X,1}^{(G_3)}(C, t + \Delta t)$.

4. **Repeat from point 2 at new time step with $t \leftarrow t + \Delta t$.**

8. THE COMPLETE NUMERICAL SCHEME

The complete numerical scheme used to numerically solve the set of reaction-diffusion equations stated in equations (24) - (27) is shown in Algorithm 2. It makes use of the RD time splitting method, to separate the collision operator from its diffusive and reactive parts and of the above mentioned grid refinement method, to refine the grid in regions where large gradients appear.
The algorithm is divided in four steps. At the first step, the initial conditions are set up, by computing the density distribution functions $f_{X,i}(\mathbf{x}, t)$ at $t = 0$. Then the numerical scheme begins until the time t_f is attained. The major steps of the numerical scheme are:

1. The collision-propagation step is applied in each subgrid G_j

2. Unknown functions are computed at adjacent subgrids

3. The distribution functions at the boundaries are updated to satisfy the boundary conditions

4. The flux at the consumed surface is computed

Finally, at the end of the computation, the concentration profiles and the flux are written on an output file.
In the next paragraph, the algorithm 2 is applied to compute the flux of a metal M and all the concentration profiles at a planar consuming surface in a chemical heterogeneous system with many ligands and complexes. Furthermore, the above mentioned algorithm 2 has been adapted to simulate environmental applications and the code MHEDYN (MultiHEterogenousDYNamics) to compute metal flux in aquatic chemically heterogeneous systems has been written. The beta version of MHEDYN is freely available in the web at `http://www.unige.ch/cabe/dynamic/`. However, a new more stable and efficient version of MHEDYN will be soon avaialable at the following web site `http://www.mhedyn-project.com/`.

9. METAL FLUXES IN MIXTURES OF ENVIRONMENTAL COMPLEXANTS

9.1. A Summary of the Physical Model and Boundary and Initial Conditions

The mathematical formulation of metal flux at a planar consuming surface in multiligand systems, precisely the problem stated by equations (24)–(27) will be solved with algorithm 2 for two applications.
In both applications, the initial conditions are the following:

- the bulk solution is homogeneous

- all species are in equilibrium in the bulk solution, and

- the bulk concentrations of any species is independent of time.

Algorithm 2 Numerical Algorithm: LBGK + Time Splitting + Grid Refinement

SET UP of INITIAL CONDITIONS
  $f_{X,i}(\mathbf{x},t)$ at $t=0$ on each subgrid $G_j \quad j=1,\ldots,s$

START of NUMERICAL SCHEME
$t \leftarrow t+\Delta t$
while $t \leq t_f$ **do**
  for all Sub-grid G_j **do**
    COLLISION 1: DIFFUSION
      Application of equation (82)
    COLLISION 2: REACTION
      Application of equation (83)
    PROPAGATION
      Application of equation (84)
  end for
  UPDATE of $f_{X,i}(\mathbf{x},t)$ at ADJACENT SUBGRIDS
    Application of flux and mass conservation
  BOUNDARY CONDITIONS
    Application of numerical boundary conditions
  COMPUTATION OF CONCENTRATION PROFILES
    $[X](\mathbf{x},t)=\sum_i f_{X,i}(\mathbf{x},t)$
  FLUX COMPUTATION at the consuming surface
  $t \leftarrow t+\Delta t$
end while

OUTPUT RESULTS
Writing of concentration profiles and flux

The boundary conditions related to each species at the consuming interface are the following:

- In the first application, M is consumed at the interface and its concentration $[M]^0$ on the solution side of the surface (subscript 0) is nil. This condition corresponds to very fast transfer rate of M through the interface, i.e. to maximum flux, in the solution phase.
- In the second application, M is consumed at the interface via the Michaelis-Menten condition (77)
- M^jL_i does not pass through the interface, equation (32).
- A complex can only be consumed after dissociation into M and ${}^jL$
- ${}^jL$ does not pass through the interface, equation (33).
- The flux is computed under conditions of finite planar diffusion, i.e. the concentrations of all species are assumed to be equal to those in the bulk solution, at a finite distance (around 20μm) of the consuming interface. This corresponds to conditions valid in stirred solution, equation (30).

Unless otherwise stated, the results of the simulations are at steady-state.
The total flux of metal through the interface is given by

$$J_{\mathrm{tot}} = \frac{1}{\mathrm{A}}\left(\frac{dN}{dt}\right)_{x=0} \tag{93}$$

where A is the surface area of the interface and N the number of mole of metal ion. The contribution of each individual complex to the flux can be expressed in terms of its degree of lability ${}^j\xi_i$. For the global system the lability degree is defined [45] as the ratio of the flux provided by the complexes over the same contribution if all the complexes were fully labile (i.e. able to dissociate completely at the interface). In the case of ligand excess under steady-state conditions, the particular degree of lability ${}^j\xi_i$ has been shown [4] to be computed with the following equation:

$${}^j\xi_i = \frac{1 - \frac{[\mathrm{M}^j\mathrm{L}_i]^0}{[\mathrm{M}^j\mathrm{L}_i]^*}}{1 - \frac{[\mathrm{M}]^0}{[\mathrm{M}]^*}} \tag{94}$$

where $[\mathrm{M}^j\mathrm{L}_i]^0$ is the surface concentration of $\mathrm{M}^j\mathrm{L}_i$, provided at any time by the numerical simulation.
The derivation of the lability degree given in [4] is new because, for the first time, mixtures of ligands and their successive metal complexes have been considered together. The derivation of equation (94) follows the same line of reasoning made in [45, 44] for simple complexes and successive complexations, separately.
Under steady-state conditions, the individual flux of $\mathrm{M}^j\mathrm{L}_i$ is related to the degree of lability by:

$$J_{\mathrm{M}^j\mathrm{L}_i} = \frac{D_{\mathrm{M}^j\mathrm{L}_i}[\mathrm{M}^j\mathrm{L}_i]^*}{\delta}\left(1 - \frac{[\mathrm{M}]^0}{[\mathrm{M}]^*}\right){}^j\xi_i \tag{95}$$

The values of the association rate constants, for complexes with simple ligands and fulvics/humics, are computed as discussed in [19]. They are given by:

$$k_{\mathrm{a}}^{\mathrm{ML}} = \frac{k_{\mathrm{a}}^{\mathrm{OS}} k_{-w}}{k_{\mathrm{d}}^{\mathrm{OS}} + k_{-w}} \tag{96}$$

where k_{-w} is the rate constant for elimination of a water molecule from the inner shell of the free hydrated metal ion. The expressions for $k_{\mathrm{a}}^{\mathrm{OS}}$ and $k_{\mathrm{d}}^{\mathrm{OS}}$ are the following:

$$k_{\mathrm{a}}^{\mathrm{OS}} = \frac{4\pi N_{\mathrm{av}} a (D_{\mathrm{M}} + D_{\mathrm{L}}) \frac{\mathrm{U}(a)}{k_{\mathrm{B}}T}}{e^{\frac{\mathrm{U}(a)}{k_{\mathrm{B}}T}} - 1} \tag{97}$$

$$k_{\rm d}^{\rm OS} = \frac{3(D_{\rm M}+D_{\rm L})\frac{{\rm U}(a)}{k_{\rm B}T}}{a^2(e^{\frac{{\rm U}(a)}{k_{\rm B}T}}-1)} e^{\frac{{\rm U}(a)}{k_{\rm B}T}} \tag{98}$$

where $N_{\rm av}$ is the Avogadro number, $k_{\rm B}$ is the Boltzmann constant, T is the temperature ($T = 273$K), a is the distance of closest approach of the metal and the ligand ($a = 0.5$nm), $D_{\rm M}$ and $D_{\rm L}$ are the diffusion coefficients of the metal M and the ligand L respectively and ${\rm U}(a)$ is the electrostatic interaction energy of M and the complexing site L, at their distance of closest approach a.
If ${\rm U}(a)= 0$ then the above equations simplify in:

$$k_{\rm a}^{\rm OS} = 4\pi N_{\rm av} a(D_{\rm M}+D_{\rm L}) \tag{99}$$

$$k_{\rm d}^{\rm OS} = \frac{3(D_{\rm M}+D_{\rm L})}{a^2} \tag{100}$$

For simple ligands [28]:

$${\rm U}(a) = \frac{z_{\rm M}z_{\rm L}e^2}{4\pi\varepsilon_0\varepsilon a}\left(1-\frac{\kappa a}{1+\kappa a}\right) \tag{101}$$

where

$$\kappa = \sqrt{\frac{2N_{\rm av}e^2 I}{\varepsilon_0\varepsilon k_{\rm B}T}} \tag{102}$$

and ε is the relative dielectric constant of water, ε_0 is the vacuum permittivity and e is the electric charge.
For humic substances [19]:

$${\rm U}(a) = z_{\rm M}e\psi \tag{103}$$

where ψ is the electric potential at the surface of the humics, due to their negative charge. Its values is discussed in [26].
Environmental "particles" are usually aggregates with wide size distribution. It has been shown that [19], as a first approximation, the effective association rate constants of their metal complexes, can be computed by [17, 40]:

$$k_{\rm a} = \frac{k_{\rm a}^{\rm ML}}{1+{\rm B}} \tag{104}$$

where

$${\rm B} = \frac{k_{\rm a}^{\rm ML}|{\rm L}|_{\rm tot}\frac{\Delta S_b}{S_{b,\rm tot}}}{4\pi r D_{\rm M} c_p} \tag{105}$$

where $|{\rm L}|_{\rm tot}$ is the total concentration of complexing sites, $\frac{\Delta S_b}{S_{b,\rm tot}}$ is the fraction of surface area, for particle with radius r and c_p is the number concentration of particle aggregates. The potential ${\rm U}(a)$ is also given by equation (103), where ψ is the surface potential at the site L (see [19] for values).

9.2. Flux Through Consuming Interfaces in Ligand Mixtures: Perfect Sink Condition for the Metal

The code MHEDYN has been used to compute the relative contribution to the total copper flux, of each copper complex in environmental mixtures of hydroxide, carbonate, fulvic and aggregate ligands. Fluxes have been computed in two waters with the typical realistic compositions given in Table **4**. Note that overall, 60 metal species with D values ranging from 2×10^{-13} up to $9\times10^{-10}{\rm m}^2{\rm s}^{-1}$ and $k_{\rm a}$ values ranging from 7×10^{2} up to $2.5\times10^{8}{\rm m}^3{\rm mol}^{-1}{\rm s}^{-1}$ have been treated by MHEDYN in each case. The fractal aggregates are assumed to be formed by the assembly of non complexing solid subparticles on which a complexing minor component which can be regarded as amorphous FeOOH is adsorbed. In the two tested cases, the roles of fulvics and aggregates are expected to change, since i) the copper to fulvic ratio increases from water n°1 to n°2 i.e. the complexing strength of Cu^{2+} by fulvic significantly decreases and ii) the overall concentration (and thus the complexing strength) of aggregate complexing sites increases in the same order.

Table 4: Main parameters used for flux computations in waters n°1 and n°2.

Compound	Water n°1	Water n°2
pH	8	8
$[CO_3^{2-}]_{tot}$	2×10^{-3}M	2×10^{-3}M
{FS}	2 mg/L	2 mg/L
{Solid concentration}	3 mg/L	3 mg/L
{FeOOH}	30 μg/L	300 μg/L
$[Cu]_{tot}$	10^{-8}M	3.5 $\times10^{-7}$M

The time evolutions of the total fluxes confirm that reaction diffusion with aggregates is also the slow step in mixtures of aggregate and soluble ligands, and that the overall time to reach a constant total flux should not be too much larger than seconds to minutes in environmental systems. Steady-state for large size complexes may take $\sim$ 30 minutes, but their contribution to the flux is negligible.

In the mixtures, at steady-state, the distribution of lability degree ${}^j\xi$ and individual fluxes ${}^jJ$ of the fulvic complexes with respect to ${}^jK$, is similar to those obtained under the same conditions with only fulvics in solution. The same observation is made for the distributions of ${}^j\xi$ and ${}^jJ$ with respect to ${}^jr$ (or ${}^jD$) for complexes with aggregates. The complete analysis of these results is reported in [4]. On the other hand, Fig. **4** shows the cumulative fluxes, $\Sigma^jJ = f({}^jr)$ of complexes with aggregates only, for the two tested waters. The curves water n°1 and water n°2 (Fig. **4**) correspond to increasing complexing strengths of aggregates

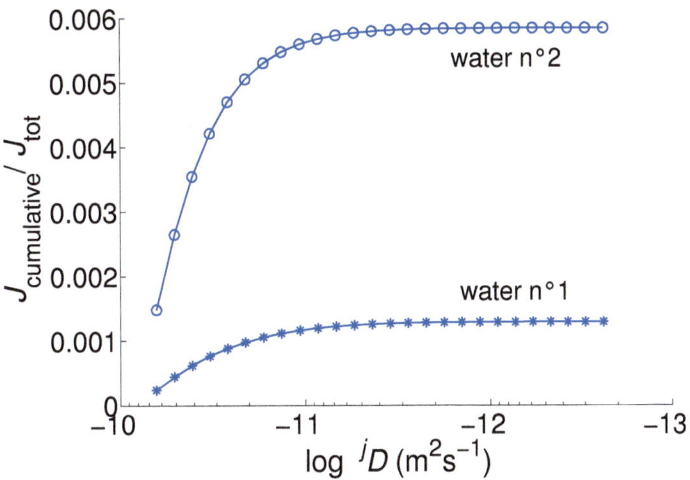

Fig. **4**: Cumulative fluxes of complexes with aggregate only as a function of the diffusion coefficients of the aggregate class, for waters n°1 and n°2. Curve water n°1 corresponds to {FeOOH}=30 μg/L, while curve water n°2 corresponds to {FeOOH}=300 μg/L. Parameters: Water 1: $[Cu]_{tot} = 10^{-8}$M, $\{FA\}_{tot} = 2$mg C/l, $\log K^* = 12.5$, $[CO_3^{2-}]_{tot} = 2$mM; Water 2: $[Cu]_{tot} = 3.5^{-7}$M, $\{FA\}_{tot} = 2$mg C/l, $\log K^* = 9.5$, $[CO_3^{2-}]_{tot} =$ 2mM.

for Cu^{2+}. It can be seen that the aggregates which contribute the most to the total flux are always those with ${}^jr \leq 10$nm, irrespective of complexation strength. Thus, even though more data are desirable, these results strongly suggest that the contribution of metal-aggregate complexes with size larger than 10nm can be neglected on flux computations.

Tables **5**, **6** provide the steady-state values of free Cu^{2+}, carbonato complexes ($CuCO_3^{2-}$ and $Cu(CO_3^{2-})_2^{2-}$), hydroxo-complexes ($CuOH^+$ and $Cu(OH)_2$), fulvic complexes and aggregate complexes, and their proportion to the total flux, J_{tot}. It also provides for comparison, the proportion of the various types of complexes

Table **5**: Steady-state values of fluxes and bulk concentrations normalised with Cu^{2+}_{tot} of the species present in waters 1. Other parameters: see Fig. **4**

Species X	**Water 1**		
	$[X]/[Cu]_{tot}\%$	J_X $(molm^{-2}s^{-1})$	$J_X/J_{tot}\%$
Cu^{2+}	0.003	1.04×10^{-14}	0.53
CuOH and $Cu(OH)_2$	0.065	6.03×10^{-14}	3.05
$CuCO_3$ and $Cu(CO_3)_2$	0.128	1.83×10^{-13}	9.28
$\sum$Cu-FS	97.50	1.70×10^{-12}	86.9
$\sum$Cu-Aggregate	2.317	2.54×10^{-15}	0.19
J		1.97×10^{-12}	100
$J_{tot}/J_{lab}\%$		1.44	

Table **6**: Steady-state values of fluxes and normalised bulk concentrations with respect to Cu^{2+}_{tot} of the species present in waters 2. Other parameters: see Fig. **4**

Species X	**Water 2**		
	$[X]/[Cu]_{tot}\%$	J_X $(molm^{-2}s^{-1})$	$J_X/J_{tot}\%$
Cu^{2+}	0.097	1.21×10^{-12}	0.45
CuOH and $Cu(OH)_2$	0.164	8.19×10^{-12}	3.02
$CuCO_3$ and $Cu(CO_3)_2$	0.305	2.09×10^{-11}	7.68
$\sum$Cu-FS	33.14	2.38×10^{-10}	87.5
$\sum$Cu-Aggregate	66.49	1.59×10^{-12}	0.58
J		2.72×10^{-10}	100
$J_{tot}/J_{lab}\%$		14.9	

in the mixture and the ratio of J_{tot}/J_{lab} where J_{lab} is the total flux which would be observed if all complexes were fully labile. Tables **5** and **6** show that, for the two cases, the overall flux is largely dominated by the Cu-fulvic complexes. Amongst inorganic complexes, the carbonate complexes, $CuCO_3$ and $Cu(CO_3)_2$, are the most important under the conditions used, but their contribution to the total flux remains marginal. Indeed, their bulk concentrations, in both water, are much less than 1% that $[Cu]_{tot}$ and their contribution to the total flux is less than 10%.

The contribution of aggregate complexes to the total flux is negligible in both cases. The negligible contribution of metal aggregates is worth emphasizing, since complexing strength of their sites have been increased by 10 and the total Cu concentration is increased by 35, in water 2. As a consequence, the total concentration of aggregate complexes, with respect to the total concentration of Cu, is 2.17% for water 1 and 66.49% for water 2. Simultaneously, the complexation strength of the fulvics decreases by a factor of $35^2 = 1225$ due to the increase of the metal/fulvic ratio by a factor of 35. Even though all these conditions should favour the contribution of aggregates to the overall flux compared to fulvics, these latter still dominate. This is due to their respective degree of lability: this latter decreases for aggregates in water 2 compared to water 1, due to the increase of complexing site concentration. On the opposite, the degree of lability of fulvic complexes drastically increases in water 2, due to the increase of the metal to ligand ratio.

The results showed that there is a complex relationship among site concentrations, the metal to ligand ratio, the equilibrium constants of the complexes, the lability and the mobility of complexes with the flux contribution of each species.

9.3. Flux Through Consuming Interfaces in Ligand Mixtures: Michaelis-Menten Condition for the Metal

In a mixture of metal ions and complexes, the understanding of how each metal species contributes to the biouptake of a metal is essential to predict ecological risk. For microorganisms, the rate of uptake (internalization flux) which obeys the Michaelis-Menten equation, has not only a major influence on the total metal flux but also on the bioavailability of the various metal species and their relative contributions to the total flux. The contribution of each metal complex to the overall metal flux, in relation to its lability, has been investigated for a number of important boundary parameters (the equilibrium constant K_a of metal with transport sites, internalization rate constant k_{int} and total transport sites concentration R_{tot}) [53]. Computations were performed for Cu(II) complexes, in a multicomponent culture medium for microorganisms and the computation of metal flux was performed with MHEDYN for ligand mixtures. Under the condition of ligand excess, as often found in the natural environment, the contribution of each metal species to the total flux is shown to be independent of the Michaelis-Menten boundary conditions.

The independence of the degrees of lability of the metal complexes on the boundary conditions, found in [53], provided that the free metal ion is the only species consumed at the interface has great utility in practice when comparing the signals given by bioanalogical sensors and metal uptake by microorganisms immersed in the same external solution. Indeed, it implies that, despite the fact that the boundary conditions of an analytical sensor and a microorganism are different, the sensor can be used to determine the individual degrees of lability of the various metal complexes which can ultimately enable prediction of the role of each complex on metal biouptake. Thus such measurements are fully relevant for prediction of the bioavailabilities of metal species and to make credible dynamic ecotoxicological risk assessments.

However, in some cases (e.g. in biofilm layers) the complex metal-ligand ML and the ligand L are consumed at the interface following the Michaelis-Menten process [24, 21]. In such conditions, it is not clear yet whether or not the contribution of each metal flux is independent of the boundary conditions.

CONCLUSIONS

In this chapter several applications were studied that allowed to test the capabilities and the major characteristic of MHEDYN. MHEDYN is a reliable numerical code that can compute the flux and profile concentrations of any species in the transient and steady-state regime, in any geometry, with a very large number of ligands (limited only by computer time), with rate constants and diffusion coefficients varying over many orders of magnitude, without requirement of ligand excess compared to metal. At the present time no equivalent code is available, hence MHEDYN can be a useful tool for the community of chemists in order to better understand the behaviour of chemical complex systems.

One of the most interesting and useful applications that can be investigated with the model presented here is in the field of biofiltration processes in biofilm which are used to treat waters and waste-waters. A successful model is the so-called transient-state multispecies biofilm model [43] which is a synthesis of key modeling features needed to describe multiple-species biofilms that experience time-varying conditions, including periodic detachment by backwashing. Recently, a three-dimensional model has been developed to numerical simulate biofilm growth [31] wherein, for the first time, a Lattice Boltzmann simulation platform complemented with an individual-based biofilm model has been implemented to perform computer simulations of 3-D pore-scale model of biofilm growth in porous media. However, the chemical reaction among nutrient species were limited to simple Michaelis-Menten kinetics and without considering metal complexation reactions. The inclusion of metal complexation reactions into the model developed in [31] could represent a significant advance in understanding the role of metal bioavailability in biofilms (e.g. diffusion and transport processes may limit the chemical activity of nutrients, toxic compounds and medicines [52]).

A preliminary first step on this direction was started in [4] with the computer code BIODYN which is an extension of MHEDYN designed for biofilm applications. The plan is to pursue in this direction by making BIODYN more robust and efficient by implementing it in the new multiscale environment (the MAPPER project) designed to deal with multiscale processes.

ACKNOWLEDGEMENT

Declared none.

CONFLICT OF INTEREST

The author confirms that this chapter content has no conflict of interest.

REFERENCES

[1] Albuquerque P, Alemani D, Chopard B, Leone P. Coupling a Lattice Boltzmann and a Finite Difference Scheme. In. Bubak M, van Albada GD, Sloot PMA, Dongarra JJ (eds.). Computational Science - ICCS 2004: 4th International Conference, Krakow, Poland, June 6-9, 2004, Proceedings, Part IV, vol. 3039, page 540. Springer Berlin / Heidelberg, 2004.

[2] Alemani D, Chopard B, Galceran J, Buffle J. Time splitting and grid refinement methods in the Lattice Boltzmann framework for solving a reaction-diffusion process. In: Alexandrov VN, van Albada GD, Slot PMA, Dongarra JJ (eds.). Proceedings of ICCS 2006, Reading, LCNS 3992, pages 70-77. Springer, 2006.

[3] Alemani D, Chopard B, Galceran J, Buffle J. Two grid refinement methods in the Lattice Boltzmann framework for reaction-diffusion processes in complex systems. Phys Chem Chem Phys 2006; 8: 4119-4130.

[4] Alemani D. A Lattice Boltzmann numerical approach for modelling reaction-diffusion processes in chemically and physically heterogeneous environments. PhD thesis, Thèse de doctorat, University of Geneva, 2007.

[5] Alemani D, Buffle J, Zhang Z, Galceran J, Chopard B. Metal flux and dynamic speciation at (bio)interfaces. Part III: MHEDYN, a general code for metal flux computation; Application to simple and fulvic complexants. Environment Sci Techn 2008; 42: 2021-2027.

[6] Alemani D, Buffle J, Zhang Z, Galceran J, Chopard B. Metal flux and dynamic speciation at (Bio)Interfaces. Part IV: MHEDYN, a general code for metal flux computation; Application to particulate complexants and their mixtures with the other natural ligands. Environment Sci Techn 2008; 42: 2028-2033.

[7] Allison JD, Brown DS, Novo-Gradac KJ. Minteqa2/prodefa2, a geochemical assessment model for environmental systems: version 3.0, user's manual, 1991.

[8] Bieniasz LK. Use of Dynamically Adaptative Grid Techniques for the Solution of Electrochemical Kinetic Equations. Part 1. J Electroanal Chem 1993; 360: 119.

[9] Bieniasz LK. Use of Dynamically Adaptative Grid Techniques for the Solution of Electrochemical Kinetic Equations. Part 2. J Electroanal Chem 1994; 374: 1.

[10] Bieniasz LK. Use of Dynamically Adaptative Grid Techniques for the Solution of Electrochemical Kinetic Equations. Part 3. J Electroanal Chem 1993; 374: 23.

[11] Bieniasz LK. A Fourth-Order Accurate, Numerov-Type, Three-Point Finite-Difference Discretization of Electrochemical Reaction-Diffusion Equations on Non-Uniform (Exponentially Expanding) Spatial Grids in One-Dimensional Space Geometry. J Comput Chem 2004; 25: 1515.

[12] Bouillot P. Méthode Lattice Boltzmann avec régularisation, appliquée à la diffusion. Master's thesis, University of Geneva, 2005.

[13] Buffle J, De Vitre R. Chemical and biological regulation of aquatic systems. Lewis Publishers, 1994.

[14] Buffle J. Complexation Reactions in Aquatic Systems. Ellis Horwood, Chichester, 1998.

[15] Buffle J, Horvai G. In situ monitoring of aquatic systems. Wiley, 2000.

[16] Buffle J, Tercier-Waeber ML. Voltammetric environmental trace metal analysis and speciation. From laboratory to in situ measurements. TrAC-Trends in Analytical Chemistry 2005; 24: 172-191.

[17] Buffle J, Startchev K, Galceran J. Computing steady-state metal flux at microorganism and bioanalogical sensor interfaces in multiligand systems. A reaction layer approximation and its comparison with the rigorous solution. Phys Chem Chem Phys 2007; 9: 2844-2855.

[18] Buffle J, Zhang Z, Alemani D. Metal Flux and Dynamic Speciation at (Bio)interfaces. Part II: Evaluation and Compilation of Physicochemical Parameters for Complexes with Particles and Aggregates. Environ Sci Technol 2007; 41: 7621-7631

[19] Buffle J, Zhang Z, Startchev K. Metal Flux and Dynamic Speciation at (Bio)interfaces. Part I: Critical Evaluation and Compilation of Physicochemical Parameters for Complexes with Simple Ligands and Fulvic/Humic Substances for simple ligands, fulvics and particles. Environ Sci Technol 2007; 41: 7609-7620.

[20] Chen S, Dawson SP, Doolen GD, Janecky DR, Lawniczak A. Lattice Methods and their Applications to Reacting Systems. Comput Chem Engng 1995; 19: 617-646.

[21] Chen-Charpentier BM, Dimitrov DT, Kojouharov HV. Numerical simulation of multi-species biofilms in porous media for different kinetics. Math Comput Simul 2009; 79: 1846-1861.

[22] Chopard B, Droz M. Cellular automata modeling of physical systems. Cambridge University Press, 1998.

[23] Dawson SP, Chen S, Doolen GD. Lattice Boltzmann computations for Reaction-Diffusion equations. J Chem Phys 1993; 98: 1514-1523.

[24] Deshusses MA, Hamer G, Dunn IJ. Behavior of biofilters for waste air biotreatment. 1. dynamic model development. Environ Sci Techn 1995; 29: 1048-1058.

[25] Dupuis A, Chopard B. Theory and applications of an alternative Lattice Boltzmann grid refinement algorithm. Phys Rev E 2003; 67: 066707.

[26] Duval J, Wilkinson KJ, van Leeuwen HP, Buffle J. Humic substances are soft and permeable: evidence from their electrophoretic mobilities. Environ Sci Technol 2005; 39: 6435-6445.

[27] Filippova O, Hanel D. Grid Refinement for Lattice-BGK Models. J Comput Phys 1998; 147: 219-228.

[28] Fuoss RM. Ionic Association. III. The Equilibrium between Ion Pairs and Free Ions. J Ann Chem Soc 1958; 80: 5059.

[29] Galceran J, Puy J, Salvador J, Cecilia J, Mas F, Garces JL. Lability and mobility effects on mixtures of ligands under steady-state conditions. Phys Chem Chem Phys 2003; 5: 5091-5100.

[30] Galceran J, van Leeuwen HP. Dynamics of Bioupatke Processes: the Role of Transport, Adsorbtion and Internalization, Chapter 4. John Wiley and Sons, Ltd, 2004.

[31] Graf von der Schulenburg DA, Pintelon TRR, Picioreanu C, Van Loosdrecht MCM, Johns ML. Three-dimensional simulations of biofilm growth in porous media. AIChE Journal 2009; 55: 494-504.

[32] He X, Luo LS, Dembo M. Some Progress in Lattice Boltzmann Method. Part 1. Nonuniform Mesh Grids. J Comput Phys 1996; 129: 357-363.

[33] Heyrovsky J, Kuta J. Principles of Polarography. Academic Press, New York, 1966.

[34] Hundsdorfer W, Verwer J. Numerical Solution of Time-Dependent Advection-Diffusion-Reaction Equations. Springer-Verlag, Berlin, 2003.

[35] Junk M. A finite Difference Interpretation of the Lattice Boltzmann Method. Numer Meth Part Diff Eqs 2001; 17: 383-402.

[36] Koster W, van Leeuwen HP. Physicochemical Kinetics and Transport at the Biointerface: Setting the Stage, Chapter 1. John Wiley and Sons, Ltd, 2004.

[37] Koutecky J, Koryta J. The general theory of polarographic kinetic currents. Electrochim Acta 1961; 3: 318-339.

[38] Li Q, Zheng C, Wang N, Shi B. LBGK Simulations of Turing Patterns in CIMA Model. J Sci Comput 2001; 16: 121-134.

[39] Mongin S, Uribe R, Puy J, Cecilia J, Galceran J, Zhang H, Davison W. Key role of the resin layer thickness in the lability of complexes measured by dgt. Environment Sci & Techn 2011; 45: 4869-4875.

[40] Pinheiro JP, Minor M, van Leeuwen HP. Metal Speciation Dynamics in Colloidal Ligand Dispersion. Langmuir 2005; 21: 8635-8642.

[41] Puigdomenech I. Medusa: make equilibrium diagrams using sophisticated Algorithms. Windows Program, 2001.

[42] Puy J, Cecilia J, Galceran J, Town RM, van Leeuwen HP. Voltammetric lability of multiligand complexes: the case of ML_2. J Electroanalyt Chem 2004; 571: 121-132.

[43] Rittmann BE, Stilwell D, Ohashi A. The transient-state, multiple-species biofilm model for biofiltration processes. Water Research 2002; 36: 2342-2356.

[44] Salvador J, Puy J, Galceran J, Cecilia J, Town RM, van Leeuwen HP. Lability Criteria for Successive Metal Complexes in Steady-State Planar Diffusion. J Phys Chem B 2006; 110: 891-899.

[45] Salvador J, Garces JL, Galceran J, Puy J. Lability of a mixture of metal complexes under steady-state planar diffusion in a finite domain. J Phys Chem B 2006; 110: 13661-13669.

[46] Stumm W (ed.). Aquatic chemical kinetics. Wiley Interscience, New York, 1990.

[47] Toffoli T, Margolus N. Invertible cellular automata: A review. Physica D 1990; 45: 229-253.
[48] van Leeuwen HP. Revisited: the conception of lability of metal complexes. Electroanal 2001; 13: 826-830.
[49] van Leeuwen HP, Galceran J. Physicochemical Kinetics and Transport at chemical-biological surfaces. IUPAC series on Analytical and Physical Chemistry of Environmental Systems, Chapter 3. John Wiley and Sons, Chichester, U.K., Vol. 9, 2004.
[50] Wilkinson KJ, Buffle J. Critical Evaluation of Physicochemical Parameters and Processes for Modelling the Biological Uptake of Trace Metals in Environmental (Aquatic) Systems, chapter 10. John Wiley and Sons, Ltd, 2004.
[51] Wolf-Gladrow DA. Lattice-Gas Cellular Automata and Lattice Boltzmann Models. An Introduction. Springer-Verlag, Berlin Heidelberg, 2000.
[52] Zhang Z, Nadezhina E, Wilkinson KJ. Quantifying diffusion in a biofilm of streptococcus mutans. Antimicrob Agents Chemother 2011; 55: 1075-1081.
[53] Zhang Z, Alemani D, Buffle J, Town RM, Wilkinson KJ. Metal flux through consuming interfaces in ligand mixtures: boundary conditions do not influence the lability and contributions of metal species. Accepted for publication in Phys Chem Chem Phys 2011.

Progress in Computational Physics, Vol. 3, 2013, 184-198

CHAPTER 7

A Lattice Boltzmann Method for Coupled Fluid Flow, Solute Transport, and Chemical Reaction

Qinjun Kang *, Peter C. Lichtner

[1] *Earth and Environmental Sciences Division, Los Alamos National Laboratory, Los Alamos, New Mexico, United States*

Abstract: In this chapter we present a lattice Boltzmann method (LBM) for modeling coupled fluid flow, solute transport, and chemical reaction at a fundamental scale where the flow is governed by continuum fluid equations. Our numerical model accounts for multiple processes, including fluid flow, diffusion and advection of species, ion-exchange and mineral precipitation/dissolution reactions, as well as the evolution of pore geometry due to dissolution/precipitation. Homogeneous reactions are described either kinetically or through local equilibrium mass action relations. Heterogeneous reactions are incorporated into the LBM through boundary conditions imposed at the mineral surface. The LBM can provide detailed information on local fields, such as fluid velocities, solute concentrations, mineral compositions and amounts, as well as the evolution of pore geometry due to chemical reactions. Simulation examples include flow in a channel coupled with different reactions (linear kinetics for a single component, nonlinear kinetics for multi-components, and ion exchange reaction with constant Kd), crystal growth from supersaturated solution, and injection of CO_2 into a limestone rock.

Keywords: Lattice Boltzmann method, fluid flow, reactive solute transport, homogeneous reaction, heterogeneous reaction, dissolution, precipitation, sorption, physicochemical transport and interfacial processes, porous media, evolution of pore geometry, pore scale, continuum scale.

1. INTRODUCTION

Fluid flow coupled with solute transport and chemical reaction in natural and man-made porous media is ubiquitous, particularly in various energy, earth, and environment systems. Examples include development of petroleum and geothermal reservoirs, geological storage of carbon dioxide and nuclear wastes, fate and transport of underground contaminants, and electrochemical energy conversion devices (fuel cells and batteries). In these examples, the multiple physicochemical transport and interfacial processes interacting with the inherently complex morphology of the porous media across different length scales make this problem extremely difficult and consequently pose several open questions.

On one hand, most of the key processes, including fluid mobility, chemical transport, adsorption, and reaction, are ultimately governed by the pore-scale interfacial phenomena, which occur at scales of microns; and at this scale, mathematical governing equations describing different processes do exist. On the other hand, because of the wide disparity in length scales, it is virtually impossible to solve the pore-scale governing equations at the scale of interest.

As a result, a continuum formulation (macroscopic approach) of reactive transport in porous media based on spatial averages and empirical parameters is often employed. As the spatial averaging is taken over length scales much larger than typical pore and grain sizes, spatial heterogeneities at smaller scales are unresolved. These unresolved heterogeneities, together with the empirical parameters often unrelated to physical properties, lead to significant uncertainties in reactive flow modeling at the larger scale. Therefore, to reduce uncertainties in the numerical modeling of reactive transport processes at the scale of interest, it is imperative to better understand these processes at the pore scale and to incorporate pore-scale effects in the continuum scale [26, 25].

*Address correspondence to: Qinjun Kang, Mail Stop T003, Computational Earth Science Group (EES-16), Earth and Environmental Sciences Division, Los Alamos National Laboratory, Los Alamos, New Mexico 87545, United States; Tel: ++1 505 665 9663; Fax: ++1 505 665 8737; E-mail: qkang@lanl.gov

Matthias Ehrhardt (Ed.)

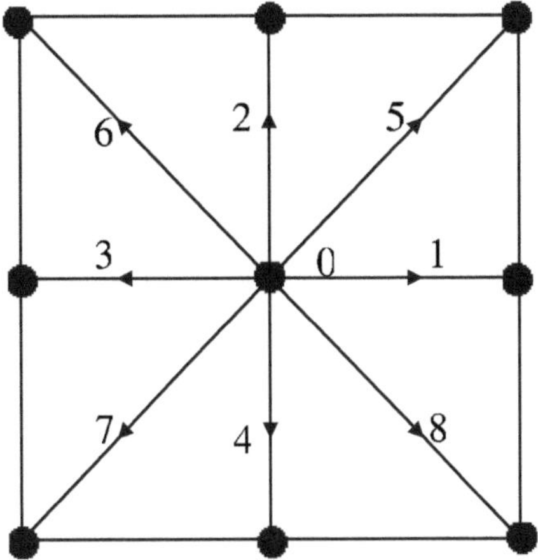

Fig. **1**: Schematic illustration of a D2Q9 lattice.

The problem of reacting flows in porous media has been studied extensively at the pore scale using various approaches under different simplifying conditions [6, 11, 35, 30, 21, 1, 4, 2, 9, 8, 17, 16, 18, 15, 13, 14, 12]. However, only recently has the full complexity of the strongly coupled flow, transport, and precipitation/dissolution reactions in realistic geochemical systems been considered [13, 14, 23, 32]. In this chapter, we present a lattice Boltzmann method (LBM) we recently developed for modeling coupled fluid flow, solute transport, and multi-component chemical reaction.

The LBM is a relatively new numerical method in computational fluid dynamics [5, 31]. However, it has undergone great advances and developed into a powerful numerical tool for simulating complex fluid flows and modeling physics of fluids since its appearance in the late 1980s. Owning to its advantage in handling nonequilibrium dynamics, especially in fluid flow applications involving interfacial dynamics, and its ease to treat complex boundaries (geometries), the LBM offers a promising approach for investigating pore-scale phenomena involving reacting flows in porous media.

In Section 2, the LBM for fluid flow is briefly reviewed. In Section 3, the LBM for multi-component reactive solute transport is presented. In Section 4, different methods to update the solid phase as a result of dissolution/precipitation is discussed. Simulation examples are given in Section 5 and Section 6 concludes this chapter.

2. LATTICE BOLTZMANN METHOD FOR FLUID FLOW

The flow of a single aqueous fluid phase in the pore space of a porous medium can be simulated by the following evolution equation (the so-called LBGK equation) [3, 29]:

$$f_\alpha(\mathbf{x}+\mathbf{e}_\alpha\delta t, t+\delta t) = f_\alpha(\mathbf{x},t) - \frac{f_\alpha(\mathbf{x},t) - f_\alpha^{\mathrm{eq}}(\mathbf{x},t)}{\tau}. \qquad (1)$$

In the above equation, δt is the time increment, f_α the distribution function along the α direction in velocity space, f_α^{eq} the corresponding equilibrium distribution function, and τ the dimensionless relaxation time. For the commonly used two-dimensional, nine-speed LB model (D2Q9) as shown in Fig. **1**, the discrete velocities $\mathbf{e}_\alpha$ have the following form:

$$\mathbf{e}_\alpha = \begin{cases} 0, & (\alpha=0), \\ (\cos\frac{(\alpha-1)\pi}{2}, \sin\frac{(\alpha-1)\pi}{2})c, & (\alpha = 1-4), \\ \sqrt{2}(\cos[\frac{(\alpha-5)\pi}{2}+\frac{\pi}{4}], \sin[\frac{(\alpha-5)\pi}{2}+\frac{\pi}{4}])c, & (\alpha=5-8), \end{cases} \qquad (2)$$

and f_α^{eq} is given by:

$$f_\alpha^{\mathrm{eq}}(\rho,\mathbf{u}) = \rho\omega_\alpha F(\mathbf{u}), \qquad (3)$$

where

$$F(\mathbf{u}) = 1 + \frac{3\mathbf{e}_\alpha \cdot \mathbf{u}}{c^2} + \frac{9(\mathbf{e}_\alpha \cdot \mathbf{u})^2}{2c^4} - \frac{3\mathbf{u} \cdot \mathbf{u}}{2c^2}. \tag{4}$$

In the above equations, the lattice speed, $c = \delta x/\delta t$, where δx is the space increment, and ω_α are weight coefficients with $\omega_0 = 4/9$, $\omega_\alpha = 1/9$ for $\alpha = 1,2,3,4$, and $\omega_\alpha = 1/36$ for $\alpha = 5,6,7,8$. It has been shown that equation (1) can be proved to recover the continuity and Navier-Stokes (NS) equations in the nearly incompressible limit [3, 29]:

$$\frac{\partial \rho}{\partial t} + \nabla \cdot (\rho \mathbf{u}) = 0, \tag{5}$$

$$\frac{\partial (\rho \mathbf{u})}{\partial t} + \nabla \cdot (\rho \mathbf{u}\mathbf{u}) = -\nabla p + \nabla \cdot \left[\rho \nu \left(\nabla \mathbf{u} + (\nabla \mathbf{u})^\top\right)\right]. \tag{6}$$

The fluid density and velocity are calculated using

$$\rho = \sum_\alpha f_\alpha, \tag{7}$$

$$\rho \mathbf{u} = \sum_\alpha \mathbf{e}_\alpha f_\alpha, \tag{8}$$

and the viscosity are pressure are determined by

$$\nu = (\tau - 0.5) {c_s}^2 \delta t, \tag{9}$$

$$p = {c_s}^2 \rho. \tag{10}$$

In the above equations, ${c_s}^2$ is speed of sound, which is related to the lattice speed as ${c_s}^2 = c^2/3$ for the D2Q9 lattice. Apparently, the LBM fluid satisfies an equation of state for ideal gases.

3. THE LATTICE BOLTZMANN METHOD FOR MULTI-COMPONENT REACTIVE SOLUTE TRANSPORT

The LB models for chemically reacting fluid flows were first introduced by [22, 7]. In their models, the LB equations for transport have a similar form as the flow equation with the addition of a source/sink term representing chemical reactions. The chemical reactions used in [7] represent the Selkov model. In a more general case, homogeneous chemical reactions taking place in an aqueous fluid can be written in the following form [13]:

$$0 \rightleftharpoons \sum_{k=1}^{N} \nu_{kr} A_k, \quad (r = 1, \ldots, N_R), \tag{11}$$

where N is the total number of solute species, N_R is the number of reactions, A_k denotes the kth species, and ν_{kr} is the stoichiometric coefficient. If the concentrations of the aqueous species are assumed to be sufficiently low so that their effect on the density and velocity of the solution is negligible, then the reactive transport of solute species can be described using another set of distribution functions, $g_{\alpha k}$, which satisfies a similar evolution equation as f_i:

$$g_{\alpha k}(\mathbf{x} + \mathbf{e}_\alpha \delta t, t + \delta t) = g_{\alpha k}(\mathbf{x}, t) - \frac{g_{\alpha k}(\mathbf{x}, t) - g_{\alpha k}^{\mathrm{eq}}(C_k, \mathbf{u})}{\tau_k} + \omega_\alpha \sum_{r=1}^{N_R} \nu_{kr} I_r. \quad (k = 1, \ldots, N), \tag{12}$$

In the above equation, I_r is the reaction rate of the rth reaction, C_k is the solute concentration of the kth species, τ_k is the relaxation time related to the diffusivity by

$$\mathscr{D}_k = (\tau_k - 0.5) {c_s}^2 \delta t, \tag{13}$$

and $g_{\alpha k}^{\mathrm{eq}}$ is the equilibrium distribution function of the kth species, which has the following form:

$$g_{\alpha k}^{\mathrm{eq}}(C_k, \mathbf{u}) = C_k \omega_\alpha F(\mathbf{u}). \tag{14}$$

The weight coefficients ω_α, the form of $F(\mathbf{u})$, as well as the relation between c_s and c all depend on the specific lattice model. For the D2Q9 lattice, they are the same as those for flow. For a D2Q4 lattice, $\omega_\alpha = 1/4$ for $\alpha = 1,2,3,4$, $c_s{}^2 = c^2/2$, and $F(\mathbf{u}) = 1 + (2\mathbf{e}_\alpha \cdot \mathbf{u})/c^2$. The discrete velocities $\mathbf{e}_\alpha$, are the same as those in (2) for $\alpha = 1,2,3,4$. The concentration C_k is defined in terms of the distribution function by the following equation:

$$C_k = \sum_\alpha g_{\alpha k}, \tag{15}$$

similar to the density in the flow equation. Using the Chapman-Enskog expansion technique, one can prove that the above LB equation (12), recovers the pore-scale advection-diffusion-reaction equation for an incompressible flow field [7]:

$$\frac{\partial C_k}{\partial t} + (\mathbf{u} \cdot \nabla) C_k = \nabla \cdot (\mathscr{D}_k \nabla C_k) + \sum_{r=1}^{N_R} \nu_{kr} I_r. \tag{16}$$

Therefore, one can solve the reactive transport problem by solving Equation (12) for each species, assuming all reaction rate constants are known. However, this may be very computationally expensive for a chemical system with many species. Moreover, the reaction rates I_r, as functions of the species concentrations through kinetic rate laws, are often unknown. For many reactions, however, their intrinsic rates are sufficiently rapid that the reactions may be assumed to be in instantaneous equilibrium. Their actual rates are then controlled by the rate of transport of species to and from the site of reaction.

For these reactions it would be desirable if the conservation equations could be formulated in such a way that rates corresponding to these fast reactions could be replaced by conditions of local equilibrium in the form of appropriate mass action equations. Such a system in which both local equilibrium and kinetic reactions take place simultaneously, is referred to as being in a state of local partial equilibrium [26, 27].

3.1. Canonical Form of Chemical Reactions

Mathematically, any linearly independent set of reactions can be rewritten in terms of a suitable set of species, referred to as primary species, with a single species, referred to as a secondary species (for aqueous species) or mineral species, appearing on the right-hand side with unit stoichiometric coefficient [27]. For homogeneous reactions in the bulk fluid,

$$\sum_{j=1}^{N_C} \nu_{ji} A_j \rightleftharpoons A_i, \quad (i = 1, \ldots, N_R), \tag{17}$$

where N_C is the number of primary species, N_R is the number of the secondary species, equal to the number of homogeneous reactions in the bulk fluid, and ν_{ji} are the stoichiometric coefficients for these reactions. For mineral reactions,

$$\sum_{j=1}^{N_C} \nu_{jm} A_j \rightleftharpoons A_m, \quad (m = 1, \ldots, N_m), \tag{18}$$

where N_m is the number of minerals, or the number of mineral reactions at the solid/fluid interface and ν_{jm} are the stoichiometric coefficients for the mineral reactions.

3.2. Homogeneous Reactions in the Bulk Fluid

For homogeneous reactions in the form of equation (17), the LB equations for the primary and secondary species are

$$g_{\alpha j}(\mathbf{x} + \mathbf{e}_\alpha \delta t, t + \delta t) = g_{\alpha j}(\mathbf{x}, t) - \frac{g_{\alpha j}(\mathbf{x}, t) - g_{\alpha j}^{\text{eq}}(C_j, \mathbf{u})}{\tau_{\text{aq}}} - \omega_\alpha \sum_{i=1}^{N_R} \nu_{ji} I_i, \quad (j = 1, \ldots, N_C), \tag{19}$$

and

$$g_{\alpha i}(\mathbf{x} + \mathbf{e}_\alpha \delta t, t + \delta t) = g_{\alpha i}(\mathbf{x}, t) - \frac{g_{\alpha i}(\mathbf{x}, t) - g_{\alpha i}^{\text{eq}}(C_i, \mathbf{u})}{\tau_{\text{aq}}} + \omega_\alpha I_i, \quad (i = 1, \ldots, N_R), \tag{20}$$

respectively, according to equation (12). In the above equations, a single relaxation time τ_{aq} is used for all the aqueous species. Different diffusion coefficients can be obtained by varying δx or δt in equation (13), though in this study only species-independent diffusion is considered, guaranteeing conservation of charge in the aqueous phase. More information on LBM simulation of electrochemical systems that includes species-dependent diffusion can be found in [10]. I_i is the reaction rate for the ith homogeneous reaction. Solving for $\omega_\alpha I_i$ from equation (20) and substituting it into (19), we have

$$g_{\alpha j}(\mathbf{x}+\mathbf{e}_\alpha\delta t, t+\delta t)+\sum_{i=1}^{N_R}\nu_{ji}g_{\alpha i}(\mathbf{x}+\mathbf{e}_\alpha\delta t, t+\delta t)=g_{\alpha j}(\mathbf{x},t)+\sum_{i=1}^{N_R}\nu_{ji}g_{\alpha i}(\mathbf{x},t)$$
$$-\frac{g_{\alpha j}(\mathbf{x},t)+\sum_{i=1}^{N_R}\nu_{ji}g_{\alpha i}(\mathbf{x},t)-\left(g_{\alpha j}^{\mathrm{eq}}(C_j,\mathbf{u})+\sum_{i=1}^{N_R}\nu_{ji}g_{\alpha i}^{\mathrm{eq}}(C_i,\mathbf{u})\right)}{\tau_{\mathrm{aq}}}. \tag{21}$$

If we introduce

$$G_{\alpha j}=g_{\alpha j}+\sum_{i=1}^{N_R}\nu_{ji}g_{\alpha i} \tag{22}$$

and use (15), then we have

$$\sum_\alpha G_{\alpha j}=C_j+\sum_{i=1}^{N_R}\nu_{ji}C_i\equiv\Psi_j, \tag{23}$$

where Ψ_j denotes the total concentration of the jth primary species. This quantity may be positive or negative depending on the selection of primary species [27]. From equations (14) and (23), we have

$$g_{\alpha j}^{\mathrm{eq}}(C_j,\mathbf{u})+\sum_{i=1}^{N_R}\nu_{ji}g_{\alpha i}^{\mathrm{eq}}(C_i,\mathbf{u})=\omega_\alpha C_jF(\mathbf{u})+\sum_{i=1}^{N_R}\nu_{ji}\omega_\alpha C_iF(\mathbf{u})$$
$$=\omega_\alpha\left(C_j+\sum_{i=1}^{N_R}\nu_{ji}C_i\right)F(\mathbf{u})=\omega_\alpha\Psi_jF(\mathbf{u})=G_{\alpha j}^{\mathrm{eq}}(\Psi_j,\mathbf{u}). \tag{24}$$

Therefore, (21) can be rewritten as

$$G_{\alpha j}(\mathbf{x}+\mathbf{e}_\alpha\delta t,t+\delta t)=G_{\alpha j}(\mathbf{x},t)-\frac{G_{\alpha j}(\mathbf{x},t)-G_{\alpha j}^{\mathrm{eq}}(\Psi_j,\mathbf{u})}{\tau_{\mathrm{aq}}},\quad (j=1,\ldots,N_C). \tag{25}$$

The above equation can be proved to recover the pore-scale advection-diffusion equation for total concentration Ψ_j, the flux of which is given by

$$\sum_\alpha G_{\alpha j}\mathbf{e}_\alpha=\Psi_j\mathbf{u}-\mathscr{D}\nabla\Psi_j. \tag{26}$$

Assuming the homogeneous reactions are in instantaneous equilibrium, we have the following mass action equations:

$$C_i=(\gamma_i)^{-1}K_i\prod_{j=1}^{N_C}(\gamma_jC_j)^{\nu_{ji}}, \tag{27}$$

where K_i is the equilibrium constant of the ith homogeneous reaction and γ_i is the activity of the ith secondary species, which is the product of this particular homogeneous reaction. Therefore, by rewriting the homogeneous reactions in the canonical form, formulating a LB equation for total concentration Ψ_j, and replacing the rates of these reactions with mass action equations, we reduce the number of unknowns and evolution equations from N_C+N_R to N_C. The reduction can be significant for a system with many aqueous species.

3.3. Heterogeneous Reactions at Fluid-Solid Interfaces

In this study, homogeneous nucleation in the bulk fluid is not considered. Heterogeneous reactions at fluid-solid interfaces are represented by boundary conditions for the primary and secondary species given by

$$\mathscr{D}\frac{\partial C_j}{\partial n} = \sum_{i=1}^{N_R} \nu_{ji} I_i^* + \sum_{m=1}^{N_m} \nu_{jm} I_m^* \tag{28}$$

and

$$\mathscr{D}\frac{\partial C_i}{\partial n} = -I_i^*, \tag{29}$$

respectively, where n is the direction normal to the interfaces pointing toward the fluid phase, and I_i^* and I_m^* are the reaction rates for the ith homogeneous and mth mineral reactions at the mineral interfaces, respectively. Following a similar procedure to the one above, we eliminate the rates I_i^* for homogeneous reactions and obtain the boundary conditions for Ψ_j:

$$\mathscr{D}\frac{\partial \Psi_j}{\partial n} = \sum_{m=1}^{N_m} \nu_{jm} I_m^*, \tag{30}$$

where I_m^* is assumed to have the form derived from transition state theory:

$$I_m^* = -k_m\left(1 - K_m Q_m\right). \tag{31}$$

In the above equation, k_m and K_m are the reaction rate and equilibrium constants, respectively, and the ion activity product Q_m is defined by

$$Q_m = \prod_{j=1}^{N_C} (\gamma_j C_j)^{\nu_{jm}}. \tag{32}$$

The treatment of heterogeneous reactions through boundary conditions in the form of equations (30)-(32) has the advantage over the continuum formulation which does not resolve the pore scale in that diffusion of solute species to and from the mineral surface is accounted for in the current method but not in the continuum formulation.

As a result, if Q_m becomes very large due to supersaturation, the reaction rate I_m^* in the continuum formulation becomes unreasonably large as there is no constraint mechanism. However, in the current formulation, the reaction rate is limited by diffusion and remains a realistic value.

The formulation of the unknown distribution functions at the fluid/solid interface from (30) depends on the orientation of the surface. For a wall node shown in Fig. **2**, the unknown distribution function G_{2j} can be calculated as the following. After each streaming process of the distribution function $(G_{\alpha j})$, G_{4j} corresponding to node Q is known.

G_{1j} and G_{3j} do not affect the fluid domain and hence there is no need to calculate their values. In contrast, G_{2j} enters the fluid domain in the next streaming step and hence must be determined using the boundary conditions. Applying (26) at node Q along the y direction gives

$$G_{2j} - G_{4j} = -\frac{\mathscr{D}}{c}\frac{\partial \Psi_j}{\partial y} = \frac{1}{c}\sum_{m=1}^{N_m} \nu_{jm} k_m\left(1 - K_m Q_m\right). \tag{33}$$

By applying the relation derived from [14] at the $\mathbf{e}_2$ and $\mathbf{e}_4$ directions, we have

$$G_{2j}^{\text{neq}} = -G_{4j}^{\text{neq}}. \tag{34}$$

From the above equation we have

$$G_{2j} + G_{4j} = G_{2j}^{\text{eq}} + G_{4j}^{\text{eq}} = \frac{\Psi_j}{2}. \tag{35}$$

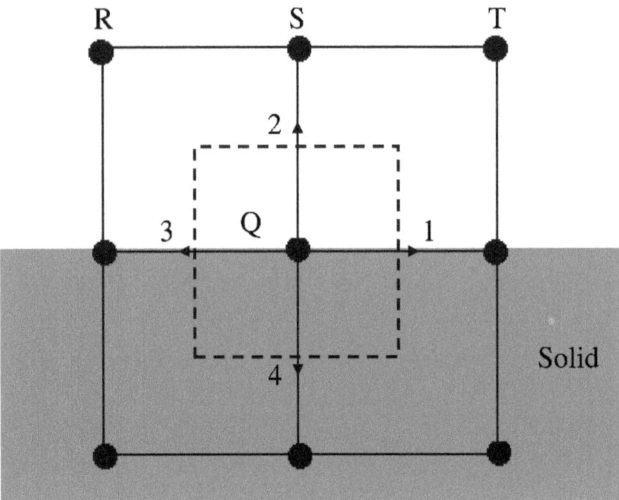

Fig. **2**: Schematic illustration of a D2Q4 lattice at a wall node.

The last equality in the above equation follows from (14), together with the no-slip velocity boundary condition. From equations (33) and (35), we can eliminate the unknown G_{2j} and obtain the following equation:

$$2G_{4j} = \frac{\Psi_j}{2} - \frac{1}{c}\sum_{m=1}^{N_m} \nu_{jm} k_m (1 - K_m Q_m). \tag{36}$$

Generally the above equation is a nonlinear algebraic equation for C_j and must be solved numerically. In the current study, Newton-Raphson iteration is used. After solving the above equation for C_j, with Ψ_j determined from equation (23), the unknown distribution functions can then be calculated by equation (35):

$$G_{2j} = \Psi_j/2 - G_{4j}. \tag{37}$$

Because $\sum_\alpha G_{\alpha j} = \Psi_j$, and G_{2j} is what enters the aqueous phase and G_{4j} is what leaves the solution, the left-hand side of the first equality in equation (33) is the net increase of Ψ_j in the aqueous solution. The right-hand side of the second equality in equation (33) is the total reaction rate for the jth primary species from heterogeneous reactions. Therefore, use of (33) guarantees solute conservation due to heterogeneous reactions.

There are some simple cases where (36) has an analytic solution, e.g., the single-aqueous-component single-mineral case with unit stoichiometric coefficient, i.e., $N_C = N_m = \nu_{11} = 1$ and $N_R = 0$. In this case, both the total concentration (Ψ) and ion activity product (Q) reduce to the concentration of the solute (C) from equations (23) and (32), respectively (Here $\gamma = 1$ has been used). Therefore, (36) reduces to

$$C = \frac{2G_4 + k_r C_s}{k_r + 0.5} \tag{38}$$

for linear reaction kinetics

$$\mathscr{D}\frac{\partial C}{\partial n} = k_r(C - C_s); \tag{39}$$

and to

$$C = \frac{2G_4 + kS}{kK_d + 0.5} \tag{40}$$

for ion exchange reaction with constant K_d formulation

$$\mathscr{D}\frac{\partial C}{\partial n} = k(K_d C - S). \tag{41}$$

4. UPDATE OF THE SOLID PHASE

To accurately model chemical reactions at fluid-solid interfaces, it is necessary to account for the time evolution of the solid phase, especially when significant mass transfer between solids and fluids is involved due to dissolution and/or precipitation.

Verberg and Ladd [33, 34] designed an algorithm for simulation of chemical erosion in rough fractures. An optimized LB scheme is used to solve the time-independent Stokes flow equations. A continuous bounce-back scheme allows for the boundary to be located anywhere between two grid nodes. The new solid structure is determined based on the local flux of tracer particles across the solid surface, where the assumption is made that the reaction kinetics are instantaneous and dissolution is therefore diffusion controlled.

In Verhaeghe's work, the amount of species injected in the system was calculated from the difference between populations leaving and entering the system. The amount of dissolved solid was then made equal to the amount of injected species divided by the difference between the solid concentration and the actual concentration in the cell. Kang et al. [12, 19] proposed two methods to update the solid phase. In both methods, the volume of the stationary solids satisfies the following equation:

$$\frac{\partial V_m}{\partial t} = \overline{V}_m a_m I_m^*, \tag{42}$$

where V_m, $\overline{V}_m$, and a_m are dimensionless volume, molar volume, and specific surface area of the mth mineral, respectively, and I_m^* is the reaction rate for the mth mineral reaction at the mineral interface. Solute diffusion in the solid phase is neglected and mineral reactions are assumed to only occur at the fluid-solid interface. Each interface node represents a control volume (a control area in the 2D case) with a size of 1×1 (in lattice units) and is located at the center of this volume. As can be seen from Fig. **2**, node Q is the center of the control volume surrounded by dashed lines. The initial control volume is given a dimensionless volume V_m^0. The volume is updated at each time step explicitly according to the equation

$$V_m(t+\delta t) = V_m(t) + \overline{V}_m a_m I_m^* \delta t, \tag{43}$$

where δt is the time increment. In this study both δt and a_m equal unity in lattice units.

When V_m reaches certain threshold values, the pore geometry needs to be updated. For dissolution, the solid node associated with V_m can be simply removed (i.e., changed to a pore node), when V_m reaches zero. For precipitation, however, there are multiple ways to add a solid node when V_m reaches a certain threshold value. A random-growth method was proposed by Kang et al. for both single-species [16, 18] and multi-component systems [13]. In that method, the growth has no preference in any particular direction and the method has been shown to be free of lattice effects.

In the above methods, the grid size is assumed to be small enough that each node is only represented by one mineral at one time, and the effect of both dissolution and precipitation is recorded at that node through equation (43). In reality, changes of solid morphology can involve scales much smaller than the lattice or pore size used in the simulations.

In the other approach, we assume that each node can be represented by multiple minerals whose initial total volume fraction sums to unity. The volume fraction of each mineral is still updated by equation (43) and the amount dissolved recorded in the solid node. Mass accumulation due to precipitation, however, is recorded at the neighboring pore nodes. As seen in Fig. **2**, when

$$\sum_m V_m(Q) = 0, \tag{44}$$

node Q is changed to a fluid node. When

$$\sum_m V_m(S) = 1, \tag{45}$$

node S becomes a solid node composed of multiple minerals. In contrast to previous methods [18, 13] for updating pore geometry, this method can account for coexistence of multiple minerals at the same node and the growth is deterministic rather than random. Clearly, the LBM can be used to simulate a variety of reacting flow problems when combined with appropriate methods to update the solid phase that account for the underlying physical properties/processes [28].

5. SIMULATION EXAMPLES AND DISCUSSION

5.1. Reactive Transport in a Channel

5.1.1. *Linear Kinetics*

To validate the LB model, we first simulate the transport of a reactive solute using first order linear kinetics described by (39) with the reaction:

$$A \rightleftharpoons A_{(s)}, \tag{46}$$

in a two dimensional channel. The simulation domain is 400×64 in lattice units, corresponding to a physical domain of 5 centimeters in length and 0.75 centimeter in width, with two layers of solids placed at the top and bottom. The steady-state velocity field has a parabolic distribution across the channel. Initially, the pore space is filled with a solution of species A in equilibrium with the channel wall consisting of solid $A_{(s)}$. A pure solvent is introduced at the entrance, which triggers the reaction. In the numerical simulations, a very low concentration (1.0×10^{-8}mol/L) is used at the entrance. A zero concentration gradient is enforced at the right boundary.

For the same reaction kinetics, Kang et al. [13] did a simulation of diffusion and reaction at a boundary in an open rectangular domain and compared the results with the analytic solution. In this study, due to the introduction of the advection term, there is no analytic solution for the finite domain under consideration. To validate our model, we simulate the same 2D problem at the pore scale using the computer code FLOTRAN developed by Lichtner [24] with a prescribed parabolic velocity profile.

Fig. **3** shows concentration contours of the 2D FLOTRAN and LB simulations for equilibrium constant $\log(K) = 1$ and for three different reaction rate constants at $t = 1.95 \times 10^4$s. Red lines denote the 2D FLOTRAN simulations and green lines the LB simulations. Fig. **4** shows relative error in concentrations between these two methods. As shown in Fig. **4** the agreement between these two methods are excellent. The larger values of relative error near the entrance are caused by the low concentration there. Another source of error is due to the different ways these two methods use to treat the heterogeneous reactions. The LB model treats the heterogeneous reactions as boundary conditions at the mineral interface, but in the FLOTRAN simulations, these reactions are volume averaged over the two nodes (one pore and one solid node) adjacent to the interface.

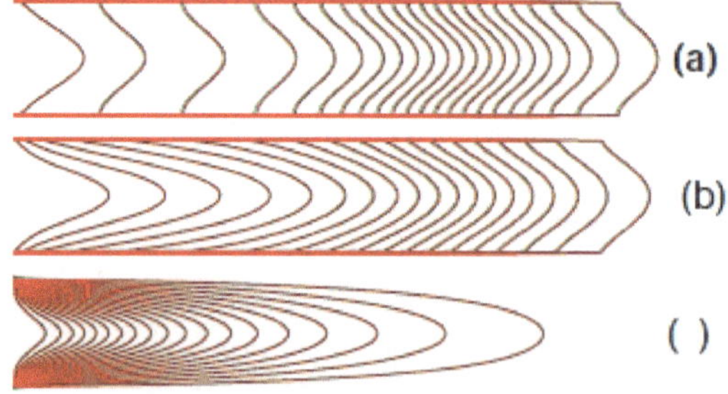

Fig. **3**: Concentration contours of 2D FLOTRAN and LB simulations for linear reaction kinetics at $t = 1.95 \times 10^4$s. Red lines denote 2D FLOTRAN simulations and green lines the LB simulations, with a linear first order reaction kinetics. Equilibrium constant $\log(K) = 1$ and reaction rate constants are (a) 1.0×10^{-12} mol/cm^2/s; (b) 1.0×10^{-9} mol/cm^2/s; (c) 1.0×10^{-6} mol/cm^2/s.

5.1.2. *Three-Component System*

In this section, a multi-component system with solute species A, B, C, and solids $AB_{(s)}$ and $AC_{(s)}$ undergoing the following reactions

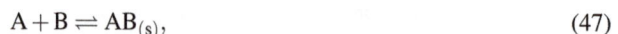

$$A + B \rightleftharpoons AB_{(s)}, \tag{47}$$

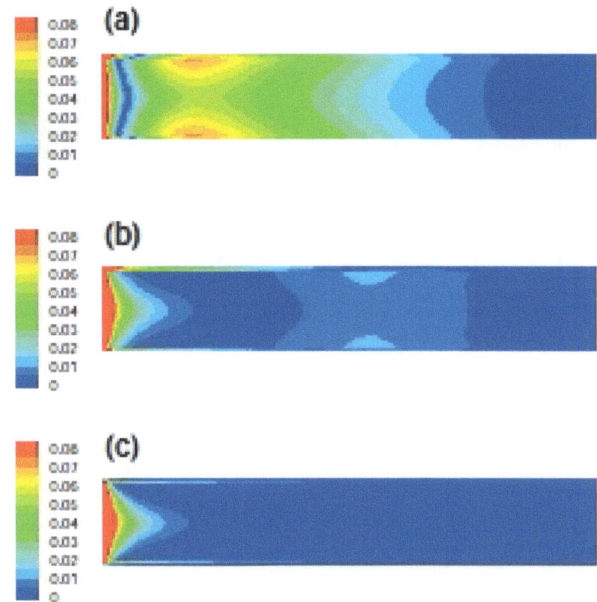

Fig. **4**: Distribution of relative error in concentrations between the 2D LB and FLOTRAN simulations for the same conditions in Fig. **3**. The error is the largest at the inlet.

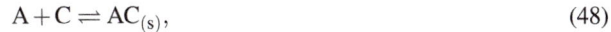

$$A + C \rightleftharpoons AC_{(s)}, \tag{48}$$

with rate law given by (31) is considered. The channel wall consists of solid $AB_{(s)}$ in equilibrium with A and B initially present in the pore space. A solution including only A and C (in equilibrium with $AC_{(s)}$) is introduced at the entrance. Solid $AB_{(s)}$ dissolves, causing precipitation of $AC_{(s)}$. In this study, we focus on the effect of the heterogeneous reactions taking place at the channel walls on solute transport by turning off the homogeneous reactions occurring in the bulk of the fluid and by neglecting changes in geometry of the walls. The LBM results are vertically averaged and compared with single continuum calculations using FLOTRAN [24], in which both velocity, specific surface area, and volume fraction are averaged across the channel. By fitting the pore-scale LB with the continuum model results, the condition at which the continuum scale formulation is valid can be determined, and macroscopic parameters needed in the continuum scale simulations can be derived [25].

Fig. **5** shows spatial distribution of the concentrations along the channel of the 1D FLOTRAN simulations (solid lines) and averaged LB results (dashed lines) for equilibrium constant $\log(K) = 1$ and for different reaction rate constants at $t = 1.95 \times 10^4$s, for the multi-component system. Dispersion corresponding to a non-reactive tracer (0.43mm) is used in the FLOTRAN simulations. Here we use the same equilibrium constant and reaction rate constant for both reactions. Red, blue, and black lines denote the concentration of solute A, B, and C, respectively. The simulation results from FLOTRAN and the LBM are in good agreement for slow reactions, as shown in Fig. **5**(a) and Fig. **5**(b), but differ significantly for very fast reactions. As shown in Fig. **5**(c), the concentrations of B and C of the continuum-scale simulations become equal after a short distance from the entrance and are in local equilibrium with solid $AB_{(s)}$ and $AC_{(s)}$, respectively. This means that the heterogeneous reactions are so fast that they can be treated in instantaneous equilibrium and the advection-diffusion-reaction process is controlled by the transport along the channel. In the LB simulation, however, because of the lateral diffusion, the averaged concentrations are not in equilibrium with the solid.

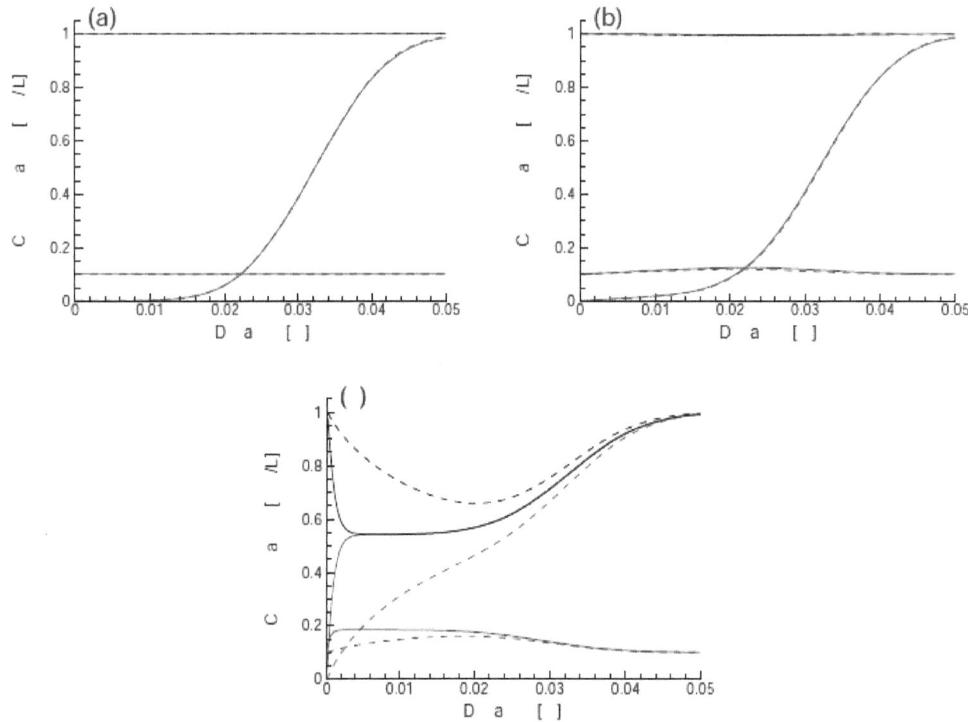

Fig. **5**: Spatial distribution of the concentrations along the channel for the 1D FLOTRAN simulations (solid lines) and averaged LB results (dashed lines) for a multi-component system at $t = 1.95 \times 10^4$s. Red, blue, and black lines denote the concentration of solute A, B, and C, respectively.

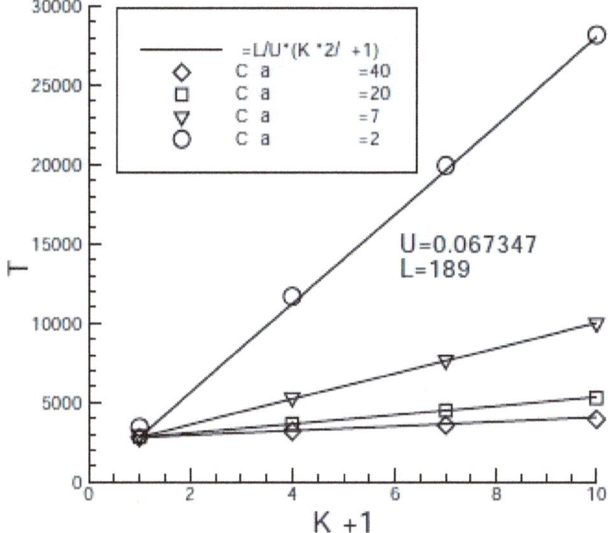

Fig. **6**: Dependence of half-concentration time on retardation factor K_d+1 for different channel widths using the kinetic rate constant $k = 6.25 \times 10^{-2}$.

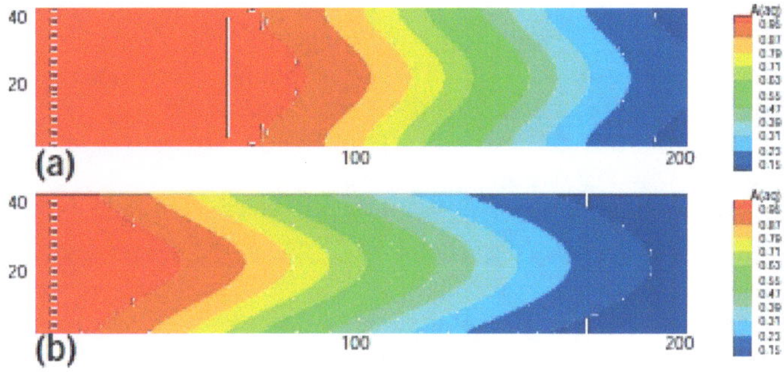

Fig. **7**: Concentration fields at t=2000 (LB units) for the channel of 200×40 at $K_d = 0$ (a) and $K_d = 9$ (b).

5.1.3. *Sorption Reaction with Constant* K_d

We then consider the sorption reaction with constant K_d described by (41) at the channel walls. As mentioned previously, in this case, the concentration has an analytical solution described by (40). Note that when the reaction rate constant (k) equals zero, we have $C = 4G_4$ from (40). Then from equation (35), we have $G_2 = G_4$, corresponding to the zero flux boundary condition. On the other hand, when $k \to \infty$, we have $S = K_d C$. According to the continuum formulation of reactive transport [26], the mean velocity with which the reactive solute moves (U^*) is retarded by a factor of $K_d + 1$ compared to the mean solvent velocity (U), i.e., $U^*/U = 1/(K_d + 1)$. Fig. **6** shows dependence of half-concentration time on retardation factor K_d+1 for different channel widths. Clearly, the time is linearly proportional to the retardation factor $K_d + 1$. This means that even for fast reaction rates the LBM agrees with generalized Taylor dispersion theory and the continuum formulation for retardation. Fig. **7** shows concentration field for $K_d = 0$ and $K_d = 9$. Clearly for the case of $K_d = 9$, the transport of solute is retarded and slower than that in the case of $K_d = 0$.

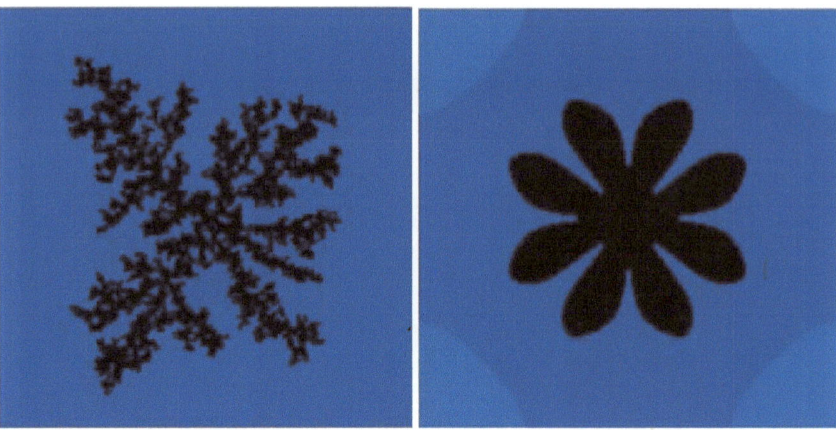

Fig. **8**: Crystal structures and solute concentration for Da=600 and solute saturation 1.2 obtained using random growth (left) and deterministic growth (right) methods.

5.2. Crystal Growth from Supersaturated Solution

We then simulate crystal growth from a supersaturated solution in a closed system. The simulation geometry is a two-dimensional cell of size $h \times h$, where h equals 200 (in lattice units) in this study. Initially, the domain is filled with a supersaturated solution with a concentration C_0. At time zero, a stable nucleus is introduced at the center of the domain. Crystal begins to grow subsequently.

The same linear reaction kinetics described by (39) is used. Simple dimensional analysis suggests that there are two important dimensionless parameters, which control the processes. They are the relative concentration or saturation ($\psi = C_0/C_s$) and Damkohler number (Da). The Da number is defined as $k_r h/D$ and describes the effect of reaction relative to that of diffusion.

The effect of saturation and Damkohler were investigated in a previous study [20]. In this study, we focus on the effect of different rules to update the solid phase on the crystal morphology.

Fig. **8** shows crystal structures at $Da = 600$ and $C_0/C_s = 1.2$ with the random-growth method (left) and deterministic method (right) to update the solid phase. At this high Da number, the process is diffusion-controlled. As a result, the crystal structure is not compact. However, while the crystal on the left is an open cluster-type structure, consistent with that of the multiparticle diffusion-limited aggregation (MPDLA) simulation of solidification structures of alloy melt, the crystal structure on the right is highly symmetric and regular. Interestingly, although the two structures are quite different, they both have a fractal dimension around 1.75.

5.3. Injection of CO_2 into a Limestone Rock

In this section, simulation results for the injection of a CO_2(g) saturated brine into a limestone rock at the pore scale are presented. The chemical system of Na^+, Ca^{2+}, Mg^{2+}, H^+, SO_4^{2-}, CO_2(aq), and Cl^- with the reaction of calcite to form dolomite and gypsum is considered. Secondary species included in the simulation are: OH^-, HSO_4^-, H_2SO_4(aq), CO_3^{2-}, HCO_3^-, $CaCO_3$(aq), $CaHCO_3^+$, $CaOH^+$, $CaSO_4$(aq), $MgCO_3$(aq), $MgHCO_3^+$, $MgSO_4$(aq), $MgOH^+$, NaCl(aq), $NaHCO_3^-$(aq), and NaOH(aq). For this system, $N_C = 7$, but $N_C + N_R = 23$. Initial fluid composition is pH 7.75 and 2.69 m NaCl brine, in equilibrium with minerals calcite, dolomite and gypsum at 25°C.

Initial rock composition is calcite. Secondary minerals include dolomite and gypsum. For boundary conditions, a constant pressure gradient is imposed across the domain for flow. When flow reaches steady state, a fluid with a pH of 3.87 and in equilibrium with 179 bars CO_2(g) and minerals dolomite and gypsum is introduced at the inlet. Zero gradient boundary conditions are imposed at the outlet. Two different cases are considered with different mineral reaction rates to show their effects on solution concentration, mineral

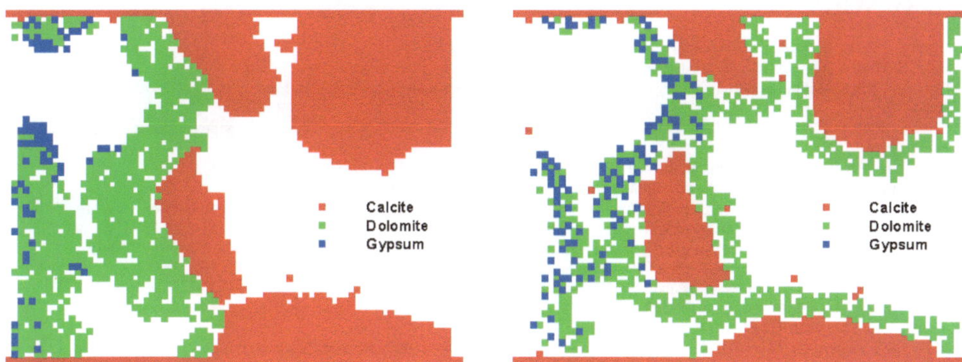

Fig. **9**: Resulting geometries at time=15625 seconds for two different mineral reaction rate constants: a) large reaction rate constants; b) small reaction rate constants.

deposition and change in geometry. Resulting geometries at time=15625 seconds for two different mineral reaction rate constants are plotted in Fig. **9**. Damkohler is 7.375 for calcite and gypsum and 0.7375 for dolomite for the faster mineral reactions and 7.375×10^{-2} for calcite and gypsum and 7.375×10^{-3} for dolomite for slower reactions. Here the random-growth method is used to update the solid phase.

As can been seen from the figures, for the case of smaller reaction rate constants, the precipitation of dolomite is more uniform surrounding the dissolving calcite grains. Only a small amount of gypsum forms on top of dolomite. At some point in the simulation, the major pores for flow become blocked halting further fluid flow through the medium.

CONCLUSIONS

In this chapter we have presented a LBM for modeling coupled fluid flow, solute transport, and chemical reaction in complex porous media. The model has been applied to the simulation of flow in a channel coupled with reactions between the solutes and channel walls, crystal growth from supersaturated solution, and injection of CO_2 into a limestone rock. It is shown that the LBM is able to provide detailed information on fluid velocity, solute concentration, mineral composition, and reaction rates, as well as the evolution of the porous media geometry. For some cases, the upscaled LBM results are compared with single continuum calculations. It is found that the strong pore-scale concentration gradients, caused by very fast reactions, lead to significant discrepancies between the upscaled LBM results and the continuum scale simulations, indicating the breakdown of the continuum models in these cases.

Therefore the LBM is a useful tool to understand the fundamental physics occurring at the pore scale for reactive transport in porous media, and to validate the continuum models. We would like to emphasize the flexibility and ease of this method in handling multiple coupled transport and interfacial processes in complex media. For example, it can be easily combined with various methods to update the solid phase that account for different underlying physical properties/processes.

Applications of this method to problems in the areas of energy and environment, such as geologic storage of carbon dioxide and nuclear wastes, subsurface contaminant migration, bioremediation etc. are ongoing and will be presented in future publications.

ACKNOWLEDGEMENT

This work was supported by LDRD projects 20100025DR and 20110009DR sponsored by Los Alamos National Laboratory, and by UC Lab Fees Research Project UCD-09-15.

CONFLICT OF INTEREST

The authors confirm that this chapter content has no conflict of interest.

REFERENCES

[1] Bekri S, Thovert JF, Adler PM. Dissolution of porous media. Chem Engrg Sci 1995; 50: 2765-2791.
[2] Bekri S, Thovert JF, Adler PM. Dissolution and deposition in fractures. Engrg Geology 1997; 48: 283-308.
[3] Chen H, Chen S, Matthaeus WH. Recovery of the Navier-Stokes equations using a lattice-gas Boltzmann method. Phys Rev A 1992; 45: R5339-R5342.
[4] Chen S, Dawson SP, Doolen GD, Janecky DR, Lawniczak A. Lattice methods and their applications to reacting systems. Comput & Chem Engrg 1995; 19: 617-646.
[5] Chen S, Doolen GD. Lattice Boltzmann method for fluid flows. Ann Rev Fluid Mech 1998 30: 329-364.
[6] Daccord G. Chemical dissolution of a porous medium by a reactive fluid. Phys Rev Lett 1987; 58: 479-482.
[7] Dawson SP, Chen S, Doolen GD. Lattice Boltzmann computations for reaction-diffusion equations. J Chem Phys 1993; 98: 1514-1523.
[8] Dijk P, Berkowitz B. Precipitation and dissolution of reactive solutes in fractures. Water Resour Res 1998; 34: 457-470.
[9] Fredd CN, Fogler HS. Influence of transport and reaction on wormhole formation in porous media. AIChE J 1998; 44: 1933-1949.
[10] He X, Li N. Lattice Boltzmann simulation of electrochemical systems. Comput Phys Commun 2000; 129: 158-166.
[11] Hoefner ML, Fogler HS. Pore evolution and channel formation during flow and reaction in porous media. AIChE J 1988; 34: 45-54.
[12] Kang QJ, Lichtner PC, Viswanathan HS, Abdel-Fattah AI. Pore scale modeling of reactive transport involved in geologic CO_2 sequestration. Transp Porous Media 2010; 82: 197-213.
[13] Kang QJ, Lichtner PC, Zhang DX. Lattice Boltzmann pore-scale model for multicomponent reactive transport in porous media. J Geophys Res Solid Earth 2006; 111: B05203.
[14] Kang QJ, Lichtner PC, Zhang DX. An improved lattice Boltzmann model for multicomponent reactivetransport in porous media at the pore scale. Water Res Res 2007; 43: W12S14.
[15] Kang QJ, Tsimpanogiannis IN, Zhang DX, Lichtner PC. Numerical modeling of pore-scale phenomena during CO_2 sequestration in oceanic sediments. Fuel Process Techn 2005; 86: 1647-1665.
[16] Kang QJ, Zhang DX, Chen SY. Simulation of dissolution and precipitation in porous media. J Geophys Res Solid Earth 2003; 108: 2505.
[17] Kang QJ, Zhang DX, Chen SY, He XY. Lattice Boltzmann simulation of chemical dissolution in porous media. Phys Rev E 2002; 65: 036318.
[18] Kang QJ, Zhang DX, Lichtner PC, Tsimpanogiannis IN. Lattice Boltzmann model for crystal growth from supersaturated solution. Geophys Res Lett 2004; 31: L21604.
[19] Kang QJ, Lichtner PC, Janecky DR. Lattice Boltzmann method for reacting flows in porous media. Adv Appl Math Mech 2010; 2: 545-563.
[20] Kang YJ, Yang C, Huang XY. Analysis of the electroosmotic flow in a microchannel packed with homogeneous microspheres under electrokinetic wall effect. Int J Engrg Sci 2004; 42: 2011-2027.
[21] Kelemen PB, Whitehead JA, Aharonov E, Jordahl KA. Experiments on flow focusing in soluble porous media, with applications to melt extraction from the mantle. J Geophys Res 1995; 100: 475-496.
[22] Kingdon RD, Schofield P. A reaction-flow lattice Boltzmann model. J Phys A: Math Gen 1992; 25: L907-L910.
[23] Li L, Peters CA, Celia MA. Upscaling geochemical reaction rates using pore-scale network modeling. Adv Water Res 2006; 29: 1351-1370.
[24] Lichtner PC. Flotran users manual. Technical report, Los Alamos National Laboratory, 2001.
[25] Lichtner PC, Kang Q. Upscaling pore-scale reactive transport equations using a multiscale continuum formulation. Water Res Res 2007; 43: W12S15 (19 pages).
[26] Lichtner PC, Steefel CI, Oelkers EH. Reactive transport in porous media. Rev Mineralogy 34. Mineralogical Society of America, Washington, D.C., 1996.
[27] Lichtner PC. Continuum model for simultaneous chemical reactions and mass transport in hydrothermal systems. Geochi Cosmochi Acta 1985; 49: 779-800.
[28] Lu GP, DePaolo DJ, Kang QJ, Zhang DX. Lattice Boltzmann simulation of snow crystal growth in clouds. J Geophys Res Atmos 2009; 114: D07305.
[29] Qian YH, Dhumieres D, Lallemand P. Lattice BGKs models for Navier-Stokes equation. Europhys Lett 1992; 17: 479-484.
[30] Salles J, Thovert JF, Adler PM. Deposition in porous media and clogging. Chem Engrg Sci 1993; 48: 2839-2858.
[31] Succi S. The Lattice Boltzmann Equation for Fluid Dynamics and Beyond. Numerical Mathematics and Scientific Computation. Oxford University Press, 2001.
[32] Tartakovsky AM, Meakin P, Scheibe TD, West RME. Simulations of reactive transport and precipitation with smoothed particle hydrodynamics. J Comput Phys 2007; 222: 654-672.
[33] Verberg R, Ladd AJC. Lattice-Boltzmann model with sub-grid-scale boundary conditions. Phys Rev Lett 2000; 84: 2148-2151.
[34] Verberg R, Ladd AJC. Simulation of chemical erosion in rough fractures. Phys Rev E 2002; 65: 056311.
[35] Wells JT, Janecky DR, Travis BJ. A lattice gas automata model for heterogeneous chemical reactions at mineral surfaces and in pore networks. Physica D: Nonlin Phen 1991; 47: 115-123.

CHAPTER 8

A Lattice Boltzmann Approach for Distributed Three-dimensional Fluid-Structure Interaction

Sebastian Geller*, **Christian Janssen, Manfred Krafczyk**

Institute for Computational Modeling in Civil Engineering, TU Braunschweig, Germany

Abstract: This work investigated the validity and efficiency of the coupling of the Lattice Boltzmann method with finite element schemes as well as rigid body approaches to model fluid-structure interaction (FSI). The results on two- and three-dimensional benchmark configurations are very promising and show that an explicit coupling scheme is able to produce accurate results which agree with reference solutions very well. The underlying fluid solver VIRTUALFLUIDS is based on adaptive hierarchical grids and component technology for parallelization. One target application is the simulation of a jet induced by a ship propeller which poses a substantial challenge to both the numerical scheme as well as to the software parallelization concept. Furthermore, the coupling to a rigid body dynamics engine (PhysicsEngine-pe) leads to the possibility to compute FSI problems with a huge number of particles which is the basis for numerical simulations in geothermal drilling, where the particles eroded by the drill head influence the fluid dynamics. A free surface Lattice Boltzmann approach coupled with rigid body motions shows further potential and demonstrates the broad applicability of the developed algorithms.

Keywords: Adaptive hierarchical grids, explicit coupling scheme, finite element scheme (FEM), fluid-structure interaction (FSI), geothermal drilling, Lattice Boltzmann Method (LBM), PhysicsEngine-pe, rigid body approaches, software parallelization concepts, VirtualFluids.

1. INTRODUCTION

The realistic simulation of fluid structure interaction processes is a challenging task. Engineers need to determine fluid loads on structures to study the transient behaviour of bridges and buildings in turbulent wind flows, the influence of bridge piers and immersed structures such as ships on water flow patterns to mention a few examples. To compute the turbulent characteristic of such interaction problems, a parallel solver is necessary on today's computer architectures to handle more than a billion degrees of freedom, which are necessary for such a problem.

In general, two approaches to solve fluid-structure interaction (FSI) problems exist. The monolithic approach [17, 18] discretizes the two separate domains with a similar discretization scheme and solves the resulting, coupled system of equations within one solver. The compatibility conditions at the interface are treated inherently within this system of equations. In contrast, the partitioned approach [33] uses separate solvers for the fluid and the structural system. Strong coupling methods [26, 47] as well as loose coupling methods [10, 11, 30, 38, 39, 40] exist. In the partitioned solution, the solvers need to communicate physical properties of their mutual boundaries to fulfill the interface conditions. Each domain may utilize any type of discretization considered efficient for its field. For the fluid, mainly two approaches are used in the context of partitioned FSI for large deformations. The fluid is either described on an arbitrarily moving grid (arbitrary Lagrangian Eulerian (ALE) formulation), or on a fixed Cartesian grid (Euler approach).

For the flow, the Lattice Boltzmann approach has matured as an alternative model to the direct solution of the Navier-Stokes equation. The discretization is based on fixed Eulerian grids, which allows to handle large deformations without extensive remeshing during the coupling process. The first LB algorithm for interaction problems between a fluid and rigid obstacles has been developed by Ladd [27, 28] for the simulation of

Address correspondence to: Sebastian Geller, iRMB – Institute for Computational Modeling in Civil Engineering, TU Braunschweig, Mühlenpfordtstrasse 4-5, 38106 Braunschweig, Germany; Tel: ++49 531 391 7582; Fax: ++49 531 391 7518; E-mail: geller@irmb.tu-bs.de

Matthias Ehrhardt (Ed.)

particulate suspensions. A special treatment of boundary conditions and the activation/deactivation of fluid nodes have been developed in this work.
Using the LB method, the fluid-structure boundary is represented as a sharp interface and the boundary conditions are directly imposed on the structure and on the fluid at this discrete interface. As such, this contribution is a continuation of the work presented in [15, 23, 43]. In [23], this methodology was layed out in detail, benchmarked and validated against the two-dimensional, numerical examples proposed in [44, 45]. Algorithms to handle interface meshes created on the basis of the computational meshes of the structural solver were implemented and validated in the scope of this work. By contrast, the present paper shows the extension of the method to simulate flow problems with rigid body motion, such as the jet created by a rotating ship propeller, drill cutting transport as well as free surface rigid body interactions.

2. THE LATTICE BOLTZMANN METHOD FOR SINGLE-PHASE FLOWS ON NON-UNIFORM GRIDS

In essence, the LBM is a finite difference discretization of the discrete Boltzmann equation in space and time on equidistant, Cartesian grids. The basic, physical notion of the method is to let fluid-mass fractions (i.e. particles) collide on the nodes formed by this lattice and propagate them among the links formed between the nodes of this lattice. To introduce the method formally, the following abbreviations are used: $\mathbf{x}$ represents a 3D vector in space and $\mathbf{f}$ is a b-dimensional vector, where b is the number of microscopic velocities along a certain number of links of the grid. The $d3q19$ model [41] with the following microscopic velocities is considered:

$$\{\mathbf{e}_i, i=0,\ldots,6\} = \left\{ \begin{array}{ccccccc} 0 & c & -c & 0 & 0 & 0 & 0 \\ 0 & 0 & 0 & c & -c & 0 & 0 \\ 0 & 0 & 0 & 0 & 0 & c & -c \end{array} \right\}$$

$$\{\mathbf{e}_i, i=7,\ldots,18\} = \left\{ \begin{array}{cccccccccccc} c & -c & c & -c & c & -c & c & -c & 0 & 0 & 0 & 0 \\ c & -c & -c & c & 0 & 0 & 0 & 0 & c & -c & c & -c \\ 0 & 0 & 0 & 0 & c & -c & -c & c & c & -c & -c & c \end{array} \right\}$$

where $c = \Delta x_l / \Delta t_l$ is a constant velocity determining the speed of sound $c_s^2 = c^2/3$. Other possible choices for the microscopic velocity space are the $d3q13$ model, introducing a reduced set of 13 velocities, or the $d3q27$ model, considering all possible links between next-neighbor nodes. The lattice Boltzmann equation on the basis of a multi-relaxation time model [8, 31] for the grid level l is then given by

$$f_{i,l}(t+\Delta t_l, \mathbf{x}+\mathbf{e}_i\Delta t_l) = f_{i,l}(t,\mathbf{x}) + \Omega_{i,l}, \quad i=0,\ldots,b-1, \tag{1}$$

where Δt_l is the time step and $\Omega_{i,l}$ the collision operator on grid level l, which is given by:

$$\Omega_l = \mathsf{M}^{-1}\mathsf{S}_l \left[(\mathsf{M}\mathbf{f}) - m^{\mathbf{eq}}\right]. \tag{2}$$

M represents the transformation matrix, which is composed of 19 basis vectors $\{\Phi_i, i=0,\ldots,b-1\}$ and which is given in the appendix. The basis vectors are orthogonal with respect to the inner product $\langle \Phi_i, \Phi_j w \rangle$ (in contrast to [7], where $\langle \Phi_i, \Phi_j \rangle = 0$, if $i \neq j$). The vector w constitutes the weights $\{w_i, i=0\ldots,b-1\}$:

$$w = (\frac{1}{3}, \frac{1}{18}, \frac{1}{18}, \frac{1}{18}, \frac{1}{18}, \frac{1}{18}, \frac{1}{18}, \frac{1}{36}, \frac{1}{36}, \frac{1}{36}, \frac{1}{36}, \frac{1}{36}, \frac{1}{36}, \frac{1}{36}, \frac{1}{36}, \frac{1}{36}, \frac{1}{36}, \frac{1}{36}, \frac{1}{36})$$

The moments $m = \mathsf{M}\mathbf{f}$ are labeled as

$$m = (\rho, e, \varepsilon, j_x, q_x, j_y, q_y, j_z, q_z, 3p_{xx}, 3\pi_{xx}, p_{ww}, \pi_{ww}, p_{xy}, p_{yz}, p_{xz}, m_x, m_y, m_z).$$

m^{eq} is the vector representing the equilibrium moments given in Eqs. (4) and $\mathsf{S}_l = \{s_{l,i,i}, i=0,\ldots,b-1\}$ is the diagonal collision matrix. The nonzero collision parameters $s_{l,i,i}$ (the eigenvalues of the collision matrix

$\mathsf{M}^{-1}\mathsf{S}_l\mathsf{M}$) are:

$$\begin{aligned} s_{l,1,1} &= s_{l,a} \\ s_{l,2,2} &= s_{l,b} \\ s_{l,4,4} &= s_{l,6,6} = s_{l,8,8} = s_{l,c} \\ s_{l,10,10} &= s_{l,12,12} = s_{l,d} \\ s_{l,9,9} &= s_{l,11,11} = s_{l,13,13} = s_{l,14,14} = s_{l,15,15} = -\frac{\Delta t_l}{\tau_l} = s_{l,\omega} \\ s_{l,16,16} &= s_{l,17,17} = s_{l,18,18} = s_{l,e}, \end{aligned}$$

The relaxation time τ_l is defined as:

$$\tau_l = 3\frac{\nu}{c^2} + \frac{1}{2}\Delta t_l, \tag{3}$$

where ν is the kinematic viscosity. The parameters s_a, s_b, s_c, s_d and s_e can be chosen in the range $[-2,0]$ and tuned to improve stability [31]. The optimal values depend on the specific characteristics of the system (considering geometry, initial and boundary conditions) and cannot be computed in advance. Some reasonable values for these parameters are given in [7]. We choose $s_a = s_b = s_c = s_d = s_e = \max\{s_{l,\omega}, -1.0\}$. The nonzero equilibrium distribution functions $\{m_i^{eq}, i = 0, \ldots, 18\}$ are given by:

$$m_0^{eq} = \rho, \tag{4a}$$
$$m_3^{eq} = \rho_0\, u_x, \tag{4b}$$
$$m_5^{eq} = \rho_0\, u_y, \tag{4c}$$
$$m_7^{eq} = \rho_0\, u_z, \tag{4d}$$
$$\tag{4e}$$
$$m_1^{eq} = e^{eq} = \rho_0\,(u_x^2 + u_y^2 + u_z^2), \tag{4f}$$
$$m_9^{eq} = 3p_{xx}^{eq} = \rho_0\,(2u_x^2 - u_y^2 - u_z^2), \tag{4g}$$
$$m_{11}^{eq} = p_{zz}^{eq} = \rho_0\,(u_y^2 - u_z^2), \tag{4h}$$
$$m_{13}^{eq} = p_{xy}^{eq} = \rho_0\, u_x u_y, \tag{4i}$$
$$m_{14}^{eq} = p_{yz}^{eq} = \rho_0\, u_y u_z, \tag{4j}$$
$$m_{15}^{eq} = p_{xz}^{eq} = \rho_0\, u_x u_z, \tag{4k}$$

where ρ_0 is a constant density and ρ a density variation. The macroscopic quantities (density and momentum) and the stress tensor are given by (omitting the index l):

$$\rho = \sum_i f_i, \tag{5a}$$
$$\rho_0 \mathbf{u} = \sum_i \mathbf{e}_i f_i, \tag{5b}$$
$$S_{xx} = -(1 - \frac{\Delta t}{2\tau})(\frac{1}{3}e + p_{xx} - \rho_0 u_x^2) \tag{5c}$$
$$S_{yy} = -(1 - \frac{\Delta t}{2\tau})(\frac{1}{3}e - \frac{1}{2}p_{xx} + \frac{1}{2}p_{ww} - \rho_0 u_y^2) \tag{5d}$$
$$S_{zz} = -(1 - \frac{\Delta t}{2\tau})(\frac{1}{3}e - \frac{1}{2}p_{xx} - \frac{1}{2}p_{ww} - \rho_0 u_z^2) \tag{5e}$$
$$S_{xy} = -(1 - \frac{\Delta t}{2\tau})(p_{xy} - \rho_0 u_x u_y) \tag{5f}$$
$$S_{yz} = -(1 - \frac{\Delta t}{2\tau})(p_{yz} - \rho_0 u_y u_z) \tag{5g}$$
$$S_{xz} = -(1 - \frac{\Delta t}{2\tau})(p_{xz} - \rho_0 u_x u_z) \tag{5h}$$

Using the Chapman-Enskog expansion [4, 21], which in essence is a gradient expansion around the local equilibrium state, it can be shown that the LB Method is a second order scheme both in space and time for the compressible Navier-Stokes equations in the low Mach number limit using the advective scaling. Using the diffusive scaling, it is a scheme of first order in time and second order in space for the *in*compressible Navier-Stokes equations.

In the grid generation process a smoothed tree-type grid (quadtrees and octrees) is used. Typical quadtree meshes are shown in Fig. **1**. In the interest of algorithmic simplicity, we employ smoothed trees i.e. the neighboring cells can only differ by at most one grid level. In Fig. **1**, a non-smoothed quadtree, a smoothed quadtree and a smoothed quadtree with a minimum width of 3 cells per grid level are shown. Regions around arbitrarily shaped objects can be refined locally. For the parallel solver a hybrid approach of blocks refined via the tree-type structure with matrices with equidistant node distance is considered, which reduces the number of special cases for the parallelization process significantly. The strategy for parallelization is described in [12].

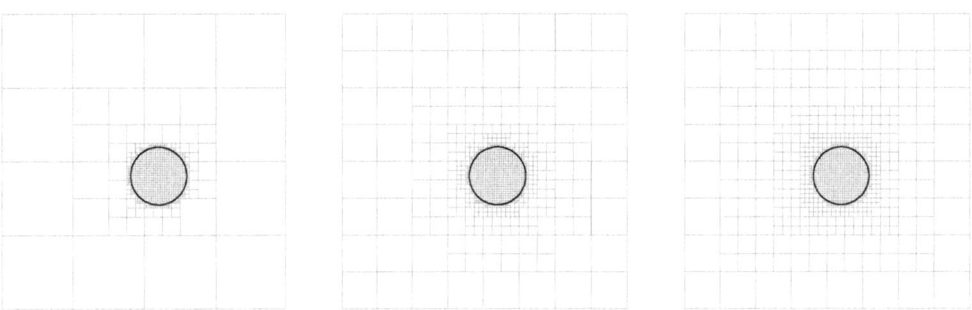

Fig. **1**: Quadtree, smoothed quadtree and quadtree with minimum width of 3 cells.

In order to model spatial refinement where one level of refinement is represented by one grid level, nested time-stepping is used. If the speed of sound, the kinematic viscosity, the Mach and the Reynolds number are assumed to be equal on all grids, then two time steps on the finer grid have to be performed during one time step on the coarser level [9].

In contrast, an approach where the Mach number is lowered on the finer grid levels to ensure a faster convergence to the *in*compressible Navier-Stokes equations can be found in [42]. This approach is called diffusive scaling and requires to perform four time steps on the finer grid during one time step on the coarser mesh.

Ensuring the continuity of pressure, the velocity, and also of their derivatives, the non-equilibrium parts (f_i^{neq} [5, 9, 48] or alternatively m_i^{neq}) have to be rescaled. Performing the rescaling after the propagation step one obtains:

$$m_{i,l-1}^{neq} = \frac{s_{l,i,i}}{s_{l-1,i,i}} \frac{\Delta t_{l-1}}{\Delta t_l} m_{i,l}^{neq} \tag{6}$$

The relaxation parameters $s_{l,a} = s_{l,b} = s_{l,c} = s_{l,d} = s_{l,e}$ can in principle be arbitrarily chosen in the range $[-2,0]$, but we used the same scaling as for $s_{l,\omega}$.

$$\frac{s_{l,a}}{s_{l+1,a}} = \frac{s_{l,b}}{s_{l+1,b}} = \frac{s_{l,c}}{s_{l+1,c}} = \frac{s_{l,d}}{s_{l+1,d}} = \frac{s_{l,e}}{s_{l+1,e}} = \frac{s_{l,\omega}}{s_{l+1,\omega}} \tag{7}$$

Different space and time interpolations have to be used because of the different mesh spacings Δx_l and time steps Δt_l. A cubic interpolation in space is used for the 'hanging' nodes. Fig. **2** shows the interpolation for the 'hanging' node P using the nodes P1, P2, P3 and P4. Linear interpolation is used in time. Details of the algorithm can be found in [6, 15, 48]. The necessary boundary conditions are described in Section 4.1..

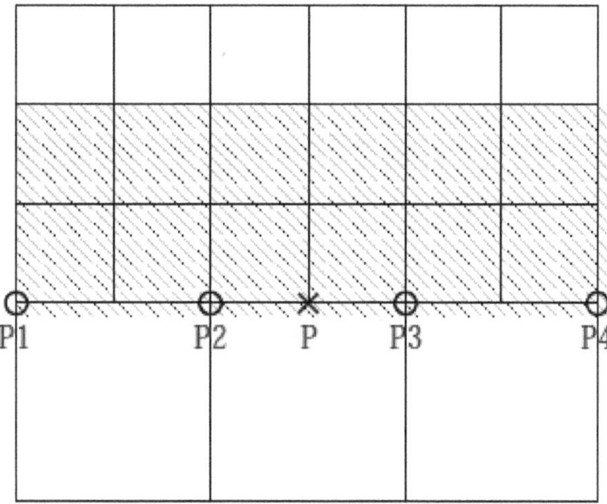

Fig. 2: Hanging node P and nodes P1-P4, which are required for the interpolation in space.

3. TRANSFORMATION OF PHYSICAL QUANTITIES

In order to reduce numerical errors and to avoid very small and/or large numerical values, the LB system is rescaled. Such a rescaling is inherent to most LB implementations, so that the procedure for the mapping between two different systems labeled as *LB* and *real* is presented below. The scaling of the forces and the time from the fluid solver system (LB) to the real system is done via the dimensionless drag coefficient, which has to be the same in both systems, yielding:

$$\mathbf{F}_{real} = \mathbf{F}_{LB} \cdot \frac{H_{real}^2 \cdot \rho_{real} \cdot u_{real}^2}{H_{LB}^2 \cdot \rho_{LB} \cdot u_{LB}^2} \tag{8}$$

with reference height H, reference density ρ and reference velocity u. By multiplying the drag formula with the square of the Reynolds number we obtain the following alternative formulation:

$$\mathbf{F}_{real} = \mathbf{F}_{LB} \cdot \frac{\rho_{real} \cdot \nu_{real}^2}{\rho_{LB} \cdot \nu_{LB}^2} \tag{9}$$

which can be applied if no reference velocity is given. Demanding equivalence of Reynolds number in both systems, and using $u = \frac{H}{\Delta T}$, the following time scaling is obtained:

$$\Delta T_{real} = \Delta T_{LB} \cdot \frac{\nu_{LB} \cdot H_{real}^2}{\nu_{real} \cdot H_{LB}^2} \tag{10}$$

The wall velocity (boundary condition) for the LB system is

$$\mathbf{u}_{LB} = \mathbf{u}_{real} \cdot \frac{\nu_{LB} \cdot H_{real}}{\nu_{real} \cdot H_{LB}} \tag{11}$$

and the formula for rescaling the gravity reads

$$g_{LB} = g_{real} \cdot \frac{u_{LB}^2 \cdot H_{real}}{u_{real}^2 \cdot H_{LB}} \tag{12}$$

or

$$g_{LB} = g_{real} \cdot \frac{\nu_{LB}^2 \cdot H_{real}^3}{\nu_{real}^2 \cdot H_{LB}^3}. \tag{13}$$

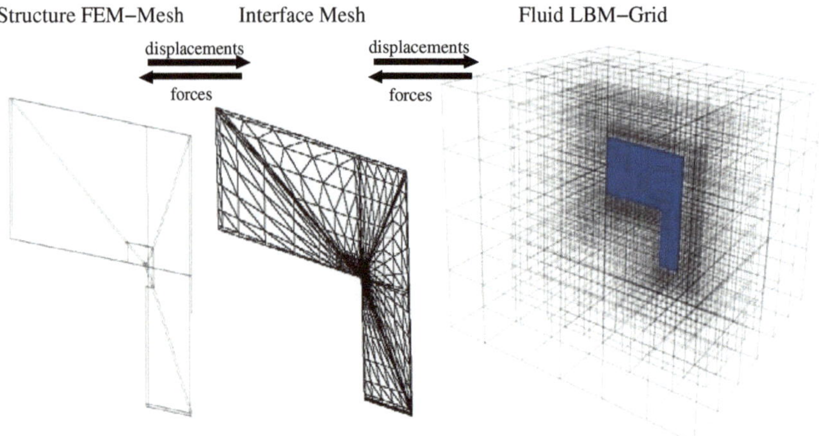

Fig. **3**: Coupling using interface mesh.

4. DESCRIPTION OF THE COUPLING APPROACH

In this work, the solution of coupled fluid-structure interaction problems is achieved by a partitioned approach, in which each domain is solved separately and information is exchanged at the common boundary. The coupling framework is described in the following. In an early version of the coupling with the high order finite element solver ADHOC, an eigenvalue approach for the structural integration in time was used [15, 22, 43], which by design could only account for geometrically linear structural deformations. The coupling algorithm was of explicit nature, since the boundary information was only exchanged once at each time step. A second, completely different and more stable implicit version which also considers geometrically nonlinear structural displacements is described in [16]. This concept proved to be quite involved from a computational point of view and was therefore not further investigated. The most efficient algorithm turned out to be a modified explicit version, which is described in the following.

4.1. Coupling Algorithm

At the fluid-structure interface, two completely different discretizations are present. The structure is described by coarse and curved *p*-elements, while the fluid is described by relatively small cubes and a locally refined octree type grid. If rigid bodies are used, the objects are described as volume bodies. We therefore introduce an interface mesh to couple the solid discretization with the fluid discretizations. The interface mesh is a moving surface mesh and consists of flat triangles. On each node of the interface mesh, the values for the fluid/structure velocity, the FE load vector and other physical quantities that are required for the exchange are stored. The mesh is constructed as follows. The nodes are defined by the Gaussian integration points on the surface of the *p*-Finite-Elements or as a set of equidistant points situated on the surface of the structure. A triangulation serves to obtain the interface mesh. In case of the structural finite element computation the interface mesh can be adapted by the *p*-FEM solver as well as by the LB solver. A Gaussian interface mesh and the coupling is shown in Fig. **3**.

The simulation framework (for details see [2]) is based on a client-server concept and realized via MPI [36]. A master process manages the exchange of data between the structural code and the fluid code. For rigid body motion, the interface mesh is directly referenced in the code. The values to be transfered (i.e. the tractions and displacements) are only allowed to vary linearly between the nodes of the interface mesh. As part of the setup, the structural solver provides the interface mesh and its initial configuration. This mesh is handed to/in the fluid solver.

The coupling algorithm with a multilevel nested time-stepping scheme is depicted in Fig. **4**. While n fluid steps and one structural step are performed, further sub-steps (internal fluid steps) need to be performed in

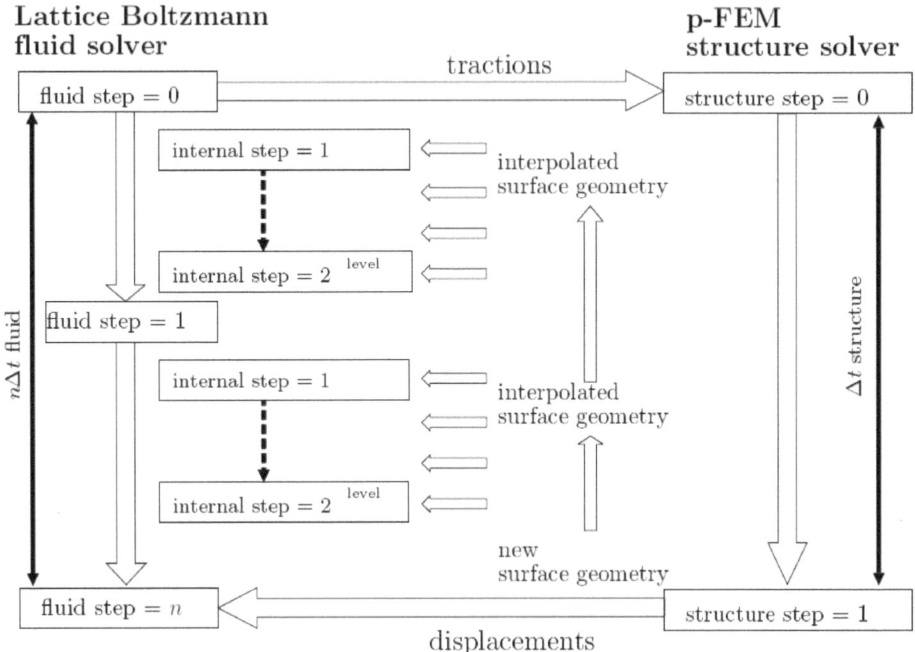

Fig. **4**: Explicit coupling algorithm with multilevel nested time-stepping.

each of these fluid steps, in order to propagate the fluid unknowns $f_l(t,\mathbf{x})$ through the hierarchical, non-uniform grid defined by the LBM solver (see Section). The overall algorithm is given as follows:

1. The fluid solver computes the traction vector at the interface mesh points.
2. The tractions are transfered to the structural solver via the interface mesh.
3. The structure integrates these tractions into the FE load vector, either in a standard way or by means of a load conservative scheme [23]. The structural displacements are calculated and then transfered back to the fluid solver via the interface mesh.
4. The fluid solver performs an interpolation of the positions of the interface mesh in time and calculates the solution for all internal fluid steps.
5. The previous step is repeated for n fluid-subcycling steps. The fluid stresses of the subcycling steps plus the fluid internal steps are averaged.

The algorithm is a staggered coupling algorithm with a sub iteration for the fluid, with the specialty that the fluid-solver needs to perform internal fluid steps on its sub grids. As such, the presented algorithm is only conditionally stable and first order accurate in time. For rigid body motion, only one force vector and, respectively, one displacement vector have to be exchanged for each rigid body.

5. BOUNDARY CONDITIONS

Kinematic boundary conditions for the flow solver are imposed consistently with the displacements and velocities at the surface of the structure. In the LBM, macroscopic flow quantities are obtained via taking low-order moments of the particle distribution functions. In turn, macroscopic boundary conditions can only be set implicitly and have to be specified directly in terms of particle distribution functions. A well known and simple way to introduce no-slip walls is the so-called bounce back scheme, in which particles

bounc ... th one
of the ... cy for
arbitr ... locity
bounc

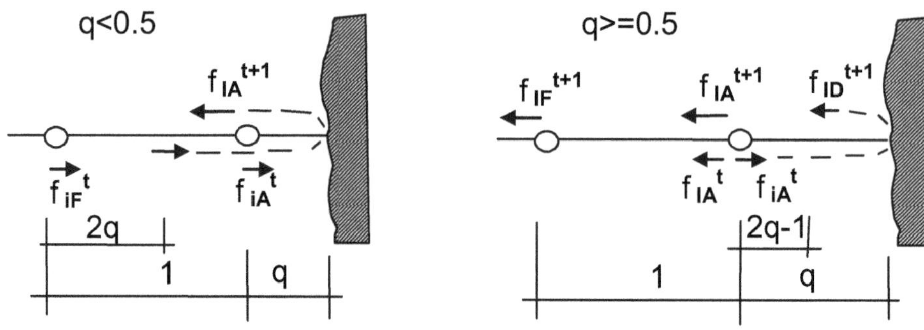

Fig. **5**: Interpolations for second order bounce back scheme.

In Figure **5** two cases along a link i can be identified:

(a) wall–node distance $q_i < 0.5\,|\mathbf{e}_i\Delta t|$ and

(b) wall–node distance $q_i \geq 0.5\,|\mathbf{e}_i\Delta t|$.

The modified bounce back scheme reads

$$f_{IA}^{t+1} = (1-2q)\cdot f_{iF}^{t} + 2q\cdot f_{iA}^{t} - 2\rho w_i \frac{\mathbf{e}_i\mathbf{u}_w}{c_s^2}, \quad 0.0 < q < 0.5 \tag{14}$$

$$f_{IA}^{t+1} = \frac{2q-1}{2q}\cdot f_{IA}^{t} + \frac{1}{2q}\cdot f_{iA}^{t} - \rho w_i \frac{\mathbf{e}_i\mathbf{u}_w}{q c_s^2}, \quad 0.5 \leq q \leq 1.0, \tag{15}$$

The recovery of second order accuracy even for curved boundaries is clearly demonstrated in [14]. A detailed discussion of LBE boundary conditions can be found in [13]. In contrast to the simple bounce-back scheme, the use of these interpolation based no-slip boundary conditions results in a notable mass loss across the no-slip lines. Yet, the results obtained with bounce-back were inferior which highlights the importance of a proper geometric resolution of the flow domain. Pressure boundary conditions are implemented by setting the missing distribution functions according to [46], such that

$$f_I(t+\Delta t,\mathbf{x}) = -f_i(t,\mathbf{x}) + f_I^{eq}(p_0,\mathbf{u}(t_B,\mathbf{x}_B)) + f_i^{eq}(p_0,\mathbf{u}(t_B,\mathbf{x}_B)) \tag{16}$$

where p_0 is the given pressure, $t_B = t + \frac{1}{2}\Delta t$, $\mathbf{x}_B = \mathbf{x} + \frac{1}{2}\mathbf{e}_i$, (f_I, f_i) are an anti-parallel pair of distribution functions with velocities $\mathbf{e}_i = -\mathbf{e}_I$, and f_I is the incoming and f_i the outgoing distribution function value. $\mathbf{u}$ is obtained by extrapolation.

If a new fluid node is created due to the moving structure, linear inter- or extrapolation (depending on the geometrical configuration) is used to compute the local velocity at the corresponding node. A local Poisson type iteration described in [34] is then applied locally to adjust the corresponding pressure and the higher order moments.

6. FORCE EVALUATION

There are basically two possibilities to evaluate forces on boundaries using the LBM, the momentum-exchange method and the pressure/stress integration method. A detailed description and comparison of both methods can be found in [35], and the details for the implementation in 2D are given in [15, 23]. The boundary of the structure is described by the interface mesh consisting of flat triangles in 3D as discussed in Section 4.

In the momentum exchange method the force acting along a link *i* on a boundary results from the momentum-exchange between the particle distributions hitting the moving object and the moving object itself [37]:

$$\Delta \mathbf{F} = \frac{\Delta x^3}{\Delta t}(\mathbf{e}_i - \mathbf{u}_B)(f_i^t + f_{\bar{I}}^{t+1}). \tag{17}$$

Here $\mathbf{u}_B$ is the velocity of the moving boundary and $\mathbf{e}_{\bar{I}} = -\mathbf{e}_i$. The momentum exchange method works well if only the integral forces acting on structural elements are of interest. However, it leads to noisy force results if the elements of the structure or the interface mesh have a size comparable to the grid distance of the fluid mesh. If too few links contribute to the force computation, incorrect local forces are obtained. In this case the stress integration method is favorable.

An advantage of the LBM compared to conventional CFD solvers is the local availability of the stress tensor. The stress tensor $S_{\alpha\beta}$ with scalar pressure is:

$$S_{\alpha\beta} = -c_s^2 \rho \delta_{\alpha\beta} + \sigma_{\alpha\beta} \tag{18}$$

δ is the Kronecker-Delta and $\sigma_{\alpha\beta}$ is computed from the non equilibrium part of the distribution functions:

$$\sigma_{\alpha\beta} = \left(1 - \frac{\Delta t}{2\tau}\right) \sum_{i=1}^{b-1} f_i^{neq} \left(e_{i\alpha} e_{i\beta} - \frac{1}{D} \mathbf{e}_i \mathbf{e}_i \delta_{\alpha\beta}\right) \tag{19}$$

In three dimensions, 256 different cases emerge of how a point may lie inside the grid. These cases conform with the Marching-Cubes (3-D) algorithm [29] which was developed for computer graphics to compute iso-surfaces from cartesian grids.

7. SHIP PROPELLER

The first very demanding test case for fluid-structure interaction is the jet stream creation of a ship propeller. It is important during the ship design process to predict the fluid load on the river bed and the river banks which eventually will lead to substantial erosion. For this purpose, the fluid velocity and stress are monitored at different probe locations in the flow (see Fig. **6**) The propeller has a speed of 3.965 rotations per second. The Reynolds number equals 31.1 Mio., related to the propeller diameter D=1.58m as characteristic length scale. The boundary condition which is applied on the propeller surface is the interpolation bounce back with an additional velocity term for the moving wall. The test case was computed with the parallel version of VIRTUALFLUIDS on a grid consisting of 50 Mio. lattice nodes. The grid was hierarchically refined for four different levels, which were computed with the nested time stepping scheme described in Section . The grid size on the coarse grid is $\Delta x_{max} = 16cm$ and on the finest grid $\Delta x_{min} = 2cm$. An LES model [25] was used to capture the turbulent structures of the flow. On the coarsest grid level the time step is $\Delta t = 0.0002353s$. Hence, to simulate one minute of real time, 255000 time steps had to be computed.

Figs. **7**, **8** and **9** show the jet which is generated from the rotation of the propeller. In Fig. **10** the time history for one minute simulation of a specified point is depicted.

For parallel computations on massively-parallel hardware, a proper load balancing is crucial. For this test case, the load balancing is done via space partitioning with the METIS library [24]. As we use a block-structured grid, the loads of the computational blocks are estimated according to the level of refinement. On top of that, during the computation, the load is determined dynamically by keeping track of the time one block needs to compute the flow equations and to update the geometry of the changing rotational structure. Subsequently, the load balancing is updated. A topology managing process redistributes the grid blocks and sends the new distribution to the calculation clients. This way, the computational time is reduced by approximately 30%. In Fig. **11**, the computed loads after the initial distribution with METIS, before the dynamic update, are depicted.

The first validation is done in comparison to jet theory for ship propellers. According to [1] the jet velocity in the central domain behind the propeller can be computed as

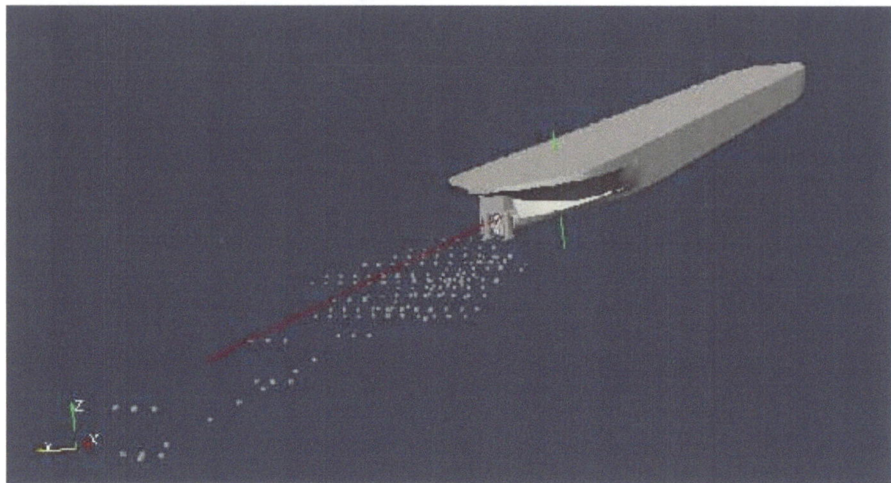

Fig. **6**: Ship hull with probe locations.

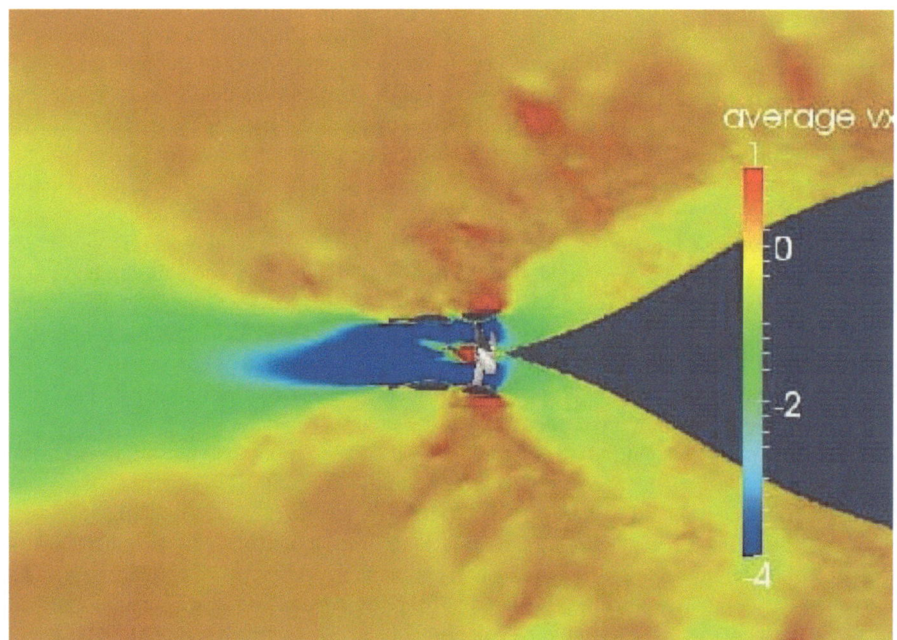

Fig. **7**: Jet behind the propeller; cut plane with averaged x-velocity (scaled).

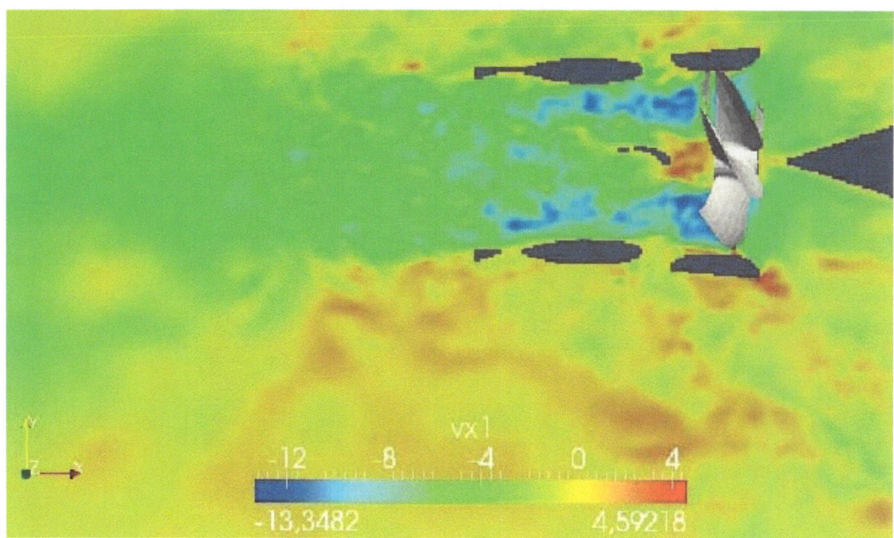

Fig. **8**: Jet behind the propeller, cut plane with immediately x-velocity.

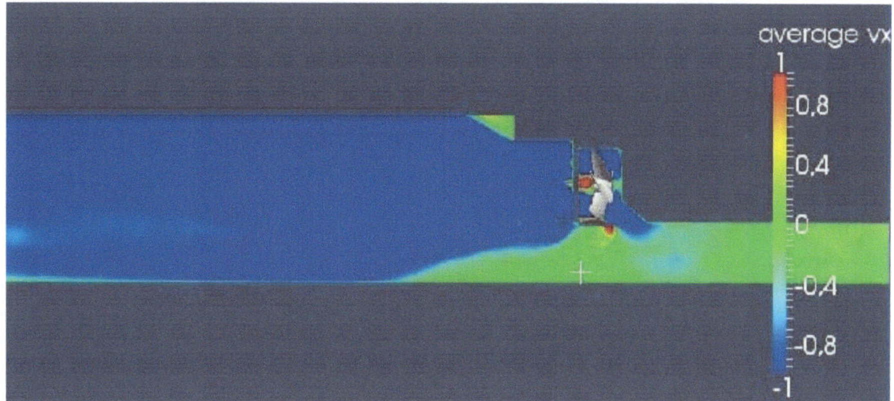

Fig. **9**: Stream created behind the propeller, side view with x-velocity (scaled).

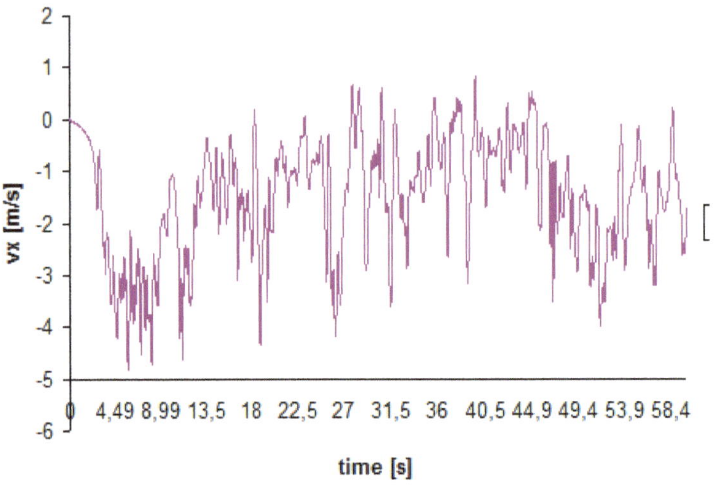

Fig. **10**: x-velocity time series at probe location $\vec{x}$=(-5.851m, 0m, -1.9291m).

Fig. **11**: Weights for redistribution of computational load.

$$v_0 = 1.60\sqrt{0.5}f_N nD\sqrt{K_{t,DP}} = 4.59m/s \tag{20}$$

with a factor for the rotation speed of $f_N = 0.75$ the rotational frequency $n = 3.965s^{-1}$, the diameter $D = 1.58m$ and $K_{t,DP} = 0.67\frac{P}{D}$ with the construction inclination $P = 1.76m$. The speed is valid for the central domain $x_0/D \leq 2.6$, so $x_0 = 4.11m$.

For the speed without impaired jet distribution the speed decays as

$$\frac{v_1}{v_0} = 2.6\left(\frac{x}{D}\right)^{-1} \tag{21}$$

for the domain $4.11m \leq x \leq 6.25m$. In the domain where the jet touches the ground the decaying velocity is

$$\frac{v_2}{v_0} = A\left(\frac{x}{D}\right)^{-a} \tag{22}$$

with exponent $a = 0.6$ and coefficient $A = 1.88e^{-0.092(h/D)}$ with the water depth $h = 3.87m$. This formula is valid for the domain $x \geq 6.25m$.

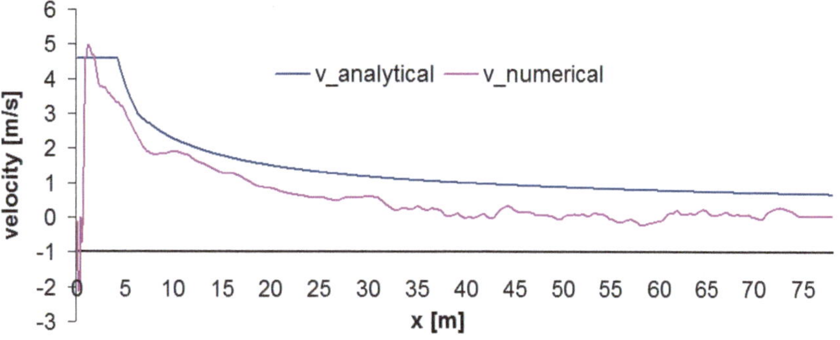

Fig. **12**: decay of averaged velocity behind propeller.

The simulated real time was one minute starting from a still fluid and propeller. It tokes three weeks on 192 CPUs of Cluster with 2.6GHz Intel Quad X5550 CPUs. The average was computed from 40 to 60 seconds.

8. DRILL CUTTINGS TRANSPORTATION

A second example is the simulation of particle transport during drilling into the earth. In geotechnical drilling, a fluid is used to pump out the cuttings caused by the drilling process. In our numerical model, the particles of the suspension are described in a Lagrangian formulation and are coupled with the Eulerian fluid part. The particles are discretized as ideal spheres with varying sphere diameter. The Newtonian motion of the rigid bodies as well the collision are computed with the rigid body dynamics engine PE - PHYSICSENGINE [19]. This rigid body solver is a sophisticated C++ framework for the simulation of rigid, inelastic bodies. One of the main focuses during the development of the PE was the massively parallel simulation of rigid bodies. The parallel version of VIRTUALFLUIDS is coupled with the parallel version of the PE, using identical computational domains for both solvers. In the vicinity of boundaries to grid blocks on other processes, the PE uses overlaps to discretize the rigid bodies. This way, the transition of the geometrical object from one process to the other can be accurately modeled. The forces which act on the particles are computed with the momentum exchange method. The structural solver computes the resulting motion of the particles and the motion is transfered back to the fluid solver via the additional velocity term in the interpolation bounce back rule.

In Fig. **13**, three screen shots of the coupled simulation are given, depicting fluid pressure, velocity with cuttings and velocity of the drill fluid. The segment is a part of the cylinder of the well. In the center is the drill string. The diameter of the bore hole is 0.5m. The fluid has an kinematic viscosity of $0.00015m^2/s$. The Reynolds number for the simulation was 3333.

The final application is the estimation of pressure losses along the bore hole and fluid force computation on the drill string. As the fluid has non-Newtonian properties, works to include power laws for the fluid are in progress.

Fig. **13**: Cutting transport in drill pipe (pressure, velocity with cuttings, velocity).

The next Fig. **14** shows the top view on a drill string. The system is periodic in vertical direction for the fluid as well as the particles. It has an diameter of $64mm$ and the drill string is $20mm$. The particles are generated randomized in position and diameter at startup and have a size of $2mm$ to $5mm$. For the fluid velocity an adaptive forcing is implemented to obtain an average velocity over the cross section. The rotation speed of the string is $-12rpm$, the average velocity is 0.5m/s, the kinematic viscosity is $\nu = 0.000083m^2/s$, the fluid density is $865kg/m^3$ and the spheres density is $1500kg/m^3$. The eccentricity of the drillstring is 50%.

9. FREE SURFACE FLOW

Finally, the combination of FSI and free surface flow is presented. In Fig. **15**, the impact of a wave on a stack of boxes is shown. The wave loading is created by a classical breaking dam scenario, in which a column of water is allowed to collapse. The box stack consists of 15 boxes, which are modeled as rigid bodies in the PE physics engine. Even at low grid resolutions ($160 \times 48 \times 48$ lattice nodes), quite realistic behavior can be observed. As the surge front impacts on the box stack, the boxes are pushed forward and advected with the flow. The simulations were run for a Reynolds number 50000 and Froude number 1.5. For the free surface capturing, a hybrid LBM-VOF-method is used (see [20]). In the long run, debris flow, consisting of billions of particles, and the generation of Tsunamis due to underwater land slides shall be addressed.

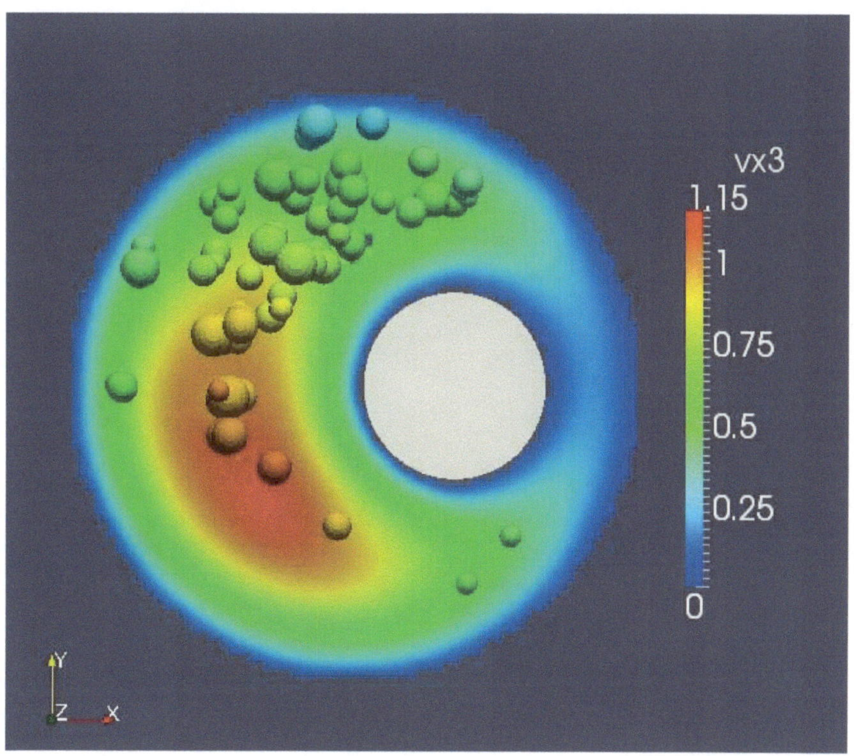

Fig. **14**: Converged state of periodic simulation with rotating drill string and cuttings transport.

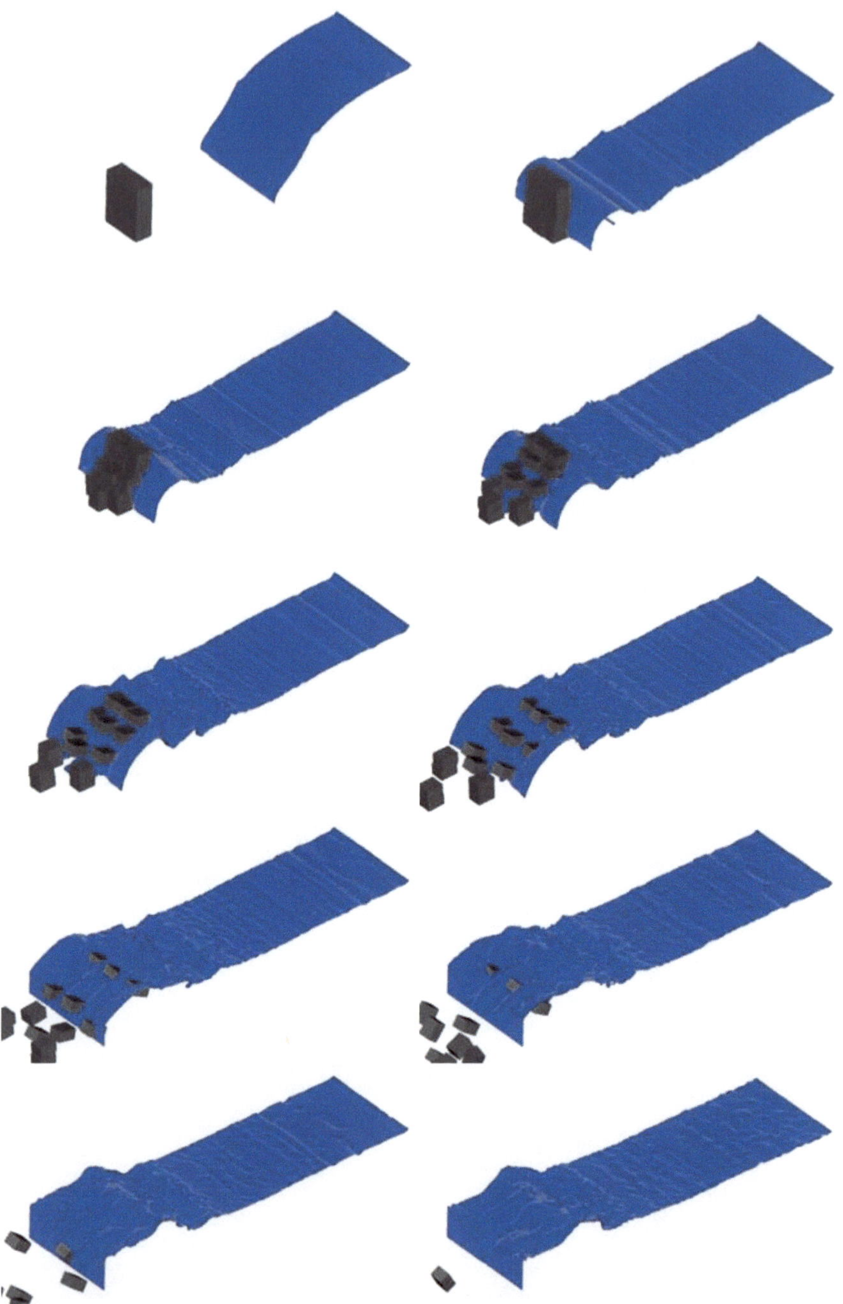

Fig. **15**: Wave impact on box stack for selected time steps.

CONCLUSIONS

In this chapter, several applications of the Lattice Boltzmann fluid solver VIRTUALFLUIDS in the field of fluid-structure interaction have been presented. A major advantage of the method is the possibility of an explicit coupling for coupled simulations leading to a high computational efficiency. Recent developments of VIRTUALFLUIDS resulted in the capability of simulating real world CFD test cases. The parallel framework is based on a Lattice-Boltzmann approach on block-structured grids and provides non-uniform and adaptive grid refinement features. VIRTUALFLUIDS has successfully been coupled to different structural displacement solvers. Due to the underlying Eulerian grid of the LBM, the handling of deforming or moving structures is straightforward and time-consuming remeshing procedures are not necessary. The applications which were shown in this contribution demonstrate that the solver is able to deal with a large amount of degrees of freedom. Further investigations have to deal with more sophisticated turbulence models and laws of the wall to improve the description of boundary layers, even at comparably low grid resolutions.

ACKNOWLEDGEMENT

The authors gratefully acknowledge support by the Ministry for Science and Culture of Lower Saxony and Baker Hughes and funding of the Geothermal Energy and High Performance Drilling (GEBO) project. We also thank Prof. Söhngen and Mr. Spitzer (Federal Waterways Engineering and Research Institute) for valuable discussions regarding the ship propeller test case.
Also we thank the DEISA Consortium (www.deisa.eu) for CPU-time and technical support from the DEISA Extreme Computing Initiative funded through EU FP7 project RI-222919.

CONFLICT OF INTEREST

The authors confirm that this chapter content has no conflict of interest.

DISCLOSURE

Part of the information included in this chapter has been previously published by the author in Geller S., Kollmannsberger S., El Bettah M., Krafczyk M., Scholz D., Düster A., Rank E. An Explicit Model for Three-Dimensional Fluid-Structure Interaction using LBM and p-FEM. in : Fluid Structure Interaction II, Lecture Notes in Computational Science and Engineering, 2010, Volume 73, 285-325.

REFERENCES

[1] BAW. Grundlagen zur Bemessung von Böschungs- und Sohlensicherungen an Binnenwasserstrassen. Bundesanstalt für Wasserbau (BAW), Karslruhe, 2004.

[2] Brenk M, Bungartz H-J, Mehl M, Neckel T. Fluid-Structure interaction on cartesian grids: flow simulation and coupling interface. In: Fluid-Structure Interaction, Modelling, Simulation and Optimisation. Lecture Notes in Computational Science and Engineering, Springer, 2006; 53: 294-335.

[3] Bouzidi M, Firdaouss M, Lallemand P. Momentum transfer of a Boltzmann-Lattice fluid with boundaries. Phys Fluids 2001; 13: 3452-3459.

[4] Chapman S, Cowling TG. The mathematical theory of non-uniform gases. Cambridge University Press, New York, 1990.

[5] Crouse B, Rank E, Krafczyk M, Tölke J. A LB-based approach for adaptive flow simulations. Int J Mod Phys B 2003; 17: 109-112.

[6] Crouse B. Lattice-Boltzmann Strömungssimulation auf Baumdatenstrukturen. PhD thesis, Lehrstuhl für Bauinformatik, Fakultät für Bauingenieur- und Vermessungswesen, Technische Universität München, 2003.

[7] d'Humières D, Ginzburg I, Krafczyk M, Lallemand P, Luo L. Multiple-relaxation-time Lattice Boltzmann models in three dimensions. Philos Trans Math, Phys Enginrg Sci 2002; 360: 437-451.

[8] d'Humières D. Generalized Lattice Boltzmann equations. In: Shizgal BD, Weave DP (eds.). Rarefied Gas Dynamics: Theory and Simulations. Prog Astronaut Aeronaut, Washington DC, 1992. AIAA 1992; 159; 450-458.

[9] Filippova O, Hänel D. Boundary-fitting and local grid refinement for LBGK models. Int J Mod Phys C 1998; 8: 1271-1279.

[10] Felippa C, Park K, DeRuntz J. Stabilization of staggered solution procedures for fluid-structure interaction analysis. In: Blytschko T, Geers T (eds.). Computational Methods for Fluid-Structure Interaction Problems. American Society of Mechanical Engineers, New York, 1977; 26: 95-124.

[11] Felippa C, Park K, Farhat C. Partitioned analysis of coupled mechanical systems. Comput Meth Appl Mech Enginrg 2001; 190: 3247-3270.

[12] Freudiger S. Entwicklung eines parallelen, adaptiven, komponentenbasierten Strömungskerns für hierarchische Gitter auf Basis des Lattice Boltzmann Verfahrens. PhD thesis, Technische Universität Braunschweig, 2009.

[13] Ginzburg I, d'Humières D. Multi-reflection boundary conditions for Lattice Boltzmann models. Phys Rev E 2003; 68: 066614.

[14] Geller S, Krafczyk M, Tölke J, Turek S, Hron J. Benchmark computations based on Lattice-Boltzmann, finite element and finite volume methods for laminar flows. Comput Fluids 2006; 35: 888-897.

[15] Geller S, Tölke J, Krafczyk M. Lattice Boltzmann methods on quadree-type grids for fluid-structure interaction. In Bungartz HJ, Schäfer M (eds.) Fluid-Structure Interaction, Modelling, Simulation and Optimisation, Lecture Notes in Computational Science and Engineering, Springer 2006; 53; 270-293.

[16] Geller S, Tölke J, Krafczyk M, Kollmannsberger S, Düster A, Rank E. A coupling algorithm for high order solids and Lattice Boltzmann fluid solvers. In: Wesseling P, Onate E, Periaux J (eds.). Proceedings of the European Conference on Computational Fluid Dynamics, ECCOMAS CFD 2006, TU Delft, The Netherlands, 2006.

[17] Hron J, Turek S. A monolithic FEM solver for ALE formulation of fluid structure interaction with configurations for numerical benchmarking. In: Proceedings of the International Conference on Computational Methods for Coupled Problems in Science and Engineering, Santorini, 2005.

[18] Hübner B, Walhorn E, Dinkler D. A monolithic approach to fluid-structure interaction using space-time finite elements. Comput Meth Appl Mech Enginrg 2004; 193: 2087-2104.

[19] Iglberger K, Rüde U. Massively parallel rigid body dynamics simulations. Comput Sci - Res Develop 2009; 23: 159-167.

[20] Janssen C, Krafczyk M. A Lattice Boltzmann approach for free-surface-flow simulations on non-uniform block-structured grids. Comput Math Appl 2009;

[21] Junk M, Klar A, Luo LS. Asymptotic analysis of the Lattice Boltzmann equation. J Comput Phys 2005; 210: 676.

[22] Kollmannsberger S, Düster A, Rank E. FSI based on bidirectional coupling of high order solids to a Lattice Boltzmann method. In: Proceedings of PVP 2006, 11th International Symposium on Emerging Technologies in Fluids, Structures, and Fluid/Structure Interactions, within the ASME Pressure Vessel and Piping Conference, Vancouver, B.C, Canada, 2006.

[23] Kollmannsberger S, Geller S, Düster A, Tölke J, Sorger C, Krafczyk M, Rank E. Fixed-grid fluid-structure interaction in two dimensions based on a partitioned Lattice Boltzmann and p-FEM approach. Int J Numer Meth Enginrg 2009; 79: 817-845.

[24] Karypis G, Kumar V. METIS – A software package for partitioning unstructured graphs, partitioning meshes, and computing fill-reducing orderings of sparse matrices, Version 4.0. Department of Computer Science, University of Minnesota, Army HPC Research Center, Minnesota, MN, 1998.

[25] Krafczyk M. Gitter-Boltzmann-Methoden: Von der Theorie zur Anwendung. Postdoctoral thesis, Lehrstuhl für Bauinformatik, Fakultät für Bauingenieur- und Vermessungswesen, Technische Universität München, 2001.

[26] Küttler U, Wall W. Vector extrapolation for strong coupling fluid-structure interaction solvers. J Appl Mech 2009; 2.

[27] Ladd A. Numerical simulations of partiuclate suspensions via a discretized Boltzmann equation. Part 1: theoretical foundations. J Fluid Mech 1994; 271: 285.

[28] Ladd A. Numerical simulations of partiuclate suspensions via a discretized Boltzmann equation. Part 2: numerical results. J Fluid Mech 1994; 271: 311.

[29] Lorensen WE, Cline HE. A discontinuous Galerkin method for the Navier-Stokes equations. ACM SIGGRAPH Comput Graph arch 1987; 21: 163-169.

[30] Löhner R, Cebral JR, Camelli FF, Baum JD, Mestreau EL, Soto OA. Adaptive embedded/immersed unstructured grid techniques. Arch Comput Meth Enginrg 2007; 14: 279-301.

[31] Lallemand P, Luo L. Theory of the Lattice Boltzmann method: dispersion, dissipation, isotropy, Galilean invariance, and stability. Phys Rev E 2000; 61: 6546-6562.

[32] Lallemand P, Luo LS. Lattice Boltzmann method for moving boundaries. J Comput Phys 2003; 184: 406-421.

[33] P. LeTallec and J. Mouro. Fluid structure interaction with large structural displacements. Comput Meth Appl Mech Enginrg 2001; 190: 3039-3067.

[34] Mei R, Luo LS, Lallemand P, d'Humières D. Consistent initial conditions for LBE simulations. Comput Fluids 2006; 35: 855-862.

[35] Mei R, Yu D, Shyy W, Lou L. Force evaluation in the Lattice Boltzmann method involving curved geometry. Phys Rev E 2002; 65: 041203.

[36] Message Passing Interface Forum. MPI: A message-passing interface standard. Int J Supercomput Appl 1994; 8.

[37] Nguyen N-Q, Ladd AJC. Sedimentation of hard-sphere suspensions at low Reynolds number. J Fluid Mech 2004; 525: 73-104.

[38] Park K, Felippa C. Partitioned analysis of coupled systems. Chapter 3 in: Blytschko T, Hughes T (eds.). Computational Methods for Transient Analysis. North-Holland,Amsterdam- New York, 1984, pp. 157-219.

[39] Piperno S, Farhat C, Larrouturou B. Partitioned procedures for the transient solution of coupled aroelastic problems – part I: Model problem, theory and two-dimensional application. Comput Meth Appl Mech Enginrg 1995; 124: 79-112.

[40] Piperno S, Farhat C. Partitioned procedures for the transient solution of coupled aroelastic problems – part II: Energy transfer analysis and three dimensional applications. Comput Meth Appl Mech Enginrg 2001; 190: 3147-3170.

[41] Qian YH, d'Humières D, Lallemand P. Lattice BGK models for Navier-Stokes equations. Europhys Letters 1992; 17: 479-484.

[42] Rheinländer M. A consistent grid coupling method for Lattice Boltzmann schemes. J Stat Phys 2005; 121: 49-74.

[43] Scholz D, Kollmannsberger S, Düster A, Rank E. Thin solids for fluid-structure interaction. In: Bungartz HJ, Schäfer M (eds.). Fluid-Structure Interaction, Modelling, Simulation and Optimisation. Lecture Notes in Computational Science and Engineering, Springer 2006; 53: 294-335.

[44] Turek S, Hron J. Proposal for numerical benchmarks for fluid-structure interaction between an elastic object and laminar incompressible flow. In: Bungartz HJ, Schäfer M (eds.). Fluid-Structure Interaction, Modelling, Simulation and Optimisation. Lecture Notes in Computational Science and Engineering, Springer 2006; 53: 371-385.

[45] Turek S, Hron J. Numerical benchmarks for fluid-structure interaction between an elastic object and laminar incompressible flow. In: Bungartz HJ, Schäfer M (eds.). Fluid-Structure Interaction, Lecture Notes in Computational Science and Engineering. Springer, 2010.

[46] Thuerey N. A single-phase free-surface Lattice Boltzmann method. PhD thesis, Lehrstuhl für Systemsimulation (Informatik 10), Universität Erlangen-Nürnberg, 2003.

[47] Wall W, Gammnitzer P, Gerstenberger A. A strong coupling partitioned approach for fluid-structure interaction with free surfaces. Comput Fluids 2007; 36: 169-183.

[48] Yu D, Mei R, Shyy W. A multi-block Lattice Boltzmann method for viscous fluid flows. Int J Numer Meth Fluids 2002; 39: 99-120.

Progress in Computational Physics, Vol. *3*, 2013, 217-267

CHAPTER 9

Lattice Boltzmann Method for MILD Oxy-fuel Combustion Research: A Potential Powerful Tool Responding to the Man-made Global Warming

Sheng Chen [1,2,3,*]

[1] *State Key Laboratory of Coal Combustion, Huazhong University of Science and Technology, P.R. China,* [2] *China-EU Institute for Clean and Renewable Energy, Huazhong University of Science and Technology, P.R. China and* [3] *Advanced Coal Technology Consortium, US-China Clean Energy Research Center, USA*

Abstract: The subject of this chapter concerns the numerical simulation/modeling of multicomponent combustion under MILD oxy-fuel operation by the lattice Boltzmann method (LBM) to deepen our understanding on this novel combustion technology and improve related methodologies, which represents a significant challenge in the combustion community to respond to the man-made global warming. The main objective is to demonstrate the opportunities to build a new modeling framework to accelerate scientific discovery in this topic by utilizing unique features of the LBM and indicate the challenges that we should overcome since until recently the LBM could not be employed in such research due to combustion's inherent complexity. The relevant preliminary explorations, including that by the present author, are reviewed. Further development of the methodology would not only provide the basis for a new platform to accelerated scientific research in combustion science, but also significantly push back the limits of LBM in general.

Keywords: Biogas combustion, entropy generation analysis, Lattice Boltzmann method (LBM), mass conservation scheme, MILD oxy-fuel combustion technology, multicomponent combustion, reaction zone structures, spatial self-copy phenomenon, surface chemistry, thermal LB models.

1. INTRODUCTION

Large-scale disintegration of Antarctic ice shelf, accelerated ice loss from Greenland, recently the recorded hottest summer and coldest winter in Europe, 2010 global deadly floods and tens of thousands of persons being killed by the extreme weather events in the latest years, etc. underscore that the remaining time for the human race to prevent global warming is limited. However, according to the latest World Energy Outlook published by the International Energy Agency (IEA), CO_2 emitted by power plants will continue to constitute the major share of greenhouse gases in the next two decades due to the sharp increasing demand on electricity power. The conflict of sharp increasing demand on electricity power and the preservation of the global warming place the human being in a serious dilemma. Up to date, it has been widely accepted that one of the most feasible approaches in response to this crisis is to develop and adopt novel combustion technologies to reduce the amount of CO_2 emitted by power plants since it is the predominant contributor to global CO_2 emissions, as emphasized in the reports of IEA and Intergovernmental Panel for Climate Change (IPCC). Among the recently emergent novel combustion technologies, the so-called oxy-fuel [8] combustion technology has attracted increasing attention. The oxy-fuel combustion is a process of burning fuel in a mixture of pure oxygen and recirculated flue gas. Compared with convectional air-fired boilers, the exhaust gases in oxy-fuel regime are composed mainly of CO_2 instead of N_2 since almost all atmospheric nitrogen is removed from input air, which makes it possible for Carbon Capture and Storage (CCS) with commercial requirements. The oxy-fuel technology was originally developed by US Argonne National Laboratory. Within two decades, it has been developed from laboratory tests to industrial applications which is an extraordinary progress as for an energy technology. Unfortunately, this novel combustion technology itself poses new challenges to combustion specialists. Through plenty of preliminary experiments a lot

Address correspondence to: Sheng Chen, State Key Laboratory of Coal Combustion, Huazhong University of Science and Technology, Wuhan 430074, P.R. China; E-mail: shengchen.hust@gmail.com

Matthias Ehrhardt (Ed.)

of new phenomena have been observed and it has been confirmed that the patterns of chemical reactions, mechanisms of heat transfer and structures of flow etc. in CO_2/O_2 atmosphere are quite different from their normal air-fired counterparts [87]. For example, applying a conventional approach for burner design and operating conditions to oxy-fuel operation will lead to unexpected distributions of temperature, species and velocity inside the combustion chamber and as a result to flame instability and poor burnout. Furthermore, there are obvious disadvantages in the original oxy-fuel combustion technology, such as local hotspots. Recently it has become clear that these hurdles must be surmounted before the oxy-fuel technology can be widely implemented with confidence in its performance. The combustion science communities, such as the International Flame Research Foundation (IFRF), have identified that the key to accomplish this urgent task is to clearly understand the intrinsic characteristics of such reactive flow for future combustion strategies, which poses substantial scientific challenges related to turbulence, multiphase/multicomponent flow, thermodynamics, combustion and techno-economic assessment.

In order to deepen our insight on oxy-fuel combustion and improve its performance for widely spreading it in practical combustion systems, Computational Fluid Dynamics (CFD) techniques have become essential tools for engineers as well as for research specialists to investigate oxy-fuel combustion because the complexity of such reactive flow often limits our ability to understand the phenomena without the help of computers. Especially, some important structures in turbulent multiphase reactive flow can not be captured even by the modern experimental instruments. With the aid of CFD, an improved variation of oxy-fuel combustion strategy named "ISOTHERM Pwr®" has been developed by ITEA in Italy. Recently, Korea electric power research institute and Korea electric power company started a conceptual design of oxy-fuel combustion systems using CFD.

The MILD (Moderate or Intense Low-oxygen Dilution) combustion technology [9], usually being characterized by both an elevated temperature of reactants and low temperature increase in the combustion process, is judged as "one of the most promising combustion technologies in 21st Century" [9] due to its intrinsic advantages such as good combustion stability, low NO_x emissions, enhanced heat transfer, high energy recovery and uniform distribution of chemical and thermodynamic variables. Through recirculated flue gas, oxy-fuel and MILD combustion technologies can be combined seamlessly and accordingly the performance of oxy-fuel combustion can be enhanced significantly. Recently, an initial key project "Theory & Equipment Development for Oxy-fuel Combustion", in which the present author participates, has been started by the US-China Clean Energy Research Center (CERC). One of goals of the project is to reveal the intrinsic characteristics of turbulent multiphase reactive flow under MILD oxy-fuel condition. Through our preliminary theoretical and experimental exploration, we have identified some special effects in the turbulent reactive flows, which may be negligible in traditional combustion systems, under MILD and/or oxy-fuel conditions:

1. Under oxy-fuel operation, because of the different thermo-physical properties between CO_2 and N_2 as compared to the air-fired counterparts, both combustion characteristics and heat transfer of reactants will be affected significantly. For example, the consumption rate of volatiles in oxy-fuel combustion is likewise expected to be slower than in air due to the lower diffusivity of small hydrocarbons in CO_2 compared to N_2, so combustion may occur in a diffusion flame around each solid fuel particle or collectively for a group of solid fuel particles. Moreover, there are several competing mechanisms in oxy-fuel combustion: Firstly, gasification (besides simple oxidation) would be presumed to increase the overall rate of char consumption, but the high endothermicity of the gasification reactions tends to cool the char particle and therefore reduce its oxidation rate, making it unclear if the overall rate of char consumption will increase or decrease from the effects of the gasification reaction; Secondly, burnout under oxy-fuel conditions would be improved due to the higher oxygen partial pressure experienced by the burning coal, possible gasification by CO_2 (and H_2O) and lower gas volumetric flows, whereas the excess CO_2 in the vicinity of the burning solid fuel particles could alter the reaction equilibrium and slow down the burning rate. The final results of these competing mechanisms sensitively depend on the instantaneous scalar quantities (concentrations and temperatures), as well as their histories, within the pores of the char, at the outer surface of the char, and within the boundary layer surrounding the char. Consequently, it is extremely important to accurately predict physical diffusion of oxidizer, products and heat through the gas film confined by complicated geometry.

2. In conventional combustion, chemistry is usually much faster than turbulence timescales to allow some simplified models, and turbulent diffusion dominates over molecular diffusion. MILD combustion, on the other hand, is characterized by low reaction rate, low heat release rate and low temperature gradients, all of

which make molecular diffusion more important. In addition, turbulence-combustion interactions are more complex due to overlapping time and length scales. These new phenomena, especially molecular diffusion in combustion, have not been studied extensively and many studies assume equi-diffusivity even when multispecies multi-step complex chemical reactions are involved. These two effects, in particular, challenge the applicability of combustion models that assume fast chemistry and neglect the effects of differential diffusion. Some essential questions remain to be answered: for instance, how does molecular diffusion affect chemical species distribution, reaction rate, temperature distribution and so on, in comparison with turbulent transport? What effects does turbulence have on flame structures in the MILD oxy-fuel combustion?

Unfortunately, although the available state-of-the-art CFD technologies for combustion research have achieved great successes in turbulent combustion concerned primarily phenomena in which the separation of scales can be justified, such as in the cases of fast chemistry and in the flamelet regime, combustion in other regimes where both chemistry and mixing are competitive as in MILD combustion and flame-based combustion are still great challenges for them [37]. Besides, to capture the fine structures of multicomponent/multiphase flows and to describe accurately the interactions between different phases are also daunting tasks for the continuum-based approaches [85]. No doubt, the popularly used CFD tools are inappropriate to pave the way to substantial progress in the field of MILD oxy-fuel combustion research. Consequently, it is still an urgent task to develop/adopt new numerical methods to accelerate scientific discovery in this field, As Freeman Dyson, a famous theoretical physicist, mathematician and Max Planck Medal winner concluded:" The great advances in science usually result from new tools rather than from new doctrines."

In the last two decades or so, the lattice Boltzmann method (LBM) has matured as an efficient alternative and promising numerical scheme for simulating and modeling complicated physical and chemical systems. Unlike conventional numerical schemes based on discretizations of macroscopic continuum equations, the LBM is based on microscopic models and mesoscopic kinetic equations. The fundamental idea of the LBM is to construct simplified kinetic models that incorporate the essential physics of microscopic or mesoscopic processes so that the macroscopic averaged properties obey the desired macroscopic equations. The kinetic nature of the LBM introduces several important features that distinguish it from other numerical methods. First, straight forward treatment of realistic boundary conditions in complicated geometries, for example multiphase/multicomponent flows through porous media and free surface flows in casting. With improvements as boundary fitting formulations, unstructured grids, and adaptive mesh refinement, now the LBM can handle moving and curved boundaries without reduction of computational accuracy. Second, the structural simplicity of the code due to the methods' explicit nature, the full parallelism of the method and its ready extension to 3D problems. The pressure field in the LBM is calculated using a simple equation of state instead of solving the Poisson equation (or an equivalent one) which is a substantial advantage because solving the Poisson problem often produces numerical difficulties requiring special treatment and can be very time consuming. In addition, the relative independence of the lattice nodes in LBM can be exploited for high parallel computational efficiency. This inherent property makes the LBM suitable for the current trend of parallel computation . For instance, the LBM has emerged as a powerful tool (either Direct Numerical Simulation (DNS) or incorporating turbulence models such as Large Eddy Simulation (LES)) to simulate thermal turbulence. Third, because the LBM can naturally incorporate interactions between different fluids and between fluids and solids, therefore it can potentially reveal novel pattern formation, such as in chemical reaction systems, liquid-solid transition, liquid-vapor boiling process, fluid-solid contact angle, charged colloidal suspensions, droplet collisions, deformation and breakdown of droplets. Especially, the LBM has been compared favourably with the spectral method, the artificial compressibility method, and the finite volume method, all quantitative results further validate excellent performances of the LBM not only in computational efficiency but also in computational accuracy. No surprise, the researchers worldwide highlighted the potential of scientific discovery of LBM in their reports and began some attempts in their work with LBM. Recently a new commercial software "PowerFLOW®", which based on the LBM, has emerged which is very successfully utilized in the automotive industry. For combustion simulation under MILD oxy-fuel conditions, the following outstanding intrinsic advantages of LBM make itself the best alternative of the existing CFD technologies to address the above-mentioned challenges induced by such novel combustion strategy:

1. The assumption of scale separation, which is the basis for "continuum" CFD to build turbulent models, is not required in the LBM any longer. Accordingly, the LB equation can represent richer and more

complicated physics when projected onto hydrodynamic variables in the phase space than the "continuum" approaches.
2. The LB approach perhaps is the best available candidate to describe equation-free multi-scale nonlinear dynamical systems in which a closed macroscopic description in terms of macroscopic variables is unavailable. Consequently, the complicated turbulence-combustion interactions due to overlapping time and length scales in MILD oxy-fuel regime can be treated more straightforwardly in the framework of LBM.
Yet, the prospect of applying the LBM to combustion research under MILD oxy-fuel condition is not entirely clear. As the proverb said: " Every coin always has two sides". Although the LBM possesses obviously outstanding advantages over traditional CFD methods on multi-scales multi-physics problems, hampered by its unsatisfactory performance for thermohydrodynamics, combustion modelling perhaps becomes the Achilles' heel of the traditional LBM due to the puzzle of its numerical instability in combustion simulation. Although the efforts on this topic began emerging soon after the LBM was proposed, the progress in this field is the slowest one in the LBM community during the past two decades. In the latest review paper on LBM published in "Annual Review of Fluid Mechanics" [1], this issue is recognized as a great challenge. Besides in the field of combustion modelling and simulation, some further efforts are required to retrofit the existing LB-based solvers so make them suitable for turbulent reactive flow research in MILD oxy-fuel regime. The objective of this chapter is to demonstrate the opportunities to build a new modeling framework to accelerate scientific discovery in this topic by utilizing unique features of the LBM and indicate the challenges that we should overcome. In order to achieve the goal, firstly the latest progress of LBM in thermal flow, chemical reaction and boundary treatment are reviewed because they all together make up of the basis of numerical combustion research for practical systems. For clarity and avoiding redundance, the related issues that have been analyzed comprehensively in previous review papers and books are just briefly presented here so to provide necessary clue for the readers who are interesting in further reading. Then its applications to complicated energy conversion systems, such as combustion under MILD oxy-fuel condition are presented. Prospects and possible challenges for future advanced research are discussed in succession.

2. ADVANCEMENT IN MODELLING

[illegible]

Combustion is a kind of oxidation reaction with intensive heat release, so a simple and robust mathematic algorithm to treat thermal flow forms the first footstone of numerical combustion research, not only for the LBM but also for traditional CFD technologies.

[illegible]

According to the popular classification used in the LBM community, the existing strategies for constructing thermal LB models may fall into three categories, i.e., the multispeed approach, the double-distribution-function (DDF) approach, and the hybrid approach. The multispeed approach is a straightforward extension of its athermal counterpart, in which only the velocity distribution functions (DFs) are used. Since the multispeed models usually suffer from severe numerical instability and the temperature variation simulated is limited to a narrow range, the DDF approach is proposed as a remedy. The DDF approach utilizes two different DFs, one for the flow field (described by the Boltzmann equation) and the other for the energy field (represented by the so-called Boltzmann energy equation). Among the existing DDF models, that proposed by He et al. [51] has attracted more attention since in their model the viscous dissipation and compression work can be incorporated for the first time. The hybrid approach is similar to the DDF approach except that the energy equation is solved by different numerical methods (e.g., finite-difference methods) rather than by solving the LB equation and accordingly some intrinsic advantages of the pure LB scheme have been lost. An excellent review on this topic can be found elsewhere [85].
The above-mentioned classification is mainly based on mathematic treatment and can not straightforwardly support necessary information for a combustion specialist to choose an appropriate approach for practical simulation. In fact, from the viewpoint of thermodynamics, the existing thermal LB schemes also can be categorized as four types according to the choice of primitive variables for the conservation law of energy

[28]: temperature-, internal-energy-, total-energy- and sensible-enthalpy-based models. Just as their names imply, the DFs to represent the evolution of temperature, internal energy, total energy and sensible enthalpy are introduced in them accordingly. The former two have been analyzed comprehensively in the book [85], so they will not be repeated here. In the rest part of this subsection, the latter two are discussed, especially on their shortcomings, which induces the fifth type of thermal LB model, the so-called total-enthalpy-based model recently proposed by Chen et al. [28]. The complete process to derive the total-enthalpy-based model is also given here to show the readers the key steps (highlighted by bold font) to construct a thermal LB model.

[illegible]

The goal of total-energy-based thermal LB model [46] is to overcome the intrinsic shortcomings of the internal-energy-based model proposed by He et al. [51] (please bear in mind there are a lot of kinds of thermal LB models based on internal energy but He's is the first one that can reflect the effects of viscous dissipation and compression work in the recovered energy equation). For instance, in the derivation of the equilibrium for the internal energy DFs, an *ad hoc* regrouping technique was employed so that some high-order terms could be neglected. The regrouping is somewhat arbitrary, and different regrouping methods may lead to different equilibria. More unsatisfactorily, in the model designed by He et al. [51], both the evolving equation for internal energy DFs and the calculation of the temperature include complicated terms involving temporal and spatial derivatives of the macroscopic variables such as velocity and temperature, which may introduce some additional errors and do harm to the numerical stability. To achieve the target, firstly, the total energy DFs

$$\mathfrak{E} = \frac{\xi^2}{2} f \tag{1}$$

were introduced [46], where $f(\mathbf{x}, \xi, t)$ is the velocity DFs of the molecules and implies the probability of finding a molecule moving with velocity ξ at position $\mathbf{x}$ and time t.
In kinetic theory, it is well known the evolution of the velocity DFs is governed by the Boltzmann equation [85]

$$\partial_t f + \xi \cdot \nabla f + \mathbf{a} \cdot \nabla_\xi f = \Omega_f \tag{2}$$

where $\mathbf{a}$ is the acceleration, and Ω_f is a collision operator that satisfies the following conservation laws:

$$\int \phi \Omega_f d\xi = 0 \tag{3}$$

where $\phi = (1, \xi, \xi^2)$ are the so-called collision invariants. Generally Ω_f is extremely complicated but fortunately can be simplified by some approximations. One widely used approximation is the so-called single-relaxation-time or Bhatnagar-Gross-Krook-BGK (BGK) model [78]

$$\Omega_f = -\frac{1}{\tau_f}[f - f^{eq}] \tag{4}$$

where τ_f is the so-called relaxation time, and f^{eq} is the local Maxwellian equilibrium distribution function (EDF) defined by

$$f^{eq}(\xi; \rho, \mathbf{u}, t) = \frac{\rho}{(2\pi RT)^{D/2}} \exp\left(-\frac{(\xi - \mathbf{u})^2}{2RT}\right) \tag{5}$$

with D being the spatial dimension, T is the local temperature and R the gas constant.
If the both sides of Eq. (2) are multiplied by $\frac{\xi^2}{2}$, then the so-called Boltzmann energy equation for total-energy DFs can be obtained as

$$\partial_t \mathfrak{E} + \xi \cdot \nabla \mathfrak{E} + \mathbf{a} \cdot (\nabla_\xi \mathfrak{E} - \xi f) = \Omega_{\mathfrak{E}} \tag{6}$$

where $\Omega_{\mathfrak{E}} = \xi^2 / 2\Omega_f$ is the collision operator characterizing the total energy change during the particle collisions.

The fluid variables in the EDF, i.e., the density ρ, velocity $\mathbf{u}$, and total energy E, are defined as the moments of f and $\mathfrak{E}$

$$\begin{pmatrix} \rho \\ \rho \mathbf{u} \\ \rho E \end{pmatrix} = \begin{pmatrix} \int f d\xi \\ \int \xi f d\xi \\ \int \mathfrak{E} d\xi \end{pmatrix} \tag{7}$$

where the definition of E reads

$$E = \mathfrak{e} + \frac{u^2}{2} \tag{8}$$

In Eq. (8), the internal energy $\mathfrak{e} = C_v T$ and C_v is the specific heat coefficient at constant volume.
The second key in Guo's thermal LB model [46] is how to construct $\Omega_{\mathfrak{E}}$ appropriately so avoiding the explicit appearance of internal-energy-related term, which will bring complicated temporal and spatial derivatives of the macroscopic variables into Eq. (6), through the collision operation. The assumption by Woods [95] was adopted in [46]. Namely $\Omega_{\mathfrak{E}}$ can be split into two parts:

$$\Omega_{\mathfrak{E}} = \Omega_i + \Omega_m \tag{9}$$

where $\Omega_i = [\frac{(\xi - \mathbf{u})^2}{2}]\Omega_f$ is the internal energy part [51] and

$$\Omega_m = \Omega_{\mathfrak{E}} - \Omega_i = (\xi \cdot \mathbf{u} - u^2/2)\Omega_f \equiv Z\Omega_f \tag{10}$$

is the mechanical energy part and it can be approximated as

$$\Omega_m = \Omega_{\mathfrak{E}} - \Omega_i = -\frac{Z}{\tau_f}(f - f^{eq}) \tag{11}$$

With the aid of this approximation and the thermodynamical relationship Eq. (8), one can straightforwardly get

$$\Omega_i = \frac{1}{\tau_{\mathfrak{E}}}(\mathfrak{E} - \mathfrak{E}^{eq}) - \frac{Z}{\tau_{\mathfrak{E}f}}(f - f^{eq}) \tag{12}$$

where $\mathfrak{E}^{eq}$ is the EDF for the total energy and

$$\frac{1}{\tau_{\mathfrak{E}f}} = \frac{1}{\tau_{\mathfrak{E}}} - \frac{1}{\tau_f} \tag{13}$$

where $\tau_{\mathfrak{E}}$ is the relaxation time for internal energy. The role of the second term in $\Omega_{\mathfrak{E}}$ is to derive the correct viscous heat dissipation in the energy equation and when $\frac{1}{\tau_{\mathfrak{E}}} = \frac{1}{\tau_f}$ the thermal model is identical to the original BGK model.
Consequently, the total-energy-based kinetic equations read

$$\partial_t f + \xi \cdot \nabla f + \mathbf{a} \cdot \nabla_\xi f = -\frac{1}{\tau_f}(f - f^{eq}) \tag{14}$$

$$\partial_t \mathfrak{E} + \xi \cdot \nabla \mathfrak{E} + \mathbf{a} \cdot \nabla_\xi \mathfrak{E} = -\frac{1}{\tau_{\mathfrak{E}}}(\mathfrak{E} - \mathfrak{E}^{eq}) + \frac{Z}{\tau_{\mathfrak{E}f}}(f - f^{eq}) + f\xi \cdot \mathbf{a} \tag{15}$$

where

$$f^{eq} = \frac{\rho}{(2\pi RT)^{D/2}} \exp\left(-\frac{(\xi - \mathbf{u})^2}{2RT}\right) \tag{16}$$

$$\mathfrak{E}^{eq} = \frac{\rho(\xi^2)}{2(2\pi RT)^{D/2}} \exp\left(-\frac{(\xi - \mathbf{u})^2}{2RT}\right) \tag{17}$$

In order to get uniquely accurate expressions of f^{eq} and $\mathfrak{E}^{eq}$, the Hermite expansions are invoked in [46] instead of the Taylor ones in [51]. The final formulas read

$$f^{eq} = \omega(\xi, T)\rho[1 + \frac{\xi \cdot \mathbf{u}}{RT} + \frac{1}{2}(\frac{\xi \cdot \mathbf{u}}{RT})^2 - \frac{\mathbf{u}^2}{2RT}] \tag{18}$$

$$\mathfrak{E}^{eq} = \omega(\xi, T)p[\frac{\xi \cdot \mathbf{u}}{RT} + (\frac{\xi \cdot \mathbf{u}}{RT})^2 - \frac{\mathbf{u}^2}{RT} + \frac{1}{2}(\frac{\xi^2}{RT} - D)] + Ef^{eq}(T), \tag{19}$$

where

$$\omega(\xi,T)=\frac{1}{(2\pi RT)^{D/2}}\exp(-\frac{\xi^2}{2RT})$$

and $p=\rho RT$ is the pressure.

In conclusion, three treatments are adopted in Guo's model [46] to remedy the shortcomings in He's [51]: Firstly, the total energy DFs are introduced instead of the internal energy ones; Secondly, the explicit appearance of internal-energy-related term in the collision operation is bypassed through the thermodynamical relationship between total energy and internal energy; Thirdly, the EDF is expanded with the aid of the Hermite expansions instead of the Taylor ones.

Despite the apparent advancement made by the total-energy-based thermal model [46], soon it is well recognized that this model still suffers from some deficiencies. Among them, the major one is that additional energy source terms are hardly naturally incorporated into the energy conservation equation formulated by total energy since the two evolution equations (the Boltzmann and so-called Boltzmann energy equations) for the pertinent DFs are not independent. In fact, similar puzzle also exists in temperature- and internal-energy-based thermal models. Recently, a sensible-enthalpy-based thermal model [10] brings the hope to remedy this shortcoming.

[illegible]

For investigation on practical energy conversion systems, enthalpy, instead of temperature, internal energy or total energy, is popularly used in the engineering community since any additional input/output energy, no matter heat or work, is convenient to be consistently incorporated into the energy conservation equation formulated by enthalpy [77]. Accordingly, Chatterjee [10] constructed a new thermal LB model through replacing the internal energy in He's model [51] by sensible enthalpy (what should be emphasized is that in [10] Chatterjee confused the definitions between sensible enthalpy and total enthalpy)

$$h_s=\text{[illegible]}+p/\rho \tag{20}$$

and the corresponding sensible enthalpy DFs read

$$\hbar=[\frac{(\xi-\mathbf{u})^2}{2}+RT]f \tag{21}$$

In Eq. (21) the relationship $p=\rho RT$ is invoked implicitly.

The Boltzmann energy equation for sensible enthalpy DFs can be gained through multiplying the both sides of Eq. (2) by $\frac{(\xi-\mathbf{u})^2}{2}+RT$

$$\partial_t\hbar+\xi\cdot\nabla\hbar=-\frac{1}{\tau_\hbar}(\hbar-\hbar^{eq})-f\mathscr{Q} \tag{22}$$

where the collision term $[\frac{(\xi-\mathbf{u})^2}{2}+RT]\Omega_f$ is modelled by $-\frac{1}{\tau_\hbar}(\hbar-\hbar^{eq})$ and

$$\mathscr{Q}=(\xi-\mathbf{u})\cdot[\partial_t\mathbf{u}+\xi\cdot\nabla\mathbf{u}]-R[\partial_tT+\xi\cdot\nabla T] \tag{23}$$

For simplicity in Eq. (22) it is assumed that $\mathbf{a}=0$.

The EDF for sensible enthalpy is defined as

$$\hbar^{eq}=\frac{\rho[(\xi-\mathbf{u})^2+2RT]}{2(2\pi RT)^{D/2}}\exp(-\frac{(\xi-\mathbf{u})^2}{2RT}) \tag{24}$$

and there is

$$h_s=\int\hbar d\xi \tag{25}$$

The disadvantages of Chatterjee's thermal LB model are obvious: it inherits almost all shortcomings of its ancestor, He's thermal LB model [51], such as the evolving equation Eq. (22) includes complicated terms involving temporal and spatial derivatives of the macroscopic variables. As demonstrated by Chen et al. [15, 16], such complicated terms will seriously hamper the numerical stability of LBM for thermal flow simulation. Indeed Chatterjee [10] has recognized that combustion simulation is beyond the scope of his sensible-enthalpy-based thermal model. No doubt, if $\mathbf{a}\neq 0$, Eq. (22) will become more complicated and the numerical stability will become worse accordingly.

[illegible]

Choice of primitive variable. In order to remedy the deficiencies in aforementioned thermal LB models, a total-enthalpy-based model was recently proposed by Chen et al. [28], in which the authors introduced the following total enthalpy DFs:

$$h = (\frac{\xi^2}{2} + RT)f \tag{26}$$

From Eq. (26) the total enthalpy H can be defined as

$$\rho H = \rho(\frac{\mathbf{u}^2}{2} + h_s) = \int h d\xi \tag{27}$$

where $h_s = \int C_p dT$ is the sensible enthalpy and $C_p = (D+2)R/2$ is the specific heat coefficient at constant pressure.

The evolution of h can be obtained from the Boltzmann equation Eq. (2) as

$$\partial_t h + \xi \cdot \nabla h + \mathbf{a} \cdot (\nabla_\xi h - \xi f) = \Omega_h \tag{28}$$

where $\Omega_h = (\xi^2/2 + RT)\Omega_f$ is the collision operator characterizing the energy change during the particle collisions.

As mentioned above, the key point to design a kinetic model based on the total enthalpy DFs h is to specify the collision term Ω_h in Eq. (28) with sound physics. Inspired by the idea proposed by Guo et al. [46], we also can decompose Ω_h into these two parts:

$$\Omega_h = \Omega_{hs} + \Omega_m \tag{29}$$

where $\Omega_{hs} = [\frac{(\xi - \mathbf{u})^2}{2} + RT]\Omega_f$ is the sensible enthalpy part and

$$\Omega_m = \Omega_h - \Omega_{hs} = (\xi \cdot \mathbf{u} - u^2/2)\Omega_f \equiv Z\Omega_f \tag{30}$$

is the mechanical energy part and it can be approximated as

$$\Omega_m = \Omega_h - \Omega_{hs} = -\frac{Z}{\tau_f}(f - f^{eq}) \tag{31}$$

With this approximation, one can straightforwardly get

$$\Omega_{hs} = \frac{1}{\tau_h}(h - h^{eq}) - \frac{Z}{\tau_{hf}}(f - f^{eq}) \tag{32}$$

where h^{eq} is the local EDF for the total enthalpy and

$$\frac{1}{\tau_{hf}} = \frac{1}{\tau_h} - \frac{1}{\tau_f} \tag{33}$$

Consequently, the total-enthalpy-based kinetic equations read

$$\partial_t f + \xi \cdot \nabla f + \mathbf{a} \cdot \nabla_\xi f = -\frac{1}{\tau_f}(f - f^{eq}) \tag{34}$$

$$\partial_t h + \xi \cdot \nabla h + \mathbf{a} \cdot \nabla_\xi h = -\frac{1}{\tau_h}(h - h^{eq}) + \frac{Z}{\tau_{hf}}(f - f^{eq}) + f\xi \cdot \mathbf{a} \tag{35}$$

where

$$f^{eq} = \frac{\rho}{(2\pi RT)^{D/2}} \exp(-\frac{(\xi - \mathbf{u})^2}{2RT}) \tag{36}$$

$$h^{eq} = \frac{\rho(\xi^2 + 2RT)}{2(2\pi RT)^{D/2}} \exp(-\frac{(\xi - \mathbf{u})^2}{2RT}) \tag{37}$$

The fluid variables are defined as

$$\begin{pmatrix} \rho \\ \rho \mathbf{u} \\ \rho H \end{pmatrix} = \begin{pmatrix} \int f d\xi \\ \int \xi f d\xi \\ \int h d\xi \end{pmatrix} \tag{38}$$

Through the standard Chapman-Enskog expansion [85], we can obtain the following hydrodynamic equations at the Navier-Stokes level [77]:

$$\partial_t \rho + \nabla \cdot (\rho \mathbf{u}) = 0 \tag{39}$$

$$\partial_t (\rho \mathbf{u}) + \nabla \cdot (\rho \mathbf{u}\mathbf{u}) = -\nabla p + \nabla \cdot \tau + \rho \mathbf{a} \tag{40}$$

$$\partial_t (\rho H) + \nabla \cdot [(p + \rho H)\mathbf{u}] = \nabla \cdot (\lambda \nabla T) + \nabla \cdot (\tau \cdot \mathbf{u}) + \rho \mathbf{u} \cdot \mathbf{a} \tag{41}$$

where

$$\tau = \mu[\mathbf{S} - 2/D(\nabla \cdot \mathbf{u})\mathbf{I}], \qquad \text{where} \quad S_{\alpha\beta} = \partial_\alpha u_\beta + \partial_\beta u_\alpha$$

is the viscous stress tensor, and the viscosity and thermal conductivity are given by

$$\mu = \tau_f p, \lambda = C_p \tau_h p.$$

The Prandtl number of the system, $Pr = \mu C_p/\lambda = \tau_f/\tau_h$, can be made arbitrary by tuning the two relaxation times. It is obvious that there is not any complicated term involving temporal and spatial derivatives of the macroscopic variables either in the evolving equations or in the calculation of the temperature, so the potential inducement leading to numerical instability and additional unexpected errors in Chatterjee's enthalpy-based model [10] is avoided.

Expansion of EDFs. For isothermal flows, both the Hermite and the Taylor expansions can give the same results. For thermal flows, however, the two methods will result in different formulations. It prefers to use the former method because the expansion coefficients obtained in this way are just the velocity moments of the DFs, and the truncation of higher-order terms do not directly alter the lower-order moments of the DFs [46]. Consequently, the Hermite expansion proposed by [46] is employed to expand f^{eq} and h^{eq} in Eqs. (36)–(37) as

$$f^{eq} = \omega(\xi, T) \sum_n \frac{\mathbf{A}_\alpha^{(n)}(\mathbf{x}, t)}{n!} H_\alpha^{(n)}(\hat{\xi}) \tag{42}$$

$$h^{eq} = \omega(\xi, T) \sum_n \frac{\mathbf{B}_\alpha^{(n)}(\mathbf{x}, t)}{n!} H_\alpha^{(n)}(\hat{\xi}) \tag{43}$$

where

$$\omega(\xi, T) = \frac{1}{(2\pi RT)^{D/2}} \exp(-\frac{\xi^2}{2RT})$$

and $\hat{\xi} = \xi/\sqrt{RT}$; $H_\alpha^{(n)}$ are the nth-order tensor Hermite polynomials. The expansion coefficients $\mathbf{A}_\alpha^{(n)}$ and $\mathbf{B}_\alpha^{(n)}$ are given by

$$\mathbf{A}_\alpha^{(n)} = \int f^{eq} H_\alpha^{(n)}(\hat{\xi}) d\xi$$

$$\mathbf{B}_\alpha^{(n)} = \int h^{eq} H_\alpha^{(n)}(\hat{\xi}) d\xi$$

As demonstrated in [46], the derivation of the hydrodynamic equations at the Navier-Stokes level from the kinetic model proposed in Eqs. (34)–(35) requires the zeroth- through third-order moments of f^{eq} and zeroth-through second-order moments of h^{eq}. Therefore, in order to obtain the same equations at the Navier-Stokes level, it is necessary to keep the terms up to third order in the Hermite expansion of f^{eq}, and to second order in the expansion of h^{eq}. With these coefficients, the truncated Hermite expansions of f^{eq} and h^{eq} can be written as

$$f^{eq,3} = \omega(\xi, T)\rho\{1 + \frac{\xi \cdot \mathbf{u}}{RT} + \frac{1}{2}(\frac{\xi \cdot \mathbf{u}}{RT})^2 - \frac{\mathbf{u}^2}{2RT} + \frac{\xi \cdot \mathbf{u}}{6RT}[(\frac{\xi \cdot \mathbf{u}}{RT})^2 - \frac{3\mathbf{u}^2}{RT}]\}$$

$$h^{eq,2} = \omega(\xi, T))\{\rho H + (p + \rho H)\frac{\xi \cdot \mathbf{u}}{RT} + \frac{p}{2}(\frac{\xi^2}{RT} - D) + (p + \frac{\rho H}{2})[(\frac{\xi \cdot \mathbf{u}}{RT})^2 - \frac{\mathbf{u}^2}{RT}]\}$$

As Guo et al. [46] pointed out that for low Mach flows, the third-order term in $f^{eq,3}$ can be neglected, and we can use the truncated expansions of f^{eq} and h^{eq} up to the second order, i.e.,

$$f^{eq,2} = \omega(\xi,T)\rho[1+\frac{\xi\cdot\mathbf{u}}{RT}+\frac{1}{2}(\frac{\xi\cdot\mathbf{u}}{RT})^2-\frac{\mathbf{u}^2}{2RT}] \tag{44}$$

$$h^{eq,2} = \omega(\xi,T))p[\frac{\xi\cdot\mathbf{u}}{RT}+(\frac{\xi\cdot\mathbf{u}}{RT})^2-\frac{\mathbf{u}^2}{RT}+\frac{1}{2}(\frac{\xi^2}{RT}-D)]+Hf^{eq,2} \tag{45}$$

Accordingly, the terms associated with the external force $\mathbf{a}$ in the kinetic model, $\mathbf{a}\cdot\nabla_\xi f$ and $\mathbf{a}\cdot\nabla_\xi h$, should also be projected on to the Hermite basis. Guo et al. [46] indicated that, in order to obtain the exact Navier-Stokes equations, it is adequate to truncate the Hermite expansions of the two terms up to the second order and first order, respectively. With this in mind, after some standard manipulations we obtain

$$\mathbf{a}\cdot\nabla_\xi f = -\omega(\xi,T)\rho[\frac{\xi\cdot\mathbf{a}}{RT}+\frac{(\xi\cdot\mathbf{u})(\xi\cdot\mathbf{a})}{(RT)^2}-\frac{\mathbf{u}\cdot\mathbf{a}}{RT}] \tag{46}$$

$$\mathbf{a}\cdot\nabla_\xi h = -\omega(\xi,T)\rho H\frac{\xi\cdot\mathbf{a}}{RT} \tag{47}$$

Discretization of velocity phase. The discrete velocity phase can be obtained by choosing the abscissae of a suitable Gauss-Hermite quadrature with the weight function $\omega(\xi,T)$ so that the required velocity moments of the truncated EDFs can be exactly evaluated. It is well-known that in the DDF framework the main challenge in this step resulting from the inconsistency between the continuous particle velocity ξ in the continuous kinetic equations and the discrete particle velocity ξ_i in the corresponding discrete velocity model since ξ is independent of $\mathbf{x}$, t and T while ξ_i not [46].

In order to overcome this difficulty, He et al. [51] designed a method in which the local temperature T in the truncated EDFs was replaced with a reference temperature T_0. With this replacement, Eqs. (44)–(47) become

$$f^{eq,2}(T_0) = \omega(\xi,T_0)\rho[1+\frac{\xi\cdot\mathbf{u}}{RT_0}+\frac{1}{2}(\frac{\xi\cdot\mathbf{u}}{RT_0})^2-\frac{\mathbf{u}^2}{2RT_0}] \tag{48}$$

$$h^{eq,2}(T_0) = \omega(\xi,T_0)p_0[\frac{\xi\cdot\mathbf{u}}{RT_0}+(\frac{\xi\cdot\mathbf{u}}{RT_0})^2-\frac{\mathbf{u}^2}{RT_0}+\frac{1}{2}(\frac{\xi^2}{RT_0}-D)]+Hf^{eq,2}(T_0) \tag{49}$$

where

$$\omega(\xi,T_0) = \frac{1}{(2\pi RT_0)^{D/2}}\exp(-\frac{\xi^2}{2RT_0})$$

and $p_0 = \rho RT_0$. Accordingly, the local temperature T appearing in the forcing terms is also replaced with T_0

$$\mathbf{a}\cdot\nabla_\xi f = -\omega(\xi,T_0)\rho[\frac{\xi\cdot\mathbf{a}}{RT_0}+\frac{(\xi\cdot\mathbf{u})(\xi\cdot\mathbf{a})}{(RT_0)^2}-\frac{\mathbf{u}\cdot\mathbf{a}}{RT_0}] \tag{50}$$

$$\mathbf{a}\cdot\nabla_\xi h = -\omega(\xi,T_0)\rho H\frac{\xi\cdot\mathbf{a}}{RT_0} \tag{51}$$

But notice that the temperature appearing in the total enthalpy H is still the local value: $H = C_pT + u^2/2$. Similar with that in [46], it is easy to verify that the zeroth- and first-order moments of $f^{eq,2}(T_0)$ and the zeroth-order moment of $h^{eq,2}(T_0)$ are the same as those of the EDFs with the local temperature T given by Eq. (21) but the higher moments required in the derivation of the Navier-Stokes equations, however, are different because of the replacement of T with T_0

$$\int \xi_\alpha\xi_\beta f^{eq,2}(T_0)d\xi = \rho u_\alpha u_\beta + p_0\delta_{\alpha\beta}$$

$$\int \xi_\alpha\xi_\beta\xi_\gamma f^{eq,2}(T_0)d\xi = p_0(u_\alpha\delta_{\beta\gamma}+u_\beta\delta_{\alpha\gamma}+u_\gamma\delta_{\alpha\beta})$$

$$\int \xi_\alpha h^{eq,2}(T_0)d\xi = (p_0+\rho H)u_\alpha$$

$$\int \xi_\alpha\xi_\beta h^{eq,2}(T_0)d\xi = p_0(RT_0+H)\delta_{\alpha\beta}+(2p_0+\rho H)u_{\alpha\beta}$$

where δ is the Kronecker delta with two indices.
Now the integral of a function of ξ, say ψ, can be evaluated as

$$\int \omega(\xi, T_0)\psi(\xi)d\xi = \sum_i w_i \psi(\mathbf{c}_i) \tag{52}$$

where $\mathbf{c}_i$ is the abscissa and w_i is the quadrature weight. Therefore, if we define

$$f_i = \frac{w_i f(\mathbf{x}, \mathbf{c}_i, t)}{\omega(\mathbf{c}_i, \quad T_0)}, h_i = \frac{w_i h(\mathbf{x}, \mathbf{c}_i, t)}{\omega(\mathbf{c}_i, T_0)}$$

we can evaluate the integrals in Eq. (38) using the quadrature and determine the fluid variables as

$$\rho = \sum_i f_i, \rho\mathbf{u} = \sum_i \mathbf{c}_i f_i, \rho H = \sum_i h_i \tag{53}$$

The evolution equations for the reduced distribution functions f_i and h_i can be easily derived from the kinetic equations (14)-(15) for f and h, which lead to the following discrete velocity model:

$$\partial_t f_i + \mathbf{c}_i \cdot \nabla f_i = -\frac{1}{\tau_f}(f_i - f_i^{eq}) + F_i \tag{54}$$

$$\partial_t h_i + \mathbf{c}_i \cdot \nabla h_i = -\frac{1}{\tau_h}(h_i - h_i^{eq}) + \frac{Z_i}{\tau_{hf}}(f_i - f_i^{eq}) + \chi_i \tag{55}$$

where $Z_i = \mathbf{c}_i \cdot \mathbf{u} - u^2/2$ and

$$f_i^{eq} = w_i \rho[1 + \frac{\mathbf{c}_i \cdot \mathbf{u}}{RT_0} + \frac{1}{2}(\frac{\mathbf{c}_i \cdot \mathbf{u}}{RT_0})^2 - \frac{\mathbf{u}^2}{2RT_0}] \tag{56}$$

$$h_i^{eq} = w_i p_0[\frac{\mathbf{c}_i \cdot \mathbf{u}}{RT_0} + (\frac{\mathbf{c}_i \cdot \mathbf{u}}{RT_0})^2 - \frac{\mathbf{u}^2}{RT_0} + \frac{1}{2}(\frac{\mathbf{c}_i^2}{RT_0} - D)] + H f_i^{eq} \tag{57}$$

F_i and χ_i are two terms related to the external force:

$$F_i = w_i \rho[\frac{\mathbf{c}_i \cdot \mathbf{a}}{RT_0} + \frac{1}{2}\frac{(\mathbf{c}_i \cdot \mathbf{a})(\mathbf{c}_i \cdot \mathbf{u})}{(RT_0)^2} - \frac{\mathbf{u} \cdot \mathbf{a}}{RT_0}] \tag{58}$$

$$\chi_i = w_i \rho H \frac{\mathbf{c}_i \cdot \mathbf{a}}{RT_0} + f_i \mathbf{c}_i \cdot \mathbf{a} \tag{59}$$

Through the standard Chapman-Enskog analysis, the thermohydrodynamic equations corresponding to the discrete velocity model Eq. (28) can be derived at the Navier-Stokes level as:

$$\partial_t \rho + \nabla \cdot (\rho \mathbf{u}) = 0 \tag{60}$$

$$\partial_t(\rho \mathbf{u}) + \nabla \cdot (\rho \mathbf{u}\mathbf{u}) = -\nabla p_0 + \nabla \cdot \tau + \rho \mathbf{a} \tag{61}$$

$$\partial_t(\rho H) + \nabla \cdot [(p_0 + \rho H)\mathbf{u}] = \nabla \cdot (\lambda \nabla T) + \nabla \cdot (\tau \cdot \mathbf{u}) + \rho \mathbf{u} \cdot \mathbf{a} \tag{62}$$

where $p_0 = \rho RT_0$, $\tau = \mu \mathbf{S}$, $\mu = \tau_f p_0$ and $\lambda = C_p \tau_h p_0$.
It is obvious in the derived Eq. (64), the expression of Pr is consistent to that in Eq. (41), namely the Prandtl number depends only on τ_f and τ_h. However, in Guo's model [46], besides τ_f and τ_h, Pr is also determined by the specific heat ratio $\gamma = C_p/C_v$ because in Guo's model the thermal conductivity is given by $\lambda = C_v \tau_h p_0$ instead of $C_p \tau_h p_0$ in its corresponding kinetic equation, which is not sound in physics. Another potential damage induced by this inconsistency is that for low Pr fluid the disparity between τ_f and τ_h will be enhanced in Guo's model because in it the effective Prandtl number is $Pr_e = Pr/\gamma$, which will cause unexpected numerical instability. These two significant shortcomings in Guo's model [46] are straightforwardly remedied by the total-enthalpy-based model [28].
Time discretization. Because in the LB framework the lattice is closely dependent on the discrete velocity set, therefore the spatial discretization is coupled with the velocity discretization [85] and has been completed in the above paragraphs. The last step to construct a thermal LB model is how to discretize time.

Similar with that in [46], in order to guarantee the accuracy of the thermal LB model, the time discretization for Eq. (54) is made by integrating the equation along the characteristic line and the trapezoidal rule [51] is used to integrate the right-hand side of Eq. (54):

$$f_i(\mathbf{x}+\mathbf{c}_i\delta t,t+\delta t)-f_i(\mathbf{x},t)=\frac{\delta t}{2}[\Omega_f(\mathbf{x}+\mathbf{c}_i\delta t,t+\delta t)+F_i(\mathbf{x}+\mathbf{c}_i\delta t,t+\delta t)]+\frac{\delta t}{2}[\Omega_f(\mathbf{x},t)+F_i(\mathbf{x},t)] \quad (63)$$

where δt is the time step and $\Omega_f=-(f_i-f_i^{eq})/\tau_f$.

As demonstrated by He et al. [51], the implicitness of the above scheme can be eliminated by introducing the following DFs:

$$\overline{f}_i=f_i-\frac{\delta t}{2}(\Omega_f+F_i) \quad (64)$$

so Eq. (63) becomes

$$\overline{f}_i(\mathbf{x}+\mathbf{c}_i\delta t,t+\delta t)-\overline{f}_i(\mathbf{x},t)=-\omega_f[\overline{f}_i(\mathbf{x},t)-f_i^{eq}(\mathbf{x},t)]+\delta t(1-\frac{\omega_f}{2})F_i \quad (65)$$

where $\omega_f=2\delta t/(2\tau_f+\delta t)$ and

$$\rho=\sum_i\overline{f}_i,\rho\mathbf{u}=\sum_i\mathbf{c}_i\overline{f}_i+\frac{\delta t}{2}\rho\mathbf{a} \quad (66)$$

Similarly, the final formulation of the time-discrete scheme for Eq. (55) can be written as

$$\overline{h}_i(\mathbf{x}+\mathbf{c}_i\delta t,t+\delta t)-\overline{h}_i(\mathbf{x},t)$$
$$=-\omega_h[\overline{h}_i(\mathbf{x},t)-h_i^{eq}(\mathbf{x},t)]+\delta t(1-\frac{\omega_h}{2})\chi_i+(\omega_h-\omega_f)Z_i[\overline{f}_i(\mathbf{x},t)-f_i^{eq}(\mathbf{x},t)+\frac{\delta t}{2}F_i] \quad (67)$$

where $\omega_h=2\delta t/(2\tau_h+\delta t)$ and

$$\overline{h}_i=h_i-\frac{\delta t}{2}\left[\frac{h_i^{eq}-h_i}{\tau_h}+\frac{Z_i\Omega_f}{\tau_{hf}}+\chi_i\right].$$

The total enthalpy is determined by

$$\rho H=\sum_i\overline{h}_i+\frac{\delta t}{2}\rho\mathbf{u}\cdot\mathbf{a} \quad (68)$$

The planar thermal Poiseuille flow driven by a constant force a has been simulated by the total-enthalpy-based thermal LB model [28]. If the gravity is neglected, the analytical velocity and the temperature profiles of planar thermal Poiseuille flow can be described as [46]

$$u(y)=4u_0\frac{y}{h}(1-\frac{y}{h}) \quad (69)$$

$$T^*=\frac{y}{h}+\frac{Pr}{Ec}[1-(1-2\frac{y}{h})^4] \quad (70)$$

where h is the channel height, $u_0=\rho_0ah^2/8\mu$, $T^*=(T-T_l)/(T_u-T_l)$ and the Eckert number

$$Ec=\frac{u_0^2}{C_p(T_u-T_l)}.$$

T_l and T_u are the temperature of the bottom and top walls of the channel, respectively. The Reynolds number in the thermal Poiseuille flow is defined as $Re=\rho_0hu_0/\mu$.

Figure **1** illustrates the temperature profiles for $Pr=\mu C_p/\lambda=0.71$ as Ec varies from 0.1 to 100 and $Re=20$ while Fig. **2** shows that for a fixed Eckert number $Ec=10$ as Pr varies from 0.1 to 4. In these simulations

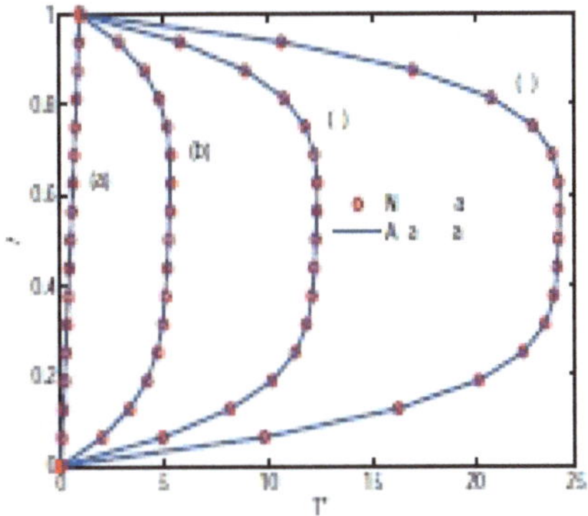

Fig. 1: Temperature variation T^* of the thermal Poiseuille flow at $Re = 20$ and $Pr = 0.71$. (a)–(d) $Ec = 0.1, 20, 50,$ and 100. Solid lines are the analytical solutions, and the symbols are the numerical results.

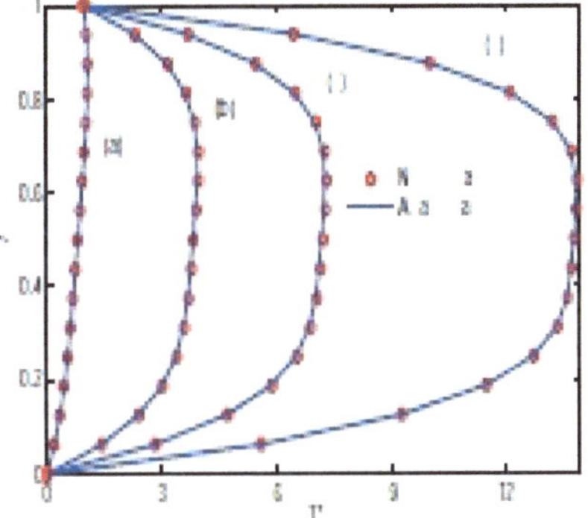

Fig. **2**: Temperature variation T^* of the thermal Poiseuille flow at $Re = 20$ and $Ec = 10$. (a)–(d) $Pr = 0.1, 1.0, 2.0,$ and 4.0. Solid lines are the analytical solutions, and the symbols are the numerical results.

the relaxation parameter $\omega_f = 0.8$ and the global Mach number $u_0/\sqrt{3RT_0}$ is about 0.008. It is clearly seen that the numerical results are in excellent agreement with the analytical solutions. The viscous heat effects are successfully captured by the total-enthalpy-based thermal LB model over a wide range of the product of Pr and Ec.

The benefits of the total-enthalpy-based thermal LB scheme are clear since it keeps the advantages in the previous models and remedies their shortcomings at the same time. The summary is listed in Table **1**, where "Applicability" means "Convenience to treat any additional energy source term"; "Stability" denotes "Without any complicated gradient term"; "Reliability" and "Consistency" represent "Avoiding arbitrary regrouping technique" and "Physically sound expressions of Pr and λ", respectively. As will be shown in Section 2.3.2, besides above benefits, there is another significant gain due to the introduction of total enthalpy: The non-Dirichlet-type heat boundaries, which are challenges in the LB community, can be addressed straightforwardly in the framework of total-enthalpy-based thermal LB scheme.

[illegible]

The simulation of combustion phenomena is one of the most challenging frontiers of computational physics / engineering, and the LBM is no exception. The key to construct an applicable LB combustion model is how to treat the stiffness of the governing equations for combustion.

[illegible]

Generally, there are two types of chemical reactions in combustion: heterogeneous and homogeneous reactions. For practical combustion systems, usually the heterogeneous reactions occur on the surfaces of solid/liquid fuels while homogeneous ones take place in gaseous combustible mixture. Kang et al. [58] perhaps are the pioneers to extend the LBM to heterogeneous reactions investigation. Later, Verhaeghe et al. [92, 93] and Walsh et al. [94] continued the discussion on the feasibility of LBM for heterogeneous reactions. Their results demonstrated the obvious advantages of LBM over traditional CFD on capturing the fine structures caused by heterogeneous reactions, competing with diffusion mechanisms. But all of them are limited in mild heterogeneous reactions. Namely the heat release due to chemical reactions is assumed so low that the flow field is not affected by them. It is obvious that such assumption can not hold water in practical combustion systems. The difficulty lies on the numerical instability induced by the significant disparity of time scales between flow and chemistry under intensive exothermic reaction (combustion) conditions. Simulation for homogeneous reactions in LB framework also meets the same puzzle but it provides a much simpler scenario for developing advanced combustion models. No doubt, the existing efforts to design robust combustion LB models all take homogeneous reactions as benchmark tests to validate their design.

According to whether the influence of combustion on fluid flow being considered, the existing combustion LB models may fall into two categories: non-coupled and coupled models. In the former, the influence of combustion on fluid flow is neglected. In succession, the latter was proposed to remedy this drawback.

[illegible]

The first non-coupled combustion LB model (NCLB) was proposed by Succi et al. [84] who are the pioneers to adopt the LBM for combustion research. In the limiting case of irreversible infinite fast chemistry reactions and "cold" flames with weak heat release the authors have described the methane-air reactive coflow

Table **1**: Comparison between several representative thermal LB models

	[51]	[10]	[46]	[28]
Applicability		√		√
Stability			√	√
Reliability			√	√
Consistency	√	√		√

dynamics in two dimensions with the 24 speeds FCHC model including two passive scalars as mixture fraction of the fuel and temperature. Although their model can only be used to simulate diffusion flame where the reaction front is infinite thin, it possesses a outstanding advantage: conserved scalars which are important in combustion simulation can be obtained straightforwardly from its extra freedoms, without obviously additional computational cost. More important, this publication [84] laid a milestone in the development of the LBM since the authors pointed out the future direction of the LBM in combustion research. The subsequent progress in this field all can be thought stemming from the vision presented in it indeed. Later, Yamamoto et al. [96] developed another non-coupled LB model to simulate counterflow propane-air flame. In fact, compared with Succi's combustion model [84], except that combustion with finite-rate reactions can be modelled, there is no significant advance in Yamamoto's LB model [96]. Then Yamamoto and his cooperators successively simulated turbulent combustion [97] and combustion in 3D porous media [98] with the aid of their non-coupled combustion LB model. However, as the authors showed in their work [97, 98], due to the flow field being compelled to be insensitive to variable density by combustion, it is not surprising that those numerical results will deviate from the real physical phenomena. The latest progress on non-coupled combustion LB model was conducted by Karlin's group [29]. But the emphasis of their work [29] is to show how to reduce the stiffness caused by disparate time scales of detailed chemical kinetics by invoking the method of invariant grids and Yamamoto's non-coupled combustion model was adopted in their publication. In this chapter, no further discussion would be made on the non-coupled combustion LB models due to their obvious shortcomings in combustion research.

[illegible]

For non-coupled combustion LB models, the fully compressible Navier-Stokes equations are their targeted governing equations. However, because only the pressure fluctuation instead of the "real" thermodynamic pressure can be identified by the popularly used LB models [53], it is hard, perhaps impossible, to construct a stably coupled combustion LB model to match the fully compressible Navier-Stokes equations. If the influence of thermodynamic pressure could be excluded (or neglected) in the macroscopic momentum equation, it would become possible to implement a robust combustion model in the framework of traditional LBM. Consequently, the question appears: does there exist such scenario where the variation of thermodynamic pressure of reactive flow could be negligible?

Fortunately, for most practical combustion systems in energy industry, because flame speeds are small compared to the sound speed (typical values for flame speeds range from 0.1 to 5 m/s while the sound speed varies between 300 and 600 m/s in most combustion chambers), accordingly, the so-called low Mach number approximation of Navier-Stokes equations (LMNA) can be derived from the full set of conservation equations for mass, momentum, energy and mass fractions of species by expanding the variables in a series of ζ (where $\zeta = \vartheta M^2$, ϑ is the ratio of specific heats and M is the Mach number) and neglecting terms of second order in ζ compared to dominant terms [77]. If the following assumptions are further made:

(1) There are no external forces.

(2) The diffusion obey the Fick's law of diffusion.

(3) Viscous energy dissipation and radiative heat loss are neglected.

Then when the Mach number approaches zero, for a mixture of N species, the resulting equations read

$$\partial_t \rho + \nabla_\alpha \rho u_\alpha = 0, \tag{71}$$

$$\partial_t \rho u_\alpha + \nabla_\beta \rho u_\alpha u_\beta = -\nabla_\alpha p_1 + \nabla_\beta \mu (\nabla_\alpha u_\beta + \nabla_\beta u_\alpha), \tag{72}$$

$$\rho C_p (\partial_t T + u_\alpha \nabla_\alpha T) = \nabla_\alpha C_p \kappa \rho \nabla_\alpha T + \sum_{i=1}^{N} h_i \omega_i + \partial_t p_0, \tag{73}$$

$$\rho (\partial_t Y_i + u_\alpha \nabla_\alpha Y_i) = \nabla_\alpha D_i \rho \nabla_\alpha Y_i + \omega_i, \quad i = 1, \ldots, N-1, \tag{74}$$

$$\sum_{i=1}^{N} Y_i = 1, \qquad \rho = \frac{p_0 \overline{W}}{RT}, \tag{75}$$

where ρ, $\mathbf{u}$ and T are the density, velocity and temperature of the mixture, μ is the dynamic viscosity, κ is the coefficient of thermal diffusivity, and C_p is the constant pressure specific heat capacity of the mixture.

Y_i and D_i are the mass fraction and diffusivity of the ith species in the mixture, and ω_i and h_i are the rate of production and heat of formation of the ith species respectively. In addition, R is the universal gas constant and $\overline{W}$ is the mean molecular weight of the mixture given by

$$\overline{W} = 1 \Big/ \sum_{i=1}^{N} \frac{Y_i}{W_i}. \tag{76}$$

W_i is the molecular weight of the ith species. Note that, in Eqs. (71)–(74) the subscripts α and β represent Cartesian coordinates and the summation convention is applied to these subscripts.

The main assumption in the LMNA is that the total pressure P can split in two parts: the "thermodynamic pressure" p_0 and the "hydrodynamic pressure" p_1. All of these equations, except Eq. (72), are derived from the balance of ζ^0 terms. The leading order momentum equation reduces to $\nabla_\alpha p_0 = 0$; therefore, p_0 can only be a function of time. In an open system, where the pressure has to approach a constant value at infinity (i.e., atmospheric pressure P_0) the last term in the temperature equation (73) vanishes. All quantities that appear in these equations, are the leading order terms ζ^0 of their corresponding ζ expansion, except for the hydrodynamic pressure $p_1(x_\alpha, t)$, which appears on the right side of the momentum equation (72) and is a first order ζ^1 quantity. Consequently the assumption that the flow is dynamically incompressible and density depends only on temperature is also reasonable. In absence of temperature gradients the low Mach number approximation of Navier-Stokes equations reduces to the system of Navier-Stokes equations for incompressible flows [81, 88]. The reaction front that can be described by the LMNA is well-known in the combustion community as "constant pressure flames" [77]. In such flames, reactive flows at low speeds are characterized by low Mach number ($M \ll 1$) but with significant changes of the density due to temperature changes by intensive exothermic reactions. The LMNA can hold water in most practical combustion processes since the acoustic influence on reaction zones is not of interest in such cases [77].

It is well-known that in the traditional LBM the "pressure" and the density are coupled closely through the state-equation-like relationship $P = c_s^2 \rho$, where c_s is the so-called "isothermal speed of sound" [78]. Consequently the pressure in the traditional LBM is determined by the density explicitly. However, as shown in Eqs. (71)–(72), it is obvious that in the LMNA the density ρ and hydrodynamic pressure p_1 are independent. Therefore, the key to design a LB combustion model for the LMNA is how to treat the density and hydrodynamic pressure independently but simultaneity through the same evolving equation. As will be shown below, the differences between the existing LB combustion models for LMNA reactive flows all result from the choices how to decouple the density and hydrodynamic pressure.

Filippova and Hänel [40, 41, 42] are the pioneers to introduce the LMNA into the LBM to simulate combustion. Because of the differences between the macroscopic target equations, Filippova and Hänel found that it was possible to construct their model basing on the pressure-like DFs developed for incompressible isothermal flows [50] instead of the standard particle density DFs [78] used in the NCLB [84, 97]. The EDF in their model reads

$$f_i^{eq} = \frac{w_i}{\rho_0}\left[\frac{\mathscr{P}}{c_s^2} + \frac{\rho u_\alpha c_{i\alpha}}{c_s^2} + \frac{\rho u_\alpha u_\beta}{2c_s^2}\left(\frac{c_{i\alpha}c_{i\beta}}{c_s^2} - \delta_{\alpha\beta}\right) + \frac{\nu u_\gamma \partial_\delta \rho}{c_s^2}\left(\frac{c_{i\gamma}c_{i\delta}}{c_s^2} - \delta_{\gamma\delta}\right)\right] \tag{77}$$

where $\nu = \mu/\rho$ is the kinematic viscosity. The last term in Eq. (77) is used to cancel non-Galilean invariant terms in the recovered macroscopic equations [41]. With the aid of such DFs, the pressure-like term $\mathscr{P}$, which consists of the hydrodynamic pressure and some correction terms, can be obtained through the zeroth moment (please bear in mind that the zeroth moment of standard particle density DFs denotes density). At the same time, in order to describe the variation of density with time, a source time $G = \frac{1}{\rho_0}[\rho(t) - \rho(t - \delta t)]$ was added to their stationary pressure-like DFs to recover the correct continuum equation, namely Eq. (71), in the LMNA. The second difference between the coupled model proposed by Filippova et al. and the NCLB is that their model is a hybrid scheme instead of the pure LBM scheme, in which the flow simulation is decoupled from the solution of the temperature/species equations. Specifically, the flow simulation is accomplished by using the LBM, while the temperature/species equations are solved by finite-difference schemes. The notable improvement in this hybrid scheme is that it can handle the variable density to respond to temperature changes. But, there still exist some shortcomings in this scheme. On one hand, the simplicity property of the pure LBM scheme has been lost [96]. On the other hand, one must pay additional special attention to handle its pressure-like term at curved walls [41]. To appropriately initialize the pressure-like DFs is also not a easy task. Moreover, the numerical stability of Filippova's combustion model [41] may be seriously hampered by the complicated derivative terms in its EDF and evolving equation [64]. In fact, this hybrid scheme [40, 41, 42] was verified only through artificial reactions with small temperature variation because the time derivative of density added into this scheme to recover the continuum equation will trigger numerical instabilities when the variation of density is not small. But, such small temperature ratio can not satisfy the actual needs of most practical combustion simulations [96].

Later, Lee and Lin [64] conducted the effort to remedy the disadvantages of the hybrid scheme proposed by Filippova et al. [41]. Similar with that by Filippova and Hänel [40, 41, 42], Lee and Lin transformed the standard particle density DFs so as to describe the variation of hydrodynamic pressure directly and their EDF reads

$$f_i^{eq} = w_i\{\mathscr{P} + \rho c_s^2[\frac{u_\alpha c_{i\alpha}}{c_s^2} + \frac{u_\alpha u_\beta}{2c_s^2}(\frac{c_{i\alpha}c_{i\beta}}{c_s^2} - \delta_{\alpha\beta})]\} \tag{78}$$

In order to cancel the pseudo-thermodynamic pressure, a spatial derivation term $\partial_\alpha(c_s^2\rho - \mathscr{P})$ was added into the corresponding evolving equation. However, the variation of density is reflected by the mixture fraction Z implicitly so their model is restricted in diffusion flame with fast chemistry. In their model, the characteristic Galerkin finite element method, which was originally developed for the discrete Boltzmann equation [65], was used to compute all unknown variables of laminar jet diffusion flames. In their way the flow and scalar fields all are solved in the LB framework and the time derivative term related to density in the evolution equation is cancelled. However, these improvements are based on the cost of additional complicated source terms including spatial derivative emerging in the evolution equations. As highlighted above, the complex source terms added into LB models would hamper the numerical stability besides decreasing the computational efficiency significantly [15, 16].

Through a comprehensive analysis, one can find that the deficiencies in the above-mentioned LB models for the LMNA (either by Filippova et al. or by Lee and his cooperator) all stem from the inappropriate construction of their DFs for hydrodynamic pressure. In fact, they all adopt the standard particle density DFs [78] as their stencil to design the DFs for hydrodynamic pressure. Consequently, the complete separation between the density and hydrodynamic pressure is impossible due to the intrinsic characteristic of the standard particle density DFs, which can be seen more clearly through their recovered continuum equation (or more accurately, the macroscopic equation for pressure). For Filippova's model, the recovered continuum equation reads

$$\frac{1}{c_s^2}\partial_t\mathscr{P} + \partial_t\rho + \nabla_\alpha\rho u_\alpha = 0, \tag{79}$$

while that of Lee and Lin's reads

$$\frac{1}{\rho c_s^2}\partial_t\mathscr{P} + \nabla_\alpha u_\alpha = -\frac{1}{\rho^2}\frac{d\rho}{dZ}\nabla_\alpha(\rho\mathscr{D}\nabla_\alpha Z) \tag{80}$$

As clearly shown in Eqs. (79)–(80), the pressure-like term $\mathscr{P}$ always accompanies the density in the recovered macroscopic equation and they are obviously different from the "true" continuum equation Eq. (71). It is well-known in the combustion community that to keep mass conservation (in other words, to solve the continuum equation correctly) is critical to guarantee numerical stability for combustion simulation. Accordingly it is straightforward to understand why they are too poor to model practical combustion. Consequently, the key is to design appropriate DFs in which the hydrodynamic pressure and density can be handled fully independently.

Before to do that, in order to make it convenient for the numerical implementation, firstly we may rewrite Eqs. (71)–(74) as their dimensionless forms [13, 14]

$$\partial_{\tilde{t}}\tilde{\rho} + \tilde{\nabla}_\alpha\tilde{\rho}\tilde{u}_\alpha = 0, \tag{81}$$

$$\partial_{\tilde{t}}\tilde{\rho}\tilde{u}_\alpha + \tilde{\nabla}_\beta\tilde{\rho}\tilde{u}_\alpha\tilde{u}_\beta = -\tilde{\nabla}_\alpha\tilde{P} + \frac{1}{Re}\tilde{\nabla}_\beta\tilde{\mu}(\tilde{\nabla}_\alpha\tilde{u}_\beta + \tilde{\nabla}_\beta\tilde{u}_\alpha), \tag{82}$$

$$\partial_{\tilde{t}}\widetilde{T}+\widetilde{u}_\alpha\widetilde{\nabla}_\alpha\widetilde{T}=\frac{1}{\widetilde{\rho}RePr}\widetilde{\nabla}_\alpha\widetilde{\mu}\widetilde{\nabla}_\alpha\widetilde{T}+\sum_{i=1}^{N}\widetilde{h}_i\widetilde{\omega}_i, \tag{83}$$

$$\partial_{\tilde{t}}\widetilde{Y}_i+\widetilde{u}_\alpha\widetilde{\nabla}_\alpha\widetilde{Y}_i=\frac{1}{\widetilde{\rho}ReSc}\widetilde{\nabla}_\alpha\widetilde{\mu}\widetilde{\nabla}_\alpha\widetilde{Y}_i+\widetilde{\omega}_i, \tag{84}$$

where $Re=\rho_0u_0L_0/\mu_0$, $Pr=\mu/(\rho\kappa)$ and $Sc=\mu/(\rho D_i)$ are the Reynolds, Prandtl and Schmidt numbers respectively. All variables with tildes are nondimensional, with the reference values of density ρ_0, velocity u_0, temperature T_0, length L_0 and dynamic viscosity μ_0 . One perhaps notices that in Eq. (82) the hydrodynamic pressure is replaced with the total pressure through the relationship $\nabla_\alpha p_0=0$ mentioned above [13, 14]. This transformation is very useful because it is the total pressure instead of the hydrodynamic pressure that can be measured in practical applications. The hydrodynamic pressure just acts like a parameter to satisfy the continuum equation.

The prospect of applying the LBM to combustion simulation was not entirely clear until the novel coupled LB model for the LMNA designed by Chen et al. [13, 14] appeared. In order to overcome the above shortcomings, a DDF-based combustion model was constructed in [13, 14] where the evolution equation for the flow field is described by

$$g_k(\mathbf{x}+c\mathbf{e}_k\Delta t,t+\Delta t)-g_k(\mathbf{x},t)=-\tau_u^{-1}[g_k(\mathbf{x},t)-g_k^{(eq)}(\mathbf{x},t)]. \tag{85}$$

where $\mathbf{e}_k$ is the discrete velocity direction, which depends on the lattice model. $k=0$ represents the rest fluid particle. $c=\Delta x/\Delta t$ is the fluid particle speed. Δx, Δt and τ_u are the lattice grid spacing, the time step and the dimensionless relaxation time for the flow field respectively. $g_k(\mathbf{x},t)$ is the distribution function at node x and time t with velocity $\mathbf{e}_k$, and $g_k^{(eq)}(\mathbf{x},t)$ is the corresponding equilibrium distribution depending on the lattice model used.

The EDF in Chen's model is defined by

$$g_k^{(eq)}=\chi_kP+\rho s_k(u_\alpha). \tag{86}$$

where χ_k is the parameter defined below; ζ_k are the weight coefficients such that $\zeta_k>0$, $\sum_k\zeta_k=1$; and

$$s_k(u_\alpha)=\zeta_k[\frac{ce_{k\alpha}u_\alpha}{c_s^2}+\frac{(ce_{k\alpha}u_\alpha)(ce_{k\beta}u_\beta)}{2c_s^4}-\frac{u_\alpha u_\alpha}{2c_s^2}],$$

where the parameter c_s satisfies

$$c_s^2\delta_{\alpha\beta}=\sum_k\zeta_kc^2e_{k\alpha}e_{k\beta}. \tag{87}$$

They also depend on the chosen lattice model. Through Eqs. (85)–(87), it is obvious that there is not any complicated source term in Chen's model no matter which lattice model will be adopted.

For the general form of this model, χ_k and $g_k^{(eq)}$ can be determined by the following constraints:

$$\sum_k g_k^{(eq)}=\rho,\quad \sum_k ce_{k\alpha}g_k^{(eq)}=\rho u_\alpha,\quad \sum_k c^2e_{k\alpha}e_{k\beta}g_k^{(eq)}=P\delta_{\alpha\beta}+\rho u_\alpha u_\beta.$$

The simplest choice is to take $\chi_k=\lambda\zeta_k$ for $k>0$, and λ is a coefficient determined by moments of the DFs. Then from the first and last constraints mentioned above one can have

$$\chi_0=\frac{\rho}{P}-\sum_{k\neq0}\chi_k,\lambda\sum_{k\neq0}c^2\zeta_ke_{k\alpha}e_{k\beta}P=\lambda c_s^2P\delta_{\alpha\beta}=P\delta_{\alpha\beta}$$

which leads to $\lambda=1/c_s^2$, therefore

$$\chi_k|_{k\neq0}=\zeta_k/c_s^2,\chi_0=\frac{\rho}{P}+\frac{\zeta_0-1}{c_s^2}. \tag{88}$$

The flow velocity, pressure and kinematic viscosity are given by

$$\rho u_\alpha = \sum_k c e_{k\alpha} g_k, \tag{89}$$

$$P = \frac{c_s^2}{1-\zeta_0}\left[\sum_{k\neq 0} g_k + \rho s_0(u_\alpha)\right] \ , \tag{90}$$

$$\nu = \frac{\mu(T)}{\rho} = (\tau_u - 1/2)c_s^2 \Delta t. \tag{91}$$

An interesting aspect of Chen's model is that the value of density of the mixture is not obtained explicitly through the evolution equation (85). In their model, at every new time step one can obtain the new value of density by

$$\rho^{n+1} = \frac{\rho^n T^n \overline{W}^{n+1}}{T^{n+1}\overline{W}^n} = \frac{\rho_0 T_0 \overline{W}^{n+1}}{T^{n+1}\overline{W}_0}. \tag{92}$$

according to the assumption of the LMNA mentioned in Eq. (75). Where $\overline{W}_0$ is the mean molecular weight of the mixture at the state ρ_0, T_0. Note that here and below the superscript n refers to the current time level and $n+1$ the next one. However, different from the model proposed by Filippova et al. [41], there is a degree of freedom in Chen's model due to its special structure (namely the parameter $\chi_k P$ in Eq. (86)), so one can recover the time derivative term $\partial_t \rho$ in Eq. (71) with the aid of this degree of freedom instead of the forward difference scheme proposed by Filippova and Hänel [41]. As mentioned above, the approach that using a forward difference scheme as an additional factor in the DFs to recover the term $\partial_t \rho$ in Eq. (71) will hamper the numerical stability because this additional factor, which serves as a source term, will make the rest fluid particle distribution negative (namely $g_{k=0} < 0$), especially at the early stage of combustion when variance of temperature is significant.

The symbol $\mu(T)$ in Eq. (91) represents the relationship between the dynamic viscosity and the temperature. There exists the approximation

$$\frac{\mu}{\mu_0} = \sqrt{\frac{T}{T_0}}. \tag{93}$$

The reference viscosity $\mu_0 = \mu(T = T_0)$ is directly related to the Reynolds number Re by

$$Re = \rho_0 u_0 L_0 / \mu_0. \tag{94}$$

Consequently with the aid of Eq. (92), Eq. (91) can be transformed into

$$\nu = \frac{\mu(T) T \overline{W}_0}{\rho_0 T_0 \overline{W}} = (\tau_u - 1/2)c_s^2 \Delta t. \tag{95}$$

Note that the relaxation time τ_u in Eq. (85) is now a field variable depending on the temperature and mean molecular weight of the mixture even if the dynamic viscosity does not depend on the temperature.

The value of ζ_k for two- and three-dimensional problems can be found in [78]. In absence of temperature gradients, Chen's model reduces to an incompressible model for isothermal flows proposed by Guo et al. [44], which agrees with the assumption of the LMNA [41]. The continuum equation Eq. (81) can be recovered correctly through Eq. (85) with the aid of the standard Chapman-Enskog expansion [13, 14] without any non-physical term like that in Eqs. (79)–(80).

The evolution equations for the temperature and concentration fields have the same form

$$f_{S,j}(\mathbf{x} + c\mathbf{e}_j \Delta t, t + \Delta t) - f_{S,j}(\mathbf{x}, t) = -\tau_S^{-1}[f_{S,j}(\mathbf{x}, t) - f_{S,j}^{(eq)}(\mathbf{x}, t)] + \zeta_j Q_S \Delta t. \tag{96}$$

The symbol S represents the concentration (mass fraction) field of the ith species or the temperature field (i.e., $S = Y_i, T$). In Eq. (96) τ_S is the dimensionless relaxation time for S and Q_S is the source term due to chemical reactions. In LB calculation, Q_S is given by the similarity in non-dimensional equations of temperature and concentration fields(i.e., $Q_{S=T} = \sum_{i=1}^{N} \widetilde{h}_i \widetilde{\omega}_i$, $Q_{S=Y_i} = \widetilde{\omega}_i$).

The corresponding EDF reads

$$f_{S,j}^{(eq)} = \zeta_j S[1 + \frac{ce_{j\alpha}u_\alpha}{c_s^2} + \frac{(ce_{j\alpha}u_\alpha)(ce_{j\beta}u_\beta)}{2c_s^4} - \frac{u_\alpha u_\alpha}{2c_s^2}]. \tag{97}$$

The meaning of e_j and ζ_j is the same as that of e_k and ζ_k mentioned above. The intent that the authors used different subscript for them is that they could employ different DdQq [78] lattice model for the flow, temperature, and concentration fields. For example, in [13], the authors simulated the flow field by a 2D 9-velocity lattice model while a 2D 4-velocity lattice model for the temperature field due to the governing equation of the temperature field is the advection-diffusion equation instead of the Navier-Stokes equations. The first advantage of this approach is it can improve the numerical efficiency of the DDF model observably. The second advantage is it can save numerous memory (the memory that 2D 4-velocity lattice model needs is less than half of that 2D 9-velocity lattice model needs). Especially, the second advantage is more attractive if detailed reaction mechanisms are adopted.
The temperature, T, and the mixture fraction of species i, Y_i, are obtained in terms of the DFs by

$$T = \sum_j f_{S=T,j}. \tag{98}$$

$$Y_i = \sum_j f_{S=Y_i,j}. \tag{99}$$

The thermal diffusivity, κ, and diffusion coefficient, D_i, are given by

$$\kappa = \frac{\nu}{Pr} = (\tau_{S=T} - 1/2)c_s^2\Delta t. \tag{100}$$

$$D_i = \frac{\nu}{Sc} = (\tau_{S=Y_i} - 1/2)c_s^2\Delta t. \tag{101}$$

The same as τ_u, τ_S $(S = Y_i, T)$ also becomes a field variable in this coupled combustion model.
Besides successfully decoupling the density and hydrodynamic pressure, another foremost characteristic of Chen's combustion model [13, 14] is that the fluid particle speed in their model can be adjusted dynamically, depending on the so-called "particle characteristic temperature", for the purpose of the numerical stability. If the fluid particle speed is a constant value during combustion simulations, the limitation $O(|\mathbf{u}|/c) \sim O(M) \ll 1$ perhaps won't be satisfied due to the increasing value of local velocity. In some existing schemes the fluid particle speed depends on the local temperature T [74]. In such schemes, the calculated DFs at the next time step may not reside on the grid nodes because the fluid particle speed on different grid nodes may have different value. A reconstruction step is required to compute the information on the grid nodes, so interpolations have to be used, but which will introduce undesirable numerical artifacts [60].
In order to avoid the reconstruction step besides to assure the numerical stability, in their work [13], a concept "particle characteristic temperature" was introduced. The authors named the temperature that determines the value of the fluid particle speed as the "particle characteristic temperature". For combustion simulation, the maximum temperature T_{max} of the mixture was chosen as the "particle characteristic temperature". The fluid particle speed c^n is allowed to respond to T_{max}^n changes, i.e.,

$$\frac{c^n}{c_0} = \sqrt{\frac{T_{max}^n}{T_0}}. \tag{102}$$

c_0 is the reference value of the fluid particle speed at the temperature T_0. Therefor at time n the value of the fluid particle speed on different grid nodes can be the same.
To guarantee the algorithm is a simple process of hopping from one grid point to the next, the same as the standard LBM, the particle speed c^n, the time step Δt^n, and the lattice grid spacing Δx must satisfy the following relationship

$$\Delta x = c^n \Delta t^n. \tag{103}$$

Because Δx is constant after the grid number has been determined, therefore now Δt^n has to vary at every iterative step, according to Eq. (103). It must be emphasized that in all existing LB schemes Δt^n cannot be adjusted dynamically, which is another major reason that hampers the numerical stability in combustion indeed. It is well known that in conventional numerical methods, Δt^n can be adjusted freely to assure numerical stabilities [77]. Although the LMNA can filter out them when the whole system approaches to the steady state, the high frequency pressure waves generated at the forepart of ignition, will hamper numerical stabilities. This phenomenon, named "overshoot", is well known in combustion simulations by direct numerical simulations (DNS). In DNS, this difficulty can be overcome easily by adjusting Δt^n. Considering the fact that the LBM starts from the kinetic theory and has been derived to conserve high-order isotropy, the LBM should be more sensitive than conventional methods in capturing pressure waves [52], Δt^n also must be adjusted reasonably. If Δt^n is not small enough, especially in the early stage, the heat of formation due to chemical reactions will be released into the mixture relatively fast, which will result in pressure P increasing sharply. That P increases sharply will immediately cause **u** increasing too quickly, then the restriction $O(|\mathbf{u}|/c) \sim O(M) \ll 1$ will break down. Consequently it is not surprising that numerical divergence appears. However, there had not been studies that try to adjust Δt^n until Chen's publication appeared [13].

Low Mach combustion described by either fast chemistry or finite-rate reaction kinetics both can be simulated straightforwardly by Chen et al. model. For the former, a two-dimensional coflow methane-air laminar diffusion flame in a plane domain, which is a benchmark test in the combustion science [77], has been simulated successfully [11]. Fig. **3** illustrates the computational domain of such case. The fuel enters the domain within an inner region centered about the midline and width $H/8$. The mixture fraction Z is adopted to simplify the calculation for the flame front with fast chemistry assumption and at the inlet $Z = 1.0$. Outlet is zero cross-flow and zero axial derivatives of the streamwise velocity. Upper and lower boundary are kept at a constant wall temperature T_0 (reference temperature $T_0 = 300K$). The inlet temperature is also T_0. The aspect ratio of the whole domain is $L/H = 4.0$, the same with the computational domain adopted by Succi et al. [84]. The *D2Q9* lattice model [78] is employed for the simulation, i.e., $e_k = (0,0)$ for $k = 0$, $e_k = (\cos[(k-1)\pi/2], \sin[(k-1)\pi/2])$ for $k = 1-4$ and $e_k = (\cos[(k-5)\pi/2] + \pi/4, \sin[(k-5)\pi/2 + \pi/4])$ for $k = 5-8$, respectively. The weight coefficients are $\zeta_0 = 4/9$, $\zeta_k = 1/9$ for $k = 1-4$ and $\zeta_k = 1/36$ for $k = 5-8$. Consequently the relationship between the parameter c_s and the particle speed c is $c_s^2 = c^2/3$. The burned plane in the plot of Z versus mass fraction of species and temperature is shown in Fig. **4**, where the following single step global reaction is considered:

$$CH_4 + 2O_2 \rightarrow CO_2 + 2H_2O$$

The stoichiometric ratio $Z_{st} = 0.054$ indicates the flame position in the Z-space. Figs. **5–6** illustrate the temperature and mixture fraction distributions when flame achieves its steady status. The fine structure of Burke-Schumann flame can be observed from the numerical results which agree well with the results obtained by the multigrid method [67]. The flame length is a very important quantity in this case. For a given set of physical-geometrical parameters, the flame length L_f, depends both on Z at the inlet (i.e. the degree of mixture richness) and on the inlet velocity (i.e. *Re*). In the simulation, Z at the inlet is a constant so L_f is just a function of *Re*. Quantitative comparisons of the flame length are presented in Fig. **7** together with theoretical predictions (the broken line in Fig. **7**)by Roper [80]. One can see that the numerical results obtained by Chen's model also can accurately reflect the variance tendency of flame length. The small differences perhaps result from the simplifications adopted in the theoretical predictions [80]. Together with Figs. **5–6**, it is obvious that the fine realistic flame structure that can hardly be captured by the NCLB [84] has been described clearly by Chen's combustion model.

The so-called "counter-flow" premixed propane-air flame [96], which is also a benchmark test in the combustion science [77], has been adopted to validate the performance of Chen's model for finite-rate premixed combustion. Fig. **8** illustrates the computational domain and boundary conditions of such premixed flames. Two parallel stationary walls are located at $y = \pm L_0$. The half-length of the distance between walls, L_0, is 20 $[mm]$. The combustible mixture is uniformly ejected from the top and bottom walls, and it reacts in the reaction zone. Then, the twin stagnation flames are formed in this counter flow. The burned gas flows outward along the x-direction. A heat source is placed with temperature $T = 5T_0$ to ignite the mixture and the reaction can be expressed with an over-all single step reaction [96]

$$C_3H_8 + 5O_2 \rightarrow 3CO_2 + 4H_2O$$

Figs. **9–10** show the temperature and concentration distributions at $x = 0, y \geq 0$ when the reactive flow

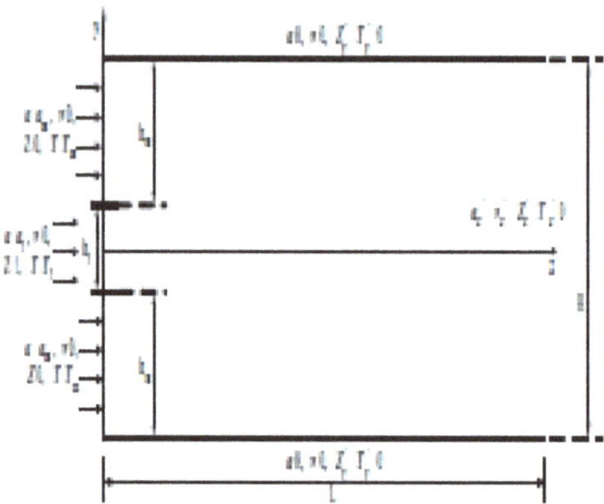

Fig. 3: Computational domain and boundary conditions of coflow methane-air laminar diffusion flame.

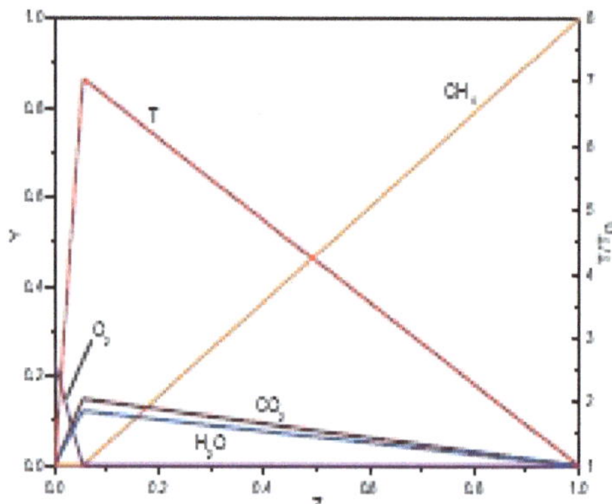

Fig. **4**: Temperature and mass fraction of species as functions of mixture fraction for coflow methane-air laminar diffusion flame.

achieves the steady state. The mass rate of production for propane ω_{N,C_3H_8} is also shown. One can see that, as the center is approached, the temperature starts to increase at $y/L \simeq 0.6$, and steeply increases at $y/L \simeq 0.2 - 0.3$. The reaction zone is located in this region, where the large heat release occurs to cause the temperature increase sharply. Then, temperature becomes constant in the burned gas region. Although there is a long preheated zone, the reactants, C_3H_8 and O_2, do not decrease obviously until the temperature rises to about 2.58 ($y/L \simeq 0.3$) (note that the ignition temperature of propane T_k is 766.48 $\sim$ 822.05 $[K]$, namely $T_k/T_0 \simeq 2.56 \sim 2.74$), then react in the reaction zone to form the products, CO_2 and H_2O. The fine structure of counter-flow flame can be clearly observed by the numerical results. In the figure Y_f is the flame stagnation position where ω_{N,C_3H_8} reaches its peak value. The differences between the concentration fields obtained by Chen's LMNA model [13] and that by Yamamoto's NCLB model [96] are not very obvious. But, there are significant differences in the flow field. Fig. **11** shows the distributions of y-direction velocity u_y of the reactive flow. One can see that in Chen's model [13] the phenomenon that the velocity is accelerated due to the flow expansion caused by the increase of temperature can be captured exactly: at the flame stagnation position Y_f where the flame propagation speed $u_f = 0$, one can find the local flow velocity u_y (where $u_x = 0$) approximates to the burning velocity S_L. It agrees very well with the relationship $S_L = u_y - u_f$ because along this line ($x = 0$) it can be approximated to a one-dimensional problem [77]. Therefore, compared with the temperature field obtained by Yamamoto's model [96], the high temperature region obtained by Chen's model is wider, with lower maximum temperature, which agrees very well with the results obtained by conventional compressible finite-difference method (FDM) [77], as Fig. **12** shows.

The oscillation of the maximum temperature T_{max} is noticeable in the beginning and gradually dissipates

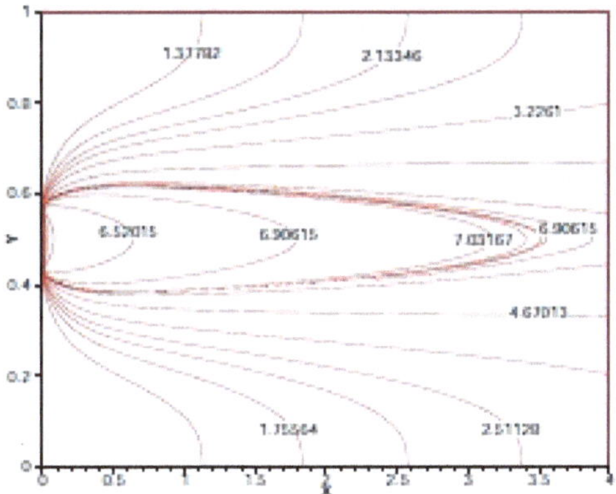

Fig. 5: Dimensionless temperature distributions of coflow methane-air laminar diffusion flame at $Re = 50$.

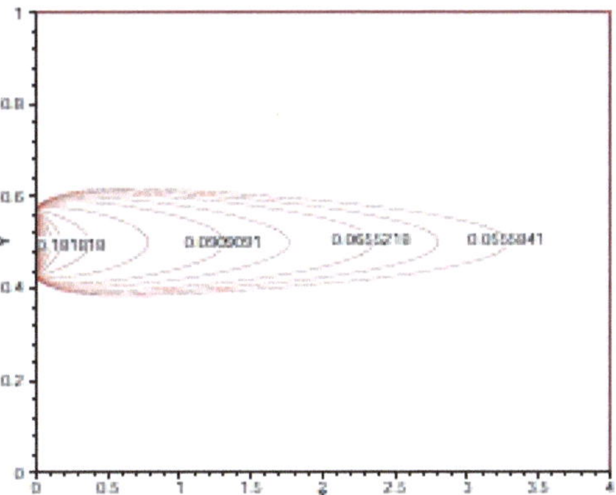

Fig. **6**: Mixture fraction distributions of coflow methane-air laminar diffusion flame at $Re = 50$.

as time t increases. The fluctuation of T_{max} is bigger than 2% before $t \approx 95$. For clarity Fig. **13** only plots oscillations of the maximum temperature in the early stage after ignition. It is very clear that the maximum temperature will sharply increase in a minute and then decrease slowly. In conventional numerical methods, in order to assure numerical stabilities, one must adjust the time step during this period to guarantee the heat of formation due to chemical reactions is released into the mixture relatively slowly, otherwise the numerical results will diverge. In Chen's model the fluid particle speed changes depending on the maximum temperature, therefore the time step will firstly decrease and then increase slowly. The variation tendency of time step in Chen's model is consistent with that in conventional numerical methods. Consequently, the numerical stability besides the numerical efficiency can be guaranteed. Another phenomenon also can be illustrated from this figure. One can see that there exists a quasi periodic behavior, which results from the traveling of waves induced by combustion. Through careful analysis, it can be found there are two different types of periods in Fig. **13**: one with the dimensionless characteristic time about 2 reflects acoustic waves bouncing back and forth between the top and bottom walls; the other represents the low frequency hydrodynamics with the dimensionless characteristic time about 16. The oscillation of the maximum temperature will last until it reaches the steady-state solutions. Similar phenomenon has also been reported by He et al. [52]. The combustion-induced waves will cause serious numerical instability and has not been solved successfully yet [11] (It will be discussed again in Section 2.3.3 as an open problem).

The burning velocity S_L and the flame temperature T_f with different equivalence ratio are very important quantities in combustion research and they are predicted in Figs. **14–15**. One can see that the numerical results obtained by Chen's model agree well with the experimental data [7], too.

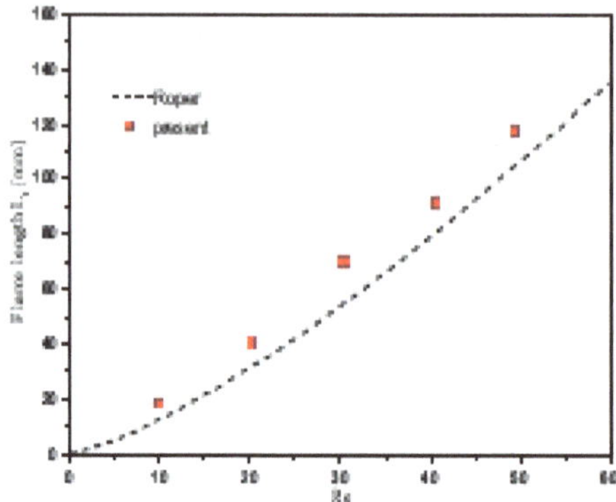

Fig. 7: Comparison of flame length of coflow methane-air laminar diffusion flame with different Reynolds numbers.

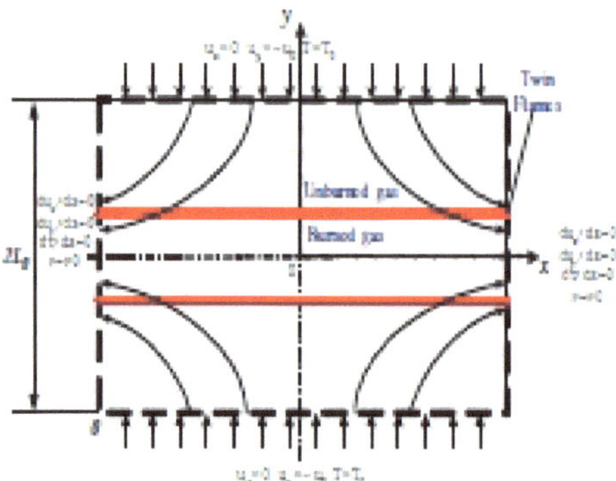

Fig. 8: Computational domain and boundary conditions of counter-flow premixed propane-air flames.

[illegible]

Initial and boundary conditions paly critical roles in the complete solution process of the LBM as well as in partial differential equations (PDE). The discussions on robust and physically sound initial and boundary strategies compose one of the most important parts in the progress of the LBM. It is well-known that for practical systems only the macroscopic quantities such as velocity and pressure are measurable while in the LBM the DFs are adopted to describe the evolving of a system. Consequently, a well-designed initial/boundary scheme in the framework of the LBM denotes a good algorithm for the mapping from the macroscopic fluid variables to the DFs. The first study on LBM-tailored initial strategy was conducted by Skordos [82]. The author tried to improve the initialization of DFs by computing the initial density by directly solving a Poisson equation to obtain the pressure and the density via the equation of state and then the nonequilibrium part of DFs by using Chapman-Enskog procedure [82]. In order to avoid the cumbersome computations in the initializing step proposed in [82], Mei et al. [72] published an iterative initialization procedure to generate consistent initial conditions for DFs along the general idea: During the initialization procedure the density is the only conserved variable in the system, while the flow momentum is relaxed to the state prescribed by the initial velocity field. The comprehensive review on boundary value problems within the framework of the LBM can be found in [85, 5, 12]. Here three important issues in this field, which have been omitted in previous review publications since obvious advancements on them are made only in the latest years, will be discussed below. These issues include: Mass conservation scheme for complicated boundary, treatment for non-Dirichlet-type heat boundary conditions and open boundary effects. In fact, the interaction between the

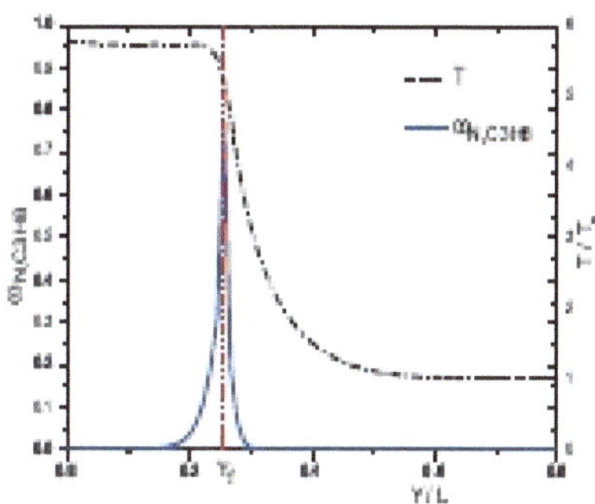

Fig. **9**: Distributions of temperature and mass rate of production for propane in counter-flow premixed propane-air flames along $x = 0, y \geq 0$.

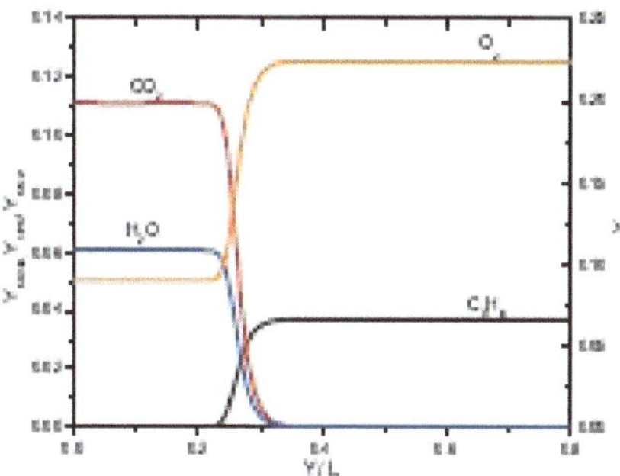

Fig. **10**: Distributions of mass fraction of species in counter-flow premixed propane-air flames along $x = 0, y \geq 0$.

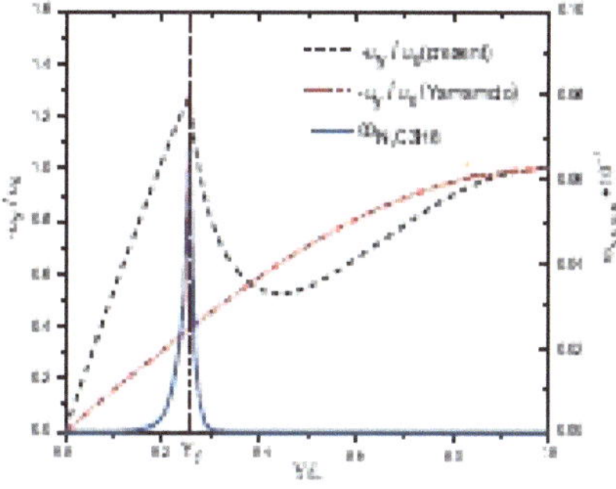

Fig. **11**: Distributions of velocity of u_y , mass rate of production for propane in counter-flow premixed propane-air flames along $x = 0, y \geq 0$.

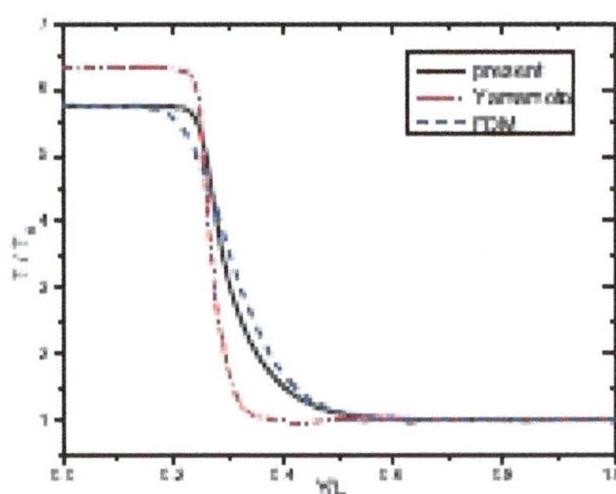

Fig. 12: Distributions of temperature in counter-flow premixed propane-air flames.

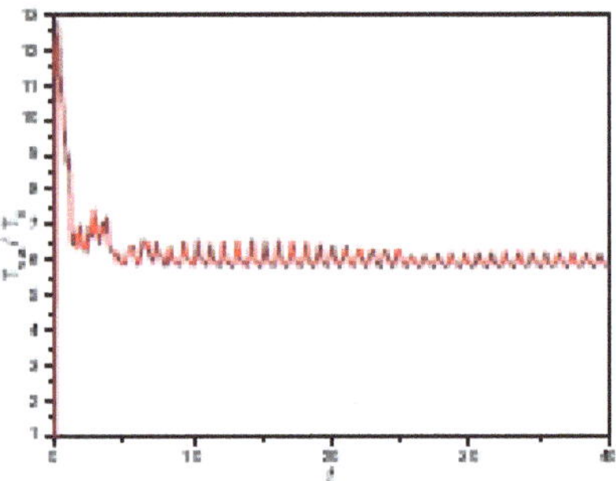

Fig. 13: Oscillations of the maximum temperature after ignition.

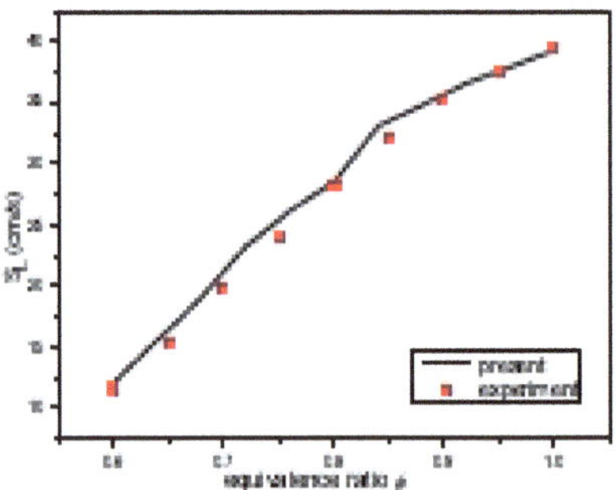

Fig. **14**: Burning velocity with different equivalence ratios in counter-flow premixed propane-air flames.

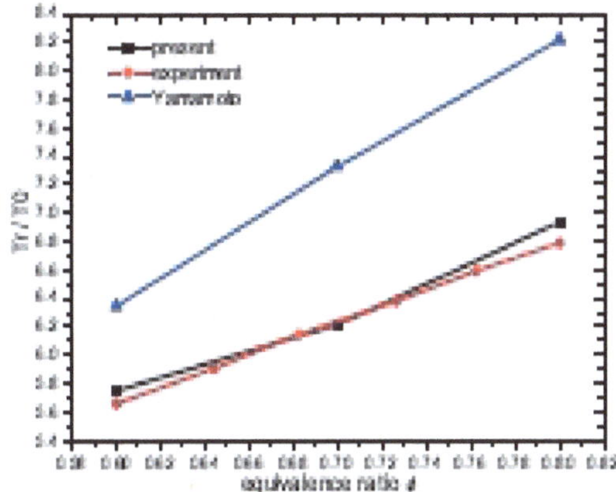

Fig. **15**: Flame temperature with different equivalence ratios in counter-flow premixed propane-air flames.

open boundaries and the fluid domain itself is still an "open" problem up to date and poses a great challenge for combustion research in the LBM community.

[illegible]

Traditionally, the LBM employs the bounce-back treatment for a wall boundary condition. The implementation of the bounce-back boundary scheme is very simple, so it is popularly used not only for macroflows but also for microflows. However, this treatment requires that the boundaries should be located midway through the last fluid node and the first outside node for second-order accuracy. Otherwise, the bounce-back treatment gives only first-order accuracy at boundaries. This constraint makes the bounce-back treatment insufficient for complex boundaries.

To overcome the limit of this boundary treatment, many modified boundary treatments have been designed for complex boundaries. The first approach is to use a body-fitted mesh and to execute the DFs throughout the entire computational domain [49, 48]. The second strategy also applies an interpolation-based approach, but under a uniform Cartesian mesh, to track the position of boundary [39, 71, 61, 99, 43, 45]. The third way is based upon the utilization of the immersed boundary treatments [38]. The fourth technique introduces unstructured grids to treat complicated geometry [90, 89, 76]. The final and latest one is the so-called "interpolation-free" schemes [59, 63]. Due to their characteristics of a superior numerical accuracy, an intuitive approach and an inherent reliability, the interpolation type schemes are the most commonly employed technique for resolving curved boundary problems in LB simulations [1]. However, there are two obvious disadvantages in the interpolation boundary schemes published in [39, 71, 61, 99, 43, 45]: First, the bounce-back rule requires only one grid spacing between the surfaces, but the linear interpolation requires at least two grid spacings, while the quadratic interpolation and multireflection require three; Second, previous studies [61, 3] have indicated that the interpolation routines used in these schemes to solve the DFs near the curved boundary may cause a significant loss of mass conservation, which reduces the accuracy of the computed momentum transfer at the boundary and therefore results in a net mass flux. Moreover, it is well-known that mass leakage may cause seriously numerical instability. The first shortcoming restrains the intrinsic advantage of local computing of the LBM and implies that they can not be used to simulate flow in narrow gaps where only one fluid node available [30]. In order to overcome this difficulty, Chun and Ladd [30] developed an equilibrium distribution interpolated boundary scheme. In Chun's boundary scheme, the interpolation operation is executed for the equilibrium distribution instead of the full velocity distribution. The unknown non-equilibrium part of the distribution at the boundary node is approximated with the corresponding non-equilibrium part of the distribution at the nearest fluid node. Chun's boundary scheme is second-order accuracy although in which only one fluid node is required. An additional advantage of Chun's boundary scheme is that by interpolating just the equilibrium distribution, one can still tune a LB model to provide a viscosity-independent boundary rule since any viscosity dependence of the hydrodynamic boundary comes from the non-equilibrium distribution. Unfortunately, the second shortcoming of interpolation treatments still remains in Chun's boundary scheme. Although the authors claimed that in the incompressible limit the mass leakage did not affect the velocity field, this conclusion is questionable in low Mach number reacting flow [13, 11]. Furthermore, there is another obvious drawback in Chun's boundary scheme. Namely the locations of the solid surfaces are strictly required if there is only one fluid node between two solid walls (please see [30, Fig. 2(c)]). While in common scenes this requirement can hardly be satisfied, which significantly limits the applicable scope of Chun's scheme. Later, in order to remedy the second drawback, Kao and Yang [59] developed an interpolation-free scheme for curved boundary. Kao's boundary scheme treats curved boundaries using an appropriate local refinement grid technique based on the work published in [39] with a bounce-back scheme at the solid surface. The non-equilibrium part of the particle distribution directed toward the curved boundary is treated as the value from the "coarse" grid transferring into the "fine" grid in accordance with the distance between the fluid node and the solid wall. The relaxation time for the "fine" grid must be modified accordingly. It should be noted that even though Kao's scheme was proposed to eliminate errors in the mass flux input near the wall boundary, the mass non-conservation effect still inevitably exists in this scheme because the modification of the unknown DFs near the curved boundary is required based upon a grid transformation technique. Another shortcoming of Kao's scheme is that when the solid wall overlaps with the boundary node, it reduces to the standard bounce-back scheme with only first-order accuracy. In [63], the volumetric LB technique [79] was used instead of the standard pointwise LB manner. Near the curved solid wall, the authors adjusted the relaxation time of the LB equation according to the ratio between the volume occupied by fluid and the whole control volume. This boundary scheme is second-order accuracy but the problem of mass leakage at the solid wall is not solved yet. Mass conserving interpolations have been developed in [79, 91] based on the fraction of fluid in a control volume, but they are too complicated to be implemented and do not give second-order accuracy in general. Peng's group [3] investigated the mass leakage in the popularly used interpolation boundary schemes in detail. In their work, the authors convincingly demonstrated that the numerical instability of the interpolation boundary schemes mainly results from the mass non-conservation effect in them, although in common configurations this effect is quite minor. Meanwhile, Peng's group proposed a mass conserving solid wall boundary scheme [3]. The difference between Peng's scheme and that designed in [39, 71] is that Peng et al. tried to calculate the fluid density at the solid wall explicitly while in [39, 71] the authors used the density at the nearest fluid node as an approximation. In order to evaluate the fluid density at the solid wall, Peng et al. assumed that the particle distributions coming from the nearest fluid nodes were equal to the EDFs at the boundary node. It is obvious this assumption can not hold water in common scenarios. The second drawback of Peng's scheme is that at least two fluid nodes are required.

Indeed, all above mentioned boundary schemes were built from the viewpoint of fluid dynamics. Recently, Chen et al. developed a boundary scheme from another viewpoint [21], i.e. a heuristic scheme for curved boundaries inspired by the idea in surface chemistry [62]. Their new scheme is very simple and overcomes the above two disadvantages synchronously. Consequently it will be discussed in detail below.

From the viewpoint of fluid dynamics, the role of an impenetrable solid wall is just to redistribute the momentum according to the laws. However, in surface chemistry, the solid wall also serves as a mass sink [62]. In this approach the fluid molecules can be adsorbed onto the solid surface, and then desorbed after some time lag. This mechanism of the deposition of a layer with a thickness of one or more molecules onto the surface is known as adsorption in the literature of surface chemistry [62]. Based on this theory, Chen et al. designed a heuristic scheme for curved boundaries according to the value of $\eta = \delta/\Delta$, where δ is the distance of the fluid node from the boundary surface and Δ is the distance between two lattice grids [21], as shown in Fig. **16**, where $\mathbf{c}_{i'} = -\mathbf{c}_i$. What should be noted is that in their scheme, motivated by [33], they introduced "virtual nodes" which contain "virtual fluid". The nodes inside the solid wall and adjacent to the wall surface are referred to as "virtual nodes" in their scheme [21], as illustrated in Fig. **16**.

Following the line proposed in their previous publications [18, 19], one may assume

$$f_{i'} = (1-\alpha) f_{i'}(\mathbf{w},t) + \alpha f_i(\mathbf{x},t) \tag{104}$$

In Eq. (104) $\mathbf{w}$ indicates the location of solid surface and α is a weight parameter depending on η.

When $0 < \eta \leq 0.5$, the post-collision distribution f_i which leaves from the boundary node $\mathbf{x}_j$ at the instant t will arrive at the solid wall surface at the instant $t + \eta\Delta t$. One can assume that f_i will be adsorbed onto the solid surface for a period of $(1-2\eta)\Delta t$ and then be desorbed as $f_{i'}$. $f_{i'}$ can reach the boundary node $\mathbf{x}_j$ at the

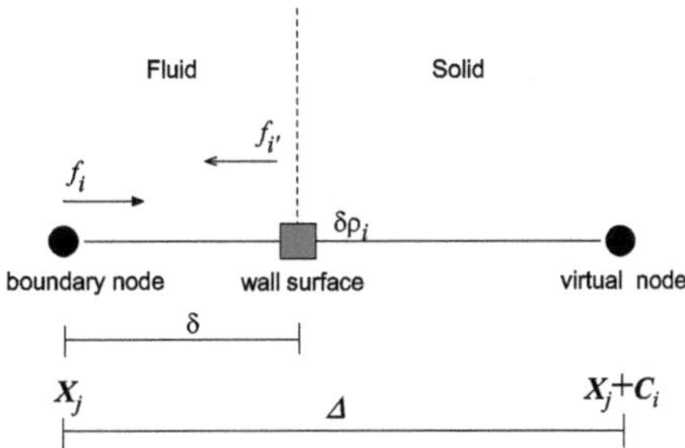

Fig. **16**: Schematic diagram of the boundary treatment rule: $\mathbf{x}_j$ is the fluid boundary node and $\mathbf{x}_j + \mathbf{c}_i$ is the virtual node in solid wall. The square indicates the position of the solid wall surface.

instant $t+\Delta t$, ready for the next collision occurring at the boundary node $\mathbf{x}_j$. The interpretation of Eq. (104) for this scene, from the viewpoint of surface chemistry, is very clear: during the period of being adsorbed on the wall surface, due to the interaction between the fluid molecules and the solid atoms, a fraction of fluid molecules changes from the state $f_i(\mathbf{x},t)$ to the state $f_{i'}(\mathbf{w},t)$ and $f_{i'}(\mathbf{w},t)$ is determined by the solid wall. Because in the LBM, the distribution can be split into two parts: $f_i = f_i^{eq} + f_i^{neq}$, Eq. (104) can be written as

$$f_{i'} = f_{i'}^{eq} + f_{i'}^{neq} \tag{105}$$

where

$$f_{i'}^{eq} = (1-\alpha) f_{i'}^{eq}(\mathbf{w},t) + \alpha f_i^{eq}(\mathbf{x},t) \tag{106}$$

and

$$f_{i'}^{neq} = (1-\alpha) f_{i'}^{neq}(\mathbf{w},t) + \alpha f_i^{neq}(\mathbf{x},t) \tag{107}$$

Since $f_{i'}^{neq}(\mathbf{w},t)$ can not be obtained exactly, usually it is assumed that $f_{i'}^{neq}(\mathbf{w},t) = \frac{2\eta w^{(c)}(1-w^{(f)})}{w^{(f)}(1-w^{(c)})} f_i^{neq}(\mathbf{x},t)$, where $w^{(c)} = \frac{1}{\tau_s}$, $w^{(f)} = \frac{4\eta}{2\eta+(2/w^{(c)}-1)}$ and τ_s is determined by fluid kinematic viscosity [59]. Consequently, Eq. (104) reads

$$f_{i'} = (1-\alpha) f_{i'}^{eq}(\mathbf{w},t) + \alpha f_i^{eq}(\mathbf{x},t) + ((1-\alpha)\frac{2\eta w^{(c)}(1-w^{(f)})}{w^{(f)}(1-w^{(c)})} + \alpha) f_i^{neq}(\mathbf{x},t) \tag{108}$$

It is straightforward that the fraction of fluid molecules changing from the state $f_i(\mathbf{x},t)$ to the state $f_{i'}(\mathbf{w},t)$ is proportional to the adsorption time, i.e. $(1-\alpha) \propto (1-2\eta)\Delta t$. The simplest choice is to set $\alpha = 2\eta$ according to the analog in this proportion. So the final expression of Eq. (104) reads

$$f_{i'} = (1-2\eta) f_{i'}^{eq}(\mathbf{w},t) + 2\eta f_i^{eq}(\mathbf{x},t) + ((1-2\eta)\frac{2\eta w^{(c)}(1-w^{(f)})}{w^{(f)}(1-w^{(c)})} + 2\eta) f_i^{neq}(\mathbf{x},t) \tag{109}$$

It is very interesting that when $\eta = 0.5$, Eq. (109) reduces to the midway bounce-back scheme, although Chen et al. constructed their boundary scheme from an entirely different way.
Another key in equation (109) is how to evaluate $f_{i'}^{eq}(\mathbf{w},t)$. Usually, $f_i \neq f_{i'}$, namely the fluid mass arriving at the solid surface is not equal to that leaving. Consequently $\delta\rho_i = f_i - f_{i'} \sim \mathcal{O}(M^2)$ [45] accumulates on the solid surface which serves as a mass sink in surface chemistry, where M denotes the Mach number. In Chen's scheme, there is

$$f_{i'}^{eq}(\mathbf{w},t) = f_i^{(eq)}(\rho_0 + \delta\rho_i, \mathbf{u}_w) \tag{110}$$

where ρ_0 is the density of the virtual node, whose initial value is the same as that of the real fluid and updated by $\rho_0(t+\Delta t) = \rho_0(t) + \sum_i \delta\rho_i$. $\mathbf{u}_w$ is the velocity of the solid surface. For simplicity, here taking a stationary wall $\mathbf{u}_w = 0$ for example to show the role of $\delta\rho_i$, namely $f_{i'}^{eq}(\mathbf{w},t) = \omega_i[\rho_0(t) + \delta\rho_i]$. If $f_i > f_{i'}$, i.e. $\delta\rho_i > 0$,

$f_{i'}^{eq}(\mathbf{w},t)$ will increase accordingly to reduce the difference between $f_{i'}$ and f_i. When $f_i < f_{i'}$, denoting $\delta\rho_i < 0$, $f_{i'}^{eq}(\mathbf{w},t)$ will decrease and the difference between $f_{i'}$ and f_i is still reduced. So $\delta\rho_i$ acts like a feedback to prevent mass leakage at the solid surface, which is analog to the effect of solute concentration on a solid wall in surface chemistry (Langmuir CR 1933) [62].

When $0.5 < \eta \leq 1$, f_i leaving from the boundary node $\mathbf{x}_j$ at t can not return back at $t+\Delta t$. According to the conclusions in [18, 19], α should equal zero in this scene. In [18, 19] the authors have demonstrated that in order to eliminate velocity slip on the wall surface when the Knudsen number $Kn \to 0$, $\alpha = 0$ is the necessary condition for $\eta > 0.5$. So Eq. (104) becomes

$$f_{i'} = f_{i'}(\mathbf{w},t) = f_{i'}^{eq}(\mathbf{w},t) + f_{i'}^{neq}(\mathbf{w},t) \tag{111}$$

where $f_{i'}^{eq}(\mathbf{w},t)$ is obtained by Eq. (110) and the non-equilibrium part can be assumed as $f_{i'}^{neq}(\mathbf{w},t) = \frac{\eta w^{(c)}(1-w^{(f)})}{w^{(f)}(1-w^{(c)})} f_{i'}^{neq}(\mathbf{x},t)$ [59]. The treatment of the non-equilibrium part in Chen's scheme is consistent with that in [45, 59, 30], which guarantees the deviatoric stress to be continuous.

It is interesting that when $\eta = 1$, equation (111) reduces to the non-equilibrium extrapolation scheme [45], which is a second-order accuracy boundary treatment for curved wall, although the start-points of these two schemes are quite different.

The simplest yet very useful case to verify whether a boundary scheme keeps mass conservation is the static flow under big intensity of gravity [3]. As shown in Fig. **17**, the flow is restricted by the top and bottom walls and periodic boundary conditions are employed for the left and right boundaries. With a mass conservation boundary scheme, the mass of this system will fluctuate at the beginning of computing iterations and then go to a constant value while durative mass leakage can be observed if a mass non-conservation treatment is adopted [3]. Chen et al. have validated their boundary scheme by this case: The initial density is equal to unity and the intensity of gravity $g = 0.001$ with a grid resolution 33×17. The analytical value of total mass is 561. As shown in Fig. **18**, the numerical results approach quickly to a constant with tiny deviation from the analytical value. Even at the foremost time steps, the fluctuation is very small. Through numerical experiments, it is found that the simulation is stable even when $\tau_s \approx 0.503$, which demonstrates the robustness of Chen's boundary scheme.

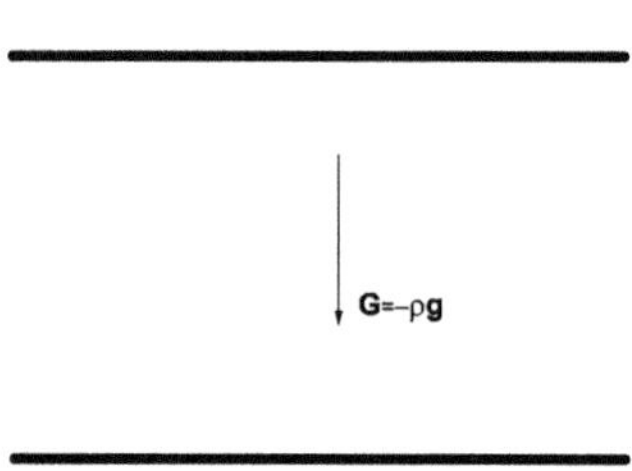

Fig. **17**: Schematic of static flow under gravity.

To demonstrate the capability of their boundary treatment for complex geometries, Chen et al. also simulated the Couette flow between two circular cylinders. In such Couette flow, the inner cylinder with radius r_1 rotates with a constant angular velocity ω and the outer cylinder with radius r_2 is kept stationary. Fig. **19** illustrates the velocity profiles of the Couette flow with different values of the radius ratio $\beta = r_1/r_2$. Their numerical results agree well with the analytical solutions [45]. The convergence rate of their scheme was tested for different values of Reynolds number $Re = (r_2 - r_1)u_0/\nu$, where $u_0 = \omega r_1$. The relative global L_2 norm errors in the velocity filed are measured and shown in Fig. **20**. Asymptotical quadratic convergence is clearly observed from this figure, and the second-order accuracy of this treatment for curved walls is confirmed.

Furthermore, Chen et al. validated their scheme by laminar flow past single cylinder. Table **2** lists the numerical results of drag coefficient (C_d), lift coefficient(C_f) and Strouhal number (St) for laminar flow past

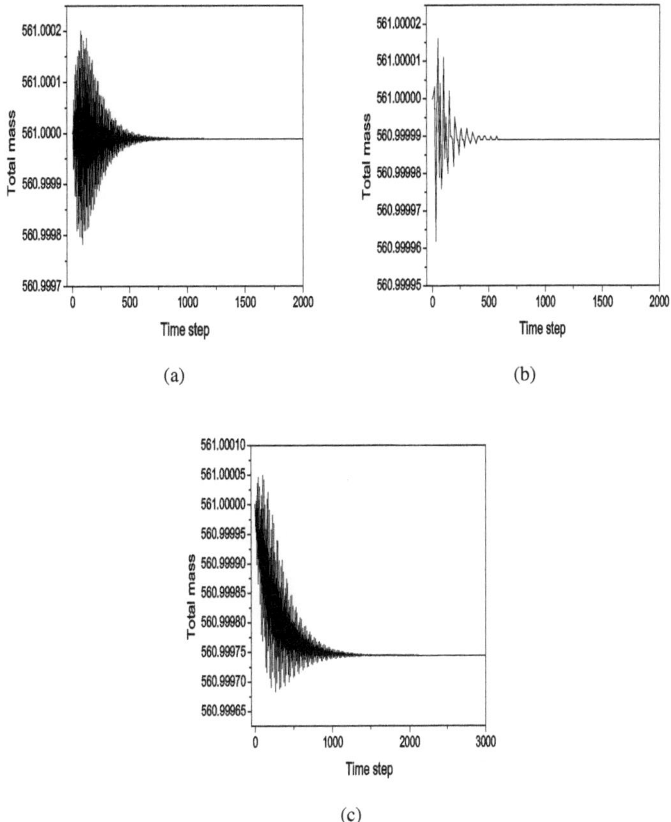

Fig. **18**: System mass changes with time step using Chen's boundary scheme(a) $q = 0.25$ (b) $q = 0.5$ (c) $q = 0.75$.

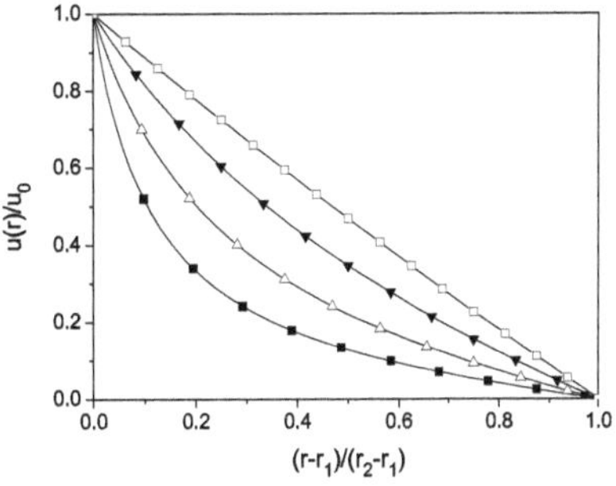

Fig. **19**: Velocity profiles of the Couette flow with different values of the radius ratio β: black square, $\beta = 0.1$; open triangle, $\beta = 0.2$; black triangle , $\beta = 0.4$; open square, $\beta = 0.8$; solid line, analytical solution.

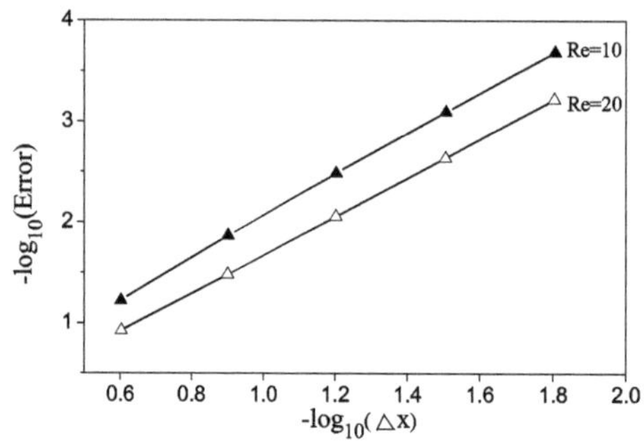

Fig. **20**: Convergence behavior with different grid resolution.

Table **2**: Numerical results for laminar flow past single cylinder with $Re = 200$.

	C_d	C_f	St
[21]	1.329	0.605	0.195
[34]	1.348	0.659	0.196
[47]	1.32	0.602	0.192

single cylinder, together with data in [34, 47]. The momentum exchange method [1] was used to evaluate the drag and lift forces imposed by the fluid on the solid body. The detailed formulae for C_d, C_f and St can be found in [3]. Their results agree well with that in previous literature.

Finally, they simulated the turbulent flow past a circular cylinder with $Re = 3900$ [73], as illustrated in Fig. **21**. It is obvious that their scheme is stable enough for high Reynolds number flow simulation.

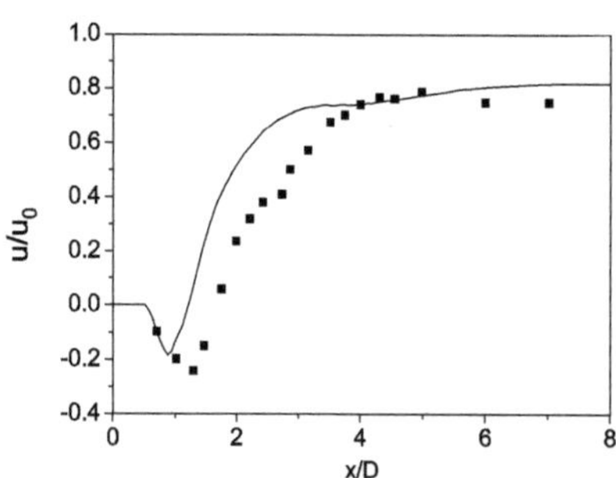

Fig. **21**: Time-averaged velocity along the centerline: solid line, numerical results; squares, experimental data.

In conclusion, besides preventing mass leakage, Chen's boundary scheme [21] possesses several obvious advantages: First, its implementation is much simpler than previous mass conserving treatments to deal with complex solid boundaries; Second, in their scheme, whenever only one fluid node is required to construct the unknown distribution from the solid surface with second-order accuracy and no restriction on the locations of the solid surfaces if there is only one fluid node between two solid walls; Third and the most attractive, following the way proposed in [18, 19, 20], Chen's mass-conservation boundary scheme for macroscopic flow can be extended to micro-flow simulation straightforwardly.

[illegible]

In the above paragraphs we focus on how to handle arbitrary curved athermal boundaries. However, in all practical energy conversion systems, the boundaries of the investigated domains are not isothermal and non-Dirichlet-type heat boundary conditions (e.g. heat-flux boundary conditions) are the most commonly found. However, the treatment for non-Dirichlet-type heat boundaries has posed a great challenge in the LBM community. Until now, almost all publications on thermal boundary conditions within the framework of LBM are restricted in the Dirichlet boundary type, namely, the temperature at boundaries being given instead of heat flux. The first popularly used thermal LB boundary scheme for Dirichlet boundary constraints was designed by He et al. [51]. In [51] the bounce-back rule of the nonequilibrium distribution was extended to the thermal boundary distribution. Later, D'Orazio et al. [35] proposed that the incoming unknown thermal populations are assumed to be equilibrium distributions with a counter-slip thermal energy. While in [86], the unknown energy distribution at the boundary node is decomposed into equilibrium and nonequilibrium parts. The nonequilibrium part is approximated with a first order extrapolation of the nonequilibrium part of populations at the neighboring domain nodes. For Neumann (heat flux) boundary conditions, D'Orazio and Succi [36], who extended their boundary scheme for Dirichlet-type thermal boundaries [35] accordingly, are the pioneers on this topic. To date, there is no doubt that the model of assuming a counter slip thermal energy density in [36] is of the highest accuracy since it can guarantee the fixed velocity and temperature or heat flux at the boundaries exactly, but the counter-temperature assumption may cast doubt on its convenient applicability to arbitrary boundary conditions or complicated geometries. Consequently, Tang et al. [86] conducted an attempt to remedy this gap. Unfortunately, their boundary treatment is an "indirect" strategy (one has to convert the given heat flux at the boundaries to the boundary temperature because their boundary scheme explicitly depends on the temperature instead of the heat flux) and loses the intrinsic advantage of local computing in the traditional LB method besides quite complicated implementation of their scheme for the Neumann-type boundaries. In fact, for a complicated curved boundary, their scheme can not work since it may be impossible to calculate the unknown boundary temperature through the heat flux boundary condition (cf. Eq. (17) in [86]). Recently, Liu et al. [68] designed a thermal boundary treatment for the Neumann thermal boundary conditions. Their way is similar with that in [86] indeed because in their scheme the heat flux on the boundary also has to be converted to the boundary temperature firstly. The difference between [86] and [68] is that such conversion is implemented by using the characteristics of DFs in the former while by invoking the finite difference algorithm in the latter. So the drawbacks in [86] remain in [68]. Moreover, the existing thermal LB boundary schemes all can not deal with the Cauchy boundary conditions, in which the temperature T_w and heat flux $\vec{q}_w$ at the boundaries both are given. Please bear in mind that a heat flux is a vector quantity.

Recently, Chen et al. [28] observed that the above deficiencies all can be overcome straightforwardly by the introduction of total enthalpy H and the implementation of their boundary treatment is very easy. The main idea in [28] is outlined below:

(1) For a Dirichlet boundary, namely the boundary temperature T_w has been given, the total enthalpy at the boundary H_w can be obtained easily by

$$H_w = C_p T_w + u_w^2/2 \tag{112}$$

where the subscript w means the location of the boundary.

(2) For a Neumann boundary, the heat flux $\vec{q}_w$ is given instead of T_w, so we can not evaluate H_w through Eq. (112) directly. According to the existing strategy in previous works, such as [86, 68], one has to convert $\vec{q}_w$ into T_w firstly and then invoke Eq. (112) to evaluate H_w with the calculated T_w. As mentioned above, this kind of "indirect" method has many obvious drawbacks. Different from them, Chen et al. tried to design a "direct" method in which the heat flux can be treated explicitly. Recalling the well-known relationship between the change of sensible enthalpy h_s and the heat flux $\vec{q}$ in the thermodynamics

$$\delta h_s = |\vec{q}| \tag{113}$$

where $|\vec{q}|$ denotes the absolute value of $\vec{q}$. Eq. (113) is attractive due to its simplicity, without any additional argument. But bear in mind that only for a constant pressure process, the equality Eq. (113) holds strictly as a reduction of the first law in thermodynamics

$$\delta h_s = |\vec{q}| + \int V dp \tag{114}$$

where V is the volume of the fluid element. Although Eq. (114) is a universal equality, it is useless for constructing an applicable boundary scheme since it is extremely difficult to determine $\int V dp$ in a complicated system.

The key is to establish an Eq. (113)-like relationship in the framework of LBM. Recalling that in the LBM with the low Mach number M assumption, there exists $\delta p \sim \mathscr{O}(M^2)$, namely a "quasi-constant-pressure" process, so Eq. (114) can be rewritten as

$$\delta h_s = |\vec{q}| + V\delta p = |\vec{q}| + \mathscr{O}(M^2) \tag{115}$$

Please notice the difference between Eq. (113) and Eq. (115).

With the aid of Eq. (115), we have

$$h_s^n - h_s^{n-1} = |\vec{q}|\delta t + \mathscr{O}(M^2) \tag{116}$$

where the superscript n refers to the current time level.

According to Eq. (27), Eq. (116) can be transformed straightforwardly as

$$H^n = |\vec{q}_w|\delta t + H^{n-1} + (u_w^n)^2/2 - (u_w^{n-1})^2/2 + \mathscr{O}(M^2) \tag{117}$$

One can see that the accuracy of Eq. (117) is second order of the Mach number, consistent to that of the traditional LBM. For a rest wall, Eq. (117) can be reduced further as

$$H^n = |\vec{q}_w|\delta t + H_s^{n-1} + \mathscr{O}(M^2) \tag{118}$$

It is clear in Chen's "direct" treatment for the Neumann boundary, the heat flux $\vec{q}_w$ can be handled explicitly and the operation is strictly local.

(3) For a Cauchy boundary, the boundary temperature T_w and the heat flux $\vec{q}_w$ on the boundary both are given. It also can be treated easily by a combined scheme of Eqs. (112) and (117)

$$H = C_p T_w + |\vec{q}_w|\delta t + u_w^2/2 + \mathscr{O}(^2) \tag{119}$$

It is obvious that now the Cauchy boundary can be handled because T_w and $\vec{q}_w$ both can be treated explicitly and independently in Eq. (119). In previous related works, such as [86, 68], the influence of $\vec{q}_w$ is realized by a converted temperature T_c on the wall. However, under the Cauchy boundary condition the wall temperature T_w has been given and generally $T_c \neq T_w$, so the strategy in [86, 68] is not sound in physics.

According to the Chapman-Enskog method, any DFs can be decomposed into its equilibrium and non-equilibrium parts. Consequently, in Chen's boundary treatmen, the interpolation operation is executed for the EDFs instead of the full DFs. The unknown non-equilibrium part of the distribution from the virtual solid node is approximated with the corresponding non-equilibrium part of the distribution at the nearest fluid node. Accordingly, their treatment can be formulated as

$$h_{i'} = h_{i'}^{eq} + h_{i'}^{neq}, \tag{120}$$

where the superscript *neq* denotes the non-equilibrium part of the energy distribution $h_{i'}$.

When $0 < \eta \leq 0.5$, inspired by [21] one may assume

$$h_{i'}^{eq} = (1-2\eta)h_{i'}^{eq}(\mathbf{w},t) + 2\eta h_i^{eq}(\mathbf{x},t) \tag{121}$$

and

$$h_{i'}^{neq} = h_i^{neq}(\mathbf{x},t) \tag{122}$$

Therefore there is

$$h_{i'} = (1-2\eta)h_{i'}^{eq}(\mathbf{w},t) + 2\eta h_i^{eq}(\mathbf{x},t) + h_i^{neq}(\mathbf{x},t) \tag{123}$$

It is very interesting that when $\eta = 0.5$, Eq. (123) reduces to the second-order accuracy thermal boundary scheme proposed in [51], although Chen et al. constructed their boundary scheme from an entirely different way.
When $0.5 < \eta \le 1$, following the way in [21], one can obtain

$$h_{i'} = h_{i'}(\mathbf{w},t) = h_{i'}^{eq}(\mathbf{w},t) + \frac{\eta w^{(c)}(1-w^{(f)})}{w^{(f)}(1-w^{(c)})} h_{i'}^{neq}(\mathbf{x},t) \tag{124}$$

where $w^{(c)} = \omega_h$, $w^{(f)} = \frac{4\eta}{2\eta + (2/w^{(c)} - 1)}$. Obviously, when $\eta = 1$, Eq. (124) reduces to the non-equilibrium extrapolation scheme designed in [46, 86], which is a second-order accuracy boundary treatment for curved boundaries, although their start-points are quite different.
In Eqs. (123) and (124), the EDFs (i.e. $h_{i'}^{eq}$ and h_i^{eq}) are evaluated by Eq. (49) . The non-equilibrium part $h_{i'}^{neq}(\mathbf{x},t)$ can be calculated easily by $h_{i'}^{neq}(\mathbf{x},t) = h_{i'}(\mathbf{x},t) - h_{i'}^{eq}(\mathbf{x},t)$ since $h_{i'}(\mathbf{x},t)$ and $h_{i'}^{eq}(\mathbf{x},t)$ both are available straightforwardly.
But for an in-depth exploration, there is still one question should be answered: why the great difficulties in previous boundary treatments can be overcome straightforwardly by the introduction of the total enthalpy DFs? Or is there any possibility to construct another scheme with other types of DFs to achieve the same success. Perhaps we can answer these questions from the viewpoint of thermodynamics.
To reply the above question, firstly we should dig out the rootstock from which the advantages of Chen's scheme originates. According to the above paragraphs, it is clear that such rootstock is Eq. (115). Through Eq. (115), the heat flux $\overrightarrow{q}$, a vector quantity, is degraded to a scalar quantity, the so-called enthalpy h_s. It is obvious that for numerical simulation the treatment for a scalar quantity is much easier than a vector one. Consequently, the aforementioned questions can be converted to: may we find a similar relationship if other DFs used?
Firstly we discuss whether there is such a possibility for the internal energy DFs [51, 86, 68] because up to date such kind of DFs have been used the most popularly. According to the first law of thermodynamics, we have

$$\delta\varepsilon = |\overrightarrow{q}| - \int p dV \tag{125}$$

It is hard to evaluate $\int p dV$ in a complicated system and only for a constant volume or "quasi-constant-volume" process, we can establish a strictly/approximately simple relationship between $\delta\varepsilon$ and $\overrightarrow{q}$ as

$$\delta\varepsilon = |\overrightarrow{q}| \tag{126}$$

It is obvious such restrictions can not be met in most practical simulations, even in the framework of the LB method. Similarly, it is also impossible for the total internal energy DFs [46] to establish such simple relationship.
After a first glance, it seems that a simple boundary scheme like Chen's treatment also can be developed based on the sensible enthalpy DFs [10] since equation (115) is built on the change of the sensible enthalpy indeed. However, please recall that if the sensible enthalpy DFs are used, the corresponding LB equation for the evolution of energy field will depend on temperature explicitly (cf. Eq. (11) in [10]). Consequently, in the scenarios where the collision step should be implemented on boundaries to guarantee numerical accuracy/stability [21, 63], one has to know the bondary temperature even under the Neumann boundary conditions. The induced shortcoming in such treatment has been explained clearly in the above paragraphs. Therefore, the intrinsic drawback of sensible enthalpy DFs [10] restricts the applicable range of Eq. (115).
Chen's thermal boundary scheme has been validated by thermal flow past a fix cylinder. The computational domain is illustrated in Fig. **22**, in which $L_u = L_d = L_y = 30.5$ and the diameter of the fixed cylinder $D = 1.0$. The boundary conditions for this problem read: (1) At the inlet: uniform flow condition; (2) At the upper and bottom solid walls: slip flow condition; (3) At the exit boundary: outflow condition and (4) On the surface of the fixed circular cylinder: no slip condition with uniform heat flux $|\overrightarrow{q}_w| = 1$.

Fig. **23** shows the isotherms and streamlines for $Re = 10$ and 45 at $Pr = 0.7$. It is seen that the front surface has the maximum clustering of temperature isotherms which indicates high temperature gradients, as compared to the other points on the surfaces of the cylinder. The clustering of isotherms also increases with an increase in the Reynolds number. This is due to the increase in the recirculation region with Reynolds number. The observations agree well with that in [6].

To take a quantitative comparison, Table **3** lists the average Nusselt numbers [6] over the surface of the fixed cylinder for different Re at $Pr = 0.7$. The deviation is less than 2% for all case considered.

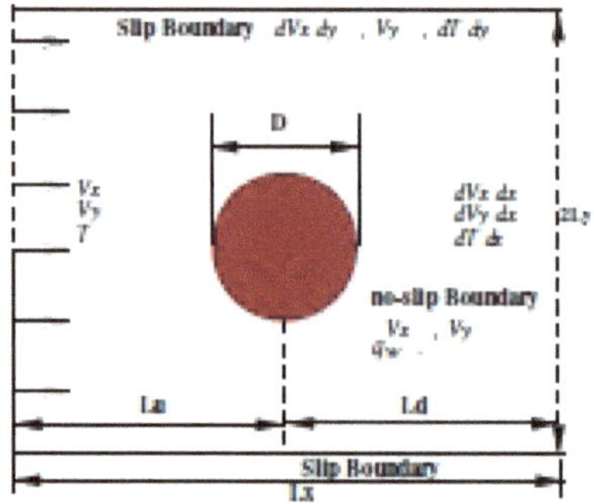

Fig. **22**: Schematics of the unconfined flow around a circular cylinder.

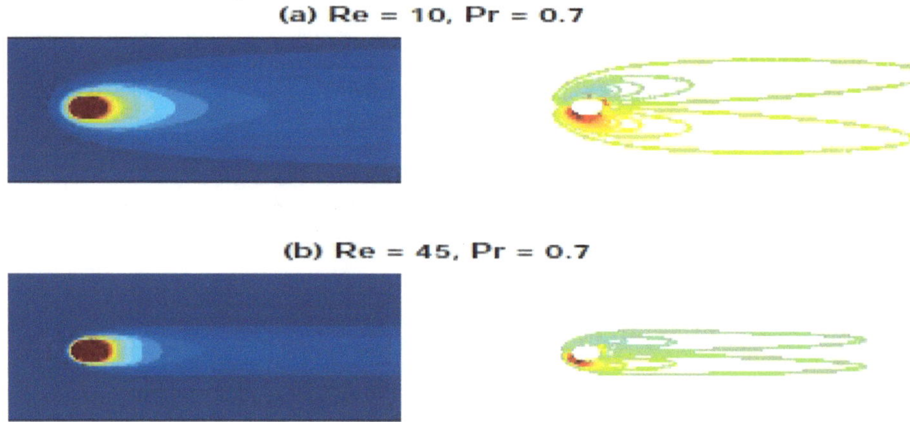

Fig. **23**: Isotherm (left) and streamline (right) plots of flow past a fixed cylinder with uniform heat flux condition for $Re = 10$ and 45 at $Pr = 0.7$.

[illegible]

Open boundaries are essential in practical fluid flow computations. Research in numerical methods for compressible fluid flow computations over the last 30 years has shown that typical boundary conditions are reflective and they may have a significant influence on the solution of unsteady compressible flows even at low Mach and Reynolds numbers [56]. In the LBM, although a significant research effort has been made to characterize the accuracy of different boundary-condition implementations [85, 1, 12], little attention has been paid so far to the study of the interaction of open boundaries with the fluid domain. However, this

Table 3: Comparison of the average Nusselt numbers over the surface of the cylinder

Re	[6]	[28]
10	1.97	1.9681
20	2.73	2.7304
30	3.32	3.3411
40	3.75	3.8092
45	4.01	3.9900

interaction will become important due to the artificial bounding of the computational domain when the real domain is in fact infinite. Yu et al. [100] conducted the first study on the pressure interaction between an inlet boundary and the interior of the flow field when the bounce-back scheme is specified at the inlet. The authors found that the bounce-back treatment would reflect most of the pressure waves back into the flow field and resulted in a poor convergence towards the steady state or a noisy flow field. Accordingly, they designed an interpolation-based superposition scheme as a remedy. While inspired by the nonreflecting boundary conception in traditional CFD for boundary treatments [32], Izquierdo and Fueyo developed a characteristic boundary [55] to prevent the reflecting of pressure wave near open boundaries. A detailed comparison between the available LB boundary schemes to reduce the impact of the boundary-interior interaction on the unsteady development of the flow field was presented in [56]. The authors showed that the interpolation-based superposition scheme proposed by Yu et al. is good enough for athermal open boundaries.

However, the relevant knowledge in the scenarios of combustion is still blank for the LBM. To the best knowledge of the present author, until now there is only one publication [11] discussing this topic. In [11], taking the coflow methane-air laminar diffusion flame illustrated in Fig. **3** as the example, Chen et al. analyzed the interaction between open boundaries and fluid domain when Yu's boundary scheme [100] is used for thermal reactive flow. It is well known there exist two different kinds of numerical waves in combustion simulations: the first is shock wave which caused by the sharp discontinuity in the initial status, the second is contact wave which results from the movement of the interface of different states, for example different temperature in this case. Figs. **24** and **25** illustrate the propagation process of shock wave. One can see that the shock wave travels with very quick speed (approximates to particle speed $c \sim O(20)$), and the pressure post the shock wave rises obviously(see the lines in Fig. **24**). When the shock wave impacts on the outlet, it will be partially reflected back to the interior of the flow field (the line $t = 0.137$ in Fig. **25**) and moves back towards the inlet (the lines between $t = 0.137$ and $t = 0.206$ in Fig. **25**) then causes a new reflection when it impacts on the inlet (the line $t = 0.229$ in Fig. **25**). But this kind of reflected wave caused by the shock wave will dissipate relatively quickly. In their study, they found that the shock wave just reduces the rate of convergence and will not cause numerical instability. Figs. **26** and **27** illustrate the propagation process of contact wave. The contact wave travels with relatively slow speed and the pressure post the contact wave changes relatively small. But when the contact wave approaches to the outlet, there will appear visible oscillations (the line $t = 0.686$ in Fig. **26**). When the contact wave impacts on the outlet, the oscillations of the reflected wave are more obvious (Fig. **27**). Such oscillation also will reduce the rate of convergence. Furthermore, if the oscillation is too large, numerical results will diverge (see the lines at $t = 0.688$, $t = 0.693$ and $t = 0.697$ in Fig. **27**). And the new reflected wave yielded by the contact wave when it returns to the inlet is more significant than that caused by the shock wave(see the line at $t = 0.718$ in Fig. **27**). The unphysical transient behavior with erratic oscillations yielded by the interaction between the open boundaries and the interior of the flow field will hamper the rate of convergence, the numerical stability, and the quality of the overall solution. In their work [11], the authors tried to increase the aspect ratio L/H to 8.0 to reduce the effect of open boundary, but such phenomena still appear.

Through the observations presented in [11], we can find that although the interpolation-based superposition scheme [100] is a well-designed boundary scheme for isothermal flow, its capability is too poor to handle contact wave because there exist significant temperature differences between the two sides of the interface.

3. APPLICATION IN COMPLICATED ENERGY CONVERSION SYSTEM

In this section, the latest progresses to extend LBM to investigation on advanced combustion technologies

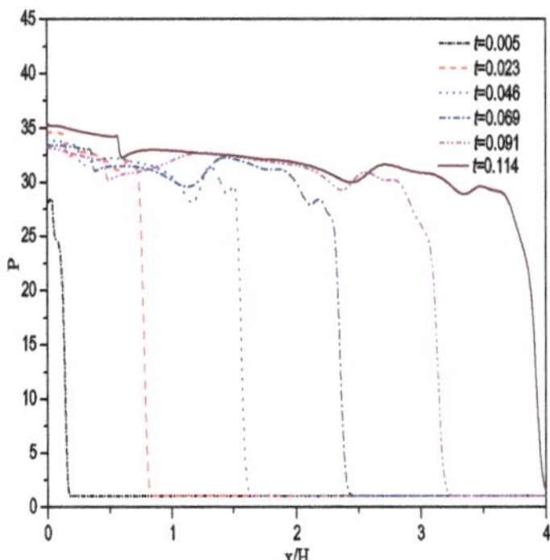

Fig. **24**: Propagation process of shock wave towards outlet;$x = 0$.

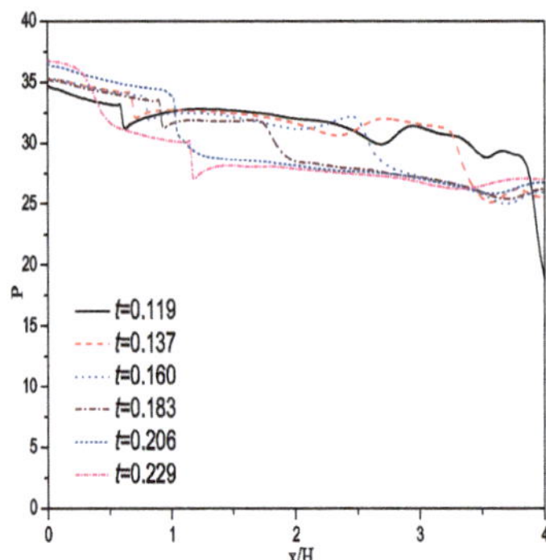

Fig. **25**: Propagation process of shock wave reflected by outlet/inlet;$x = 0$.

and energy systems are presented. The goal of this section is to try to show the readers the potential fields where the LBM can serve as a powerful tool.

[illegible]

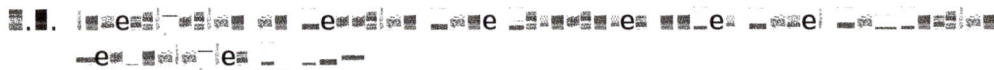

Biogas, which typically originates from the anaerobic digestion of biomass and organic wastes by microorganisms, is a renewable and biodegradable energy source that can be used for heating, lighting, transportation, small-scale power generation, and large gas turbines as a complementary fuel. The benefits of biogas are generally similar to those of natural gas. In addition, burning biogas reduces greenhouse gas emissions; it reduces the net CO_2 release and prevents CH_4 release. Thus, biogas combustion is a potential means to satisfy various legislative and ecological constraints [57]. Up to date there are numerous studies on the combustion characteristics of biogas. However, until now the research on the performance of biogas under MILD oxy-fuel combustion operations is still quite sparse, although the potential benefits to utilize biogas in MILD oxy-fuel regime are obvious. It is well-known that there are generally two drawbacks to utilize biogas in conventional modes: its small low calorific value (LCV) and induced high cost of upgrading which means removal of CO_2 to raise the LCV of biogas. However, under MILD oxy-fuel operation, these two disadvantages of biogas disappeared naturally: the MILD combustion mode can work well even using the fuels with extremely small LCVs and in the oxy-fuel regime the removal of CO_2 from biogas is unnecessary because additional CO_2 is required to dilute the hot reactants. In spite of it is a promising way to utilize biogas more efficiently and economically, the necessary knowledge on this novel strategy is highly desired before it could be widely implemented with confidence in its performance. To achieve the goal, the LBM was employed by Chen et al. [27] to investigate the reaction zone structure of counterflow diffusion flame

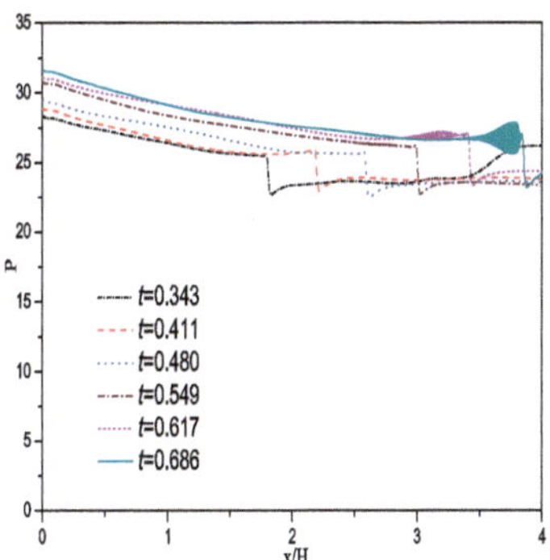

Fig. **26**: Propagation process of contact wave towards outlet;$x = 0$.

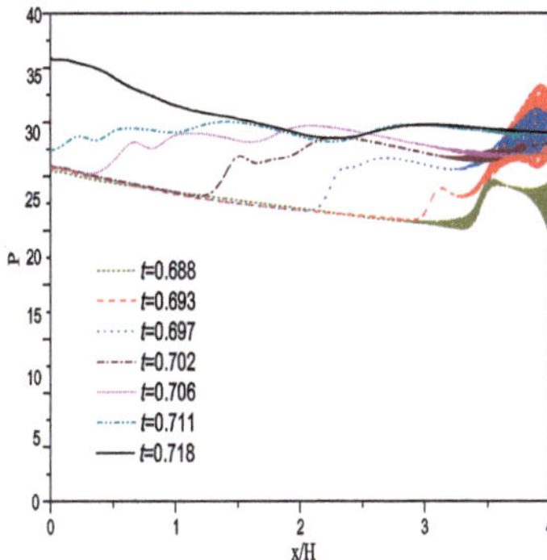

Fig. **27**: Propagation process of contact wave reflected by outlet/inlet;$x = 0$.

of hydrogen-enriched biogas under MILD oxy-fuel operation.

The problem domain and and boundary conditions are summarized in Fig. **28**. Two-dimensional rectangular coordinates are used. The origin of coordinates is located at the domain geometric center. Two parallel stationary walls are located at $y = \pm L$. The aspect ratio $A = \frac{L}{W} = 0.6$. The fuel mixture, which consists of hydrogen and biogas (being composed of 60% CO_2 and 40% CH_4 by volume ratio, namely typical compositions for biogas from active or recently closed landfills), is uniformly ejected from the bottom wall with temperature $T_{fuel} = T_0$ and the preheated oxygen diluted by CO_2 is uniformly ejected from the top wall with temperature T_{oxi}. The counter flow impacts and reacts in the reaction zone. Then, the diffusion stagnation "flame" is formed. The burned gas flows outward along the x-direction.

The volume percentage of hydrogen in fuel blends X_h is defined as [22, 25]

$$X_h = \frac{V_h}{V_h + V_b} \tag{127}$$

where V_h and V_b are the volume fraction of hydrogen and biogas in the mixtures, respectively. Similarly, for the oxidizer mixtures there is

$$X_o = \frac{V_o}{V_o + V_c} \tag{128}$$

where V_o and V_c are the volume fraction of oxygen and carbon dioxide in the mixtures, respectively.

Figure **29** illustrates the profile of temperature along line $x = 0$ with various X_h. From this figure, it can be observed that since $X_h > 0.05$ an obvious temperature peak due to the exothermic reaction emerges. The temperature peak reduces with the decrease of X_h and cannot be seen when $X_h < 0.05$, although the moderate chemical reactions are still going on for those conditions, as shown in Fig. **30**.

The variation of the maximum temperature T_{max} in the reactive flow versus X_h is plotted in Fig. **31**. It is straightforward that T_{max} increases with more hydrogen being added into the fuel blends since the calorific

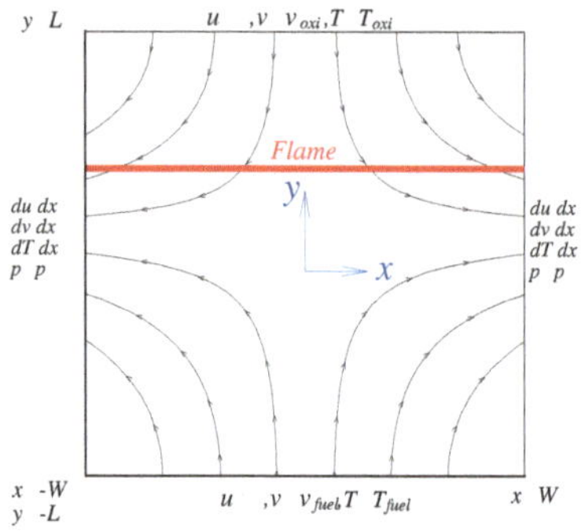

Fig. **28**: Schematic configuration and coordinate system of the computational domain of counterflow diffusion flame.

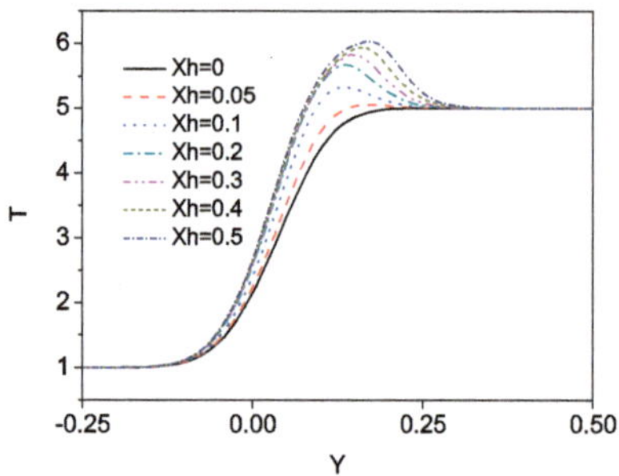

Fig. **29**: Distributions of temperature of various X_h along line $x = 0$ at $Re = 200$, $T_{oxi} = 5T_0$, $T_{fuel} = T_0$ and $X_o = 0.03$.

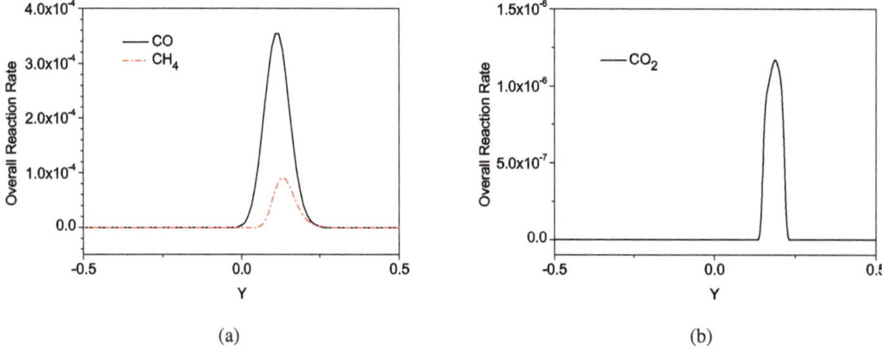

Fig. **30**: Overall Reaction Rate of (a) CH_4 and CO (b) CO_2 along line $x = 0$ at $Re = 200$, $X_h = 0$, $T_{oxi} = 5T_0$ and $X_o = 0.03$.

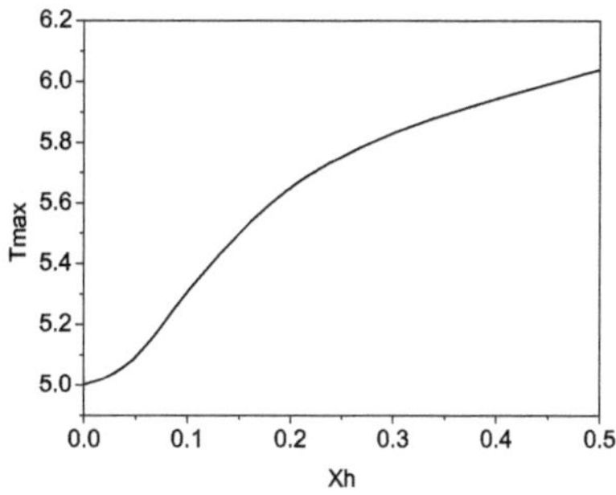

Fig. **31**: The maximum temperature versus X_h at $Re = 200$, $T_{oxi} = 5T_0$, $T_{fuel} = T_0$ and $X_o = 0.03$.

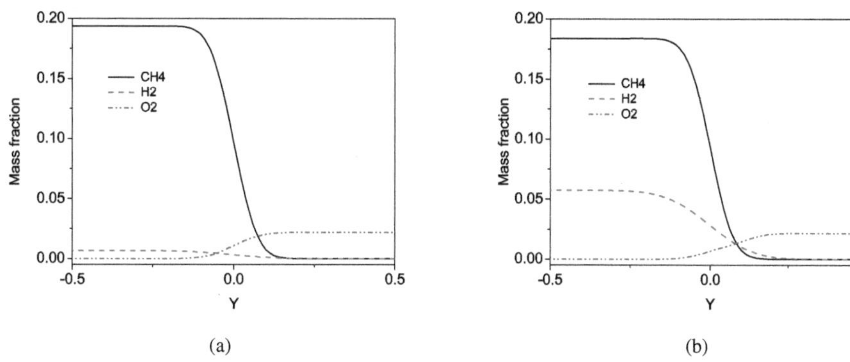

Fig. **32**: Distributions of mass fraction of CH_4, H_2 and O_2 with (a) $X_h = 0.1$ (b) $X_h = 0.5$ at $Re = 200$, $T_{oxi} = 5T_0$ and $X_o = 0.03$.

value of hydrogen is much higher than that of biogas investigated in this work. When $X_h < 0.05$, the increment of T_{max} is relatively small while significant since $X_h > 0.05$, consistent to Fig. **29**. However, the increasing speed of T_{max} will decrease against X_h and there is an approximate power-function-like relationship between T_{max} and X_h which reads $T_{max} = 6.39582{X_h}^{0.07791}$ since $X_h > 0.05$. The reduction of increasing speed of T_{max} versus X_h mainly results from two mechanisms: Firstly, the volume of reaction zone with higher temperature will increase with more H_2 being added (see Fig. **29**). However, it is well-known that the radiative heat loss is proportional to the volume of reaction zone with higher temperature, which will reduce the peak temperature of "flame" [75]. Secondly, the endothermic reactions will be enhanced due to the increase of flow temperature, which also will reduce the peak temperature of "flame". Consequently, the influence of the endothermic reactions on reduction of peak temperature can be neglected when X_h is very small.

The distributions of mass fraction of CH_4, H_2 and O_2 at various X_h are presented in Fig. **32**. Even X_h up to 0.5, the investigated combustion cases in this subsection are always in MILD regime, so the reaction zone structures with different X_h all are very similar to the so-called *Liñán's* premixed flame regime [70] in spite of the diffusion combustion mode being adopted in the present study: O_2, H_2 and CH_4 concentration distributions cross over, which implies that the fuel and oxidizers are not consumed at the thin "flame" region and so quite different from their conventional air combustion counterpart [25, 23]. It may be an intrinsic feature of MILD combustion [70].

According to the above discussion, one can concludes: even for an extremely low oxygen concentration (e.g. $X_o = 0.03$ in the oxidizer stream), there is no necessity to add too much hydrogen to sustain MILD oxy-fuel combustion of small LCV biogas which mostly consists of CO_2. This discovery supports the commercial

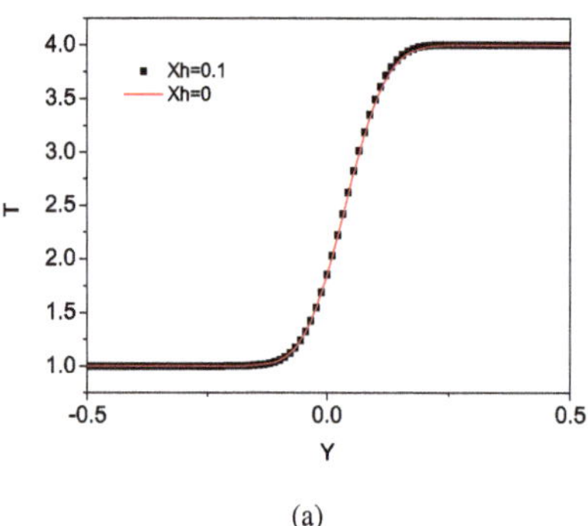

(a)

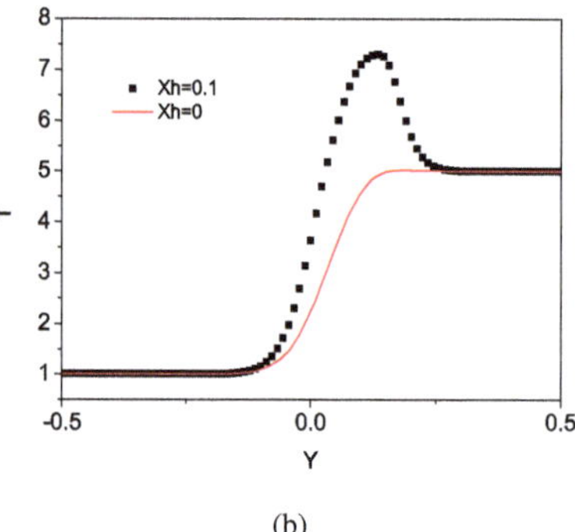

(b)

Fig. **33**: Distributions of temperature of various X_h with (a) $T_{oxi} = 4T_0$ and (b) $T_{oxi} = 5T_0$ along line $x = 0$ at $Re = 200$, $T_{fuel} = T_0$ and $X_o = 0.06$.

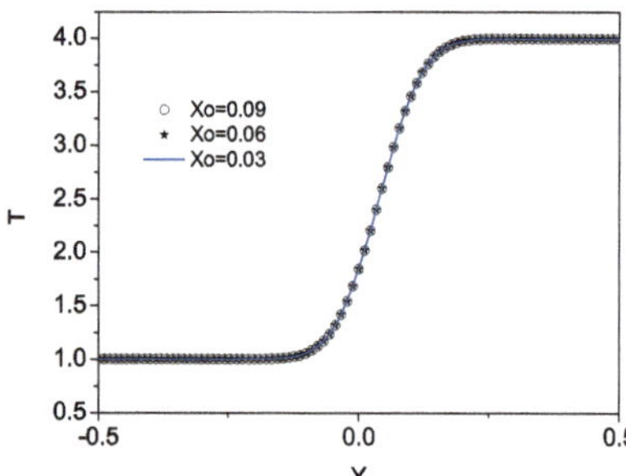

Fig. **34**: Distributions of temperature of various X_o along line $x = 0$ at $Re = 200$, $T_{oxi} = 4T_0$, $T_{fuel} = T_0$ and $X_h = 0$.

possibility to utilize biogas under MILD oxy-fuel conditions in practical applications due to the high cost of manufacturing hydrogen.

As concluded in previous studies [9], the preheated temperature of oxidizer stream plays a very important role to form MILD combustion. Through their simulation, Chen et al. [27] found that there will be no reaction if the preheated temperature of oxidizer stream T_{oxi} is less than $4T_0$ (corresponding to $1200K$) and the influences of hydrogen addition in the fuel stream and oxygen concentration in the oxidizer stream on reaction zone structures, temperature, etc. all depend closely on T_{oxi}.

Figure **33** plots the distributions of temperature of various X_h at $T_{oxi} = 4T_0$ and $T_{oxi} = 5T_0$. For the cases with $T_{oxi} = 4T_0$, the profiles of temperature at $X_h = 0$ and $X_h = 0.1$ are nearly same. Namely at this level of preheated temperature the reaction zone structures are insensitive to hydrogen addition. However, with higher preheated temperature of oxidizer stream (e.g. $T_{oxi} = 5T_0$), the distributions of temperature of $X_h = 0$ and $X_h = 0.1$ are quite different even the added hydrogen is relatively small, and there appears a significant temperature jump when $X_h = 0.1$, which denotes the MILD regime has broken down.

The profiles of temperature of various X_o at $T_{oxi} = 4T_0$ are illustrated in Fig. **34**. At relatively low preheated temperature, these three curves almost overlap each other, namely at $T_{oxi} = 4T_0$ the reaction structure are insensitive to X_o. While as above mentioned, at $T_{oxi} = 5T_0$ the temperature distributions depend sensitively on X_o, especially when X_o is big enough (cf. Fig. **35**).

Based on the above comparison, the crucial role of T_{oxi} on MILD oxy-fuel combustion is obvious. In previous study [70] it is claimed that the flameless combustion in methane-air counterflow can be sustained only when $T_{oxi} \geq 1500K$, so under oxy-fuel conditions, the MILD combustion is more easily sustained than

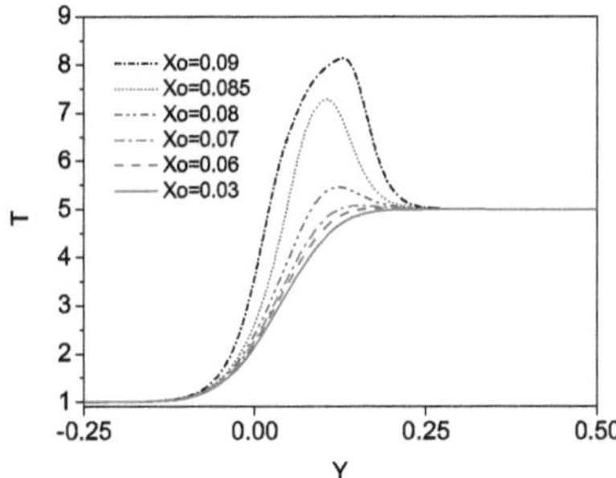

Fig. **35**: Distributions of temperature of various X_o along line $x = 0$ at $Re = 200$, $T_{oxi} = 5T_0$, $T_{fuel} = T_0$ and $X_h = 0$.

its air combustion counterpart, agreeing with the previous conclusion in [66]. Moreover, according to the aforementioned discussion, it is straightforward that a lower T_{oxi} is safer than a higher one to sustain the MILD regime for hydrogen-enriched biogas combustion under fluctuant input/operation.

[illegible]

Since Hirschfelder et al. [54] made the pioneering work of analysis on entropy generation in systems involving heat and mass transfer, a lot of open literature dealing with entropy generation in combustion systems have emerged. A latest comprehensive account of entropy analysis in combustion systems is available in the review by Som and Datta [83]. Up to date entropy generation analysis has matured as a powerful tool for system optimization [4].

Entropy generation in combustion can be described by the entropy generation equation which reads [24]

$$S = \frac{\prod : \nabla \vec{u}}{T} + \frac{k\nabla T \cdot \nabla T}{T^2} + \sum_i \frac{\rho D_i}{x_i} \nabla y_i \cdot \nabla x_i - \sum_i \frac{\mu_i \omega_i}{T} \tag{129}$$

The first term on the right-hand side of Eq. (129) is due to fluid friction (referred to as S_{vis}), the second term is due to heat transfer (referred to as S_{cond}), the third term pertains to mass transfer (referred to as S_{mix}) and the fourth term is due to chemical reaction (referred to as S_{chem}). The last two terms have summation over all the species and for all the reactions. Because there is no external body force in the present situation, the entropy generation induced by body force vanishes in Eq. (129). The entropy generation term due to coupling between heat and mass transfer also can be ignored in the above equation since it usually makes rare contribution to the local entropy generation rate unless the Soret and Dufour effects have significant influence. In Eq. (129), $\prod$ is the viscous stress, $\vec{u}$ is the velocity vector, ρ is the density of the mixture and k is the thermal diffusivity. y_i, x_i, ω_i, D_i and μ_i are the mass fraction, the mole fraction, the production rate, diffusion coefficient and chemical potential of species i respectively. The total entropy generation number is defined as [24]

$$S_{total} = \int_{\Omega} S \partial \Omega \tag{130}$$

where Ω means the global computational domain. Similar expressions can be written for $S_{vis,total}$, $S_{cond,total}$, $S_{mix,total}$ and $S_{chem,total}$.

As clearly shown in Eq. (129), the macroscopic quantities, such as temperature and velocity, are the necessary inputs for entropy generation analysis. Being aware of the outstanding advantages of the LBM over traditional CFD for complicated physical and/or chemical systems, Chen and Krafczyk [17] are the pioneers to introduce the LBM into the field of entropy generation analysis and the computation of Eq. (129) can be simplified since the first term on the right-hand side of Eq. (129) which contains complicated spatial derivatives can be straightforwardly obtained in the framework of LBM.

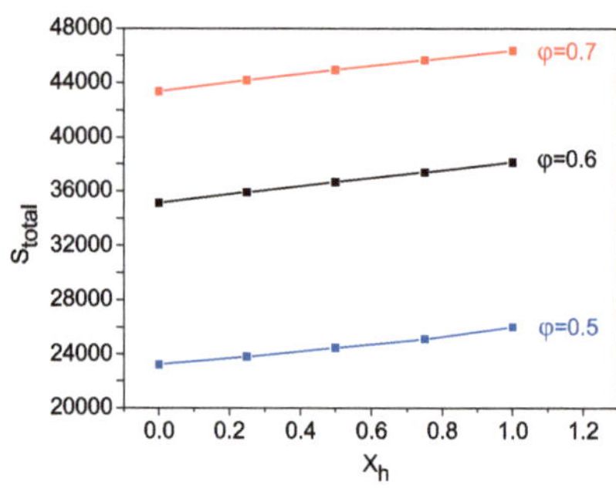

Fig. **36**: Total entropy generation number versus variable φ and X_h.

With the aid of LBM, entropy generation in hydrogen enriched ultra-lean counter-flow methane-air non-premixed combustion confined by planar opposing jets in the MILD regime has been investigated by Chen et al. [23] for the first time. The configuration adopted in [23] is similar with that illustrated by Fig. **28**. Figure **36** illustrates the variation of the total entropy generation number S_{total} versus different equivalence ratio φ and volume percentage of hydrogen in fuel blends X_h. It can be seen that S_{total} increases linearly with X_h for a given φ and these lines are nearly parallel. The relationship among them can be approximated as a linear function reading:

$$S_{total} = (-0.2726 + 1.0194\varphi + 0.0290X_h) \cdot 10^5 \tag{131}$$

According to Eq. (131) it is straightforward that the reduction in excess air level but the enhancement in hydrogen enriched level both can increase the entropy generation in a combustion process, which agrees well with previous work [83]. From this figure, it can be seen that φ plays a more important role than X_h to influence S_{total}.

In previous investigations, not only for pure hydrocarbon diffusion combustion fuel but also for hydrogen enriched non-premixed flames [83], it was claimed that in non-premixed combustion the contribution of $S_{vis,total}$ to S_{total} usually can be ignored since the magnitude of S_{vis} is too small. However, through their numerical analysis, Chen et al. [23] found that it is not true for the opposing jets counter-flow non-premixed hydrogen enriched methane-air combustion in the MILD regime. Figures **37**-**38** illustrate the variations of the relative total entropy generation rate due to heat transfer ($\gamma_{cond,total} = S_{cond,total}/S_{total}$), chemical reaction ($\gamma_{chem,total} = S_{chem,total}/S_{total}$), fluid friction ($\gamma_{vis,total} = S_{vis,total}/S_{total}$) and mixing ($\gamma_{mix,total} = S_{mix,total}/S_{total}$) versus different φ and X_h. As shown in these figures, the contribution of $S_{vis,total}$ to S_{total} is the largest no matter what φ and X_h are. The order of the predominant irreversibilities in the entire investigated domain is $\gamma_{vis,total} > \gamma_{chem,total} \approx \gamma_{cond,total} \gg \gamma_{mix,total}$. It is an obvious difference between the opposing jets counter-flow non-premixed hydrogen-air combustion and the co-flow hydrocarbon non-premixed combustion with/without hydrogen addition [83]. This characteristic also can be found in its premixed counterpart [22]. From Figs. **37**–**38**, one also can observe that $\gamma_{vis,total}$ and $\gamma_{cond,total}$ both are monotonic increasing functions of φ while monotonic decreasing functions of X_h; $\gamma_{chem,total}$ is a monotonic increasing functions of X_h but a monotonic decreasing functions of φ; $\gamma_{mix,total} \approx 0.02$ is the smallest one. Moreover, $\gamma_{vis,total}$, $\gamma_{cond,total}$, $\gamma_{chem,total}$ and $\gamma_{mix,total}$ all are nearly linear functions of φ. The variation trends of $\gamma_{vis,total}$, $\gamma_{cond,total}$ and $\gamma_{mix,total}$ are similar with those in their premixed counterpart [22] except $\gamma_{chem,total}$: in ultra-lean counter-flow methane-air premixed combustion, $\gamma_{chem,total}$ decreases against φ when $X_h < 0.5$ while it increases with φ for $X_h > 0.5$, however for the non-premixed scenarios $\gamma_{chem,total}$ always decrease against φ no matter what X_h is.

Through their numerical experiment, Chen et al. [23] observed at least five new phenomena in MILD combustion through entropy generation. More important, as will be shown below, the combination of LBM and entropy generation analysis also provide a promising way to construct turbulent models.

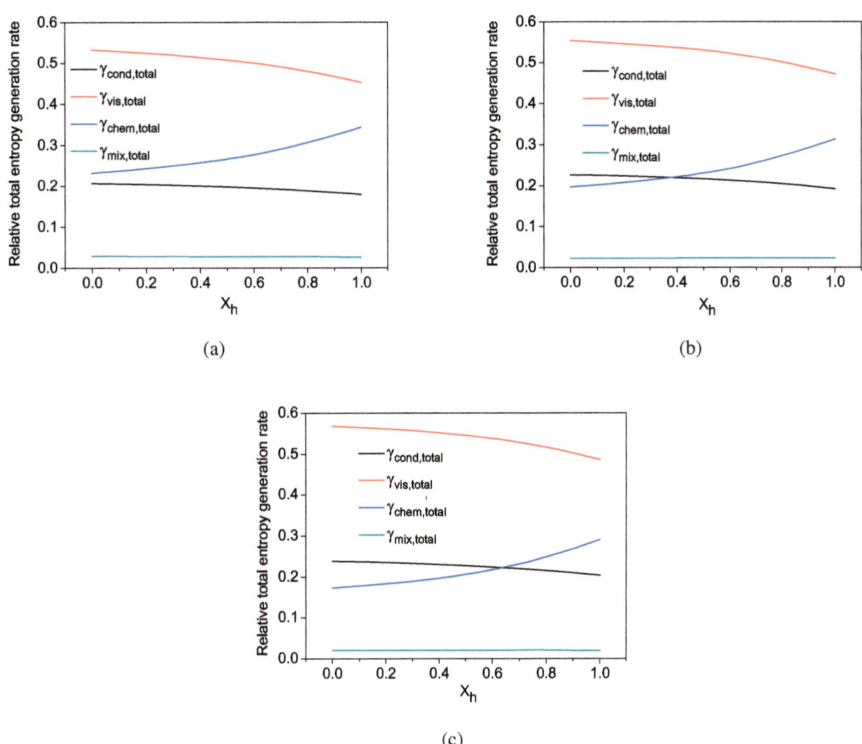

Fig. **37**: Variations of relative total entropy generation rate versus X_h at (a) $\varphi = 0.5$ (b) $\varphi = 0.6$ (c) $\varphi = 0.7$.

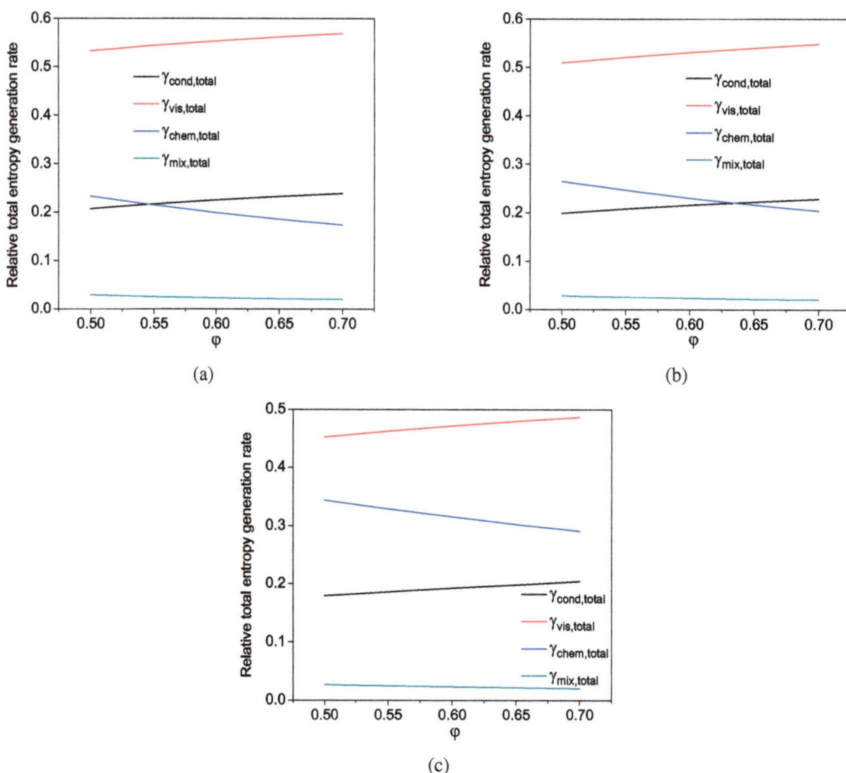

Fig. **38**: Variations of relative total entropy generation rate versus φ at (a) $X_h = 0$ (b) $X_h = 0.5$ (c) $X_h = 1$.

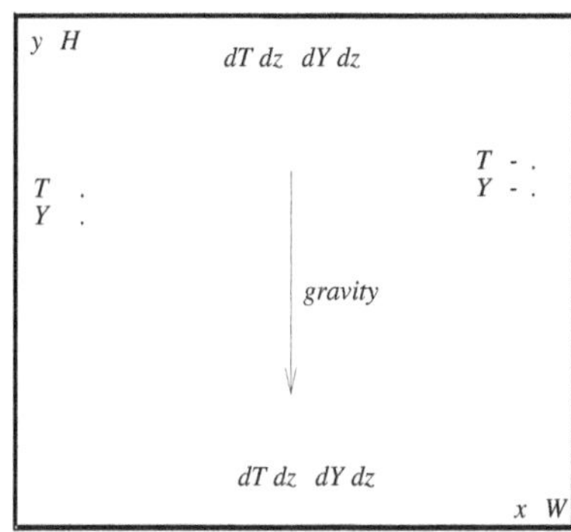

Fig. **39**: Configuration of the computational domain and boundary conditions of turbulent double-diffusive natural convection in a rectangle cavity.

[illegible]

About two decades ago it had been pointed out that an appropriate definition for the microscale in turbulence can help us to build improved turbulent models and that of entropy generation is thought as very helpful not only for turbulent model closure [2] but also for extraction of coherent structures [31]. Some characteristics of a system, which are hardly exhibited by velocity, temperature etc., can be shown clearly in the pictures of entropy generation. For example, with the aid of the LBM, recently Chen and Du [26] investigated the entropy generation of turbulent double-diffusive natural convection in a rectangle cavity and they discovered a new kind of self-organizing phenomenon by analyzing of the specific entropy generation.

The configuration of the computational domain is illustrated in Fig. **39**. The aspect ratio $A = H/W$, where H is the height of the cavity and W is the width.

It is well-known that with the turbulence intensity being enhanced, the maps of scalar fields will become more irregular due to the stochastic motion. However, through the picture of entropy generation, one can observe a very interesting regular phenomenon in turbulent double-diffusive natural convection: for a tall cavity with $A > 1$, the irregular distributions of irreversibility due to heat transfer will form regular periodical maps along the vertical direction. In [26] this phenomenon was named "spatial self-copy". It can be explained with the aid of Fig. **40**. It is clear that in Fig. **40**(a), the contours within $0 \leq y \leq 1$ can fully overlap with that in $1 \leq y \leq 2$ while in Fig. **40**(b) the contours within $0 \leq y \leq 1$, $1 \leq y \leq 2$ and $2 \leq y \leq 3$ are completely same.

Furthermore, one can observe that the periodical length in space is unity, as shown in Fig. **41**. The maps for different A are quite dissimilar. In Fig. **41**(a) the contours within $1 \leq y \leq 1.25$ fully overlap with that in $0 \leq y \leq 0.25$; in Fig. **41**(b) the contours within $1 \leq y \leq 1.5$ fully overlap with that in $0 \leq y \leq 0.5$; in Fig. **41**(c) the contours within $1 \leq y \leq 1.75$ fully overlap with that in $0 \leq y \leq 0.75$ and in Fig. **41**(d) the contours within $1 \leq y \leq 2$ fully overlap with that in $0 \leq y \leq 1$.

The "spatial self-copy" phenomenon implies that large-scale regular patterns could emerge through small-scale irregular and stochastic distributions, which is a little similar with the coherent structure [69] in shear turbulence but their mechanism is quite different. Moreover, this phenomenon was observed only for entropy generation due to thermal irreversibility. Other scalar fields investigated in [26] did not have such characteristic. What is its physical meaning hidden behind this phenomenon? Or may it be a new indicator to describe turbulent double-diffusive natural convection? It is an open question and requires further investigation in future research.

CONCLUSION

As concluded in the latest book on turbulent combustion modelling [37], although the available state-of-the-art CFD technologies have achieved great successes in combustion research concerned primarily phenomena

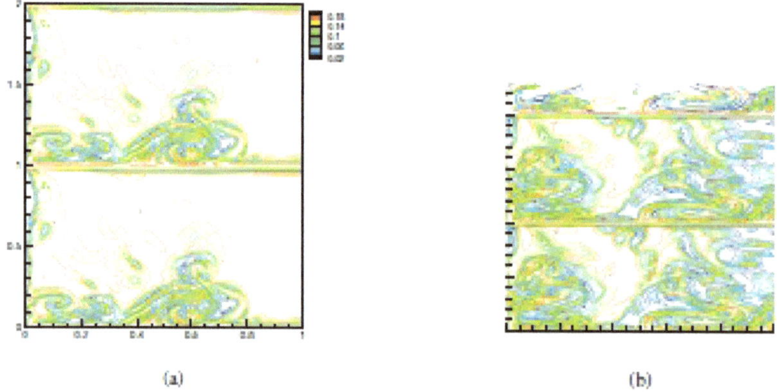

Fig. **40**: Entropy generation number due to thermal irreversibility at (a) $A = 2$ (b) $A = 3$.

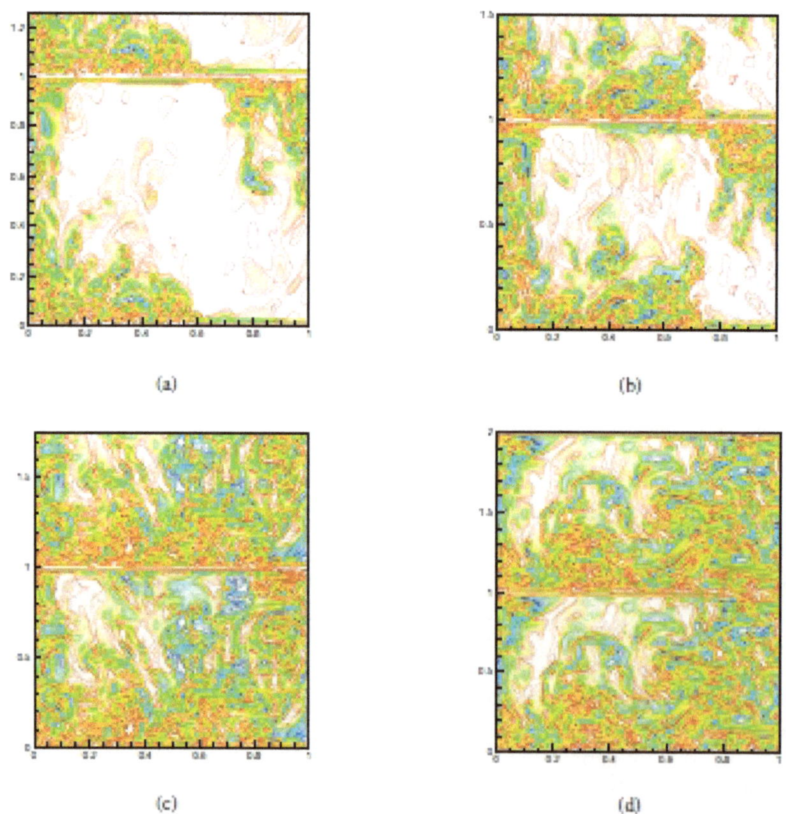

Fig. **41**: Relative total entropy generation rate due to thermal irreversibility at (a) $A = 1.25$ (b) $A = 1.5$ (c) $A = 1.75$ (d) $A = 2$.

in which the separation of scales can be justified, their applicability is seriously challenged by combustion in the regimes where both chemistry and mixing are competitive and/or turbulence-combustion interactions become more complicated due to overlapping time and length scales. However, as mentioned above, MILD oxy-fuel combustion is characterized by these regimes, so the popularly used "continuum" CFD tools based on macroscopic models have great difficulties to meet these challenges. Unlike conventional numerical methods based on direct discretizations of macroscopic continuum equations, the LBM is based on microscopic models and mesoscopic kinetic equations. The fundamental idea of the LBM is to construct simplified kinetic models that incorporate the essential physics of microscopic or mesoscopic processes so that the macroscopic averaged properties obey the desired macroscopic equations. Consequently it becomes a natural candidate to serve as a framework to tackle the challenges. Many famous combustion specialists also have recognized the LBM as a promising tool to advance turbulent combustion research in the future [37]. In this chapter, the present author tries to provide a comprehensive picture of the latest progresses of the LBM associated with combustion research: both the relevant advancement in basic modelling and potentially applicable fields are discussed. According the discussions, there are at least two theoretical puzzles should be solved before the LBM can be extended to simulate practical systems:

1. A robust combustion LB strategy for turbulent combustion simulation.
2. An appropriate boundary treatment to reduce the open boundary effects in reactive flows.

In order to achieve the goal, one should absorb the nutrition from the traditional CFD because there are many mature treatments in computational combustion science to overcome above shortcomings, for example, the nonreflecting boundary strategy for reactive flow presented in [77] to reduce influences of open boundaries on combustion. The progress of the LBM can be accelerated by the achievement of traditional CFD technologies and the investigation on MILD oxy-fuel combustion technology also can be advanced by the introduction of the LBM.

ACKNOWLEDGEMENT

This work was supported by the State Key Development Programme for Basic Research of China (Grant Nos. 2010CB227004 and 2011CB707301), and the National Natural Science Foundation of China (Grant Nos. 50936001, 51021065 and 51006043). The present author would also acknowledge the support from the Research Foundation for Outstanding Young Teachers, HUST (Grant No. 2010QN027) and the Research Fund for the Doctoral Program of Higher Education of China (Grant No. 20100142120048).

CONFLICT OF INTEREST

The authors confirm that this chapter content has no conflict of interest.

REFERENCES

[1] Aidun CK, Clausen JR. Lattice-Boltzmann method for complex flows. Annu Rev Fluid Mech 2010; 42 : 439-472.
[2] Arpaci VS. Microscales of turbulent combustion. Prog Energy Combust Sci 1995; 21: 153-171.
[3] Bao J, Peng Y, Schaefer L. A mass conserving boundary condition for the Lattice Boltzmann equation method. J Comput Phys 2008; 227: 8472-8487.
[4] Bejan A. Entropy generation through heat and fluid flow, 2nd ed. Wiley, New York:Wiley; 1994.
[5] Benzi R, Succi S, Vergassola M. The Lattice Boltzmann equation: theory and applications. Phys Report 1992; 222: 145-197.
[6] Bharti RP, Chhabra RP, Eswaran V. A numerical study of the steady forced convection heat transfer from an unconfined circular cylinder. Heat Mass Transfer 2007; 43: 639-648.
[7] Bosschaart KJ, Goey LPH. The laminar burning velocity of flames propagating in mixtures of hydrocarbons and air measured with the heat flux method. Combust Flame 2004; 136; 261-269.
[8] Buhre BJP, Elliott LK, Sheng CD, Gupta RP, Wall TF. Oxy-fuel combustion technology for coal-fired power generation. Prog Energy Combust Sci 2005; 31: 283-307.
[9] Cavaliere A, de Joannon M. Mild combustion. Prog Energy Combust Sci 2004; 30: 329-366.
[10] Chatterjee D. An enthalpy-based thermal Lattice Boltzmann model for non-isothermal systems. Europhys Lett 2009; 86: 14004.
[11] Chen S, Liu Z, He Z, Zhang C, Tian Z, Zheng C. A new numerical appraoch for fire simulation. Int J Mod Phys C 2007; 18: 187-202.
[12] Chen S, Doolen GD. Lattice Boltzmann method for fluid flows. Annu Rev Fluid Mech 1998; 30: 329-364.
[13] Chen S, Liu Z, Zhang C, He Z, Tian Z, Shi B, Zheng C. A novel coupled Lattice Boltzmann model for low Mach number combustion simulation. Appl Math Comput 2007; 193: 266-284.
[14] Chen S, Liu Z, Tian Z, Shi B, Zheng C. A simple Lattice Boltzmann scheme for combustion simulation, Comput Math App 2008; 55: 1424-1432.
[15] Chen S, Tolke J, Krafczyk M. Numerical simulation of fluid flow and heat transfer inside a rotating disk-cylinder configuration by a Lattice Boltzmann model. Phys Rev E 2009; 80: 016702.
[16] Chen S, Tolke J, Krafczyk M. Simulation of buoyancy-driven flows in a vertical cylinder using a simple Lattice Boltzmann model. Phys Rev E 2009; 79: 016704.
[17] Chen S, Krafczyk M. Entropy generation in turbulent natural convection due to internal heat generation. Int J Thermal Sci 2009; 48: 1978-1987.

[18] Chen S, Tian ZW. Simulation of microchannel flow using the Lattice Boltzmann method. Physica A 2009; 388: 4803-4810.
[19] Chen S, Tian ZW. Simulation of thermal micro-flow using Lattice Boltzmann method with Langmuir slip model. Int J Heat Fluid Flow 2010; 31: 227-235.
[20] Chen S. Lattice Boltzmann method for slip flow heat transfer in circular microtubes: Extended Graetz problem. Appl Math Comput 2010; 217: 3314-3320.
[21] Chen S, Bao S, Liu Z, Li J, Yi C, Zheng CG. A heuristic curved boundary treatment in Lattice Boltzmann method, Europhys Lett 2010; 92: 54003.
[22] Chen S, Li J, Han HF, Liu ZH, Zheng CG. Effects of hydrogen addition on entropy generation in ultra-lean counter-flow methane-air premixed combustion. Int J Hydrogen Energy 2010; 35: 3891-3902.
[23] Chen S, Liu ZH, Liu JZ, Li J, Wang L, Zheng CG. Analysis of entropy generation in hydrogen-enriched ultra lean counter-flow methaneeair non-premixed combustion. Int J Hydrogen Energy 2010; 35: 12491-12501.
[24] Chen S. Analysis of entropy generation in counter-flow premixed hydrogen-air combustion. Int J Hydrogen Energy 2010; 35: 1401-1411.
[25] Chen S, Han HF, Liu ZH, Li J, Zheng CG. Analysis of entropy generation in non-premixed hydrogen versus heated air counterflow combustion. Int J Hydrogen Energy 2010; 35: 4736-4746.
[26] Chen S, Du R. Entropy generation of turbulent double-diffusive natural convection in a rectangle cavity. Energy 2011; 36: 1721-1734.
[27] Chen S, Zheng CG. Counterflow diffusion flame of hydrogen-enriched biogas under MILD oxy-fuel condition. Int J Hydrogen Energy (revised)
[28] Chen S, Bao S, Liu Z, Zheng CG. Simple enthalpy-based Lattice Boltzmann scheme for complicated thermal systems. Phys Rev E (submitted)
[29] Chiavazzo E, Karlin IV, Gorban AN. Coupling of the model reduction technique with the Lattice Boltzmann method for combustion simulations. Combust Flame 2010; 157: 1833-1849.
[30] Chun B, Ladd AJC. Interpolated boundary condition for Lattice Boltzmann simulations of flows in narrow gaps. Phys Rev E 2007; 75: 066705.
[31] Coats CM. Coherent structures in combustion. Prog Energy Combust Sci 1996; 22: 427-509.
[32] Colonius T. Modeling artificial boundary conditions for compressible flow. Annu Rev Fluid Mech 2004; 36: 315-345.
[33] Ding EJ, Aidun CK. Extension of the Lattice-Boltzmann method for direct simulation of suspended particles near contact. J Stat Phys 2003; 112: 685-708.
[34] Ding H, Shu C, Yeo KS, Xu D. Numerical simulation of flows around two circular cylinders by mesh-free least square-based finite difference methods. Int J Numer Meth Fluids 2007; 53: 305-332.
[35] D'Orazio A, Succi S, Arrighetti C. Lattice Boltzmann simulation of open flows with heat transfer. Phys Fluids 2003; 15: 2778-2781.
[36] D'Orazio A, Succi S. Simulating two-dimensional thermal channel flows by means of a lattice Boltzmann method with new boundary conditions. Future Generation Comput Syst 2004; 20: 935-944.
[37] Echekki T, Mastorakos E. Turbulent combustion modeling: advance, new trends and perspectives. London: Springer-Verlag; 2011.
[38] Feng ZG, Michaelides EE. The immersed boundary-lattice Boltzmann method for solving fluid-particles interaction problems. J Comput Phys 2004; 195: 602-628.
[39] Filippova O, Hanel D. Grid refinement for Lattice-BGK models. J Comput Phys 1998; 147: 219-228.
[40] Filippova O, Hanel D. Lattice-BGK model for low Mach number combustion. Int J Mod Phys C 1998; 9: 1439-1445.
[41] Filippova O, Hanel D. A novel lattice BGK approach for low Mach number combustion. J Comput Phys 2000; 158: 139-160.
[42] Filippova O, Hanel D. A novel numerical scheme for reactive flows at low Mach numbers. Comput Phys Commun 2000; 129: 267-274.
[43] Ginzburg I, d'Humieres D. Multireflection boundary conditions for Lattice Boltzmann models. Phys Rev E 2003; 68: 066614.
[44] Guo ZL, Shi BC, Wang NC. Lattice BGK model for incompressible Navier-Stokes equation. J Comput Phys 2000; 165: 288-306.
[45] Guo Z, Zheng CG, Shi B. An extrapolation method for boundary conditions in Lattice Boltzmann method. Phys Fluids 2002; 14: 2007-2010.
[46] Guo Z, Zheng C, Shi B, Zhao TS. Thermal Lattice Boltzmann equation for low Mach number flows: Decoupling model. Phys Rev E 2007; 75: 036704.
[47] Harichandan AB, Roy A. Numerical investigation of low Reynolds number flow past two and three circular cylinders using unstructured grid CFR scheme. Int J Heat Fluid Flow 2010; 31: 154-171.
[48] He X, Luo LS, Dembo M. Some progress in Lattice Boltzmann Method. Part I. Nonuniform mesh grids. J Comput Phys 1996; 129: 357-363.
[49] He X, Doolen G. Lattice Boltzmann method on curvilinear coordinates system: flow around a circular cylinder. J Comput Phys 1997; 134: 306-315.
[50] He X, Luo LS. Lattice Boltzmann model for the incompressible Navier-Stokes equation. J Stat Phys 1997; 88: 927-944.
[51] He X, Chen S, Doolen GD. A novel thermal model for the Lattice Boltzmann method in incompressible limit. J Comput Phys 1998;146: 282-300.
[52] He X, Doolen GD, Clark T. Comparison of the Lattice Boltzmann method and the artificial compressibility method for Navier-Stokes equations. J Comput Phys 2002; 179: 439-451.
[53] He X, Doolen GD, Clark T. Comparison of the Lattice Boltzmann method and the artificial compressibility method for Navier-Stokes equations. J Comput Phys 2002; 179: 439-451.
[54] Hirschfelder JC, Curtiss CF, Bird RB. The molecular theory of gases and liquids. New York: Wiley; 1954.
[55] Izquierdo S, Fueyo N. Characteristic nonreflecting boundary conditions for open boundaries in Lattice Boltzmann methods. Phys Rev E 2008; 78: 046707.
[56] Izquierdo S, Martinez-Lera P, Fueyo N. Analysis of open boundary effects in unsteady Lattice Boltzmann simulations. Comput Math Appl 2009; 58: 914-921.
[57] Jahangirian S, Engeda A, Wichman IS. Thermal and chemical structure of biogas counterflow diffusion flames. Energy Fuels 2009; 23: 5312-5321.
[58] Kang Q, Zhang D, Chen S, He X. Lattice Boltzmann simulation of chemical dissolution in porous media. Phys Rev E 2002; 65: 036318.
[59] Kao PH, Yang RJ. An investigation into curved and moving boundary treatments in the Lattice Boltzmann method. J Comput Phys 2008; 227: 5671-5690.
[60] Lallemand P, Luo LS. Theory of the Lattice Boltzmann method: Acoustic and thermal properties in two and three dimensions. Phys Rev E 2003; 68: 036706.
[61] Lallemand P, Luo LS. Lattice Boltzmann method for moving boundaries. J Comput Phys 2003; 184: 406-421.
[62] Langmuir I. Surface Chemistry. Chem Rev 1933; 13: 147-191.
[63] Lee J, Lee S. Boundary treatment for the Lattice Boltzmann method using adaptive relaxation times. Comput Fluids 2010; 39: 900-909.
[64] Lee T, Lin CL. A Lattice Boltzmann algorithm for calculation of the laminar jet diffusion flame. J Comput Phys 2006; 215: 133-152.
[65] Lee T, Lin CL. A characteristic Galerkin method for discrete Boltzmann equation. J Comput Phys 2001; 171: 336-356.
[66] Li P, Mi J, Dally BB, Wang F, Wang L, Liu Z, Chen S, Zheng C. Progress and recent trend in MILD combustion. Science China: Technological Sciences 2011; 54: 255-269.
[67] Liu C. Multigrid method and its application in CFD. Beijing: Tsinghua University Press; 1995 (in Chinese).
[68] Liu C, Lin K, Mai H, Lin C. Thermal boundary conditions for thermal Lattice Boltzmann simulations. Comput Math Appl 2010; 59: 2178-2193.
[69] Liu JTC. Coherent structures in transitional and turbulent free shear flows. Ann Rev Fluid Mech 1989; 21: 285-315.
[70] Maruta K, Muso K, Takeda K, Niioka T. Reaction zone structure in flameless combustion. Proc Combust Inst 2000; 28: 2117-2123.
[71] Mei R, Luo LS, Shyy W. An accurate curved boundary treatment in the Lattice Boltzmann method. J Comput Phys 1999; 155: 307-330.
[72] Mei R, Luo LS, Lallemand P, d'Humieres D. Consistent initial conditions for Lattice Boltzmann simulations. Comput Fluids 2006; 35: 855-862.
[73] Ong L, Wallace J. The velocity field of the turbulent very near wake of a circular cylinder. Exp Fluids 1996; 20: 441-453.
[74] Pavol P, Vahala G, Vahala L. Preliminary results in the use of energy-dependent octagonal lattices for thermal Lattice Boltzmann simulations. J Stat Phys 2002; 107: 499-519.
[75] Park J, Kim JS, Chung JO, Yun JH, Keel SI. Chemical effects of added CO2 on the extinction characteristics of H2/CO/CO2 syngas diffusion flames. Int J Hydrogen Energy 2009; 34: 8756-8762.
[76] Peng G, Xi H, Duncan C, Chou S. Finite volume scheme for the Lattice Boltzmann method on unstructured meshes. Phys Rev E 1999; 59: 4675-4682.

[77] Poinsot T, Veynante D. Theoretical and numerical combustion. Philadelphia: R T Edwards Inc; 2001.
[78] Qian Y, d'Humieres D, Lallemand P. Lattice BGK models for Navier-Stokes equation. Europhys Lett 1992; 17: 479-484.
[79] Rohde M, Derksen JJ, van den Akker HEA. Volumetric method for calculating the flow around moving objects in Lattice Boltzmann schemes. Phys Rev E 2002; 65: 056701.
[80] Roper FG. The prediction of laminar jet diffusion flame sizes: Part I. Theoretical model. Combust Flame 1977; 29: 219-226.
[81] Sivashinsky GJ. Hydrodynamic theory of flame propagation in an enclosed volume. Acta Astronaut 1979; 6: 631-645.
[82] Skordos PA. Initial and boundary conditions for the Lattice Boltzmann method. Phys Rev E 1993; 48: 4823-4842.
[83] Soma SK, Datta A. Thermodynamic irreversibilities and exergy balance in combustion processes. Prog Energy Combust Sci 2008: 34: 351-376.
[84] Succi S, Bella G, Papetti F. Lattice kinetic theory for numerical combustion. J Sci Comput 1997; 12: 395-408.
[85] Succi S. The Lattice Boltzmann equation for fluid dynamics and beyond. Oxford: Oxford University Press; 2001.
[86] Tang GH, Tao WQ, He YL. Thermal boundary condition for the thermal Lattice Boltzmann equation. Phys Rev E 2005; 72: 016703.
[87] Toftegaard MB, Brix J, Jensen PA, Glarborg P, Jensen AD. Oxy-fuel combustion of solid fuels. Prog Energy Combust Sci 2010; 36: 581-625.
[88] Tomboulides AG, Lee JCY, Orszag SA. Numerical simulation of low Mach number reactive flows. J Sci Comput 1997; 12: 139-167.
[89] Ubertini S, Bella G, Succi S. Lattice Boltzmann method on unstructured grids: Further developments. Phys Rev E 2003; 68: 016701.
[90] Ubertini S, Succi S. Recent advances of Lattice Boltzmann techniques on unstructured grids. Prog Comput Fluid Dyn 2005; 5: 85-96.
[91] Verberg R, Ladd AJC. Lattice Boltzmann model with sub-grid-scale boundary conditions. Phys Rev Lett 2000; 84: 2148-2151.
[92] Verhaeghe F, Arnout S, Blanpain B, Wollants P. Lattice Boltzmann model for diffusion-controlled dissolution of solid structures in multicomponent liquids. Phys Rev E 2005; 72: 036308.
[93] Verhaeghe F, Arnout S, Blanpain B, Wollants P. Lattice Boltzmann modeling of dissolution phenomena. Phys Rev E 2006; 73: 036316.
[94] Walsh SDC, Saar MO. Interpolated Lattice Boltzmann boundary conditions for surface reaction kinetics. Phys Rev E 2010; 82: 066703.
[95] Woods LC. An introduction to the kinetic theory of gases and magnetoplasmas. Oxford: Oxford University Press; 1993.
[96] Yamamoto K, He X, Doolen GD. Simulation of combustion field with Lattice Boltzmann method. J Stat Phys 2002; 107: 367-383.
[97] Yamamoto K. LB simulation on combustion with turbulence. Int J Mod Phys B 2003; 17: 197-200.
[98] Yamamoto K, Takada N, Misawa M. Combustion simulation with Lattice Boltzmann method in a three-dimensional porous structure. Proc Combust Inst 2005; 30: 1509-1515.
[99] Yu DZ, Mei R, Luo LS, Shyy W. Viscous flow computations with the method of Lattice Boltzmann equation. Prog Aerosp Sci 2003; 39: 329-367.
[100] Yu D, Mei R, Shyy W. Improved treatment of the open boundary in the method of Lattice Boltzmann equation. Prog Comput Fluid Dyn 2005; 5: 3-12.

Index

A

B

C

D

E

F

Matthias Ehrhardt (Ed.)

N

P

R

S

T

V

www.ingramcontent.com/pod-product-compliance
Lightning Source LLC
Chambersburg PA
CBHW050817220326
41598CB00006B/241
9781608057177